AF703274

IET POWER AND ENERGY SERIES 157

Electrical Steels

Other volumes in this series:

Volume 1 **Power Circuit Breaker Theory and Design** C.H. Flurscheim (Editor)
Volume 4 **Industrial Microwave Heating** A.C. Metaxas and R.J. Meredith
Volume 7 **Insulators for High Voltages** J.S.T. Looms
Volume 8 **Variable Frequency AC Motor Drive Systems** D. Finney
Volume 10 **SF_6 Switchgear** H.M. Ryan and G.R. Jones
Volume 11 **Conduction and Induction Heating** E.J. Davies
Volume 13 **Statistical Techniques for High Voltage Engineering** W. Hauschild and W. Mosch
Volume 14 **Uninterruptible Power Supplies** J. Platts and J.D. St Aubyn (Editors)
Volume 15 **Digital Protection for Power Systems** A.T. Johns and S.K. Salman
Volume 16 **Electricity Economics and Planning** T.W. Berrie
Volume 18 **Vacuum Switchgear** A. Greenwood
Volume 19 **Electrical Safety: A guide to causes and prevention of hazards** J. Maxwell Adams
Volume 21 **Electricity Distribution Network Design, 2nd Edition** E. Lakervi and E.J. Holmes
Volume 22 **Artificial Intelligence Techniques in Power Systems** K. Warwick, A.O. Ekwue and R. Aggarwal (Editors)
Volume 24 **Power System Commissioning and Maintenance Practice** K. Harker
Volume 25 **Engineers' Handbook of Industrial Microwave Heating** R.J. Meredith
Volume 26 **Small Electric Motors** H. Moczala *et al.*
Volume 27 **AC–DC Power System Analysis** J. Arrillaga and B.C. Smith
Volume 29 **High Voltage Direct Current Transmission, 2nd Edition** J. Arrillaga
Volume 30 **Flexible AC Transmission Systems (FACTS)** Y.-H. Song (Editor)
Volume 31 **Embedded generation** N. Jenkins *et al.*
Volume 32 **High Voltage Engineering and Testing, 2nd Edition** H.M. Ryan (Editor)
Volume 33 **Overvoltage Protection of Low-Voltage Systems, Revised Edition** P. Hasse
Volume 36 **Voltage Quality in Electrical Power Systems** J. Schlabbach *et al.*
Volume 37 **Electrical Steels for Rotating Machines** P. Beckley
Volume 38 **The Electric Car: Development and future of battery, hybrid and fuel-cell cars** M. Westbrook
Volume 39 **Power Systems Electromagnetic Transients Simulation** J. Arrillaga and N. Watson
Volume 40 **Advances in High Voltage Engineering** M. Haddad and D. Warne
Volume 41 **Electrical Operation of Electrostatic Precipitators** K. Parker
Volume 43 **Thermal Power Plant Simulation and Control** D. Flynn
Volume 44 **Economic Evaluation of Projects in the Electricity Supply Industry** H. Khatib
Volume 45 **Propulsion Systems for Hybrid Vehicles** J. Miller
Volume 46 **Distribution Switchgear** S. Stewart
Volume 47 **Protection of Electricity Distribution Networks, 2nd Edition** J. Gers and E. Holmes
Volume 48 **Wood Pole Overhead Lines** B. Wareing
Volume 49 **Electric Fuses, 3rd Edition** A. Wright and G. Newbery
Volume 50 **Wind Power Integration: Connection and system operational aspects** B. Fox *et al.*
Volume 51 **Short Circuit Currents** J. Schlabbach
Volume 52 **Nuclear Power** J. Wood
Volume 53 **Condition Assessment of High Voltage Insulation in Power System Equipment** R.E. James and Q. Su
Volume 55 **Local Energy: Distributed generation of heat and power** J. Wood
Volume 56 **Condition Monitoring of Rotating Electrical Machines** P. Tavner, L. Ran, J. Penman and H. Sedding
Volume 57 **The Control Techniques Drives and Controls Handbook, 2nd Edition** B. Drury
Volume 58 **Lightning Protection** V. Cooray (Editor)
Volume 59 **Ultracapacitor Applications** J.M. Miller

Volume 62 **Lightning Electromagnetics** V. Cooray
Volume 63 **Energy Storage for Power Systems, 2nd Edition** A. Ter-Gazarian
Volume 65 **Protection of Electricity Distribution Networks, 3rd Edition** J. Gers
Volume 66 **High Voltage Engineering Testing, 3rd Edition** H. Ryan (Editor)
Volume 67 **Multicore Simulation of Power System Transients** F.M. Uriate
Volume 68 **Distribution System Analysis and Automation** J. Gers
Volume 69 **The Lightening Flash, 2nd Edition** V. Cooray (Editor)
Volume 70 **Economic Evaluation of Projects in the Electricity Supply Industry, 3rd Edition** H. Khatib
Volume 72 **Control Circuits in Power Electronics: Practical issues in design and implementation** M. Castilla (Editor)
Volume 73 **Wide Area Monitoring, Protection and Control Systems: The enabler for smarter grids** A. Vaccaro and A. Zobaa (Editors)
Volume 74 **Power Electronic Converters and Systems: Frontiers and applications** A.M. Trzynadlowski (Editor)
Volume 75 **Power Distribution Automation** B. Das (Editor)
Volume 76 **Power System Stability: Modelling, analysis and control** B. Om P. Malik
Volume 78 **Numerical Analysis of Power System Transients and Dynamics** A. Ametani (Editor)
Volume 79 **Vehicle-to-Grid: Linking electric vehicles to the smart grid** J. Lu and J. Hossain (Editors)
Volume 81 **Cyber-Physical-Social Systems and Constructs in Electric Power Engineering** S. Suryanarayanan, R. Roche and T.M. Hansen (Editors)
Volume 82 **Periodic Control of Power Electronic Converters** F. Blaabjerg, K. Zhou, D. Wang and Y. Yang
Volume 86 **Advances in Power System Modelling, Control and Stability Analysis** F. Milano (Editor)
Volume 87 **Cogeneration: Technologies, optimisation and implementation** C.A. Frangopoulos (Editor)
Volume 88 **Smarter Energy: From smart metering to the smart grid** H. Sun, N. Hatziargyriou, H.V. Poor, L. Carpanini and M.A. Sánchez Fornié (Editors)
Volume 89 **Hydrogen Production, Separation and Purification for Energy** A. Basile, F. Dalena, J. Tong and T.N. Veziroğlu (Editors)
Volume 90 **Clean Energy Microgrids** S. Obara and J. Morel (Editors)
Volume 91 **Fuzzy Logic Control in Energy Systems with Design Applications in MATLAB®/Simulink®** İ.H. Altaş
Volume 92 **Power Quality in Future Electrical Power Systems** A.F. Zobaa and S.H.E.A. Aleem (Editors)
Volume 93 **Cogeneration and District Energy Systems: Modelling, analysis and optimization** M.A. Rosen and S. Koohi-Fayegh
Volume 94 **Introduction to the Smart Grid: Concepts, technologies and evolution** S.K. Salman
Volume 95 **Communication, Control and Security Challenges for the Smart Grid** S.M. Muyeen and S. Rahman (Editors)
Volume 96 **Industrial Power Systems with Distributed and Embedded Generation** R. Belu
Volume 97 **Synchronized Phasor Measurements for Smart Grids** M.J.B. Reddy and D.K. Mohanta (Editors)
Volume 98 **Large Scale Grid Integration of Renewable Energy Sources** A. Moreno-Munoz (Editor)
Volume 100 **Modeling and Dynamic Behaviour of Hydropower Plants** N. Kishor and J. Fraile-Ardanuy (Editors)
Volume 101 **Methane and Hydrogen for Energy Storage** R. Carriveau and D.S.-K. Ting
Volume 104 **Power Transformer Condition Monitoring and Diagnosis** A. Abu-Siada (Editor)
Volume 106 **Surface Passivation of Industrial Crystalline Silicon Solar Cells** J. John (Editor)
Volume 107 **Bifacial Photovoltaics: Technology, applications and economics** J. Libal and R. Kopecek (Editors)

Volume 108 **Fault Diagnosis of Induction Motors** J. Faiz, V. Ghorbanian and G. Joksimović
Volume 110 **High Voltage Power Network Construction** K. Harker
Volume 111 **Energy Storage at Different Voltage Levels: Technology, integration, and market aspects** A.F. Zobaa, P.F. Ribeiro, S.H.A. Aleem and S.N. Afifi (Editors)
Volume 112 **Wireless Power Transfer: Theory, technology and application** N. Shinohara
Volume 115 **DC Distribution Systems and Microgrids** T. Dragičević, F. Blaabjerg and P. Wheeler
Volume 117 **Structural Control and Fault Detection of Wind Turbine Systems** H.R. Karimi
Volume 119 **Thermal Power Plant Control and Instrumentation: The control of boilers and HRSGs, 2nd Edition** D. Lindsley, J. Grist and D. Parker
Volume 120 **Fault Diagnosis for Robust Inverter Power Drives** A. Ginart (Editor)
Volume 123 **Power Systems Electromagnetic Transients Simulation, 2nd Edition** N. Watson and J. Arrillaga
Volume 124 **Power Market Transformation** B. Murray
Volume 126 **Diagnosis and Fault Tolerance of Electrical Machines, Power Electronics and Drives** A.J.M. Cardoso
Volume 128 **Characterization of Wide Bandgap Power Semiconductor Devices** F. Wang, Z. Zhang and E.A. Jones
Volume 130 **Wind and Solar Based Energy Systems for Communities** R. Carriveau and D.S.-K. Ting (Editors)
Volume 131 **Metaheuristic Optimization in Power Engineering** J. Radosavljević
Volume 132 **Power Line Communication Systems for Smart Grids** I.R.S. Casella and A. Anpalagan
Volume 905 **Power System Protection, 4 volumes**

Electrical Steels

Volume 1: Fundamentals and basic concepts

Anthony Moses, Philip Anderson, Keith Jenkins
and Hugh Stanbury

The Institution of Engineering and Technology

Published by The Institution of Engineering and Technology, London, United Kingdom

The Institution of Engineering and Technology is registered as a Charity in England & Wales (no. 211014) and Scotland (no. SC038698).

© The Institution of Engineering and Technology 2019

First published 2019

This publication is copyright under the Berne Convention and the Universal Copyright Convention. All rights reserved. Apart from any fair dealing for the purposes of research or private study, or criticism or review, as permitted under the Copyright, Designs and Patents Act 1988, this publication may be reproduced, stored or transmitted, in any form or by any means, only with the prior permission in writing of the publishers, or in the case of reprographic reproduction in accordance with the terms of licences issued by the Copyright Licensing Agency. Enquiries concerning reproduction outside those terms should be sent to the publisher at the undermentioned address:

The Institution of Engineering and Technology
Michael Faraday House
Six Hills Way, Stevenage
Herts, SG1 2AY, United Kingdom

www.theiet.org

While the authors and publisher believe that the information and guidance given in this work are correct, all parties must rely upon their own skill and judgement when making use of them. Neither the authors nor publisher assumes any liability to anyone for any loss or damage caused by any error or omission in the work, whether such an error or omission is the result of negligence or any other cause. Any and all such liability is disclaimed.

The moral rights of the authors to be identified as authors of this work have been asserted by them in accordance with the Copyright, Designs and Patents Act 1988.

British Library Cataloguing in Publication Data
A catalogue record for this product is available from the British Library

ISBN 978-1-78561-970-0 (hardback)
ISBN 978-1-78561-971-7 (PDF)

Typeset in India by MPS Limited

Contents

Acknowledgements

Many friends, associates and colleagues have helped us during the preparation of this book in informal discussions giving advice and useful comments on many aspects of its content. Others have expanded on points in their publications to enable us to do justice to their own work. It would not be appropriate to attempt to name or attribute individually in case we should miss any out! Nevertheless, we are very grateful to this anonymous group of academic and industrial experts from around the world.

Past and present colleagues at Cardiff University and Cogent Power have directly and indirectly helped shape our understanding and appreciation of the particular fields of expertise touched on in the book long before we put pen to paper.

We are grateful to individuals and publishers who freely have allowed us to reproduce or update original figures and to Hamid Shahrouzi and Ryan Davies who produced many draft figures while they were carrying out postgraduate studies at Cardiff University. We also appreciate the contribution of Dr Wojciech Pluta of Czestochowa University of Technology in preparation of several figures.

We would also like to thank Dr Christopher Harrison, Cardiff University, and Morgan Amboun, Iowa State University, for helping to make magnetic measurements on large numbers of samples for incorporation into Chapter 15 of Volume 2.

We are grateful to Doug Warne (Editor of the IET Energy Engineering book series) for carrying out a critical review of the draft of Chapter 6 of Volume 2 which helped us to cover the background to applications of electrical steels in rotating electrical machines more thoroughly. Likewise, we are grateful to Dr Filippos Marketos of GE Power Transformers Product Line for reading the draft of Chapter 5 of Volume 2 and providing useful suggestions which enabled us to take better account of the application of GO steel in power transformers from a practical, design and manufacturing point of view. Howard Smith of Cogent Power also provided a great deal of useful insight into the challenges of magnetic measurements in an industrial environment for Chapter 12 of Volume. 1.

Thanks are due to Dr Stanislaw Zurek, Head of Research & Innovation at Megger Instruments Ltd, UK, on a few accounts. First, in providing much advice in the early stages of writing based on his experience in producing his own excellent book on rotational magnetisation, published in 2017, which we refer to in chapters of this book. Second, he used his drawing skills and magnetic expertise to interpret handwritten drawings from which he produced several figures for us. Third, he redrew many other published figures (whose original sources are acknowledged in the text). In several cases, this enabled us to either update or change the originals to make them more suitable for the needs of the book. Finally, the rotational magnetisation characteristics presented in Chapter 15 of Volume 2 are taken from Excel files of measurement data which he produced while carrying out his PhD studies at the Wolfson Centre for Magnetics, Cardiff University.

We are grateful to Suzanne Yap of the International Electrotechnical Commission and Simon Merriman of the British Standards Institution for their help and guidance in producing Chapter 12 of Volume 2.

We wish to thank Dr Christoph Von Friedenburg, Senior Commissioning Editor- Power Energy and Transportation at IET Publishing for guiding us with the structure of the work from its formulation to the publication stages and to Olivia Wilkins, Assistant Editor and her colleagues at IET Publishing for their patience and continuous encouragement and prompt guidance while assisting us during all stages of preparation of the work. We are grateful to staff at MPS Ltd for their valuable contribution to the presentation of the book at the typesetting stage. We also appreciate Doug Warne's contribution to this process.

As for any work of this size, it has taken us many hours, days and months of effort to research and prepare the several iterations. We are grateful to our families for their editorial input, patience and encouragement which speeded up this process and made the experience all the more enjoyable.

Preface

The impact of electrical steels, and their importance in electrical machine design and performance, are often substantially underestimated. Despite their somewhat conventional sounding label, they should be regarded as one of the most technologically important materials in volume production today.

Indeed, anticipated advances in applications ranging from electrical power distribution to electric vehicle drives are expected to be limited by the extent to which the magnetic properties of present commercial electrical steels can be used more effectively, and improved steels can be developed.

The continued improvement of magnetic properties of the best grades of electrical steels over the past half century has made a significant positive environmental impact and it is likely that technological improvements will increase this impact in the next half century.

In this book, we have attempted to present many aspects of electrical steels technology. These range from how their magnetic properties are determined by fundamental principles of magnetism and metallurgical features to their influence on the performance of rotating electrical machines and transformers. This should cater for readers, who might be involved in material production, to appreciate how steels have to respond to large differences of localised magnetisation conditions in electrical machine cores. Equally, it should satisfy readers involved in electrical machine manufacture, or operation, who need to understand magnetic factors which should be taken into account in the choice and use of the materials.

Although the book only touches on commercial issues of the production and use of electrical steels, it should be noted that the industry is in a state of flux. For political or financial reasons, over the last few years the number of producers has dropped significantly and others are merging or restructuring. However, the demand for high quality electrical steel is set to continue to increase in the future, with increased emphasis being placed on the best grades.

Some text books referred to in this book are solely focussed on the properties of electrical steels. These are useful works in their own right, but they are generally limited in scope. However, we wish to refer to the works of one particular author, the late Professor Philip Beckley, who has been an inspiration to many people working on all aspects of electrical steels. His book entitled *Electrical Steels for Rotating Machines,* published by the IET in 2002, gives valuable insight into the interplay between magnetic properties, manufacture, physical properties and costs of electrical steels used in rotating machines. Our interest and knowledge of electrical steels has been strengthened by our experiences of working with Phil throughout our entire careers in our differing academic and industrial roles.

While writing this book, many people in the industry have asked us if we intended just to produce an update of Phil's book. The simple answer to this is that it was not the case. It was decided that the changes in the industry and the exciting opportunities which lie ahead required an approach which, in two volumes, covered the academic background to address the needs and interests of a readership in academia, the steel production industry, lamination and core production as well as in the manufacture of electrical power conversion equipment such as transformers and rotating machines. Furthermore, although an inspiration to us, Phil's book was written in a unique and entertaining format which would have been impossible for us to replicate and expand to a wider range of electrical steels and applications.

This book is intended for graduate students and other researchers engaged in any aspect of research involving the basic properties, characterisation and applications of electrical steels. It should be of value to material scientists engaged in the production and development of electrical steels, and also to engineers who are involved in any aspect of magnetic core production or design which makes use of electrical steels, or other soft magnetic strip or sheet where similar criteria apply.

It is hoped that this is a useful format for both newcomers to electrical steels and also to experienced workers wishing to find sufficient in-depth detail in their particular fields of interest. It should enable all readers to develop an understanding of important aspects of the production, characterisation and usage of the steels with which they might not be so knowledgeable.

A comprehensive work which covers the production, properties, characterisation and industrial application of electrical steels requires expertise and experience beyond the capability of a single author. Between us, we feel that we have the academic and industrial backgrounds for us to seamlessly link these topics in this book.

Tony Moses is Emeritus Professor of Magnetics at Cardiff University. He was awarded a DSc in recognition of his academic achievement in the field of soft magnetic materials. He is the author of over 500 research publications in the area and is a life member of the IEEE and also of the UK Magnetics Society. Phil Anderson is a Senior Lecturer in Magnetics in the School of Engineering at Cardiff University. He has 20 years of post-doctoral experience in research into the magnetic properties of electrical steels. He is a senior member of the Magnetics and Materials Group at Cardiff University where he has developed world standard experimental facilities for magnetic characterisation of electrical steels. After obtaining a first degree in Metallurgy, Keith Jenkins pursued a career of almost 40 years covering several technical and research functions in the electrical steel industry at the Orb Works in Newport, UK. Hugh Stanbury also spent nearly 40 years at the Orb Works after obtaining his first degree in Physics. He has been mainly responsible for industrial magnetic testing and the UKAS accreditation of the calibration systems for the measurement of the magnetic properties of electrical steels at Orb Electrical Steels. For 18 years he served as Chairman of the IEC Technical Committee 68 responsible for Magnetic Alloys and Steels. He also chairs the British Standards Institution Technical Committee ISE/108 for Magnetic Alloys and Steels. He is a life member of the UK Magnetics Society.

The book is divided into two volumes and although different members of our group of four are formal authors of each to reflect their expert contributions, all four of us have contributed to the structure of each volume.

A challenge in producing any broad work on electrical steels is to cater for the potentially wide range of prior knowledge of the related magnetism, metallurgy and engineering application content which readers may have. For example, a reader with background in the physics of magnetism should have sufficient knowledge of the magnetic domain theory which is needed to understand important magnetic characteristics of electrical steels. However, the same reader might have no prior knowledge of aspects of electrical machine core design and operation whose optimisation can be indirectly traced to the behaviour of magnetic domains. The detailed content draws upon our experience in teaching postgraduate modules and delivering short courses on the magnetic characteristics of electrical steels to practising electrical engineers, and complementary presentations on the technological role of electrical steels in electrical machines to materials technologists with limited engineering knowledge. We decided to pitch the academic level and breadth of each chapter on the magnetic theory and engineering applications at a basic level which any reader should be able to cope with, without the need to refer to more specialised text books.

The work is split into two volumes. Volume 1 comprises 18 chapters which cover background to electrical steels and also compares and contrasts their properties with the other families of magnetic materials used in electrical power applications. Volume 2 comprises 15 chapters which build on fundamental aspects presented in Volume 1 to focus on properties and applications of electrical steels. It is not possible to set a precise dividing line between the background and specific detail of electrical steels so readers are occasionally referred between chapters in the two volumes. The important topics of sensitivity of magnetic properties to mechanical stress and to temperature are dealt with for all soft magnetic materials in Volume 1. It could be argued that separate chapters on the effect of stress and temperature on the magnetic properties of electrical steels could have been included in Volume 2 but we felt it would be more concise and logical to deal with the theory and background to stress and temperature together for all groups of soft magnetic materials in Volume 1.

In Volume 1, Chapters 1 and 2 give a broad introduction to soft magnetic materials and the basic underlying concepts of related magnetic theory. These show how important magnetic properties of electrical steels compare with other soft magnetic materials, some of which are alternative choices in certain applications. Chapter 3 gives an in-depth background to the origin and characterisation of magnetic domains. We only introduce sufficient physics of magnetism to enable readers to understand phenomena in electrical steels and other soft magnetic materials used in electrical power applications. This is presented in a non-rigorous way to help a reader with little background in magnetics to appreciate the underlying factors which determine the bulk magnetic properties of electrical steels. The observation of magnetic domains helps us to understand some of the properties of bulk soft magnetic materials from a nano-dimensional level up to performance in large magnetic cores, so various techniques applicable to electrical steels are discussed in detail in Chapter 4.

Chapters 5 and 6 introduce some basic electromagnetism and waveform analysis related to electrical steels with which most readers with backgrounds in electrical engineering will be very familiar be very familiar with this. As in other chapters, much of the content here is relevant to all soft, and even hard, magnetic materials subjected to a change in magnetisation in time.

Chapters 7 and 8 introduce the important topic of losses in electrical steels. This ranges from a theoretical background to the practical impact. Much of this applies directly to thin strip with a significant amount of reference to electrical steels, some of which is expanded on in more detail in Volume 2.

Chapters 9 and 10 of Volume 1 give further background into the magnetic anisotropy of electrical steels and the basic forms of magnetic circuit in which the steels are used. These two, somewhat introductory, chapters form the foundation for later chapters in Volume 2 dedicated to the magnetic properties of the steels and their performance in magnetic cores.

Some readers will be familiar with harmful effect of mechanical stress on the magnetic properties of soft magnetic materials. This is discussed in detail in Chapter 11 of Volume 1. As with Chapters 7 and 8, by the nature of the topic there is a significant reference to effects in thin sheet or strip very relevant to all metallic soft magnetic materials designed to operate at power magnetising frequencies.

Magnetic measurements are notoriously difficult to carry out and interpret. Pitfalls and precautions relevant to such measurements on strip materials, including electrical steels, are discussed in Chapter 12. It also explains aspects of common and specialised measurements used for commercial grading of electrical steels and for providing their characteristics in a form suitable for use in computational electromagnetic analysis of electrical machine cores.

Chapters 13 and 14 cover the history and metallurgical requirements of commercial electrical steels and the production of modern steels. These highlight the complex processes and high degree of control needed to produce best quality materials.

Chapters 15–17 summarise important properties of other types of soft magnetic materials commercially used in moderate quantities. Dedicated text books cover the properties of these materials in far more depth. However, they are included in these chapters to show similarities and differences to electrical steels which may be relevant in the selection and development of soft magnetic materials for future electrical power applications.

Magnetic properties of all soft magnetic materials are liable to be temperature dependent. Today, magnetic cores of many devices are expected to operate safely and effectively over wider temperature ranges than ever before. Chapter 18 shows the general trends in the changes of magnetic properties to be expected.

Volume 2 starts with two chapters covering a wide range of specific properties and characteristics of electrical steels. These range from domain structures in individual grains to the effects of insulating coating on the bulk properties of electrical steels. Chapter 3 of Volume 2 covers specialised magnetic steels which are, or have been, produced commercially in small quantities. Some of these may become used more widely in the future.

A vast amount of the electrical steel produced each year is used in transformer and rotating machine cores. Technical interactions between producers and users have increased significantly over recent years, but they can be hindered by each party's lack of knowledge of the other's products. Chapters 5, 6, 8 and 9 of Volume 2 give an introduction to electrical machine design and performance pitched at a level which should be clear to material scientists with little prior knowledge. Electrical machine experts will note that the details in these chapters are superficial and simplified, but readers should find it interesting to see how the basic magnetic material phenomena covered in earlier chapters relate to machine performance.

Chapter 7 of Volume 2 includes a discussion of the deterioration of the magnetic performance of electrical steels under non-sinusoidal magnetisation conditions. It focusses on the effects of several types of distorted flux density increasingly found in many types of electrical machine cores today.

In Chapter 10 many aspects of magnetic core, and electrical machine, vibration and acoustic noise are discussed. This is a topic of increasing environmental concern. It might appear surprising that, even after at least 70 years of concern in the industry, the exact roles of the magnetic properties of electrical steels in causing noise and vibration are still not quantified. We hope that the content of this chapter helps readers to appreciate the wider aspects of these factors in order to more accurately assess the importance of the magnetic characteristics of a steel in any particular core design.

Chapter 11 summarises some approaches to prediction of losses and magnetic flux distribution in magnetic cores assembled from electrical steels. Some readers might routinely use commercial electromagnetic computational packages to design and predict the magnetic characteristics of electrical machine cores. These packages are suitable for most purposes, but it is shown that the way in which the basic properties of an electrical steel are characterised can have a large impact on the accuracy of the final solutions.

Commercial sales of electrical steels depend very much on all parties using recognised magnetic measurement systems and material quality standards to grade the products. Many users will routinely refer to such Standards neither realising the thoroughness and complexity of the ways in which these are produced, maintained and developed nor how new standards are produced according to calls to cover emerging, or changed, technologies. Chapter 12 gives the background to these activities, focussing on electrical steels.

Chapters 13 and 14 of Volume 2 respectively review the development of some renewable energy sources and the environmental impact of magnetic losses in electrical steels.

Chapter 15 concludes Volume 2 with a comprehensive set of magnetic characteristics of modern, commercial electrical steels. Individual producers publish data on their own products, but this is usually limited to basic loss and magnetisation characteristic under a limited range of conditions. Characteristics of a range of electrical steels, produced by various suppliers, are included. We have deliberately not included the names of the producers. This is because we wish to show general trends in properties with parameters such as temperature, stress, thickness and flux density without raising attention to any perceived differences between nominally similar products from different manufacturers.

We hope that this book clearly shows that electrical steels are structurally and magnetically very sophisticated materials subject to continual development as new applications arise. The future of electrical steels appears to be bright so we hope this book will inspire young graduates to consider careers in this exciting and stimulating area.

Anthony John Moses
Philip Ian Anderson
Keith W. Jenkins
Hugh John Stanbury
February 2019

Common acronyms, symbols and abbreviations used in the text

h_1	instantaneous value of fundamental component of magnetic field
h_n	percentage of the nth harmonic component of field with respect to H_0
h_r	instantaneous value of rotational field
h_{rx}, h_{ry}	instantaneous orthogonal components of rotational magnetic field
h_s	instantaneous surface field
h_x, h_y	component of field along x- and y-directions
$\mathcal{R}$	reluctance
$\mathcal{R}_y$, $\mathcal{R}_g$, $\mathcal{R}_p$, $\mathcal{R}_r$	circuit component reluctances
$\tan \delta$	loss angle
$\delta\, dz$	displacement due to curvature
η_{xy}	shear strain
λ_{xy}	shear magnetostriction between x- and y-directions
A_{air}	cross-sectional area of air space
$A_h(B)$	induction dependence of static B–H loop
A_c	cross-sectional area of a core
	cross-sectional area of coating
A_{cn}	cross-sectional area of stressed area
A_{coil}	cross-sectional area of a magnetising coil
A_g	cross-sectional area of air gap
A_m	cross-sectional area of an iron core
	current per unit circumferential length of a winding
A_n	sound amplitude
A_p	anisotropy factor
A_s	cross-sectional area of steel
A_w	cross-sectional area of core window
B_0	peak value of distorted flux density waveform
B_1, B_3, B_5, $\ldots$, B_{2n-1}	peak values of harmonic components of flux density

B_1	flux density near a hole; peak value of fundamental component of B
	flux density along an easy direction of magnetisation
B_2	flux density along a difficult direction of magnetisation
B_{3p}, B_{5p}	harmonic components of flux density
B_H	value of flux density when field is maximum
B_a, B_y, B_m	magnetic circuit component flux densities
B_a	flux density in air space within an iron cored coil
B_{ac}	component of a.c. flux density under rotational magnetisation
B_{actual}	true flux density in a material
B_{av}	average flux density
B_{cr}	critical flux density
B_g	air gap flux density
B_i	amplitude of ith harmonic flux density
B_{int}	internal flux density in a strip
B_{long}	component of flux density along the longitudinal direction
B_m	amplitude of flux density averaged over cross section of a lamination
B_{ma}	flux density in medium '*a*'
B_{man}	normal component of flux density in medium '*a*'
B_{mat}	tangential component of flux density in medium '*a*'
B_{max}	maximum value of step-like flux density
B_{mb}	flux density in medium '*b*'
B_{mbn}	normal component of flux density in medium '*b*'
B_{mbt}	tangential component of flux density in medium '*b*'
$B_{measurement}$	*B* calculated from B-coil output
B_n	normal flux density; peak value of nth harmonic component of B
B_{nx}, B_{ny}	peak value of the nth harmonic components of rotational field in orthogonal directions
B_p	peak value of flux density
B_r	remanent flux density
	instantaneous value of rotational flux density
B_{rrd}	resultant component of flux density along the RD
B_{rtd}	resultant component of flux density along the TD

B_s	maximum flux density; peak surface flux density
B_{sur}	surface flux density
B_t	flux density in a single turn
B_w	flux density remote from a hole
B_x, B_y, B_z	orthogonal components of flux density
B_x	amplitude of flux density at distance x beneath surface
	component of flux density in x-direction; localised flux density
	component of flux density along major axis
	instantaneous components of rotational flux density along x-axis
	peak value of rotational flux density in the x-direction
B_y	component of flux density along minor axis
	instantaneous components of rotational flux density along y-axis
	peak value of rotational flux density in the y-direction
C_1, C_2, C_3, C_4, C_5	constants related to domain width
C_{clas}	classical rotational loss coefficient
C_{cu}	cost of copper loss
$C_e^{'}$	excess loss constant
C_{fe}	cost of iron loss
C_p	specific heat capacity
C_{rex}	excess rotational loss coefficient
D_{xy}	shear magnetic modulus between x- and y-planes
E_1, E_2	free energy
E_{100}	Young's modulus along [100] direction
E_{1rms}	r.m.s. value of primary winding e.m.f.
E_2	secondary coil e.m.f.
E_h	instantaneous hysteresis energy loss
E_K	anisotropy energy
E_{RD}	Young's modulus along RD
E_T	total wall energy
E_{TR}	ratio of Young's modulus along TD to that along RD
E_{av}	average value of e.m.f.
E_c	Young's modulus of coating
E_{clas}	instantaneous classical eddy current energy loss

E_{el}	elastic strain energy
E_{ex}	exchange energy
E_g	induced e.m.f. in armature
	energy stored in an air gap
E_{me}	magnetoelastic energy
E_{ms}	magnetostatic energy
E_p	peak value of e.m.f. in primary coil
E_{ph}	e.m.f. of a phase winding
E_{rhys}	energy stored in material during rotational magnetisation
E_{rms}	r.m.s. value of e.m.f.
E_s	peak value of e.m.f. in secondary coil
	anisotropy energy
E_t	energy stored in a domain wall
	total eddy current loss
E_{total}	instantaneous total eddy current energy loss
E_u	uniaxial anisotropy energy
	energy due to external field
E_w	domain wall energy
E_x, E_y	components of electric field
E_σ	magnetostriction strain energy
F_c	core space factor
F_g	force between magnetic poles
F_n	normal force
F_p	winding space factor
F_w	force on a domain wall
G_{xy}	shear modulus
H_0	field along axis of an air solenoid
	peak value of distorted magnetic field
H_1	magnetic field along an easy direction of magnetisation
H_2	magnetic field along a difficult direction of magnetisation
$H_h(\mathrm{B})$	hysteresis component of surface field
H_X, H_Y	components of magnetic field
H_a	air field; external field
	anisotropy field

$H_a(t)$	instantaneous magnetic field during cyclic magnetisation
$H_{a1,}H_{a3,}H_{a5},\ldots,H_{a(2n-1)}$	peak values of harmonic components of magnetic field
H_{ac}	component of a.c. field under rotational magnetisation
H_{acx}	peak component of a.c. field along the x-direction
H_b	field above grain boundary
H_{cr}	critical magnetic field
H_d	demagnetising field
H_{dc}	d.c. component of magnetic field
H_{dx}	demagnetising field along the x-direction
H_{drd}	demagnetising field along the RD
H_{dtd}	demagnetising field along the TD
H_e	internal field
	effective field
H_g	air gap field
H_{in}	field at surface of a lamination
H_m	field inside a steel core
	value of field when flux density is maximum
H_p	peak value of magnetic field
H_{pk}	peak field strength
H_{ss}	field in single strip tester
H_r	circumferential field at radius r around a current-carrying conductor
	instantaneous value of rotational magnetic field
	resultant field
H_{rot}	rotational field
H_{rx}, H_{ry}	orthogonal components of rotating field
H_s	peak value of surface magnetic field
H_t	tangential component of surface magnetic field
H_x , H_y	fields parallel to x- and y-directions
$H_{x,}$, H_y, H_z	orthogonal components of field
H_x	circumferential field at radius x in a ring
	instantaneous components of rotational field along x-direction
	peak value of field in the x-direction
H_{xAC}	a.c. component of magnetic field in the x-direction

H_{xr}	instantaneous component of magnetic field along the x-direction
H_y, H_p, H_m, H_r	magnetic circuit component fields
H_y	instantaneous components of rotational field along y-axis
	peak value of field in the y-direction
H_{yr}	instantaneous component of magnetic field along the y-direction
H_β	surface field component
I_0	no-load current
I_1	primary coil current
	input current
I_2	secondary coil current
	output current
I_a	armature current
I_c	resistive component of no-load current
I_{dc}	d.c. current
I_m	magnetising current
I_p	primary current
I_{ph}	phase current
J_0	saturation polarisation (at 0 K)
J_s	saturation polarisation
J_T	torque coefficient
K_0, K_1, K_2, K_3	anisotropy constants
K_h	hysteresis loss factor
K_{cl}	classical loss factor
	dependent eddy current loss constant of ferrites
K_e	loss factor in SMCs
K_{ex}	excess loss factor
K_u, K_{u1}, K_{u2}	uniaxial anisotropy energy constants
L_I	load noise
L_f	loss factor
L_p	sound pressure
M_0	saturation magnetisation at 0 K
M_s	saturation magnetisation
M_x	magnetisation at position x
M_α	magnetisation at angle α to the RD

N_1, N_2	number of primary and secondary turns
N_{TD}	demagnetisation factor along TD
N_d	demagnetising factor
N_{dx}, N_{dy}, N_{dz}	demagnetising factors along orthogonal directions
N_p	number of turns of primary coil
N_s	number of turns of secondary coil
N_w	number of turns in phase winding
$P_{1.3T}$	power loss at 1.3 T
$P_{1.7T}$	power loss at 1.7 T
P_{3D}	total power loss under 3D magnetisation
P_h	hysteresis loss
P_h^n, P_{clas}^n, P_{ex}^n	loss components calculated numerically
P_{PWM}	loss under PWM excitation
P_a	additional core loss
P_{area}	loss per unit area
P_{av}	average loss
P_b	nominal loss of a core
P_{bias}	loss under bias field magnetisation
P_c	stator core loss
P_{clas}	classical eddy current energy loss
P_{clas}^{2d}	classical power loss under elliptical magnetisation
P_{clas}^{ac}	classical power loss under a.c. magnetisation
P_{clas}^{skin}	classical eddy current loss based on skin effect
P_{clas}^{step}	classical eddy current loss based on a step-like flux density
P_{dom}	eddy current loss in ideal Goss-oriented grain
$P_{dynamic}$	dynamic loss
P_{ec}^{dist}	eddy current loss under arbitrary B waveform
P_{ec}^{sine}	eddy current loss under sinusoidal B waveform
P_{ex}	excess eddy current energy loss
P_{iron}	no-load iron loss
P_{load}	energy loss on-load
P_n	nominal loss of steel in a wound toroid
P_{noload}	no-load energy loss
P_{oc}	open circuit power input
P_r	residual loss in a ferrite

P_{rh}	rotational hysteresis loss per cycle
ρ	resistivity
P_{rclas}	rotational classical loss per cycle
P_{rd}	loss when magnetised along RD
P_{rex}	rotational excess loss per cycle
P_{rin}	energy flow into a material under rotational magnetisation
P_{sine}	total energy loss in a lamination when no harmonics present
P_{static}	static loss
P_t	total power loss
	instantaneous power loss
	loss in an SMC particle
P_{td}	loss when magnetised along TD
P_{total}	total energy loss in a lamination
P_w	winding and friction loss
$P_{x,}$ P_y	magnetic moduli along x- and y-directions
	eddy current loss components in a burred region
P_x	component of power loss in the x-direction
P_θ	loss at temperature θ
R	resistance
R_1	primary winding resistance
R_2	secondary winding resistance
R_H	resistor used for measurement of primary current
R_a	armature resistance
R_c	resistive component of magnetising impedance
R_p	primary coil resistance
R_s	source resistance; stator winding resistance
R_{w1}	resistance of primary winding
R_w'	resistance of eddy current path
S_0	reference power of 1 MVA
S_R	transformer rating
THD_F	distortion factor referred to fundamental component of waveform
THD_R	distortion factor referred to total waveform
T_l	resultant retained stress
U_H	voltage across resistor in the primary circuit

V_0	excess loss constant
V_1	peak value of fundamental component of voltage
V_1	primary coil voltage
	input voltage
V_2	secondary coil voltage
	output voltage
V_{AB}, V_{BC}, V_{CA}	phase to phase voltages
VA_r	reactive power
V_a	active volume
V_{ab}, V_{bc}, V_{cd}, V_{da}	circuit voltages
V_{ave}	average voltage
V_{coil}	voltage across search coil
$V_{control}$	control voltage
V_g	output voltage of a generator
V_m	rectified mean voltage
V_n	peak value of nth harmonic component of voltage
V_{needle}	output voltage of needle probes
V_{out}	output voltage of a B-coil
V_p	primary voltage
V_{ph}	phase voltage
V_s	peak value of secondary voltage
$W_{1.0/400}$	loss at 1.0 T, 400 Hz magnetisation
$W_{1.5/50}$	loss at 1.5 T, 50 Hz magnetisation
W_h	hysteresis energy loss
W_a	work done on a domain wall
W_{clas}	classical eddy current energy loss
W_{ex}	excess power loss
W_l	energy component
W_{ms}	magnetostatic energy
W_r	total rotational power loss per cycle
W_{rhys}	rotational loss per cycle
X_1, X_2	winding reactances
X_m	reactive component of magnetising impedance
X_s	synchronous reactance
Y_1, Y_2, Y_3, …, Y_n	peak values of harmonic components of cyclic waveform

Z_L	load impedance
$a_1, a_2, a_3, \ldots, a_{30}$	waveform constant
a_h	peak value of hth harmonic component of flux density
a_r	waveform parameter
b_1	instantaneous value of fundamental component of flux density
b_n	percentage of the nth harmonic component of field with respect to B_0
b_r	instantaneous value of rotational flux density
b_{rx}	instantaneous component of rotational flux density along the x-direction
b_{ry}	instantaneous component of rotational flux density along the y-direction
b_s	instantaneous surface flux density
b_x, b_y	components of flux density
c_{11}, c_{12}	elastic constants
d_0	characteristic thickness
d_1	inner diameter
d_2	outer diameter
d_s	width of surface closure domain
d_ϕ	diameter of stressed area
$e_1, e_2, e_p, e_s, e_x, e_y$	induced e.m.f.s
e_{xav}, e_{yav}	average value of orthogonal components of e.m.f.
$f_{harmonic}$	sideband frequency
$f_{carrier}$	carrier frequency
$f_{control}$	control signal frequency
f_i	frequency of ith harmonic
f_{rn}	n^{th} resonant frequency
f_s	sampling frequency
i_{ave}	average current
k_1, k_2, k_3	integer constants
k_1	hysteresis loss per cycle
k_{1L}, k_{1T}	extrapolated hysteresis loss
k_2	eddy current loss constant
k_h, k_{clas}, k_{ex}	loss component constants
k_h	hysteresis loss constant
	Steinmetz coefficient

k_c	search coil design constant
k_{clas}	classical eddy current constant
k_{dc}	loss factor under bias field magnetisation
k'_{ec}	eddy current coefficient
l_1, m_1, n_1	direction cosines
l_g, l_y, l_p, l_m, l_r	magnetic circuit component lengths
l_g	air gap length
l_i	material magnetic path length
l_m	mean magnetic path length
m_a	active/effective mass of a sample
m_f	frequency modulation ratio
n_s	number of samples per cycle
p_{ref}	reference sound pressure
p_{rms}	r.m.s. sound pressure
R_t	resistance of lamination across its thickness
R_w	resistance of lamination across its width
t_0	reference thickness
t_1, t_2 , t_3, t_4	instants in time
u_1	primary voltage
$u_{2(t)}$	secondary voltage
v_m	interlaminar voltage
v_p	instantaneous primary voltage
x_t	flux transfer length
θ_0	angle between applied field and [001] direction
θ_0	reference base temperature
θ_a, θ_b, θ_c	orientation of strain gauges relative to reference direction
θ_{bh}	spatial angle between instantaneous flux density and magnetic field
θ_c	Curie temperature
θ_{ma}	angle between B_{ma} and normal direction
θ_{mb}	angle between B_{mb} and normal direction
θ_{rh}	angle between rotational magnetic field and easy direction
θ_{rh}	direction of magnetic field with respect to the *y*-direction
θ_{rb}	angle between rotational flux density and reference direction

θ_{rm}	angle between rotational magnetisation and reference direction
Φ_A	flux in the air space of an iron cored solenoid
Φ_m	magnetic flux
Φ_p	peak value of magnetic flux
Φ_t	total magnetic flux in a coil
α_{RD}	coefficient of linear expansion of steel along its RD
α_c	coefficient of linear expansion of coating
α_θ	mean value of α between θ and θ_0
β_x	retarding force on domain wall
γ_{xy}	magnetostriction shear modulus
δ_b	angle between the flux density and the RD
δ_g	elemental length of air gap
ε_h, ε_{clas}, ε_{ex}	errors in loss components
ε_a, ε_b, ε_c, ε_x, ε_y	components of strain
ε_{elast}	elastic component of strain
ε_{mag}	magnetostrictive component of strain
λ_{100}	saturation magnetostriction along [100] direction
λ_{111}	saturation magnetostriction along [111] direction
λ_e	engineering magnetostriction
λ_p	peak magnetostriction
λ_s	saturation magnetostriction
λ_x	in-plane peak to peak magnetostriction along the *x*-direction
λ_y	in-plane peak to peak magnetostriction along the *y*-direction
μ_0	magnetic constant/permeability of free space
μ_1	permeability along an easy direction of magnetisation
μ_2	permeability along a difficult direction of magnetisation
μ_{RD}	permeability when magnetised along the RD
μ_{TD}	permeability when magnetised along the TD
μ_d	differential/incremental permeability
μ_i	initial permeability
μ_m	maximum permeability
μ_{ma} μ_{mb}	permeability in medium '*a*' and medium '*b*'
μ_r	relative permeability
μ_x, μ_y	permeability along *x*- and *y*-directions

$\mu_{xx}, \mu_{yy}, \mu_{xy}$	permeability expressions as functions of μ_{RD} and μ_{TD}
$\xi_{n1}(x,t)$ $\xi_{n2}(x,t)$	sound at a point x at time t emitted from two sources
ξ_x	magnetic Poisson's ratio in x-direction
ξ_y	magnetic Poisson's ratio in y-direction
ρ_0	resistivity at temperature θ_0
ρ_s	surface value of resistivity
ρ_θ	resistivity at temperature θ
σ_{CRD}	stress in coating along RD
σ_{MRD}	shift in stress sensitivity curve measured along RD
σ_{MTD}	shift in stress sensitivity curve measured along TD
σ_{PRD}	stress induced in GO steel by differential contraction
σ_b	bending stress
σ_c	compressive stress; isotropic stress due to differential contraction
σ_{line}	line tension
σ_m	tensile stress along RD due to coating on GO steel
σ_p	planar stress
σ_{res}	residual tension
σ_{trans}	transverse stress in steel
σ_u	uniaxial stress
υ_{RD}	Poisson's ratio along RD
υ_{TD}	Poisson's ratio along TD
ψ_h, ψ_n, ψ_r	phase angles
$\psi_{h1}, \psi_{h3}, \psi_{h5}, \ldots, \psi_{h(2n-1)}$	phase angles of harmonics of H relative to fundamental value
$\psi_{b3}, \psi_{b5}, \ldots, \psi_{2n-1}$	phase angles of harmonics of B relative to fundamental value
ω_n	sound wave frequency
ω_r	speed of rotor
ω_s	synchronous speed
ϕ_1, ϕ_2	power factor angles
ϕ_a, ϕ_b, ϕ_c	instantaneous values of core flux
ϕ_{cr}	critical field
ϕ_l	leakage flux
ϕ_m	mutual flux
ϕ_θ	component of flux at angle θ to main axis

1D	one dimensional
2D	two dimensional
$2n$	number of bar domains in a grain
3D	three dimensional
A.C./a.c.	alternating current
AFM	atomic force microscope
ANN	artificial neural network
ASD	adjustable speed drive
ASST	amorphous single strip tester
ASTM	American Society for Testing and Materials
BEM	boundary integral method
BF	building factor
BLDC	brushless d.c. (motor)
BN	Barkhausen noise
CC	compilation of comments
CCW	counter clockwise
CD	committee draft
CGO	conventional grain oriented steel
CGS	centimetre–gram–second
CoFe	cobalt–iron alloy
CoFeV	cobalt–iron–vanadium alloy
CSI	current source
CSST	compensated single strip tester
CT	current transformer
CVD	chemical vapour deposition
CW	clockwise
D.C./d.c.	direct current
DFI	dark field imaging
DHM	dynamic hysteresis model
DR	domain refined grain oriented steel
DSP	digital processing hardware
e.m.f.	electromotive force
EDM	electrical discharge machine
EMC	electromagnetic compatibility
EMU	electromagnetic units
EV	electric vehicle

FDM	finite difference method
FE	finite element
FEA	finite element analysis
FEM	finite element method
GMR	giant magnetoresistance
GO	grain oriented
GPIB	general purpose interface bus
h	harmonic number
	instantaneous magnetic field
	depth of groove on DR steel
HEV	hybrid electric vehicle
HGO	high permeability grain oriented steel
HV	high voltage
ICE	internal combustion engine
IEC	International Electrotechnical Commission
ISO	International Organisation for Standards
KMO	Kerr magneto-optic
LV	low voltage
LVDT	linear variable differential transformer
MDM	magneto-dynamic model
MEE	more electric aircraft
MFM	magnetic force microscope
MI	modulation index
MKS	metre–kilogram–second
MMF	magnetomotive force
MnZn	manganese–zinc (ferrite)
MR	magnetoresistance
NiFe	nickel–iron (alloy)
NiZn	nickel–zinc (ferrite)
NO	non-oriented
NWIP	new work item proposal
ODF	orientation distribution function
OEM	original equipment manufacturer
OFGEM	Office of Gas and Electricity Markets (UK)
PC	personal computer
PCB	printed circuit board

PCl	peripheral computer bus
PD	proportional derivative controller
PM	permanent magnet; powder metallurgy
PP	purchase price
PV	photo voltaic
PVD	physical vapour deposition
PWM	pulse width modulation
RCD	residual current detector
RD	rolling direction
RE	rare earth (magnet)
RST	rapid solidification technology
SEM	scanning electron microscope
SET	single Epstein tester
SI	Système International
SiFe	silicon–iron
SMC	soft magnetic composite
SMPS	switched mode power supply
SNR	small nuclear reactor
SPWM	sinusoidal pulse width modulation
SR	switched reluctance (motor)
SST	single sheet tester
STL	statistical theory of losses
SVM	space vector modulation
TC	technical committee
TD	transverse direction
THC	total harmonic content
THD	total harmonic distortion
TOC	total ownership cost
TSM	thin sheet model
TTA	Technical Trade Association
UHDV	ultra high d.c. voltage
USB	universal serial bus
UTES	ultra-thin electrical steel
VSD	variable speed drive
VSI	voltage source

VSM	vibration sample magnetometer
WG	working group
A	cross-sectional area
	contact area
	waveform constant
B	waveform constant
	magnetic flux density
$B(S)$	flux density at time S
$B(t)$	instantaneous flux density during cyclic magnetisation
C	coefficient of surface resistance
D	density
	electric flux density
	displacement field
	current density
DF	distortion factor
E	electric field
	Young's modulus
	free energy
EF	efficiency factor
F	force
	residual transverse component of stress
$F(\gamma)$	skin effect function
Fd	force on a domain wall
G	excess loss constant
H	magnetic field
$H(B)$	value of field when cyclic flux density is maximum
$H(i)$	instantaneous field strength
$H(t)$	total surface field
I	intensity of magnetisation
	current
J	current density
	torque coefficient
	atomic molecular momentum
	magnetic polarisation
K	anisotropy constant

	anisotropy factor
	heat transmission coefficient
L	self-inductance
	width of bar domain
l	length
M	magnetisation
	mutual induction
N	number of items
	speed
NFA	not finally annealed
P	power flow
	average power
	number of pole pairs
p	ratio of length to diameter
M_m	magnetic modulus
Q	quality factor
T	period of a cyclic waveform
	torque
T, T_o	temperature at time t and t_o
U	volume
	voltage
VA	volt–ampère
W	energy stored in magnetic field
	dipole
	real power
Z	number of armature conductors
a	half thickness of a lamination
	height of burred region
	number of parallel paths
	domain width
	turns ratio of a transformer
b	instantaneous flux density
	width of bar domain in SMC particle
c	width of burr
d	equivalent depth of uniform magnetisation
	diameter

	distance
	length of a grain in a GO matrix
	width of a bar domain
d_p	length of bar domain in SMC particle;
e	ellipticity factor
f, f_1, f_2	frequency
g	average grain size
g	length of air gap
i	instantaneous current
j	instantaneous eddy current density
k	constant of proportionality
	core flux distortion factor
	eddy current loss constant
	design constant
	temperature coefficient
w_g	width of a grain in a GO matrix
l, m, n	direction cosine
m	magnetic dipole moment
	loss coefficient in SMC cores
	mass
n	hysteresis constant
	integer
	number of samples per cycle
p	eddy current function
p	instantaneous power
r	harmonic number
r	radius
s	skin depth
t	time
	sheet thickness
u	energy stored in an electric field
v	volume
v	velocity
w	grain width
	strip width
$y(t)$	instantaneous value of cyclic waveform

$\boldsymbol{A}$	vector potential
$\boldsymbol{B}$	vector flux density
$\boldsymbol{E}$	vector electric field
$\boldsymbol{H}$	vector magnetic field
Δl_{mf}	elongation due to Maxwell force
Δl_{total}	total elongation of lamination
Δl_{λ}	elongation due to magnetostriction
ΔE	change in Young's modulus due to stress
ΔP	increase in loss due to cutting stress
ΔW	percentage increase in building factor
K	texture constant
θ	angle between dipole and external field
	angle between magnetostriction and [001] orientation
	angle between groove and RD
	angle between magnetisation and easy direction
	angle between B and easy direction of magnetisation
	spatial angle between flux density and magnetic field
	temperature; fall in temperature
Φ	peak flux
Φ	phase difference between orthogonal components of flux density
Ω	density of grain boundaries
α	angle between B and H
	angle of yaw
	coefficient of linear expansion
	hysteresis loss constant
	temperature coefficient of resistance
α, β	projection angles from a point on the axis of a solenoid to its ends
α, β, γ, S	toroid magnetisation constants
α, δ	material dependent constants
γ	eddy current function
δ	angular spatial direction of rotating flux density
δ	power angle
	domain wall width
δP_h	localised hysteresis energy loss

δP_{clas}	localised classical eddy current energy loss
δh	elemental instantaneous magnetic field
δB	incremental change in flux density
δH	incremental change in magnetic field
δP	elemental eddy current loss
δi	elemental instantaneous eddy current
δl	incremental change in length
δp	incremental power flow
δx	elemental length
ε	permittivity
	strain
η	anomaly factor
	efficiency
	stray field factor
ι, κ	bias field loss parameters
λ	magnetostriction
E_θ	Young's modulus under stress
μ	permeability
ξ	loss factor
p	pole strength
ρ	resistivity
σ	conductivity
	stress
τ	time constant
	instant in time
χ	magnetic susceptibility
ψ	angle of dipole
	angle of roll
ω	angular frequency
ϕ	angle of tilt
	magnetic flux

Introduction to Volume 1

Volume 1 mainly includes background information which is useful for understanding the magnetic properties and characteristics of electrical steels which are given in detail in Volume 2. The block diagram shown below summarises the content.

Chapter numbers where most of the detail of each topic is found are shown in brackets. 'Electrical Steel' indicates chapters which include a substantial amount of content focused on aspects of electrical steel.

It is the intention of the authors to give readers, who have an engineering/science background, a full introduction to the concepts and theories applying to soft magnetic materials and, specifically, electrical steels. The authors believe that it will be particularly helpful to physics and engineering graduates, with no specialised knowledge in this field and are commencing a research or industrial career related to soft magnetic materials. Importantly, all chapters have comprehensive references to take the reader directly to the publications of eminent experts.

Much of the content of Volume 1 is directly relevant to all soft magnetic materials used in power applications. Indeed, we focus on aspects of all the families of soft magnetic materials used in large volume applications. Of course, we do not go into the depth which can be found in specialised sections of textbooks focusing,

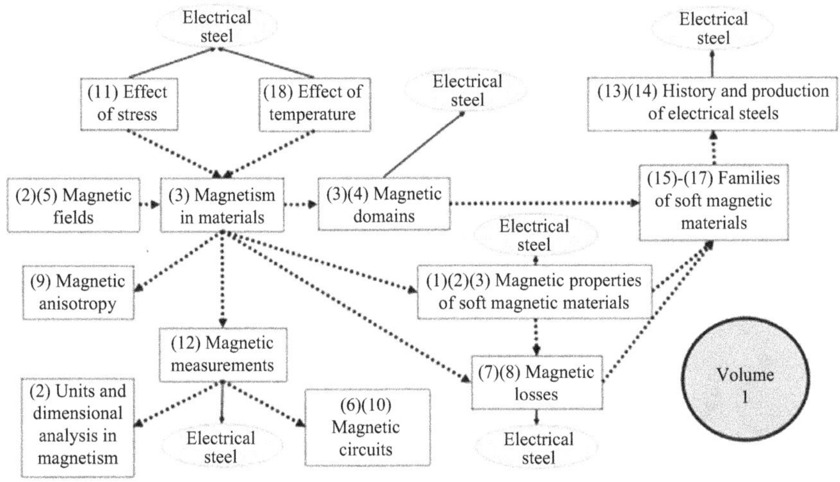

for example, on the CoFe, NiFe, soft ferrites and amorphous compositions. These are included so that clear comparisons with electrical steels can be made at any point within the book. Also, as will be seen later in this volume, there are some emerging engineering and electrical power applications where intermediate magnetic and physical properties might be needed, which none of today's commercial families of soft magnetic materials can completely satisfy. In such cases, it is important to be aware of the strengths and weaknesses of candidate materials to help focus future development in the most effective manner.

Detailed accounts of the mechanical stress and temperature dependence of electrical steels are included in Chapters 11 and 18 respectively in this volume. These broad topics could have been presented in dedicated chapters in Volume 2. However, we feel that it is more appropriate to include them here alongside discussions of their underpinning, controlling mechanisms which apply to all magnetic materials. However, more practical aspects of the effect of stress on the properties of electrical steels assembled into electrical machine cores are included in Volume 2. These are clearly cross referenced to Chapter 11 in this volume.

About the authors

Anthony John Moses is Emeritus Professor of Magnetics at Cardiff University, UK, where he was previously Director of the Wolfson Centre for Magnetics. His research focuses on aspects of the production, characterization and application of soft magnetic materials. He has authored more than 500 papers. He previously served as Chairman of the UK Magnetics Society and Chair of the International Organizing Committee of the Soft Magnetic Materials (SMM) series of conferences.

Philip Ian Anderson is a Senior Lecturer in the Magnetics and Materials Group at Cardiff University's School of Engineering. He has 25 years of research experience focussing on the development, application and characterisation of soft magnetic materials. He is a member of the British and International Standards Committees on Magnetic Alloys and Steels together with the International Organising Committees of several major magnetics conference series.

Keith W. Jenkins worked at the British Steel Electrical Steels Research Department after graduating in Metallurgy at the University of Sheffield. He also worked in Orb Electrical Steels for over 35 years in a variety of technical and research functions, where he built up a wide network of contacts with other electrical steel makers, transformer and motor manufacturers, including Nippon Steel (Japan), Posco (Korea), ThyssenKrupp (Germany), AK Steel (USA), Siemens and ABB (Europe), Rolls-Royce (UK). Recently, he has been an honorary visiting professor at the Magnetics and Materials Group at Cardiff University's School of Engineering.

Hugh John Stanbury joined the Electrical Steels Research Department of the British Steel Corporation after graduating in Physics from Imperial College. He progressed to Technical Manager of Orb Electrical Steels, Cogent Power, and was responsible for the development of their United Kingdom Accreditation Service Magnetics Standards Laboratory. He currently serves as chair of the British Standards Institution Technical Committee for Magnetic Alloys and Steels and is a former Chair of the International Electrotechnical Commission Technical Committee for Magnetic Alloys and Steels.

Chapter 1
Soft magnetic material

1.1 Range and application of commercial bulk magnetic materials

The growing reliance on renewable energy, increasing concern over energy efficiency and the drive towards hybrid and all-electric transport stimulate the need for high performance magnetic materials and their most effective use in electrical equipment. Magnetic materials are broadly classed as being either *hard* or *soft* based on how difficult it is to alter their internal magnetisation.

An efficient hard magnetic material has to be placed in a large magnetic field to become fully magnetised, but when removed from the field it retains that magnetism. In the magnetised state it is commonly referred to as a permanent magnet. On the other hand, a soft magnetic material is much easier to magnetise fully but it loses most, or even all, of its net magnetisation when the field is removed. Soft magnetic materials are sometimes referred to as temporary magnets. The first commercial permanent magnet alloys happened to be mechanically hard and the first metallic magnetic materials suitable for use in rapidly changing magnetic fields happened to be mechanically soft; this nomenclature is still used today irrespective of the mechanical state of the magnetic material.

The magnetic flux density is of major importance in any engineering application. Whether it has a constant magnitude (d.c.[1] or static) or changes with time (a.c. or dynamic), its fundamental role may be one or more of the following three basic functions:

- to produce a voltage proportional to its rate of change with time,
- to generate a force proportional to its magnitude on a current-carrying conductor or
- to create an attractive force between magnetised surfaces proportional to the square of its magnitude.

The purpose of a hard or soft magnetic material in electromagnetic devices such as electricity generators, motors, actuators or transformers is to produce and help control the magnitude and direction of the magnetic flux needed in a power transfer or conversion process. In a transformer the flux effectively acts as a medium to convert electrical energy from one voltage level to another. In a motor the magnetic flux in the air gap between the rotor and stator determines the transfer

[1]Throughout this book a.c. and d.c. have the conventional meanings of alternating current and direct current respectively.

of mechanical energy to or from electrical energy, whereas in a generator the airgap flux also has a strong effect on the wave shape of the resulting output voltage.

Hard magnetic materials are critical components in many household, transport and industrial applications often incorporating low cost, compact, high power, energy efficient actuators, motors and generators. Although hard magnetic materials are not a theme of this book, it is useful briefly to compare some properties with those of soft magnetic materials. After all, in many applications they are used alongside each other.

Figure 1.1 shows where the most common families of magnetic materials roughly lie on a saturation magnetisation versus coercivity graph. These parameters are two of the most important characteristics of any magnetic material. The commercial nickel–iron (NiFe) alloys, amorphous magnetic ribbon and soft magnetic nano-crystalline materials have the lowest coercivity indicating that they are the easiest of all soft magnetic materials to magnetise or demagnetise. Neodymium–iron–boron (NdFeB) compounds, members of the rare earth magnet family, have the highest coercivity so they can retain a state of permanent magnetisation better than any other commercial hard magnetic material. It is interesting to note that the magnitude of the coercivity varies by a factor of over 100 million between the softest and hardest material.

The physical volume or weight of magnetic components and also the quantity of magnetic energy that electromagnetic systems can store is dependent on the saturation magnetisation of the incorporated soft or hard magnetic materials. Figure 1.1 indicates that the highest saturation magnetisation in common commercial magnetic materials is achieved in cobalt–iron (CoFe) alloys and the lowest in the soft and hard

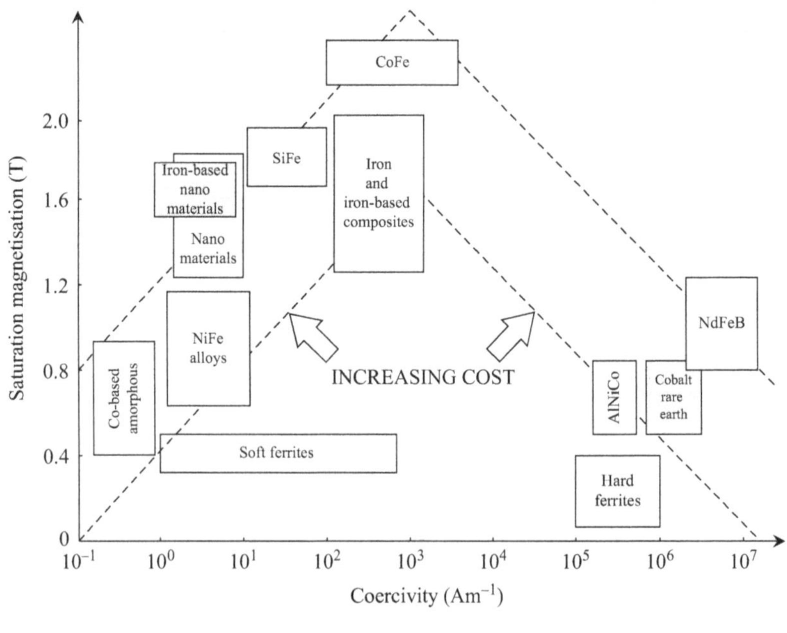

Figure 1.1 Range of coercivity, saturation magnetisation and cost trends of families of commercial soft and hard magnetic materials

ferrites. Pure iron itself has the highest saturation magnetisation of any element but its resistivity is very low so eddy current losses, which will be seen later to be crucially important in soft magnetic materials, are very high under alternating (a.c.) magnetisation. Therefore its use is mainly restricted to d.c. applications.

It is perhaps not surprising that both the best soft materials (very low coercivity together with high saturation magnetisation) and the best hard materials (very high coercivity together with high saturation magnetisation) are the most expensive. This is due to high costs of either the raw material, the production process or both. The families of soft magnetic materials included in Figure 1.1 are discussed in detail in Chapters 15–17.

In addition to saturation magnetisation and coercivity, permeability and energy losses are important properties of soft magnetic materials in many applications. The largest single application of soft magnetic materials is found in power conversion equipment where energy is transmitted via magnetic flux such as in a transformer core. High magnetic permeability of the magnetic material helps tightly link the flux between primary and secondary electrical windings and, at the same time, concentrates the flux to enable high energy transfer and to minimise core size. Soft magnetic materials used in motor and generator cores have a similar function, but an effective combination of their specific magnetic properties and core geometry helps control and optimise the magnitude and spatial profile of the magnetic flux in the air gap between the stator and rotor needed to obtain effective power transfer.

Another important application of soft magnetic materials is as magnetic shielding for protecting sensitive electrical equipment from electromagnetic radiation or, conversely, by providing a barrier to stop large magnetic fields, such as those generated in medical imaging systems, from being a potential human hazard. The basic shielding is simply a box of a suitable magnetic material designed to capture the unwanted airborne magnetic flux and divert it away from the area to be protected.

1.2 Industrially important characteristics of soft magnetic materials

In most applications the main purpose of soft magnetic materials is to control either or both the direction and magnitude of magnetic flux. However, there are many factors which determine how suitable a material might be for a specific application. The first consideration is what specific magnetic flux conditions are required in the material, e.g. its magnitude, d.c. or a.c., and, if a.c., the magnetising frequency and magnetic flux waveform. Next, the working environment must be considered to ensure that the magnetic performance is not adversely affected by factors such as high or low ambient temperature or subjection to mechanical stress.

The a.c. magnetisation will generate energy losses and acoustic noise might be generated due to magnetostriction. The determination of the most appropriate material is usually a balance between material cost, performance and application; however, the ultimate suitability of a material might be determined by the availability of a material in a physical form compatible with component design and manufacturing steps.

If the magnetising frequency is fixed in a particular application the range of possible suitable materials is reduced. Eddy currents, sometimes still referred to as Foucault currents in recognition of the discoverer,[2] Léon Foucault, are created within a magnetic material when its internal magnetisation changes and if the change is rapid, e.g. at high magnetising frequency, the effective magnetic permeability falls so the material becomes difficult to magnetise, also magnetic losses increase. It is worth clarifying what is meant by *magnetic losses*. In a magnetic material the term *loss* simply refers to energy transferred from an electromagnetic state into heat or sound (via mechanical vibration) during the magnetising process. Similar expressions are used such as iron loss, core loss, power loss, energy loss or specific total loss. Of course, energy is not being lost but it is simply being converted into another form which, in most applications, represents some form of inefficiency.

Table 1.1 separates the families of soft magnetic materials according to the range of frequencies at which they are most effectively magnetised in the most common applications. The applications and commonly used materials given in each

Table 1.1 Common high frequency applications of soft magnetic materials

Magnetisation frequency band	Typical range	Strip thickness range	Applications	Commonly used materials
D.C.	<1 Hz	Any	Shielding, relays and actuators	SiFe SiFe Ferrites Amorphous
Power	25 Hz to 1 kHz[a]	0.18–0.65 mm	Transformers, motors, generators and power conditioning	SiFe CoFe NiFe Ferrites Amorphous Powder composites
Communications and sensors	1–100 kHz	0.05–0.10 mm	Inductors, switched mode power supplies and filters	Ultra-thin SiFe Amorphous NiFe alloys Nano-materials Powder composites Ferrites
Radio	0.1–1,000 MHz	–	Inductors and antennae	Ferrites
Microwave	<1 GHz	–	Isolators, circulators and phase shifters	Ferrites

[a]These refer to the fundamental or first harmonic component of magnetising frequency.

[2]Another Frenchman, François Arago discovered the effect of eddy currents in a copper plate 30 years earlier. He called the phenomenon *magnetisation of rotation* and is sometimes credited jointly with Foucault with the discovery of eddy currents.

frequency range are not exclusive, boundaries are generally blurred by other considerations so the list only gives an indication of the spread.

Commonly used abbreviations for material groups given in the table should be defined here to avoid any misunderstanding later. SiFe refers to all silicon–iron alloys with 0.2%–6.5% silicon content (by weight); NiFe refers to nickel–iron alloys with normally around 36%, 50%, 65% or 80% nickel; ferrite refers to NiZn or MnZn ferrites; amorphous materials include iron-based compositions used at low magnetising frequencies and those which are nickel-based or cobalt-based used at high frequencies; CoFe refers to alloys normally around 49% or 35% cobalt, the balance being iron and small additions of other elements. Powder composites are restricted to iron or silicon–iron compressed powder cores and nano-materials are nano-crystalline alloys containing mainly iron, boron and silicon. The properties of all of these alloys are discussed in Chapters 15–17.

It is important to note the large magnitude of variation, by a factor of 10^9, between the lowest and highest magnetising frequencies encountered in practical applications of soft magnetic materials. It can be seen in Table 1.1 that, as the frequency increases, the practical metallic strip thickness is significantly reduced (to restrict eddy currents as will be explained in Section 3.3) but at very high frequency a limit is reached when further reduction becomes impracticable. Fortunately, soft magnetic ferrites can still be used even in the MHz region because of their extremely high electrical resistivity compared to metallic strip.

The practical strip thickness ranges included in the table largely apply to any composition of magnetic metallic strip irrespective of their magnetic performance at low frequency since the eddy current phenomenon dominates over all other magnetic properties at high frequency. It has been suggested [1] that above a magnetising frequency of around 5 kHz, the physical and microstructural factors which control the losses and permeability at lower frequency only have a small influence and the influence of sheet thickness dominates.

Today's electrical steel can only operate at the lower end of the frequency range (up to around 20 kHz in ultra-thin steel) but if it could be economically produced much thinner it could be used effectively at a much higher frequency. In contrast, amorphous and nano-materials are most conveniently produced as very thin strip which naturally makes them more suitable for high frequency magnetisation.

In Table 1.1, it can be seen that under d.c. magnetisation there is no limit on the material thickness. This is simply because there is no need to resort to thin sheet since eddy currents are not normally present. Hence soft magnetic materials used under d.c. magnetisation are commonly used in the form of bars, rods and other shapes to a large extent, just like permanent magnets. Even silicon steel in this form is not referred to as electrical steel.[3] The term was in use in the 1960s and appears

[3]The term electrical steel is used throughout this book when referring to thin iron rich alloys containing small quantities of silicon or/and aluminium and some smaller amounts of other elements which are included to tailor magnetic or physical properties. A more exact term would be iron–silicon alloys. Old terminology includes dynamo steel, magnetic steel and electrotechnical steel. Some experts in the industry maintain that the material deserves a more dedicated name which acknowledges its complex production route, unique microstructure and specialised application which makes it far different from the more well-known families such as carbon steels, alloy steels, stainless steels, structural steels and tool steels.

to have informally become reserved for cold-rolled strip of thickness up to 1.0 mm used in a.c. magnetisation applications. Much of the general content of this book is relevant to the d.c. magnetisation and application of the bulk silicon steels, but it is not specifically highlighted.

Every magnetic property of any soft magnetic material can be classified as being either *structure-insensitive* or *structure-sensitive.* The structure-insensitive properties depend only on temperature and chemical composition. This, of course, includes some influence from the presence of any impurities in a specific alloy. Structure-sensitive properties are determined as a result of the influence of microstructure, at the nano-metre scale, on the magnetic domain structure. Hence the structure dependent properties of a magnetic material are affected by the production process, mechanical stress and heat treatment. All the properties are liable to be temperature sensitive to different degrees dependent on factors explained in Chapter 3.

The most commonly encountered structure-sensitive and structure-insensitive properties are listed in Table 1.2. The table includes properties termed engineering magnetostriction, which is structure-sensitive, and the magnetostriction constants, which are structure-insensitive. These two parameters play important roles in the effective application of electrical steels and their different structural dependencies are explained in Section 3.7.

A very practical factor to consider often in material selection is the possible variation of magnetic properties from batch to batch of material of nominally the same composition. Some products are known to be difficult to produce to the tight magnetic tolerances increasingly called for in components with critical performance specifications.

In practice, the magnetic properties of a soft magnetic component depend on complex relationships between the material structure and shape, ambient environmental conditions and magnetisation conditions. Broad relationships between these parameters are illustrated in Figure 1.2. It is impossible to quantify the effect of all (or even a few) of these factors on the magnetic properties of a given material but Volume 2 will give readers an understanding of their possible occurrence and subsequent influence on the performance of electromagnetic devices incorporating any type of soft magnetic material.

Table 1.2 Common structure-insensitive and structure-sensitive properties of magnetic materials

Structure-insensitive	**Structure-sensitive**
Saturation magnetisation (M_s)	Permeability (μ)
Crystal anisotropy constants (K_1, K_2)	Coercivity (H_c)
Magnetostriction constants ($\lambda_{100}, \lambda_{111}$)	*B–H* and *M–H* relationships
Curie temperature (θ_c)	Magnetic losses
Resistivity (ρ)	Engineering magnetostriction (λ)

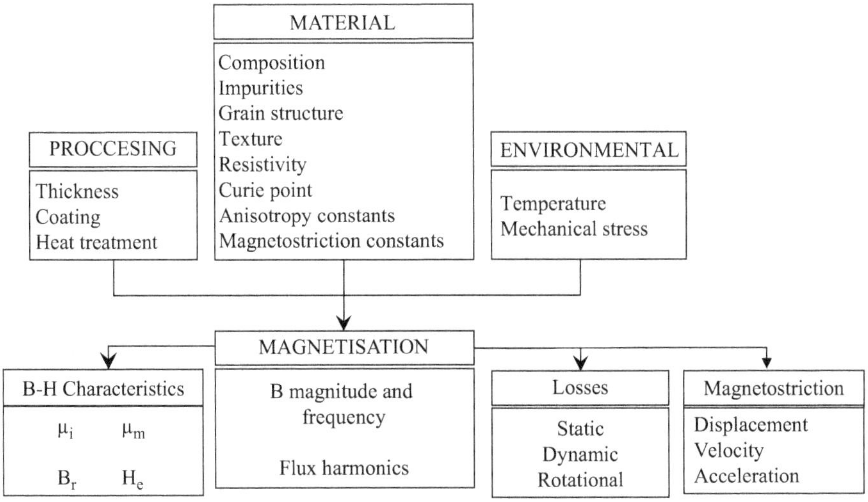

Figure 1.2 Relationships between key parameters and magnetic properties of soft magnetic materials

1.3 Families of commercial soft magnetic materials

The historic development of commercial soft magnetic materials has been driven by scientific or metallurgical breakthroughs brought about when searching for solutions to higher performance specifications that are periodically demanded by users. The need for very high permeability material to enable the expansion of worldwide telecommunication systems in the 1930s encouraged major developments in NiFe alloys and the sudden realisation of the need for low loss magnetic material in response to global energy production crises in the 1970s stimulated the commercialisation of iron-based amorphous materials together with major advances in electrical steels.

Today, families of commercially available soft magnetic materials are used either as alternatives to electrical steels in energy-sensitive applications or where new magnetic demands exceed the capability or effectiveness of standard electrical steels. The generic properties of these groups are briefly introduced here and fuller accounts of the magnetic properties and applications of the most important members of these families are given in Chapters 15–17.

Soft magnetic thin films and other specialised soft and semi-soft magnetic alloys not directly used in what can be termed *power magnetic* applications are not included in this book. (The authors coined the expression power magnetics as an all-embracing term covering aspects of properties of soft magnetic materials, mainly at power frequency magnetisation, and their application in magnetic cores ranging from grams to many kilograms in mass.)

Table 1.3 Key milestones in the development of soft magnetic materials

1900 Silicon–iron (Hadfield)
1921 Iron powder cores (Speed and Elmen)
1923 Nickel–iron alloys (Smith, Garnett, Arnold and Elmen)
1932 Iron–cobalt–vanadium alloys (White and Whal)
1934 Grain oriented silicon–iron (Goss)
1947 Soft ferrites (Snoek)
1974 Amorphous alloys (Chen and Polk)
1987 Nano-crystalline alloys (Yoshizawa, Yamauchi and Oguma)

The emergence of each of the new families of soft magnetic materials has stimulated electrical steel manufacturers to find ways of steadily improving their own product ranges. This has helped maintain their dominating share of the soft magnetic materials market, both in terms of tonnage and monetary value until the present time.

Historic milestones in the discovery and commercialisation of all of today's commercial soft magnetic materials are summarised in the Table 1.3.

The names of the main contributors to the discovery of these materials are included in the table; most of these will be referred to in later chapters. Alloy compositions have remained broadly unchanged since their infancy and the materials produced today are greatly improved versions of these original materials. All these families were discovered and commercialised over a 90-year period in history but no others have emerged during the past 30 years. As shown in later chapters, there are continual incremental advances in most of these materials. It is an open question whether a completely new family will be discovered in the foreseeable future, but new manufacturing techniques could lead to the next revolution in magnetic materials.

Iron and carbon steel alloys were the first soft magnetic materials to be produced commercially more than 100 years ago driven by the demand for magnetic cores for the first motors, generators and power transformers. Probably, the first application of thin iron sheet was even longer ago when, in about 1837, Sturgeon found that the performance of an induction coil was improved when the solid iron core was replaced by thin iron laminations [2].

It should be noted that, ever since its inception, electrical steel has consistently accounted for around 1% of global steel production; so, not surprisingly, even today, the bulk of commercial electrical steel originates from basic steel making routes not designed for the optimisation of magnetic properties. The importance of this and the sequential stages in the manufacturing cycle on the development of magnetic properties is discussed in Chapters 13 and 14.

As the telecommunications and general electronics industries became established towards the mid-twentieth century, demands for far higher permeability than could be offered by electrical steels grew. This, together with the drive towards higher magnetising frequencies, stimulated the development of soft magnetic ferrites and NiFe alloys. The soft ferrites are magnetically inferior to electrical steels but far cheaper, whereas the NiFe alloys are far more expensive but have far better high frequency properties. The ferrites are particularly suitable for power electronic systems involving non-sinusoidal waveforms, usually triangular or trapezoidal, at operating frequencies up to more than 1 MHz.

The oil crises of the 1970s caused a greater awareness of the significant proportion of electrical power generated in industrialised countries which was, and still is, being wasted as magnetic losses in the cores of electrical machines. This stimulated development of amorphous magnetic materials which have significantly lower losses than electrical steels, even at power frequencies. Today, iron-based amorphous materials are widely used in magnetic cores installed in distribution transformers in electricity distribution networks throughout the world. However, their disadvantages, discussed in detail in Chapter 15, have restricted wider take-up, particularly where the true cost of lifetime ownership is not taken into account.

Present day environmental concerns and drives for energy efficiency or conservation might provide an incentive for the use of soft magnetic nano-crystalline magnetic materials in power applications. Although far more expensive than electrical steels at present, innovative approaches to the design of electrical energy conversion systems being developed today could enable their excellent high frequency magnetic performance to be exploited, particularly in transport electrical drive applications.

Powder metallurgy processing developments and cost effective means of producing accurately dimensioned bulk net-shape soft magnetic parts have escalated interest in iron alloy powder composites. These composites can be used to produce cheap, small magnetic cores in large volumes. Also, the technology provides the capability of manufacturing complex three-dimensional core topologies which are difficult or impossible to produce as laminated steel assemblies. Such topologies might offer electromagnetic performance not possible in designs based on conventional rotating machine core geometries.

Indicative values of important magnetic parameters of commercial soft magnetic materials are summarised in Table 1.4. Non-oriented (NO) electrical steel and grain oriented (GO) SiFe included in the table are given systematically and in more detail in Chapter 15 of Volume 2 of this book. Other properties, such as thermal conductivity, thermal expansion, density, tensile strength, yield stress and hardness which are often important in applications of electrical steels are not discussed in this book. It should be noted that there are wide ranges of values of permeability and losses between grades of materials within some family groups according to material purity, heat treatment or other processing. The quoted values are simply mid-range examples.

When comparing magnetic properties, such as power loss and permeability, it is important to note carefully the magnetising frequency and peak flux density under which measurements are made in order to make meaningful comparisons. For example, the maximum (highest) relative permeability values given in Table 1.4 are obtained from d.c. magnetisation characteristics whereas data presented in Figure 1.3 are typical values at 10 kHz magnetising frequency. If the permeabilities are compared at another magnetising frequency and flux density, different relative values would be obtained. Neither case would necessarily be representative of each material under its optimum magnetising conditions.

Table 1.4 Comparison of typical values of magnetic properties of commercial soft magnetic materials[a]

Material	Saturation magnetisation (T)	Resistivity (μΩm)	Typical thickness range (mm)	Curie temperature (°C)	Saturation magnetostriction (×10^6)	Maximum relative permeability under d.c. magnetisation
Pure iron	2.15	0.15	–	770	+21	6,000–300,000
NO SiFe	2.0–2.1	0.40–0.50	0.30–0.50	750	+20	30,000
GO SiFe	2.04	0.48	0.15–0.35	750	+20	35,000
MnZn ferrite	0.25	2×10^6	-	200	–	6,000
SiFe powder composite	1.2–1.9	10,000–80,000	NA	750	–	–
49/49/2% CoFeV	2.35	0.35	0.3	950	+72	10,000
35%–80% NiFe	1.50–1.55	0.40–0.60	0.1	250–500	+1.0 to +25	10,000–400,000
Fe-based amorphous material	1.5–1.6	140–150	0.025	290–370	+20	50,000
Fe-based nano-material	1.55	140	0.02	415	27×10^{-6}	50,000 at 1 kHz

[a]In some cases characteristics are not widely quoted or appear to vary very widely. No data are entered in these cases.

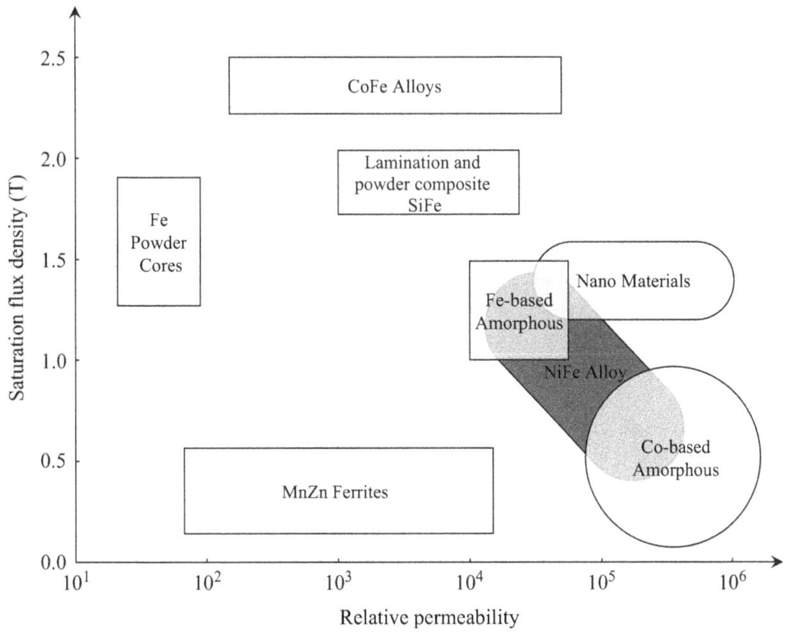

Figure 1.3 Typical ranges of common operating regimes of soft magnetic materials related to their relative permeability and saturation magnetisation

The properties of both high relative permeability and high saturation magnetisation are required in many applications. Typical permeability and saturation magnetisation properties of the groups are compared in Figure 1.3. The permeability versus magnetising frequency varies from material to material so, as a compromise, its value at 1 kHz magnetisation frequency is shown here (typical d.c. values are shown in Table 1.4 for some of these materials). More specific information is given about the range of properties of individual products within each group in Chapters 15–17. Of course, a magnetising frequency of 1 kHz is not the optimum for each group of materials, e.g. the permeability of electrical steels is far higher at, say, 50 Hz than at 1 kHz whereas nano-materials are most suitable in the 10s of kHz magnetisation regime.

Another informative way of comparing general characteristics of soft magnetic materials is to plot the relationships between their relative permeability and coercivity as shown in Figure 1.4. Using this performance measure, cobalt-based amorphous material has the best combination of low coercivity and high permeability but its poor saturation magnetisation compared with electrical steels is obvious in Figure 1.1. It should be noted that these comparisons are not made under the ideal operating conditions of several of the materials.

Typical losses of various groups of materials are compared at magnetising frequencies of 50 Hz and 1,000 Hz (at a peak flux density of 1.0 T) in Table 1.5.

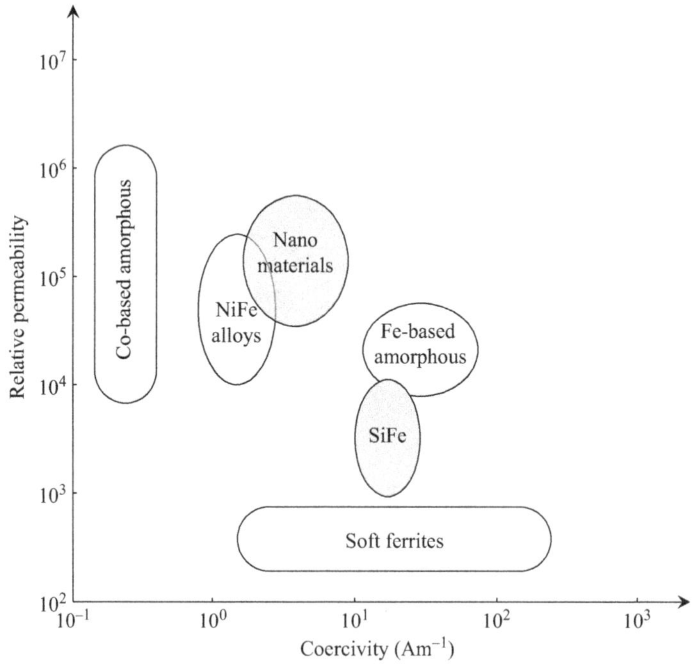

Figure 1.4 Ranges of d.c. maximum relative permeability and d.c. coercivity of soft magnetic materials (updated version of Figure 13 in [3])

Table 1.5 Typical losses of common soft magnetic materials under the same magnetising conditions

Material family	**Loss at 1.0 T, 50 Hz (W kg^{-1})**	**Loss at 1.0 T, 1,000 Hz (W kg^{-1})**
NO SiFe (0.35 mm)	0.8	70
NO SiFe (0.20 mm)	1.0[a]	50
GO SiFe (0.35 mm)	0.4	45
GO SiFe (0.20 mm)	0.2	35
SiFe powder composite (5 × 5 mm)	3–4	90–200
49/49/2% FeCoV alloy (0.35 mm)	1.5	60
50/50% NiFe (0.35 mm)	0.20	50
Fe-based amorphous ribbon	<0.1	5

[a]It might be expected that, because the loss of the thin material is lower than that of the thick material at 1,000 Hz magnetising frequency, the same should be expected at 50 Hz. It will be seen later that the variation of loss with thickness in electrical steel depends of the proportions of hysteresis loss and eddy current loss occurring at any particular magnetising frequency, so the frequency variation shown in Table 1.5 is not unexpected.

Again, the magnetising conditions are not necessarily those under which each group normally would be used in practice in order to take full advantage of their particular basic characteristics; so the figures only give a performance guide. Furthermore, the properties of laboratory produced materials can be far superior to those of commercial materials presented here. For example, laboratory produced 0.1 mm thick GO steel has losses comparable to those of commercial iron-based amorphous material [4].

Electrical steels account for some 96% of all commercially produced soft magnetic materials by weight and around 83% by monetary value as shown in Figures 1.5 and 1.6 (updated using Figure 3 in [5] and Figure 1 in [3]). Commercial

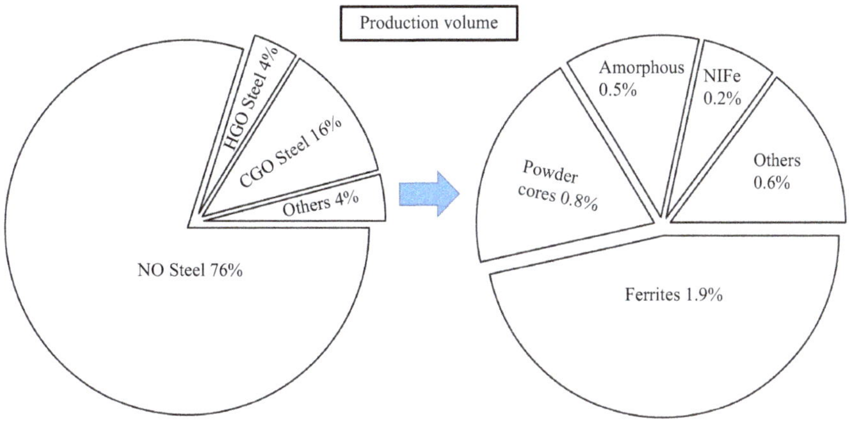

Figure 1.5 Distribution of approximate worldwide annual production of soft magnetic materials

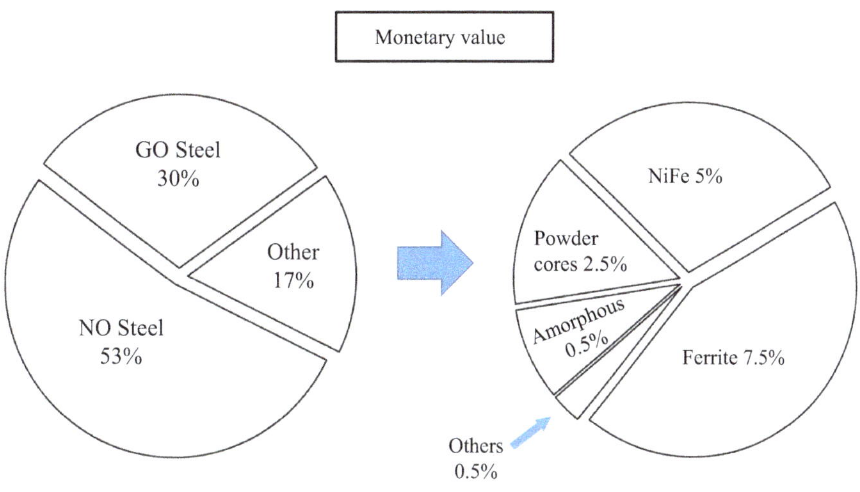

Figure 1.6 Distribution of approximate annual value of soft magnetic materials

production levels of nano-crystalline material are still too low to register in this figure so fall within the 'Other' category.

The estimates of monetary value are not very accurate since they are obtained from various sources compiled in different ways and the selling prices of all the products fluctuate widely over short periods of time in line with global production capacity and user demand.

It was pointed out earlier that magnetising frequency is often the factor which most strongly influences material selection. Other pressures follow according to the particular application. If space is restricted high saturation magnetisation becomes a critical requirement whereas instrument transformer cores operate at very low flux density where a linear *B–H* loop is essential and saturation magnetisation is of no consequence.

Where energy efficiency is a priority, low magnetic losses become a priority which usually makes high electrical resistivity an important requirement. If acoustic noise is to be minimised then the engineering magnetostriction must be as low as practicable. Table 1.6 shows ranges of parameters which are critical for common applications. Many applications are cost driven so performance is less of a priority. In general, the most effective material selection has to be based on a complex optimisation of material cost, magnetic and mechanical performance and the cost of the magnetic losses over a component's lifetime.

1.4 Electrical steels

Before 1900, wrought-iron or better quality Swedish iron were used for electrical machine cores. The birth of modern electrical steels occurred as a result of research carried out by Sir Robert Hadfield and colleagues in Sheffield, UK at the end of the nineteenth century. As early as 1882, he noticed the unexpected high mechanical hardness of an iron alloy which had been accidentally produced containing 1.5% silicon [7]. Magnetic studies followed later when it was found that the addition of 2.0%–2.5% silicon reduced the coercivity of the standard iron used in electrical machine cores [8].

The first commercial production of hot rolled silicon steel started in Germany and the USA in about 1903 and in the UK in 1909. Magnetic properties gradually improved as the understanding of the harmful effects of impurities became understood. Work, mainly by Norman P. Goss in the 1930s, led to the commercialisation of GO SiFe with a step reduction in core losses and coercivity when magnetised along the rolling direction (RD) of the sheet. The next and final major improvement to date occurred in the 1970s with the commercialisation of high permeability GO SiFe.

The 1970s was a period of great activity in the development of electrical steels, particularly GO grades aimed primarily at loss reduction in light of the global oil crisis of the time. This brought special coatings, thinner grades and magnetic domain refinement (laser scribing). These major innovations were partly prompted by iron-based amorphous materials beginning to make inroads into the traditionally GO steel dominated distribution transformer core market.

The NO grades of electrical steel have developed less slowly with focus on consistent quality and on reducing production costs. In the mid-1980s, low carbon

Table 1.6 Key requirements from soft magnetic materials for specific applications (based on Table 5 in [6])

Application	Special requirements	Potential materials
Power frequency transformer; motor and generator cores	High induction at 800 Am^{-1} Low losses at 1.5–1.7 T High M_s High permeability	SiFe Amorphous FeNi ($f > 50$ Hz)
Medium frequency transformers (400 Hz to 20 kHz)	High M_s Low losses Easy magnetisation Good high temperature performance	SiFe ($f < 1$ kHz) NiFe Amorphous Nano-crystalline ($f > 2$ kHz)
Switches, relays and circuit breakers	'R' (rounded) B–H loop High permeability Low coercivity High resistivity Mechanically hard Resistant against humidity	SiFe NiFe Nano-crystalline Ferrite
Magnetic shielding	High permeability	SiFe NiFe Amorphous Ferrite
Transducers	High permeability Temperature stability Low losses 'R' (rounded) B–H loop	SiFe NiFe Nano-crystalline Powder composites
Electronic components (e.g. power supplies, pulse transformers, chokes and HV d.c. transformers)	'F' (flat topped) loops Linear B–H loop Temperature stability Low losses High permeability	Amorphous NiFe Nano-crystalline CoFe Powder composites
Telecommunication components (e.g. cores, antennae, filters and oscillators)	High initial permeability Low loss angle (tan δ)	

Note: Throughout this book, HV refers to high voltage.

NO steels became available in response to a growing demand for energy efficient grades brought about by higher motor efficiency standards being introduced by many regulatory authorities around the world.

Table 1.7 shows the diverse range of applications of electrical steels. The list is not exhaustive or exclusive and some of the alternative soft magnetic materials listed in Table 1.4 have well established market shares of some of these application areas.

The application range is split between the two main groups of electrical steels produced in thin strip form, i.e. the NO steels where one aim is to have similar magnetic properties when magnetised in any direction in the plane of the strip, and the GO steels where the main aim is to have optimum magnetic properties when

Table 1.7 Common applications of electrical steels

Non-oriented (NO) steels	Grain oriented (GO) steels
Small and large machine cores including stepper motors, small transformers, freezers and domestic appliance motors, medium frequency motors, generators, electric vehicle drives, shielding, voltage regulators, welders, battery chargers, audio and television, standby generators and fractional horse power motors	Power transformers, distribution transformers, power inductors and filters, large generators, shielding, current transformers, voltage regulators and actuators

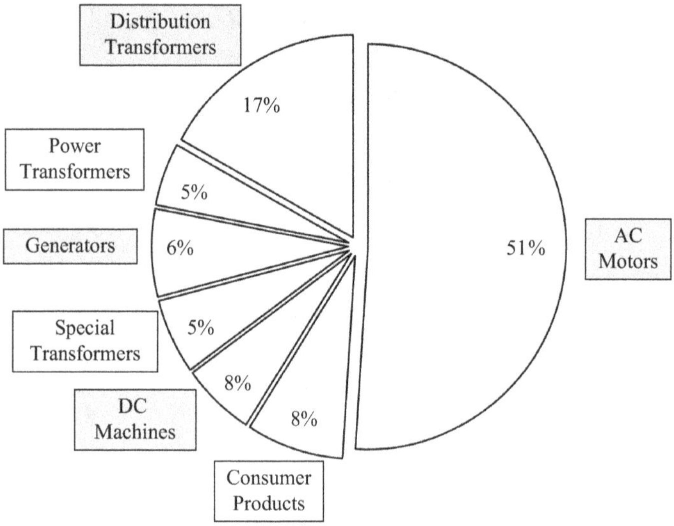

Figure 1.7 Usage of soft magnetic materials by application sector

magnetised along the RD of the strip despite it having to come at the expense of poorer properties if magnetised along other directions.

NO steel may be supplied in a ready for use, fully processed form or in a semi-processed (semi-finished) form where the customer carries out a final heat treatment to develop fully the magnetic properties. High value sub-sets produced in far smaller volumes, but with growing markets, are high silicon NO alloys and NO ultrathin strip. Figure 1.7 shows the proportions of usage of electrical steels across its main application areas. The majority of GO steel is used in transformers, taking advantage of its low losses when magnetised along its RD. Most NO material is used in rotating electrical machine cores where isotropic magnetic properties are desirable. The higher quality, more expensive NO steel grades are used where losses are an important factor, and the lower quality, lower cost grades are used when losses are not at such a premium and low cost is the most important factor.

Electrical steel is graded and sold according to the magnitude of its a.c. losses at a specified magnetising frequency and flux density. Other properties such as permeability and magnetostriction are very important in many applications but the magnetic loss is a good indicator of a steel's overall magnetic quality so it is a common factor used in evaluating new products. The historic downward trend in the average losses of best grade commercial electrical steel is shown in Figure 1.8. The initial discovery of silicon steel and then the introduction of GO material caused the two significant sudden drops in losses shown on the curve. Smaller step improvements have occurred due to the introduction of high permeability GO steel, domain refinement processes and the steady introduction of thinner materials.

Note that the trends in Figure 1.8 are shown for typical grades of 0.35 mm thick materials at a magnetisation of 1.5 T, 50 Hz. Today, thinner grades of both families are tailored for use at higher flux density or frequency so the graph is not meant to show the optimum performance of modern steels.

The loss of the best grade of electrical steel in the 1950s was around 60% of that of the first batches of silicon steel produced 50 years earlier. In the following 50 years the loss of the best grades had fallen by another 20% and it has continued to drop on average by around 3% per annum due to improved processing, better control of impurities, etc. The substantial economic and environmental benefits of this continued improvement is quantified in Chapter 14 of Volume 2 of this book.

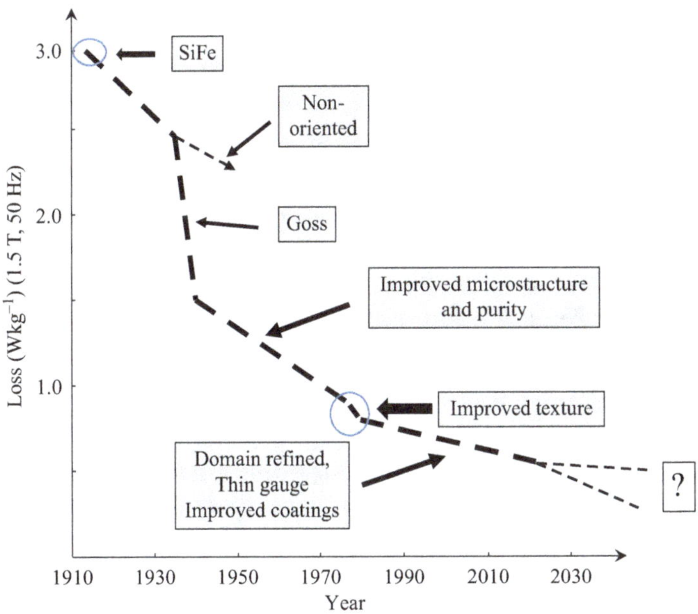

Figure 1.8 Improvement of best commercial grades of electrical steel since its discovery

It is interesting to hypothesise on a theoretical lowest possible loss which can occur in GO steel in particular. It was predicted many years ago that there was a minimum loss of 0.62 W kg^{-1} which could be attained in 0.3 mm thick GO steel magnetised at 1.7 T, 50 Hz [9]. Laboratory produced single crystals have been reported to have losses of 0.5 W kg^{-1} under the same magnetising conditions [10], more than 40% less than today's commercial best. Polycrystalline, 0.15 mm thick, 2.9% SiFe produced on a laboratory scale was reported with a loss of only 0.4 W kg^{-1} at 1.7 T, 50 Hz [11]. More recently, 0.15 mm thick development material with a loss of 0.35 W kg^{-1} at 1.7 T, 50 Hz has been produced in Japan [12]. The corresponding theoretical minimum loss in this material obtained using the approach in [9] is 0.38 W kg^{-1}. The actual loss of development material appearing to be lower than the theoretical value should not be a concern since the theoretical approach is now more widely accepted as being based on dubious grounds as explained in Chapter 4 of Volume 2 of this book.

However, the laboratory results confirm that there is still scope for considerable improvement in today's GO steel but obviously it is a massive challenge to reproduce such laboratory achievements on a commercial scale. To achieve such loss levels almost complete removal of carbon, sulphur and nitrogen (to $<0.001\%$) would be required and, although possible, the cost implications due to extended processing times would be prohibitive. In addition, the surface roughness of the steel would need to be significantly reduced to that described as a 'mirror finish', which again, although possible, would require additional processing steps and increased costs over those currently used. Improved texture control would need to be accompanied by increased domain refinement through greater optimisation of scribing and/or improvements in tension coatings.

Hadfield's discovery of the magnetic benefit of adding small amounts of silicon to iron occurred purely from the result of an accident in alloy preparation. The common adoption of around 3.0%–3.5% silicon is a compromise between higher concentrations which cause embrittlement and lower concentrations where the beneficial increased electrical resistivity due to silicon addition is not so pronounced. Furthermore, grain orientation was found to be less difficult to achieve than in other compositions. Today's electrical steels typically contain 0.2%–3.5% silicon and the balance is iron.[4,5]

The development of new soft magnetic alloys requires a systematic search of countless possible alloying elements, now possible due to developments in rapid alloy processing, and a means of determining the magnetic potential of the most promising compositions.

There is no such thing as an ideal soft magnetic material, so if there is a better option than silicon steel it will be based on an optimised combination of resistivity,

[4]Sometimes a small amount of aluminium or other elements are added for reasons which is discussed in Chapter 13.

[5]Alloys with up to 6.5% silicon are produced in small quantities for high magnetising frequency applications. They are produced in special ways as they are difficult to produce in thin strip form because they are brittle. They are discussed in Section 3.3.1 of Volume 2 of this book.

saturation magnetisation, magnetostriction constants and magnetocrystalline anisotropy constants.[6,7] These parameters were measured in a large number of single crystal iron alloys in probably the most comprehensive study of the effect of small elemental additions to iron to be reported [13]. Figure 1.9 shows a random selection of results from almost 150 alloys produced. Each point represents an alloy; for clarity exact percentage compositions are not included. It should be noted that other important properties such as magnetostriction and anisotropy constants are not given here.

Both resistivity and saturation magnetisation should be as high as possible to restrict eddy currents and at the same time give high permeability and magnetic energy storage capability. It is clearly seen that both cannot be very high in the same alloy. The main findings here are that the addition of manganese seemed to improve the key magnetic parameters of all the alloys to which it was added and that some alloys containing additions of aluminium or phosphorus have similar magnetic properties, or even simultaneously higher resistivity and saturation magnetisation than 3.2% SiFe. However, it was concluded unlikely that they could be made commercially by a simpler or cheaper process than used at that time. Germanium was one of the few feasible additions not studied in [13] and it was found later that small additions to iron resulted in alloys with similar magnetic properties to silicon–iron [14].

Because no compositions have been identified which promise to be substantially superior to the silicon steels it does not mean that there are no undiscovered or unrecognised iron-based alloys which would offer advantages under increasingly

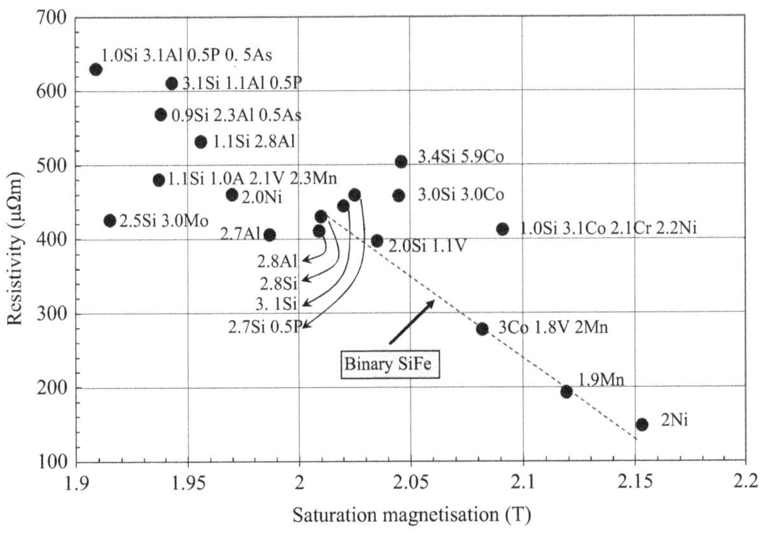

Figure 1.9 Resistivity versus saturation magnetisation trends of laboratory produced iron alloys showing random alloy compositions (data points taken from [13])

[6]These parameters are discussed in detail in Chapter 3.

[7]Obviously other factors such as mechanical properties and temperature stability need to be considered.

demanding unconventional magnetisation conditions. Apart from the basic magnetic properties related to the alloy content it is interesting to note that the addition of nickel to silicon–iron makes it easier to produce Goss texture in alloys with a high silicon content [15]. The addition of nickel does not appreciably reduce the saturation magnetisation compared with a binary Si-Fe alloy with the same silicon content so it is claimed that alloys with around 4%Ni4%Si can be competitive for high frequency applications despite the higher material cost. For example, a 4%Ni44% Si laboratory alloy containing vanadium carbide was reported to have similar properties to low silicon binary alloy at 50 Hz magnetisation but at 900 Hz its loss was around 13% lower than typically 3.2% GO SiFe [15]. However, it is unlikely that the small improvement would justify the cost of capital intensive infrastructure changes required to move from the established silicon–iron alloys.

It has been suggested that there might be future markets for low volume customised alloys for specific dedicated applications [16]. If metallurgists can develop the necessary low output, low cost, adaptable production plant which can quickly move from one alloy to another without furnace contamination, this could be feasible in the future and give a whole new prospective on customised products.

The difficulty of obtaining alloys which have both high relative permeability and low losses is a challenge which steel manufacturers have steadily worked on for the last 30 years. This combination is needed to produce high efficiency rotating machines as discussed in Chapter 6 of Volume 2 of this book. Figure 1.10 indicates

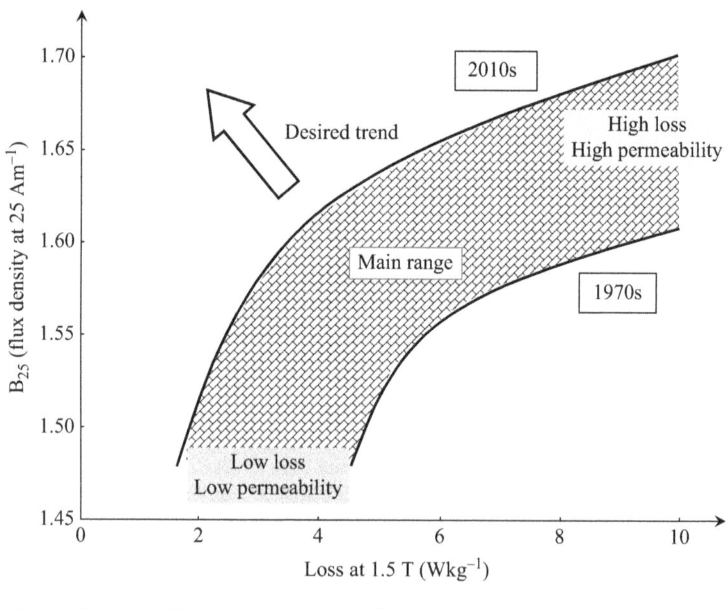

Figure 1.10 Range of loss versus permeability characteristics of typical NO electrical steels showing trend towards more effective steels achieved over the past 40 years

to some extent what has been achieved in terms of the evolution of new commercial grades of NO steel. How this is being achieved and how much further it might be improved is discussed in Chapter 2 of Volume 2 of this book.

1.5 Global impact of energy wastage in electrical steels

Figure 1.8 shows how the losses of the best grade electrical steels continue to fall on average year by year. Clearly the industry is still putting a significant effort into loss reduction despite only incremental improvements being achieved. The efficiency of large transformer and motor cores can be over 99.5% and 90.0% respectively which appears to be close to their limits, so the reasons and incentives for aiming for further reduction in magnetic losses are sometimes questioned. However, these ongoing small improvements can provide substantial economic benefits to users of motors and transformers when the life-time cost of losses is fully taken into account as shown in Chapter 14 of Volume 2 of this book.

The global impact of losses estimated from the annual worldwide production of electrical steel is shown in Table 1.8. An accurate estimate would be very difficult and time consuming to make taking all factors into account and the result would be impossible to verify anyway but the simplified approach here gives a realistic estimate of the scale of the problem.

Consider all the electrical steel produced in one year. Many grades of steel are produced and they will be used in magnetic cores operating over a wide range of flux densities and duty cycles so the best estimate values of average loss and utilisation factors are used. The calculation of the annual core loss of the steels, when installed in electrical equipment, is based on the average losses and utilisation factors from which the total loss occurring over the anticipated lifetime of the annual production is calculated. Assuming that the total core steel volume in use does not change, the annual loss in all cores is simply the annual loss in new cores multiplied by the anticipated life-time. This, of course, ignores the facts that losses in direct replacement cores will be lower than those they replace, which tend to reduce global losses, and that many cores are installed in completely new applications which add more to the global losses.

The figure given as the average loss in NO steel is partly based on a previous analysis of average global losses in motor cores [17]. The figures given in Table 1.8 show that around 11% of all generated electrical energy is consumed in electrical steels. Large approximations and assumptions have been made in arriving at this figure so it is not expected to be very accurate but it shows the scale of the energy wastage.

A large proportion of the losses in GO steel occurs as no-load losses in transformers operating on electrical transmission and distribution networks. This is widely recognised as a major contributor to electrical system losses. The total system losses on the UK networks account for around 8% of generated electricity

Table 1.8 Estimates of global iron losses in electrical steel cores (see Chapter 14 of Volume 2 of this book for further explanation)

Material	Annual production (Mt)	Average loss (W kg^{-1})	Utilisation factor (percentage time)	Average operational life (years)	Annual core loss in new equipment (TWh)	Total annual loss in all cores (TWh)	Annual loss as a percentage of electricity production (%)	Approximate annual cost of losses (at £100 per MWh)
NO steel	10	3	30	25	79	1975	7.9	200×10^9
GO steel	2	1.1	90	40	17	680	2.8	70×10^9

or around 30 TWh[8] per year [18]. Transformer losses account for around 30% of this and core losses account, in turn, for 70% of transformer losses so the no-load loss in transmission and distribution transformers is 7 TWh per annum costing £700 million.[9] Hence, even a modest percentage no-load loss reduction would result in a substantial financial saving apart from the reduced impact on the environment. Obviously, there are thousands of transformers on the network which could only be replaced gradually so such a saving could only build up gradually over several years.

UK electricity consumption is around 1.5% of world output; so, based on a simplistic linear model, we would expect 10 TWh per annum to be consumed in all installed GO steel cores in the UK which is compatible with the estimated 6 TWh consumed on the supply network.

Around 8,000 kWh of energy is required to make a tonne of electrical steel which is around 20 times less than the energy which can be dissipated within a transformer core in just one year. If it were possible to produce a far more energy efficient magnetic material how much more would the industry tolerate in added production cost? It seems as if a significant increase could be justified from this energy balance but many other factors would have to be considered.

These examples are included here to give an initial perception of the order of magnitude of the energy wastage and financial cost of losses in electrical steels which justifies the continued effort to reduce it further. A fuller explanation and justification for the approach is included in Chapter 14 of Volume 2 of this book.

References

[1] Usigami Y., Okazaki Y., Abe N., Kumano T., Kikuchi M., and Inokuchi T. 'Magnetic properties of thin gauge 3% Si-Fe with {110}<001> orientation'. *J. Mater. Eng. Perform.* 1994, vol. 4(4), pp. 435–439

[2] Brailsford F. 'Electrical sheet steels for power transformers'. *J. IEE* 1957, vol. 3(35), pp. 569–575

[3] Davies H.A., Fiorillo F., Herzer G., Guenther K., Hasegawa R., and Sievert J. 'Challenges in optimizing the magnetic properties of bulk soft magnetic materials'. *J. Magn. Magn. Mater.* 2008, vol. 320, pp. 2411–2422

[4] Kim Y.H., Ohkawa M., Ishiyama M., and Arai K. 'Iron loss of grain size controlled very thin grain-oriented silicon steels'. *IEEE Trans. Magn.* 1991, vol. 29(6), pp. 3535–3537

[5] Moses A.J. 'Opportunities for exploitation of magnetic materials in an energy conscious world'. *Interdiscipl. Sci. Rev.* 2002, vol. 27(2), pp. 100–113

[6] Kolb-Telieps A., and Hattendorf H. 'Soft magnetic materials and their applications status and trends'. *Proceedings of Workshop on Metallurgy and Magnetism*, Freiberg, Germany, 16th June 2004, paper 6

[7] Bozorth R.M. *Ferromagnetism*. IEEE Press, Piscataway, NJ, 1993, p. 67

[8] One terawatt is 10^{12} W.

[9] Based on a nominal cost of £0.1 per kWh.

[8] Barrett W.F., Brown W., and Hadfield R.A. 'Electrical conductivity and magnetic permeability of various alloys of Fe'. *Sci. Trans. R. Dublin Soc.* 1900, vol. 7, pp. 67–126

[9] Moses A.J. 'Electrical steels: past, present and future developments'. *IEE Proc.* 1990, vol. 137, Pt. A(5), pp. 233–245

[10] Ferro A., and Soardo G.P. 'Recent research progress in grain-oriented silicon steel'. *J. Magn. Magn. Mater.* 1980, vol. 19, pp. 6–14

[11] Honma K., and Nozawa T. 'Improvement of core loss in grain-oriented electrical steel'. *Magnetics Seminar*, Illinois Institute of Technology, USA, June 1984

[12] Kubota T., Fujikura M., and Ushigami Y. 'Recent progress and future trends on grain-oriented silicon steel'. *J. Magn. Magn. Mater.* 2000, vol. 215–216, pp. 69–73

[13] Foley T.F., Leak D.A., Newbury R.A., Pomeroy A.R., and Matthews A.R. 'A study of ferrous based soft magnetic materials for transformer and similar applications'. *Report 5285*, Electrical Research Association, Leatherhead, UK, June 1969

[14] Littmann M.F. 'Iron and silicon-iron alloys'. *IEEE Trans. Magn.* 1971, vol. 7(1), pp. 48–60

[15] Clark C.A., Mason J.J., and Mee P.B. 'Effect of nickel on the properties of grain-oriented silicon-iron'. *Proc. IEE* 1966, vol. 113(2), pp. 345–351

[16] Moses A.J. 'Development of alternative magnetic core materials and incentive for their use'. *J. Magn. Magn. Mater.* 1992, vol. 112, pp. 150–155

[17] Werner F.E., and Jaffee R.I. 'Energy efficient steels for motor laminations'. *J. Mater. Eng. Perform.* 1992, vol. 1, pp. 227–234

[18] Digest of UK Energy Statistics (DUKES): electricity. 'DUKES chapter 5: statistics on electricity from generation through to sales'. UK Government, https://www.gov.uk/government/statistics/electricity-chapter-5-digest-of-united-kingdom-energy-statistics-dukes (Accessed June 2018)

Chapter 2

Basic magnetic concepts

2.1 Magnetic fields, flux density and magnetisation

2.1.1 Magnetic field (H)

The simplest concept of a *magnetic field* is that it exists in a region in space where a mechanical force would be exerted on a magnetic material if it were placed there. The most common demonstration of this is the action of the earth's own magnetic field forcing the needle of the simple magnetic compass to align itself with the direction of the field which we call the magnetic north–south direction. A magnetic field can be present in a vacuum, inside any non-magnetic material, or inside a magnetic material. There are several more formal definitions based on magnetic energy gradients, electric charges in motion, force on current-carrying conductors, etc. but it is not necessary to consider these here.

A magnetic field can be produced in two ways:

- First, it can be created by the orbital and spin motion of electrons inside a magnetic material. We are conscious of the magnetic field near the ends of a bar magnet because we are aware of the magnet's ability to attract other magnetic materials to either of its ends which we refer to as north and south poles. It will be seen later that a similar sort of electron motion that causes the external magnetic field of the permanent magnet is the reason why soft magnetic materials become magnetised when placed in a magnetic field.
- Second, a magnetic field can be created as a result of electric charges in motion, such as those which form an electric current. Again this is easy to confirm, in this case just by holding a magnetic compass near to a conducting wire carrying a d.c. current and observing the effect of the field on the compass needle.

The magnetic field is a vector quantity having magnitude and direction. In the Système International (SI) terminology of magnetic units it is measured in amps/metre (Am^{-1}).

2.1.2 Magnetic dipole moment (m)

It is useful to appreciate the fundamental concept of the magnetic dipole of an atom in order to understand magnetic domains and magnetisation processes in electrical steels and other magnetic materials. First, consider the creation of a magnetic field by an electric current passing through a loop of wire as shown in Figure 2.1(a).

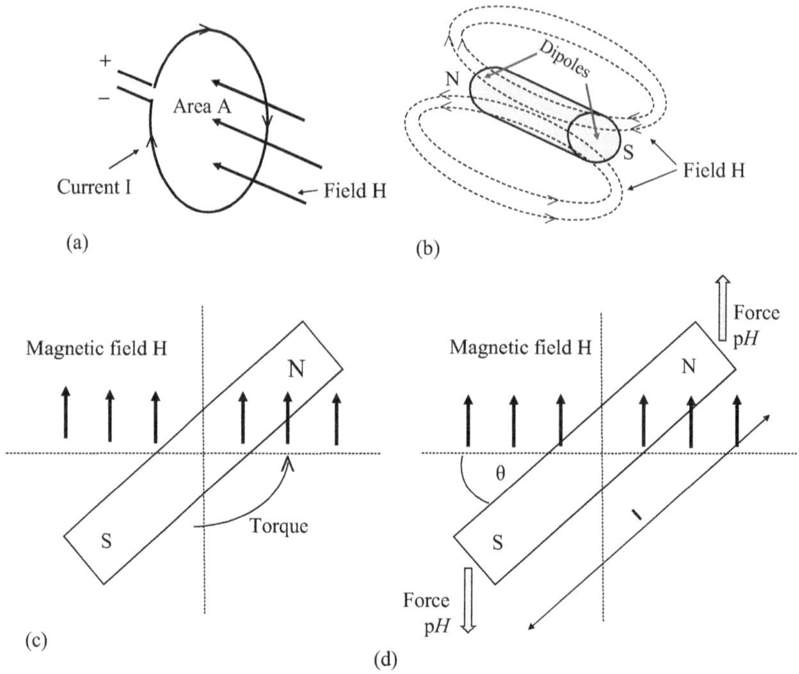

Figure 2.1 (a) Creation of a magnetic field by a current loop, (b) basic representation of a magnetic dipole, (c) the bar magnet as a dipole and (d) forces acting on a magnetic dipole at an angle to an external field

A circular magnetic field is produced around the conductor so the sum effect is a resultant field in the direction shown perpendicular to the axis of the current loop (this phenomenon is further discussed in Section 2.1.4).

An electron circulating the nucleus of an atom effectively can be thought of as an electric current flowing in the opposite direction. This, in turn, creates a magnetic field in a similar way to that created by the current loop. Hence each electron circulating the nucleus of an atom effectively generates a component of field similar to that set up by the simple current loop; together the electrons produce the atom's own fixed magnetic field or magnetisation. This atomic magnetisation is termed a magnetic dipole which is represented schematically in Figure 2.1(b). The field caused by the electron motion can be imagined as emanating from the pair of magnetic poles located at a fixed distance apart (hence the term dipole).

The simple bar magnet shown in Figure 2.1(c) is an example of a dipole. We know that if such a magnet (e.g. a compass needle) is placed in the magnetic field as shown, it experiences a mechanical torque, or moment, which tends to align the magnet in the direction of the external magnetic field. The same happens to an atomic dipole if it is placed in an external magnetic field.

We can quantify this torque at dipole level by imagining the magnet shown here as the basic form of an atomic dipole whose cross sectional area is A and the distance between its poles is l. Each pole has a pole strength p which produces an intensity of magnetisation J (tesla or T) within the dipole such that $= p/A$, i.e. the pole strength per unit area. We can also write $J = pl/Al$. The magnetic moment of the dipole is equal to pl. Hence the intensity of magnetisation is equal to the magnetic moment per unit volume.

If a single dipole is placed in an external magnetic field it will have stored magnetic potential energy which is effectively the work which would have to be done to move the dipole from a field-free region into the field. The presence of that energy is evident in the simple magnetic compass whose needle, oriented as shown in Figure 2.1(d), experiences forces in the directions shown tending to align its axis with the external field which, as is shown next, minimises the stored energy.

Suppose the dipole is moved from a field-free location, an infinite distance away, into the magnetic field and the dipole is oriented at $\theta = 90^\circ$, i.e. perpendicular to the field. Then the work done on each pole is equal and opposite so the total work done is zero and the magnetic potential energy, known as the magnetostatic energy,[1] is also zero.

We can calculate the magnetostatic energy of the dipole when it is oriented at any angle to the field from the torque on the dipole. This anti-clockwise torque is given by

$$T = 2(l/2)pH \sin\theta \,(\mathrm{Nm}) \tag{2.1}$$

The work done against this torque as the dipole is rotated in the field from the position when $\theta = 0^\circ$ to that where the direction is as shown in Figure 2.1(d) is $T \cdot d\theta$. Hence the work done (energy stored) in moving the dipole through this angle is given by

$$W = \int_{\pi/2}^{\theta} plH \sin \cdot d\theta \left(\mathrm{Jm^{-3}}\right) \tag{2.2}$$

From which the energy stored per unit volume is

$$W = -HJ \cos\theta \left(\mathrm{Jm^{-3}}\right) \tag{2.3}$$

The energy is minimum when $\theta = 0^\circ (\cos\theta = 1)$ and has the value $-HJ$. In this position, the dipole is oriented along the same direction as the field. The maximum energy is stored when the dipole is oriented in the opposite direction to the field.

A current loop generates a magnetic field as shown in Figure 2.1(a). If the cross sectional area of the loop is A and the current is I then the moment of the current loop is simply IA ($\mathrm{Am^2}$).

Every dipole has a fixed value of magnetic moment whose magnitude depends on the electron motion in the atom. To be specific, the magnetic moment of an atom is the sum of its components of orbital momentum and spin magnetic momentum.

[1]Magnetostatic energy is discussed in Section 3.7.

The magnitude of the inherent magnetisation can also be considered as being dependent on the magnetic strength of the poles and the distance between them.[2]

The different magnetic strengths of the dipoles of the common magnetic elements, iron, cobalt and nickel are due to their different combinations of orbital and spin magnetic moments. Atoms of most other elements have dipole magnetic moments but they need to be placed in an extremely high magnetic field to be forced into alignment.

The standard symbol for dipole moment is m.[3]

2.1.3 Magnetic polarisation (J)

The magnetic polarisation J is also referred to as the intensity of magnetisation. This is especially the case in the standards for measurements and specifications (see Chapter 12 of Volume 2 of this book) where the reference condition for magnetic induction is given in terms of the magnetic polarisation J. Consider a magnetic material with n dipoles per unit volume each of which has a magnetic momentum m. According to the Weiss[4] theory of magnetism, all the dipoles will be aligned parallel to each other, even with no external field present. (This phenomenon is referred to as magnetic long range order.) Hence all their moments add up and we can define the saturation polarisation J_0 at 0 K as

$$J_0 = nm(\mathrm{T}) \tag{2.4}$$

The saturation polarisation has a fixed value for a given material but if the temperature is increased above 0 K the atomic forces holding all the dipoles parallel to each other are weakened by thermal agitation of the atoms, so the saturation polarisation begins to fall over the region *a-b* in Figure 2.2(a) as the perfect alignment is disturbed.

Figure 2.2(a) shows the saturation polarisation J_s of iron versus temperature and Figure 2.2(b) shows a schematic representation of the misalignment of a group of adjacent dipoles as the temperature rises. The drop in J_s becomes steeper in the range *b-c* before eventually a temperature is reached at which the thermal agitation is so great that the alignment has completely disappeared. This point is called the Curie temperature, or Curie point, which, in pure iron, is approximately 770 °C.

The variation of J with temperature can be expressed as

$$J_s = J_0\left(1 - A_m T^{3/2}\right)(\mathrm{T}) \tag{2.5}$$

where J_0 is its value at absolute zero temperature, T is the temperature in degrees Kelvin and A_m is a constant dependent on the material structure. This expression is

[2]Note that the magnetisation of an atom can be studied in detail either in terms of its electron motion or as a magnetic dipole depending on the objective. This should not be surprising since the current carrying wire in Figure 2.1a experiences a torque which is proportional to the area of the loop and the magnitude of the current.

[3]The unit of m depends on the magnetic units used. In the Sommerfeld system of SI units, m is given in Am^2 (Amp.metre^2) whereas in the Kennelly system it is given in wbm (weber.metre). These systems of magnetic units are compared in Section 2.2.

[4]The Weiss theory is summarised in Section 3.2.

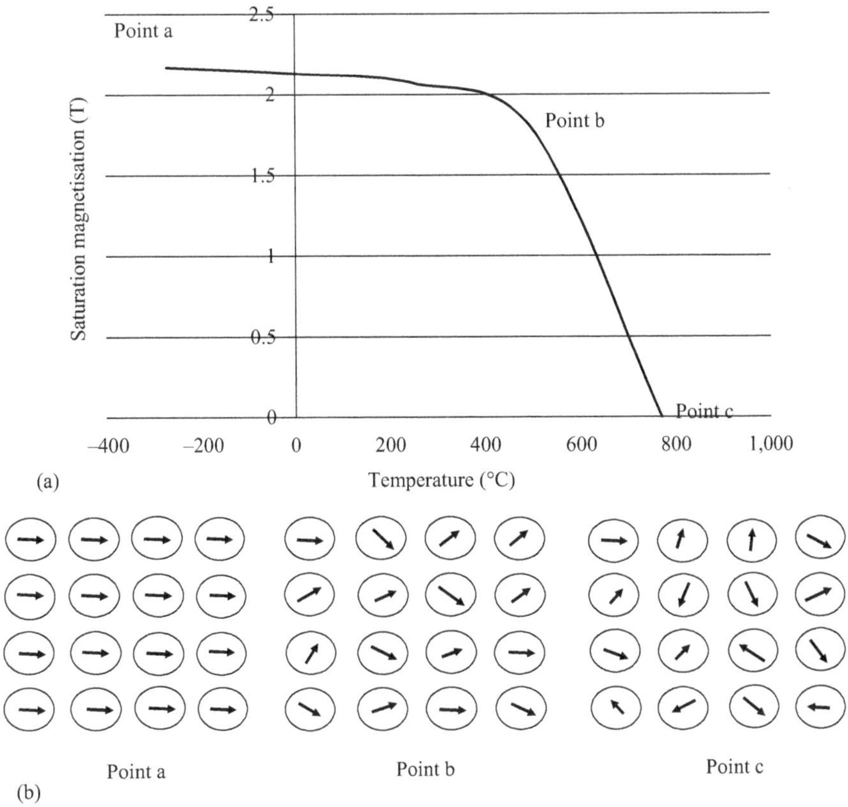

Figure 2.2 (a) Temperature variation of saturation magnetisation of iron and (b) schematic diagram of growing misalignment of dipoles with increasing temperature

derived from quantum theory [1] which takes account of the misalignment of dipole moments from the field direction caused by increasing thermal agitation. In fact, (2.5) is based on the simplified theory which gives good agreement for the $T^{3/2}$ relationship with measurements but empirical agreement for the value of the constant A is poor [2].

There is little difference between the magnitudes of J_s and J_0 at temperatures at which electrical steels usually operate as can be seen in Section 18.1.1 of this volume.

2.1.4 *Magnetic flux density (B)*

When a magnetic field is present in any medium the response is the generation of a magnetic flux density also commonly called magnetic *induction*. The flux density is a vector quantity and in the SI system it is measured in tesla (Wbm^{-2}). It can be quantified in terms of the force it produces on a current-carrying conductor placed

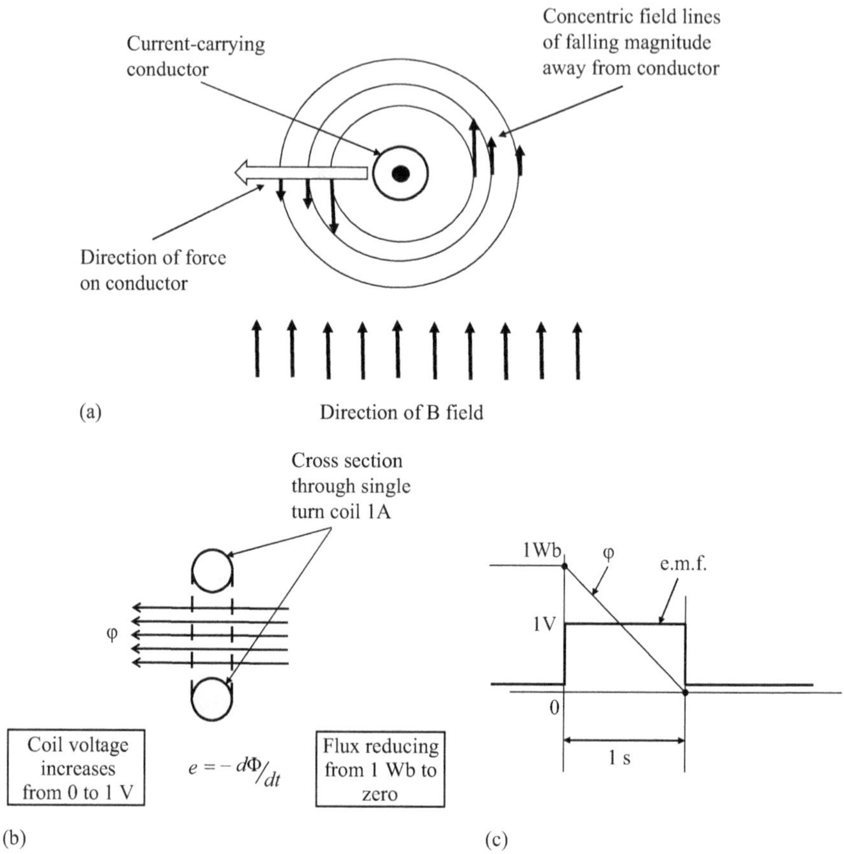

Figure 2.3 (a) Force produced on a current-carrying conductor, (b) e.m.f. generation in a fixed coil placed perpendicular to a slowly decreasing field defining the weber and (c) e.m.f. and flux variation with time as the flux drops to zero

in a uniform field.[5] In this case one tesla is defined as the value of flux density which produces a force of 1 Nm^{-1} on a conductor carrying a current of 1 A perpendicular to the direction of the field.

We can define flux and flux density in other ways. One common definition comes from Faraday's law which is discussed in Section 5.1. In this case the weber is defined as the magnetic flux which, when linking a one-turn electric circuit, produces in it an electromotive force (e.m.f.) of 1 V if it were reduced to zero at a uniform rate of 1 Wbs^{-1}. The schematic diagrams in Figure 2.3 show these two fundamental mechanisms sometimes called the motor and generator effects.

[5]In a non-magnetic medium the flux density is directly proportional to the field producing it as expressed in (2.8) so here we can say the current-carrying conductor is in a flux density of 1 tesla. Alternatively we can say it is in a magnetic field of $1/4\pi \cdot 10^{-7}$ Am^{-1}.

The stationary conductor in Figure 2.3(a) carries a constant electric current in a uniform, constant magnetic field. The current in the conductor itself produces its own magnetic field which interacts with the fixed external field as shown. We can visualise the process as field strengthening on one side of the conductor and weakening on the opposite side. This unbalance produces a force as shown. This is the fundamental way in which a motor works since if the force is large enough the conductor will move.

Figure 2.3(b) shows how we can define the weber by quantifying the role of flux in the generator effect. Here we have a fixed coil of conducting wire placed in a uniform field which is perpendicular to the plane of the coil. Suppose that the field is reduced to zero at a constant rate. By Faraday's law an e.m.f. is induced. If the induced e.m.f. is 1 V, as shown in Figure 2.3(c), then the flux passing through the coil initially was 1 Wb.

Of course all the above definitions are equivalent.

In contrast, magnetic flux (φ) is a scalar quantity (measured in weber). In order to understand this apparent contradiction it is useful to consider the magnetic field around a simple bar magnet as shown in Figure 2.4. The magnetic field is shown emerging from the north pole of the magnet and re-entering at the south pole. The magnetic field can be thought of as a flow of energy which is called magnetic flux. An equal amount of flux enters and leaves the poles but at other places the amount of flux is not uniform. For example, at point A, fewer field lines are present than at point B near the poles. This leads to our definition that the flux passing through unit area at right angles to the field lines is the flux density.

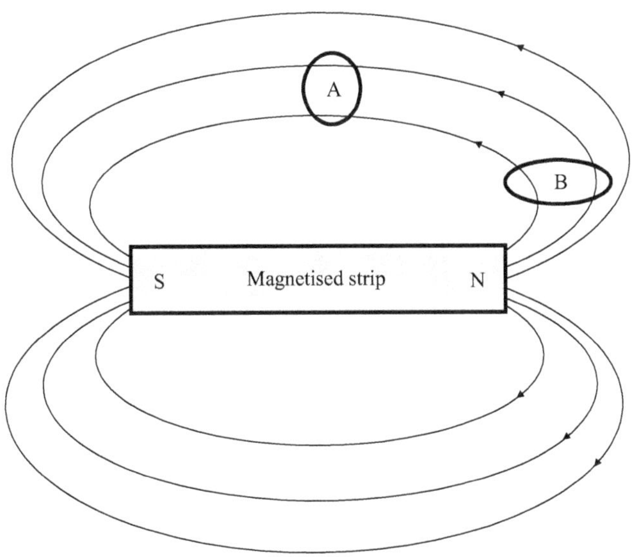

Figure 2.4 Magnetic field in the vicinity of a bar magnet

The concept of the flux entering and leaving a closed surface is expressed mathematically as

$$\int B \cdot DA = 0 \tag{2.6}$$

This is one of Maxwell's equations of electromagnetism used in computational electromagnetics.

Strictly speaking ϕ is measured across a surface and does not exist at a single point, it is effectively the component of magnetic field passing through a surface. It is normally defined mathematically based on the concept of electromagnetic fields which is beyond the scope of this book. For engineering applications it is best considered in an abstract way as a magnetic flow, sometimes pictorially presented as stream of lines of flux created by a magnetic field as shown in Figure 2.4.

Fortunately, there should not be any confusion over the concept of magnetic flux since from an engineering , or materials, standpoint it is usually treated as a vector quantity as it is in most classical electrical engineering textbooks whereas, in the physics approach to electromagnetic analysis, it is treated in a different way as a scalar quantity. From here on it is treated in its engineering context simply defined as

$$\Phi = B \cdot A(\mathrm{Wb}) \tag{2.7}$$

where B is the flux density and A is the area through which the flux passes.

Although magnetic flux is such an important quantity in magnetic measurements, it cannot be directly measured and, as will be shown later, the only way in which we can quantify it is by measuring the change of the voltage it induces in an electric circuit as it changes.

2.1.5 *Permeability* (μ)

The flux density is a measure of the degree to which a magnetic material is magnetised. As a magnetic field increases, the flux density which it produces also increases. If a field is present in a vacuum there is a simple linear relationship between it and the resulting flux density which can be written as

$$B = \mu_0 H(\mathrm{T}) \tag{2.8}$$

where μ_0 is a constant of magnitude $4\pi \cdot 10^{-7}\ \mathrm{Hm}^{-1}$. In the SI system, μ_0 is called the magnetic constant but it is still widely referred to as the permeability of free space.

If the space in which the field is acting is now filled with any material, the flux density will increase to a value given by

$$B = \mu H(\mathrm{T}) \tag{2.9}$$

where μ is termed the absolute permeability. This can also be expressed as

$$\mu = \mu_0 \mu_r (\mathrm{Hm}^{-1}) \tag{2.10}$$

where μ_r is the relative permeability of the material. Soft magnetic materials, such as electrical steels, with ferromagnetic properties, have very high values of relative

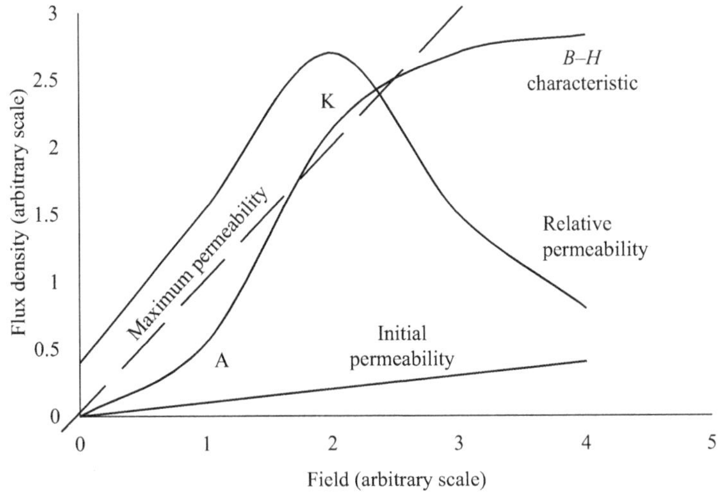

Figure 2.5 Variation of flux density and permeability under an increasing d.c. field

permeability compared with other materials.[6] However, μ_r is not a constant and tends to drop to a limiting value of unity in all magnetic materials as the magnetic field increases. From (2.9) and (2.10) we can also express the relationship between flux density and magnetic field in a material as

$$B = \mu_o \mu_r H(\mathrm{T}) \tag{2.11}$$

Figure 2.5 shows a typical *B–H* curve of a ferromagnetic material and its corresponding $\mu - H$ characteristic. The shape of the *B–H* curve is explained in Chapter 3. All that needs to be noted for now is that it is highly non-linear. *B* increases rapidly at a low field and then increases very slowly and at a constant rate at very high fields. The permeability μ, corresponding to any flux density, is simply the ratio *B/H* at that flux density. The $\mu - H$ curve shown is typical of any soft magnetic material in that the permeability is low at a very low field, it reaches a maximum value and then decreases to a very low value at a high field.

The point *K* on the curve in Figure 2.5 is referred to as its knee point. It is often defined as the point at which a 50% increase in field produces a 10% increase in flux density. It is a useful concept for transformer designers who define the same point as the point at which a 10% increase in voltage draws a 50% increase in magnetising current. This gives a measure of the effect of an overvoltage on a power distribution system. The point *A* on the curve is sometimes referred to as the ankle point. The *B–H* curve between these two points is approximately linear which is important in many applications.

[6]Ferrimagnetic materials such as ferrites have *high* permeability also but not as high as electrical steels.

The value of permeability in a very low field is an important parameter in some applications of soft magnetic materials, such as in instrument transformer cores. In practice, it is of course impossible to measure permeability at zero field. In electrical steels, it is normally measured at a field value of 4 Am^{-1}. This low field permeability is called the initial permeability μ_i.

Another useful parameter is the maximum permeability μ_m which occurs at a point on the curve where a tangent from the origin touches the curve as illustrated in Figure 2.5. The maximum permeability occurs at different values of field in different materials because of their different *B–H* curve shapes, so it should be used cautiously when comparing materials for particular applications. Instead, the flux density at a specified value of *H* is often a more useful measure. For example, B_{10} means the value of flux density at a field of 10 Am^{-1}.

It should be noted that μ in Figure 2.5 is *not* the slope of the *B–H* curve apart from at the one point when the permeability is maximum as shown in the figure. The slope at any point on the curve is referred to as the differential or incremental permeability μ_d.

The permeability characteristic in Figure 2.5 would of course be the same shape whether plotted as absolute or relative permeability since they are directly proportional to each other as seen from (2.9). The term *relative* is often omitted in technical specifications quantifying the permeability of electrical steels. For example, saying ‘the permeability of the material is 1,000’ usually means that the relative permeability is 1,000. The absolute permeability in this case is of course the product $\mu_0\mu_r$, or 12.6×10^{-4} Hm^{-1}. It is not too difficult to spot this common omission, but it is good practice to include ‘relative’ when this is what is meant.

2.1.6 Relationships between H, B and M

When a magnetic material is placed in an increasing magnetic field its flux density increases according to the following relationship[7]:

$$B = \mu_o H + J(\mathrm{T}) \tag{2.12}$$

The fundamental relationship is shown graphically in Figure 2.6.[8] This shows the *B–H* curve of a soft magnetic material initially in a demagnetised[9] state before magnetising to increase the flux density towards saturation along the magnetising direction. The flux density is made up of two parts: one is the simple $\mu_o H$ component, which increases linearly with increasing *H*, and the second is the polarisation, which increases non-linearly with *H* until it reaches its saturation value J_s.

[7]This is often seen written as $B = \mu_o(H + M)$ where *M* is the magnetisation in Am^{-1} as discussed in Section 2.2.

[8]In Figure 2.6, the material is initially *demagnetised* and becomes magnetised when a field is applied whereas in Section 2.1.3 it was inferred that a magnetic material is always magnetised to a value M_s, whether a field is present or not. This apparent contradiction is explained in Chapter 3.

[9]A magnetisable material is said to be fully demagnetised when no magnetisation can be detected outside the material.

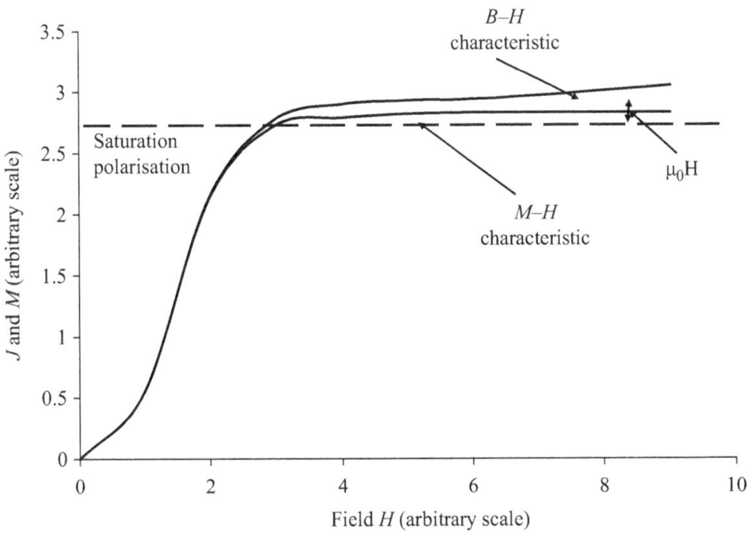

Figure 2.6 Variation of flux density with increasing field showing the approach to technical saturation

The term, saturation flux density, B_s, is usually quoted for electrical steels whereas Figure 2.6 shows that B increases indefinitely as H is increased so there is no limiting maximum flux density. However, magnetisation M does reach a definite maximum value J_0 when all magnetic dipoles are perfectly aligned. The difference in magnitude between B and J close to technical saturation[10] in electrical steel is very small, so $B_s \approx J_s$. For example, a field around 1,000 Am^{-1} is needed to bring a piece of silicon iron close to saturation and at that field the $\mu_o H$ component is less than 0.005% of the total magnetisation.

If the field applied to the material in Figure 2.6 is reduced to zero and then reversed, the well-known *hysteresis loop* is produced.[11] Figure 2.7(a) shows a complete loop which occurs over one complete cycle of magnetisation. The applied field is increased in one direction to reach the maximum value H_p at point a. During this process the flux density follows the initial magnetising curve to reach a value of B_p. The field is reversed in direction and dropped to zero when at point b some magnetisation will be found to remain in the material. The value of flux density at point b is called the remanent induction or remanent flux density B_r.

[10]Technical saturation is defined in many ways such as the point when there is no further change with increasing field in the B–H loop or that the material has become a single magnetic domain. Strictly speaking, magnetisation continues to increase, at a very small rate, in a material when it reaches its paramagnetic state. However, the definition of technical saturation magnetisation and its numerical value is of importance in the development and use of models to predict transformer core performance when subjected to high voltage transients.

[11]The hysteresis process is discussed in detail in Chapter 6. It is introduced here to help explain other basic magnetisation properties and definitions.

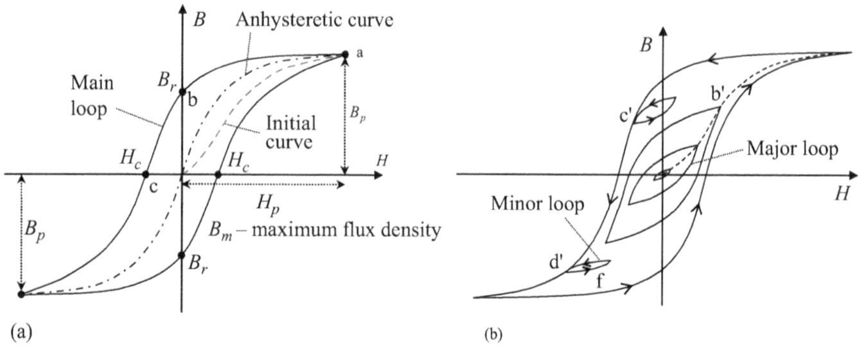

Figure 2.7 (a) Typical magnetisation curve over a full cycle of magnetisation starting in a fully demagnetised state and (b) major and minor loop development due to reversals of the magnetic field during a magnetising cycle

If the field is next increased in the opposite direction, the flux density continues to fall and becomes zero at point c. The field required to reduce the flux density to zero is called the coercive force or coercive field strength, H_c. If the field continues to be increased in this direction, and then reversed, the complete *B–H* loop will be traced.

The anhysteretic curve is the characteristic which would be obtained if no hysteresis effect occurs during cycling of the field. Each point on the anhysteretic curve corresponds to a magnetic domain structure which gives lowest possible energy. The curve is difficult to obtain by experiment. However, each point on the curve can be obtained in a similar way to the sample demagnetisation process by superposition of a slowing reducing low frequency a.c. field upon a constant d.c. bias. It is a concept used in several magnetisation models [3].

If B_p is close to saturation the coercive force is referred to as the coercivity and the remanent flux density is referred to as the remanence.

Earlier, the loop in Figure 2.7(a) was referred to as a *hysteresis* loop. It will be seen later that the *B–H* loop can attain an infinite number of shapes according to the variation of the magnetising field. The hysteresis loop, or d.c. loop, is one form of the generalised loop obtained when the field is traversed very slowly. The shape of the loop changes dramatically due to the generation of eddy currents if the field is changed more rapidly so it is not then appropriate to refer to it as a hysteresis loop. It is surprising how often this incorrect terminology appears in journal publications and text books. Hence it is more meaningful and unambiguous to refer to the characteristic as a *B–H* or *M–H* loop unless it is definitely obtained under pseudo-d. c. conditions as referred to in Chapter 12.

Figure 2.7(b) shows sub sets of a general *B–H* loop. Reversing the magnetisation at any induction such as at the point *b'* on the initial magnetisation curve produces *major loops*. If a material is being cycled through a major loop and the field is reversed by a small amount and then returned to the same values as those shown at the points *c'* and *d'*, *minor loops* are formed. These are important in producing accurate loss models and other types of characterisation of electrical steels.

Three distinct *B–H* loop shapes are widely found in practice. Their general shapes are shown in Figure 2.8. The normal or classical loop shape is shown in (a) with its 'rounded' extremes. The remanence of the rectangular loop shape shown in (b) is almost equal to its peak flux density and the flat loop in (c) has a very low remanence ratio B_r/B_s typically of 0.1. The other significant feature of the flat loop is the almost linear characteristic over much of the field range.

Demagnetisation is an important process which has to be carried out at the start of many types of d.c. measurements in order to ensure that initially the magnetisation is indeed at the origin, i.e. $B = 0$ when $H = 0$. In theory this can be achieved in a sample by applying a very low frequency a.c. field and slowly reducing it to zero. The process is illustrated in Figure 2.9. Suppose that the material is initially in some remanent state following a previous magnetisation, i.e. not in a demagnetised state. Application of the high a.c. field initially increases the flux density and then slow cycling takes the material through a series of decreasing *B–H* loops until

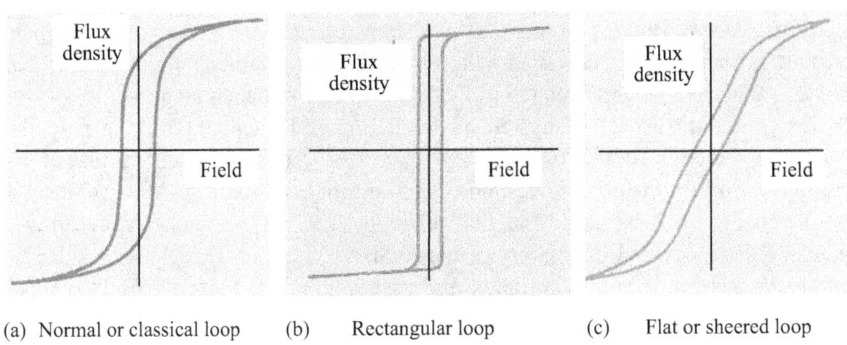

Figure 2.8 Common B–H loop shapes: (a) classical loop, (b) rectangular loop and (c) flat loop

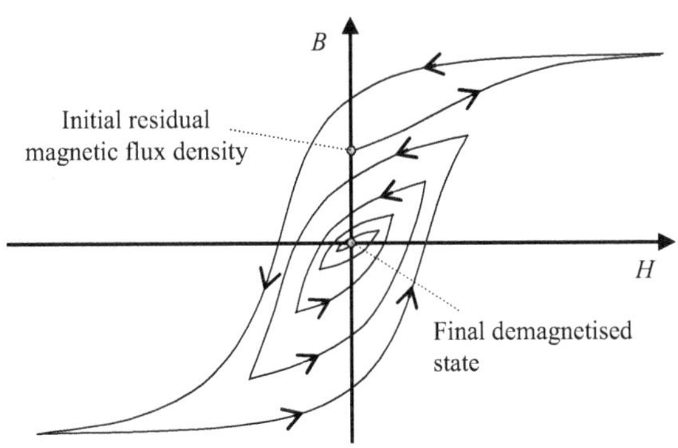

Figure 2.9 Process of demagnetisation of a sample by a decreasing a.c. field

the a.c. field has been reduced to zero. In practice full demagnetisation is sometimes difficult to achieve particularly in highly anisotropic materials. Also, there is no guarantee that residual magnetisation does not remain along other directions within a sample. Methods of demagnetisation are discussed in Chapter 12.

For completeness, another quantity, the susceptibility χ, is defined as

$$\chi = M/H \tag{2.13}$$

Hence

$$\mu_r = 1 + \chi \tag{2.14}$$

The susceptibility and relative permeability of an electrical steel are almost always practically identical[12] so the term is rarely used in a materials or engineering context.

2.1.6.1 Hysteresis models

Hysteresis modelling is one of the challenging fields of study in many engineering disciplines including electromagnetics [4]. It is widely embedded into computational electromagnetics[13] solvers to predict the electromagnetic performance of magnetic cores. It is particularly relevant to the evaluation of magnetic properties of soft magnetic materials directly related to hysteresis processes such as power loss.

An important aspect of hysteresis modelling is to account for the so-called history dependence of the magnetisation process. All this means is that if the magnetisation of a material corresponds to a certain point on a *B–H* curve, such as the one in Figure 2.7(b), and the field changes, the magnetisation will follow a trajectory dependent on the history of the field.

Hysteresis modelling is complex and many approaches are used. Two types, physical modelling and phenomenological modelling are widely applied. Physical types, as the name implies, are based on simulations of the real hysteresis processes going on in a material. Phenomenological models mathematically generate *B–H* curves following well defined rules. The phenomenological approaches are more efficient but give no insight into the physical processes going on during magnetisation of each material.

Hysteresis modelling is not dealt with in detail in this book but some of the terminology is occasionally used, so these are introduced here. Scalar hysteresis models assume that the field and flux density are unidirectional but can change continuously. In vector hysteresis models the direction of B and H can change as well as their amplitudes. Vector models are more versatile, but more complex. Scalar models are far simpler and are sufficiently accurate in many cases.

[12]From (2.11)–(2.13), it can easily be shown that $\chi = (\mu_r - 1)$.

[13]The term *computational electromagnetics* refers to the study and application of analytical and numerical approaches to calculating and field, flux, loss and mechanical force distributions associated with magnetic core. The numerical solvers use techniques such as the finite element method to obtain approximate solutions of Maxwell's equations. Computational electromagnetics relevant to electrical steels is discussed in Chapter 11 of Volume 2 of this book.

Both scalar and vector models can be static or dynamic. In static models the history of the magnetisation is taken into account but the speed at which the magnetisation changes is not accounted for. Dynamic models take into account the effect of magnetising frequency on the *B–H* loop shape. Further insight into hysteresis modelling is given in Chapter 4 of Volume 2 of this book and can be found in more detail in [5] and [6].

2.1.7 Demagnetising effects

In order to understand many aspects of the properties and applications of electrical steels it is necessary to understand the concept of demagnetising fields, which can be present inside a magnetic material, and associated demagnetisation factors which help us quantify them. Consider the iron strip shown in Figure 2.10(a). Suppose it is placed in an external field H_a which sets up a flux density B in the strip as shown. Magnetic poles are created from which the external stray flux emanates from the ends of the strip. There is a field, referred to as the demagnetising field H_d, directed from the north pole to the south pole inside the strip.

The H field and B field inside and surrounding the same magnetised strip is shown in Figure 2.10(b) and (c). It can be seen that the field and flux paths are identical outside the material but inside they are quite different because of the demagnetising field.

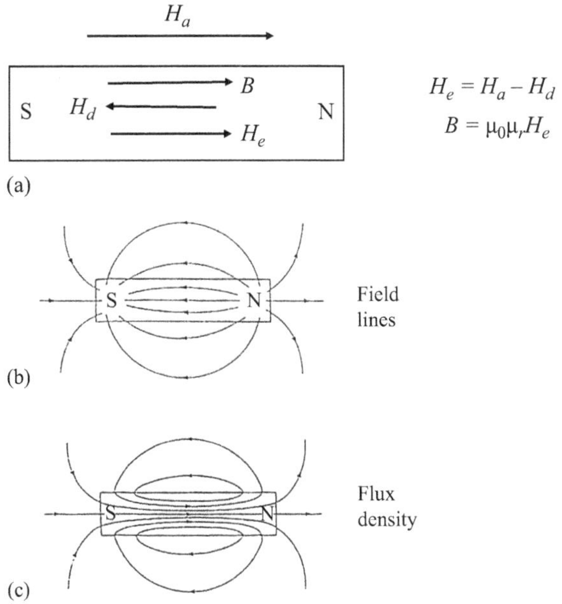

Figure 2.10 (a) Demagnetising field of a long strip of ferromagnetic material due to free north and south poles, (b) corresponding H field distribution and (c) corresponding B field distribution

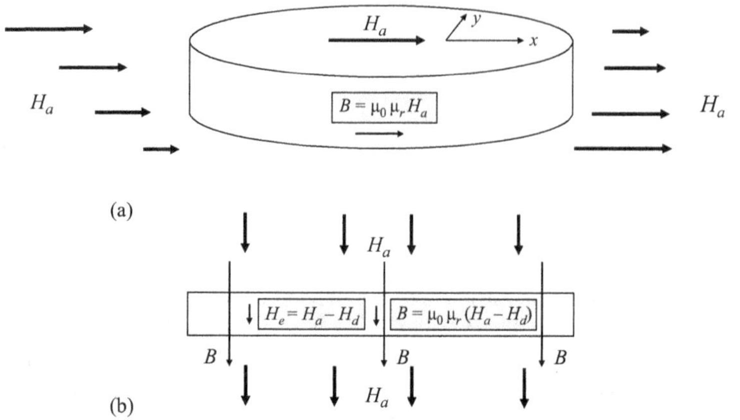

Figure 2.11 Effect of an external field applied to a thin disc of infinite diameter (a) along a direction in the plane of the disc and (b) perpendicular to the plane

The demagnetising field reduces the influence of the external field so the effective field H_e available to magnetise the strip is reduced such that

$$H_e = H_a - H_d \left(\mathrm{Am}^{-1}\right) \tag{2.15}$$

The demagnetising field is approximately proportional to the magnetisation. The proportionality constant is called the demagnetising factor N_d which is a function of the shape of the material. Depending on which system of units is used the demagnetising factor has different numerical values. Hence we can write

$$H_e = H_a - N_d M \left(\mathrm{Am}^{-1}\right) \tag{2.16}$$

Alternatively

$$H_e = H_a - N_d (J/\mu_0) \left(\mathrm{Am}^{-1}\right) \tag{2.17}$$

A demagnetising field is present in any specimen which has free poles, i.e. the specimen is not part of a closed magnetic circuit.[14] This is illustrated by considering the case of a hypothetical very thin sheet with an infinite surface area. If the material is magnetised in its plane, as shown in Figure 2.11(a), no poles are present; strictly speaking we could say they are an infinite distance apart. Hence they do not produce a demagnetising field and the demagnetising factor N_{dx} is zero as it is for a closed magnetic circuit, which also does not have poles on any part of its surface.[15]

[14]Of course a permanent magnet always has free poles. Its demagnetising field is an important feature of its characteristics.

[15]We can apply this result to an infinitely long rod or strip in the same way. In an open magnetic circuit a practical finding is that it is easier to uniformly magnetise a long strip than a short one of the same material. This is simply due to the far lower demagnetising factor of the long strip.

However, if it is magnetised perpendicular to its plane along the z direction, as in Figure 2.11(b), the flux density is the same outside and inside the material. Outside the material, $B = \mu_0 H_a$. Inside the material $B = \mu_o H_e + J$, or $B = \mu_0(H_a - N_d J) + J$ so equating these terms we obtain, $N_{dz} = 1/\mu_0$.

Note that the demagnetising factor is dimensionless, thus, $H_d = N_d M = N_d J/\mu_0$.

Practically, the demagnetising factor has to be measured experimentally, unless a sample has an ellipsoidal shape or has dimensions close to that of an oblate (flattened) or a prolate (elongated) ellipsoid. For a given ellipsoidal shape the demagnetising factor depends on the length along each principal axis and it can be shown that the sum of the three is equal to unity [7] Therefore, for a sphere $N_{dx} = N_{dy} = N_{dz} = 1/3$ but for other shapes it is difficult to exactly calculate N_d apart from that of a thin plate where $N_d = 1$ and an infinitely long bar along its length where $N_d = 0$. Values for wide ranges of dimensional ratios are widely available; typical values obtained from several experimenters are given in [8].

The demagnetising factor of a rod, whose ratio of length to diameter is p, is approximately given by

$$N_d = \frac{1}{p^2 - 1}\left\{\frac{p}{\sqrt{p^2 - 1}} \cdot \log\left(p + \sqrt{p^2 - 1}\right) - 1\right\} \tag{2.18}$$

Because of the difficulty in accurately determining the demagnetising factor, particularly its permeability dependence, whenever possible it is advisable to avoid its use in measurements by testing materials in closed magnetic circuits when $H_d = 0$.

2.2 Units in magnetism

A difficulty often encountered when interpreting magnetisation data is the confusing systems of units which have been, and are still being, used. Each system is claimed to have advantages so a single, uniform system, internationally recognised by physicists and material scientists, still seems unlikely in the foreseeable future. SI units were adopted many years ago to replace the CGS system (centimetre-gram-second)[16] partly to discourage the proliferation of unnecessary units.

However, still today, many publications include CGS terminology, so it is necessary to understand the differences between the approaches. Furthermore, poor distinction between *B–H* and *M–H* characteristics leads to mistakes in interpretation of permeability and hysteresis loops which still appear in modern text books and research publications [9].

Even the SI approach to units in magnetism itself is based on one of two approaches. The *Kennelly* and *Sommerfeld* systems were proposed in the 1940s and 1950s, respectively. Both are based on the MKS (metre-kilogram-second). They use the same definitions of magnetic field H and flux density B with units of Am^{-1} and tesla, respectively. The systems are used to similar extents today and often mixed.

[16]The CGS system is also called the Gaussian or EMU (electromagnetic unit) set of units.

A survey of the confusing terminology and inconsistent use of symbols showed that in each of eight respected text books, or in-depth journal papers, different combinations of definitions and symbols for the most common magnetic materials characteristics were used [10]. All use the basic relationship between magnetic field and flux density in a non-magnetic material given by (2.4) but there the agreement ends.

What follows is a brief comparison between important quantities defined in the three systems and relationships between them. Further background can be found in textbooks such as [11] and [12] covering wider aspects of magnetism and magnetic materials.

The CGS system is still used and the corresponding units of magnetic field, flux density, etc. are widely encountered in technical and commercial literature so the system needs to be briefly introduced here before discussing the SI systems.

In the CGS system, the relationship between magnetisation and field is expressed in the following equation.

$$B = H + 4\pi I \text{(gauss)} \tag{2.19}$$

The flux density B and the intensity of magnetisation I are measured in units of gauss.[17] The magnetic field H is measured in oersteds. Equation (2.19) effectively shows H and I contributing to lines of flux. The intensity I is proportional to the magnetic moment per unit volume which, in turn, is proportional to the magnetic pole strength. Each unit pole gives rise to a unit field (magnetisation) which spreads uniformly in all directions such as on a sphere of unit radius whose area is 4π which explains why the constant appears in this equation.

The relative permeability is not defined in the CGS system because the permeability of free space in this set of units is effectively unity; hence it does not appear in any equations.

It is not necessary to delve further into the CGS system so next we turn to the Kennelly version of SI magnetic units which complies with the following

$$\text{Kennelly } B = \mu_0 H + J \text{(T)} \tag{2.20}$$

where J is defined as the intensity of magnetisation in the same way as I in the CGS expression in (2.10). In the Kennelly system, J is measured in tesla and is often referred to as the magnetic polarisation or intrinsic magnetisation and sometimes is also given the symbol I. In some standards J replaces B but the difference between J and B is usually negligible in electrical steels.

Finally, we come to the Sommerfeld system which is built around

$$\text{Sommerfeld } B = \mu_0 (H + M) \text{(T)} \tag{2.21}$$

In (2.21), M is the magnetisation per unit volume measured in Am^{-1}. The first term, $\mu_0 H$, is the component of flux density occurring in a material due to the presence of the magnetic field and it is the same in (2.20) and (2.21). In practice the

[17]A further distraction is that flux density is also measured in lines/cm^2 where 1 gauss =1 line /cm^2.

symbol M is used indiscriminately to denote some form of magnetisation. To avoid any confusion, it should just be remembered that magnetic polarisation or intensity of magnetisation is measured in tesla and the magnetisation per unit volume is measured in $Am^{-1.}$

It should be noted that saturation magnetisation can be expressed in terms of intensity of magnetisation J or in terms of magnetisation per unit volume M. These are alternative measures of magnetisation. For example, the saturation of iron can be expressed as 1.7×10^6 Am^{-1} or, alternatively, 2.15 tesla. The magnetic saturation of soft magnetic materials is usually expressed in tesla whereas it is expressed in Am^{-1} for a hard magnetic materials. However, it is easy to convert between the two.

Of course in free space, provided that no magnetic material is present, M or J is zero so the field and flux density are the same (ignoring the constant μ_0 multiplier in the Sommerfeld system) hence either can be used as a reference source.

Table 2.1 summarises the units of the most important magnetic properties of soft magnetic materials and factors to convert from CGS units to SI units in each case. The consistency of magnetisation terms can easily be verified here. For example, in Section 2.1.3, saturation magnetisation per unit volume was introduced by multiplying the dipole moment by the number of dipoles per unit volume. The units of magnetic moment in the Kennelly and Sommerfeld systems are Wbm and Am^{-2}, respectively. Dividing each by volume, m^3 gives units of Wbm^{-2} (tesla) and Am^{-1} respectively as expected for intensity of magnetisation and magnetisation per unit volume in (2.20) and (2.21), respectively.

Table 2.1 Common magnetic units and conversion factors used with soft magnetic materials [13]

Quantity	**Common symbol**	**CGS units**	**SI units (Kennelly)**	**SI units (Sommerfeld)**	**To convert CGS to Kennelly, multiply by**
Magnetic field, magnetising force	H	Oersted (Oe)	Am^{-1}	Am^{-1}	$10^3/4\pi$
Flux density/induction	B	Gauss (G)	Tesla (Wbm^{-2})	Tesla (Wbm^{-2})	10^{-4}
Magnetisation (volume)	M	Gauss (G)	–	Am^{-1}	10^3
Intensity of magnetisation/ polarisation	I, J	$emu.cm^{-3}$-	Tesla	–	$4\pi \cdot 10^{-4}$
Magnetic moment	M	emu	weber.metre (Wbm)	Am^2	10^3
Pole strength	p	$emu.cm^{-1}$	Wb	Am	–
Magnetic flux	ϕ	Maxwell	Wb	Weber (Wb)	10^{-8}
Permeability	μ	Dimension-less	Hm^{-1}	Hm^{-1}	$4\pi \cdot 10^{-7}$
Relative permeability	μ_r	Not defined	Dimensionless	Dimensionless	–
Demagnetising factor	N_d	Dimensionless	Dimensionless	Dimensionless	$1/4\pi$
Field equation		$B = H + 4\pi M$	$B = \mu_0 H + J$	$B = \mu_0(H + M)$	–

It is interesting to note that in the preface to his classic book, written in the early days of industrial importance of magnetic materials, Sir James Alfred Ewing said [14]:

> Throughout this book the author has endeavoured to familiarise the student with the notion of intensity of magnetisation (**I**) as well as with the notion of magnetic induction (**B**). It has been urged by some writers that the alternative which is in this often is unnecessary and confusing and that if we keep ' **B**' we might dispense with '**I**'. The scientific value and the practical utility of '**B**' are so obvious that no one proposes to avoid using that. It is **I** that we are told must go. In this cry the author is by no means disposed to join. It is not too much to say that in stating the magnetic qualities of a metal the quantity '**I**' is of primary importance. The facts of saturation, the molecular theory and the phenomena of magneto-optics all demonstrate its physical reality and its fundamental interest.

Here, Ewing is referring to $\boldsymbol{B}$ and $\boldsymbol{I}$ related by (2.19). Throughout *this* book we have the benefit that, where we do have to switch between the two SI sets of units, the intensity of magnetisation and magnetic induction in practical usage are almost the same magnitude. In the Kennelly system, $B \approx J(I)$ and in the Sommerfeld system, $B \approx \mu_0 M$; so no confusion should arise.

2.3 Dimensional analysis of magnetic quantities

All physical characteristics can be derived from seven base SI quantities namely, mass, length, time, electric current, temperature, amount of substance and luminous intensity. All magnetic quantities can be related by so-called dimensional analysis to just four of these, namely mass (M), length (L), time (T) and current (I). Instead of current, other quantities such as permeability, permittivity or charge can be used, but this can lead to difficulties [15].

Dimensional analysis was developed in the nineteenth century. Renowned scientists such as Joseph Fourier and James Clerk Maxwell were prominent in defining these fundamental units and in showing how all other units can be derived from them. Dimensional equations do not convey any information about the actual value or any other information about a particular quantity.

The main use of dimensional analysis is to perform checks on the correctness of physical equations or formulae from our knowledge of the 'dimensions' of some basic quantities. A scientific equation can only be plausible if the dimensions on either side are the same. This can be tested just using simple algebra. Checking for dimensional correctness of equations can be particularly useful in magnetics. Common examples of its use are to clarify which set of SI units is used in an expression or whether a derived expression gives the absolute value of a quantity, such as power, as per unit length, volume or mass.

The dimensions of some quantities commonly encountered in magnetics are given in Table 2.2. For the sake of readers unfamiliar with the technique, it is worth considering examples of how the units are derived and used.

First, consider a simple example of acceleration of a body. Acceleration is measured in metres per second per second or, in old British imperial units, as feet

Table 2.2 Dimensions of some common quantities related to magnetics

Quantity (units)	Dimensions
Force (N)	MLT^{-2}
Pressure/stress (Nm^{-2} or pascals)	$ML^{-1}T^{-2}$
Torque (Nm)	ML^2T^{-2}
Density (kgm^{-3})	ML^{-3}
Frequency (Hz)	T^{-1}
Current (amps or coulomb.s^{-1})	I
Voltage (V)	$ML^2T^{-3}I^{-1}$
Resistance (ohms)	$ML^2T^{-3}I^{-2}$
Resistivity (ohm.metre)	$MLT^{-3}I^{-2}$
Inductance (H)	$ML^2T^{-2}I^{-2}$
Magnetic reluctance (AWb^{-1})	$M^{-1}L^{-2}T^{-2}I^2$
Magnetic field (Am^{-1})	$L^{-1}I$
Flux density (T)	$MT^{-2}I^{-1}$
Magnetic moment(Wb)	$ML^2T^{-2}I^{-1}$
Pole strength (Wb)	$ML^2T^{-2}I^{-1}$
Flux (Wb)	$ML^2T^{-2}I^{-1}$
Permeability (Hm^{-1})	$MLT^{-2}I^{-2}$
Energy (J)	ML^2T^{-2}
Power (W or Js^{-1})	ML^2T^{-3}
Electric charge (coulomb)	TI
Relative permeability	–
μ_0 (Hm^{-1})	$MLT^{-2}I^{-2}$

per second per second and it is obvious that in both cases acceleration can be said to have dimensions LT^{-2}. Now, Newton's Second Law of mechanical force states that *force* = *mass* × *acceleration*; hence, force has dimensions MLT^{-2}.

As a second example let us consider the product of magnetic field and flux density. These have dimensions $L^{-1}I$ and $MT^{-2}I^{-1}$ respectively so their product *HB* has dimensions $ML^{-1}T^{-2}$. The product *BH* must have units of energy since it is related to the area of the *B–H* loop which itself is proportional to the energy loss in a cycle of magnetisation. In Table 2.2, energy is shown as having dimensions ML^2T^{-2} and when we look more closely at the dimensions of the product *BH* we see it can be written as $ML^2T^{-2}L^{-3}$ or energy per unit volume. In this case it is a useful reminder that the area of the BH loop measured in tesla.amp.metre^{-1} is in fact the same as the energy measured in more familiar terminology as Jm^{-3}.

The dimensional form of a quantity is not the units in which it is measured. It simply shows its form with reference to the four arbitrary quantities, mass, length, time and current.[18] The more fundamental theory and wider aspects of dimensional analysis can be found in many focused publications, e.g. [16].

[18]We can deduce the dimensions of electric current in terms of *M*, *L*, *T* from its definition related to the force it produces on a pair of conductors but it is more common to include the fourth reference quantity.

2.4 Crystal planes and directions

Electrical steels have a crystalline structure which happens to have a cubic symmetry. Furthermore, both the magnetic and elastic properties of single crystals can be highly dependent on the direction of magnetisation or applied stress with respect to the crystal axes. This is the well-known phenomenon of anisotropy which simply means that the material has different properties when measured in different directions.

Silicon–iron is a very common electrical steel composition. It has what is called a body-centred cubic crystal structure. Figure 2.12 shows the smallest unit cell of the material which can exist and theoretically have the same intrinsic properties as the bulk material. It has perfect cubic symmetry with a either a silicon or an iron atom located at each corner of the cube and one at its centre, hence the term body-centred. Some nickel iron alloys have a face-centred cubic crystal structure and cobalt has a hexagonal crystal structure.

In 3% silicon–iron[19] there are only a few silicon atoms in every 100 atoms, so we would need to draw a much larger array of atoms than in Figure 2.12 to show how the silicon atoms normally are randomly distributed throughout the atomic lattice. A real crystal of SiFe would be built up from millions of such unit cells where the atoms are all fixed distances apart and spread, three-dimensionally along the crystal axes shown as dotted lines in Figure 2.12.

Silicon is called a substitutional element here because it replaces one of the iron atoms which comprise the majority of the volume. Later it will be seen that magnetic properties of electrical steels can be disturbed by the presence of interstitial elements such as carbon, these are sited randomly between the lattice sites occupied by the main and substitutional elements.

Material scientists are very familiar with the system of *Miller indices* used to define crystal planes and directions in a clear manner. Many engineers will not be familiar with the system so its basic concept for use with cubic crystalline structures

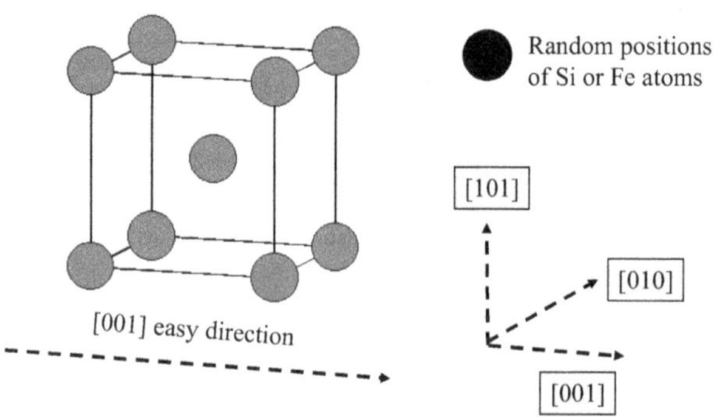

Figure 2.12 Unit crystal cell of a silicon–iron alloy

[19]GO steel contains around 3% by weight of silicon atoms which is around 6% by volume so around six atoms of Si are present in 100 atoms of 3% SiFe.

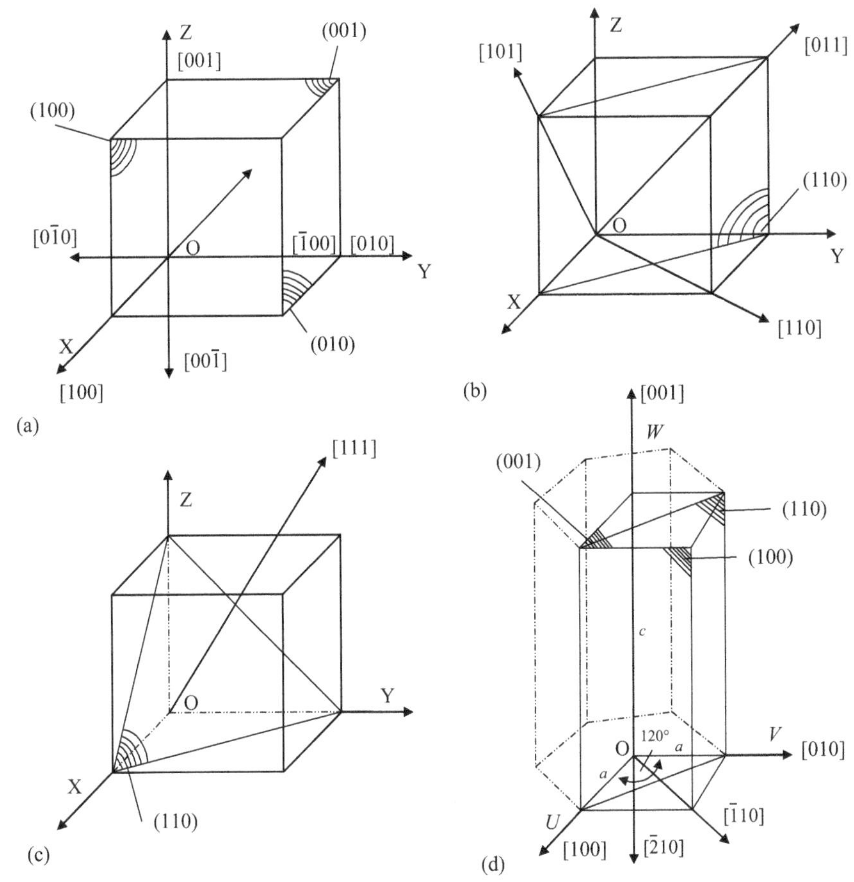

Figure 2.13 Unit cell of a body centred material showing the principal crystal directions and planes as Miller indices: (a) the six cube edge <100> directions and the (100) plane, (b) three of the twelve cube-face diagonals or <110> directions and the (110) plane, (c) one of the eight body diagonals or <111> directions and the (111) plane and (d) principal planes and directions in a hexagonal structure[20]

is introduced here. It is used later in analysis of anisotropy and magnetostriction in electrical steels. A fuller description focusing on electrical steels is given in [17].

Any direction in the cubic crystal can be uniquely defined in terms of three numbers expressed in square brackets as [*xyz*]. Likewise any crystal plane can be represented by three numbers in curved brackets as (abc).

Figure 2.13(a) shows the three cube edge directions, [100], [010] and [001] which in this case are identical to the *x*, *y* and *z* directions. The term <100> is used to mean

[20]The hexagonal form is often written in a four digit notation using Miller–Bravais indices. Hexagonal structures are not relevant to electrical steels so the indices are not discussed here.

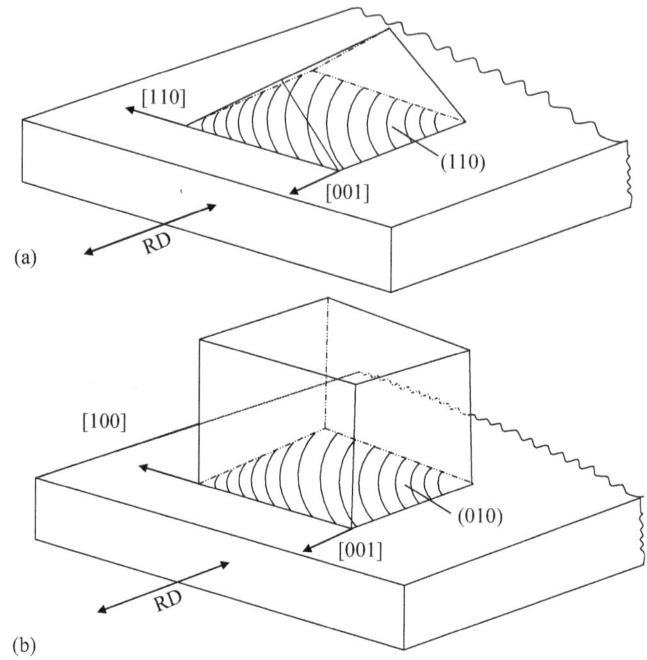

Figure 2.14 Orientation of ideal grains in (a) GO silicon–iron and (b) cube-on-edge material

any, or all, of these directions. These notations mean the positive *x*, *y* and *z* directions. Terms such as [$\bar{1}$00] mean a direction along the *x*-axis but in the negative direction.

Crystal planes perpendicular to each direction have the same number as the particular direction, e.g. the (100) (*yz*) plane is perpendicular to the [100] (*x*) direction. Likewise, the (010) and (001) planes are the *xz* and *xy* planes, respectively. Collectively, they are referred to as {100} planes.

Figure 2.13(b) and (c) shows the designations of cube face and cube body diagonals, respectively. The way in which these designations, and the more complex ones for intermediate planes and directions, are worked out is explained in detail in [13]. For completeness, Figure 2.13(d) shows the set of indices used for hexagonal crystal structures.

Grain oriented (GO) steel is sometimes referred to as {110} <001> oriented steel.[21] An ideal or perfectly oriented Goss grain (or crystal[22]) has its [001] directions parallel to the rolling direction and its (110) plane parallel with the plane of the strip as shown in Figure 2.14(a).

[21]This is often written in the more restricted form (110)[001].

[22]Crystal and grain mean the same thing when we are dealing with electrical steels. Crystal is usually used when discussing basic theory and grain is used more often when structure of the steels themselves. We often refer to the importance of grain boundaries in steels, we could equally call them crystal boundaries but although having the same meaning the term is hardly ever used.

A user can choose any set of references. Sometimes, in the literature, GO steel is treated as having an ideal (011)[100] texture.

Another texture present in some nickel–iron alloys, and also considered for silicon–iron alloys, is shown in Figure 2.14(b). This is sometimes referred to as a cube-on-face texture in which the majority of grains have a (100)[001] orientation with respect to the rolling direction of the strip. An ideal grain in cube-on-face material has easy directions of magnetisation along, and transverse to, the rolling direction. This material is discussed in Section 3.2.2 of Volume 2 of this book.

References

[1] Bozorth R.M. *Ferromagnetism*. IEEE Press, Piscataway, NJ, 1993 p. 719

[2] Bozorth R.M. *Ferromagnetism*. IEEE Press, Piscataway, NJ, 1993 p. 448

[3] Liorzou F., Phelps B., and Atherton D.L. 'Macroscopic models of magnetization'. *IEEE Trans. Magn.* 2000, vol. 36(2), pp. 418–428

[4] Zakerzadeh M., Firouzi M., Sayyaadi H., and Shouraki B. 'Hysteresis nonlinearity identification using new Preisach model-based artificial neural network approach'. *J. Appl. Math.* 2011, Art. No. 458768

[5] Dupré L., and Melkebeek J. 'Electromagnetic hysteresis modelling: from material science to finite element analysis of devices'. *Int. Compumag. Soc. Newslett.* 2003, vol. 10(3), pp. 4–15

[6] Bertotti G., *Hysteresis in Magnetism for Physicists, Material Scientists and Engineers*. Academic Press, San Diego, CA, 1998

[7] Brailsford F. *Physical Principles of Magnetism*. Van Nostrand, London, 1966, pp. 8–10

[8] Bozorth R.M. *Ferromagnetism*. IEEE Press, Piscataway, NJ, 1993, pp. 846–849

[9] Sung H.W.F., and Rudowicz C. 'Physics behind the magnetic hysteresis loop-a survey of misconceptions in magnetism literature'. *J. Magn. Magn. Mater.* 2003, vol. 260, pp. 250–260

[10] Giacoletto L.J. 'Standardized use of SI magnetic units'. *IEEE Trans. Magn.* 1974, vol. 10(4), pp. 1134–1135

[11] Crangle J. *Solid State Magnetism*. Van Nostrand, New York, 1991, pp. 12–15

[12] Bozorth R.M. *Ferromagnetism*. IEEE Press, Piscataway, NJ, 1993

[13] http://www.ieeemagnetics.org/index.php?option=com_content&view=article&id=118&Itemid=107 (Accessed July 2018)

[14] Ewing J.A. *Magnetic Induction in Iron and Other Metals*. D. Van Nostrand, London, 1900

[15] Baldwin C.T. *Fundamentals of Electrical Measurements*. G.G. Harrap & Co, London, 1961, p. 13

[16] Sonin A.A. *The Physical Basis of Dimensional Analysis*, 2nd edn. Department of Mechanical Engineering, MIT, Cambridge, MA, 2001

[17] Brailsford F. *Physical Principles of Magnetism*. Van Nostrand, London, 1966, pp. 17–21

Chapter 3

Magnetic domains, energy minimisation and magnetostriction

The magnetic domain structure of a magnetic material determines its magnetic properties. When the domain structure changes, perhaps due to factors such as temperature variation, mechanical stress or the presence of an external magnetic field, magnetic properties also change, in a beneficial or detrimental way. The physics controlling the formation and structure of domains, as well as the dynamics of so-called *domain wall motion* and *domain rotation* is quite well understood, but it is very complicated and difficult to apply to materials with complex metallurgical structure such as electrical steels.

Magnetic domains of direct relevance to electrical steels are covered in Chapter 1 of Volume 2 of this book. In this chapter their origin and basic structures are explained. Readers delving deeper into the fundamental theoretical treatments of the physical nature of magnetic domains in soft magnetic materials can refer to dedicated works such as [1]–[3].

3.1 Magnetic dipole moments and domains

To understand magnetic domains, we need to start by considering the fundamental source of magnetisation in an atom, the magnetic dipole. It was stated in Chapter 2 that the dipole basically behaves like a bar magnet. The strength, or magnetisation, of a dipole never changes but the direction in which its magnetic field is oriented is dependent on the influence of adjacent dipoles and other factors discussed later.

There are billions of dipoles per kilogram of any magnetic material. Consider a piece of iron which has all its atoms, hence dipoles, fixed in a crystal lattice. If there were no other constraints, the magnetic axis of each dipole would settle in arbitrary directions according to the attractive and repulsive forces from nearby poles. The material is then in a state of magnetic equilibrium as shown in Figure 3.1(a).[1]

It was realised, soon after Ewing built on Weber's early concept of the dipole moment in an attempt to explain ferromagnetism [4], that the large permeabilities of ferromagnetic materials could not be explained by random orientation of dipole moments depicted in Figure 3.1(a). James Ewing did not realise that there are large

[1]In fact this sort of distribution occurs in other types of magnetism but the effect is so small compared with ferromagnetism that it is not discussed here.

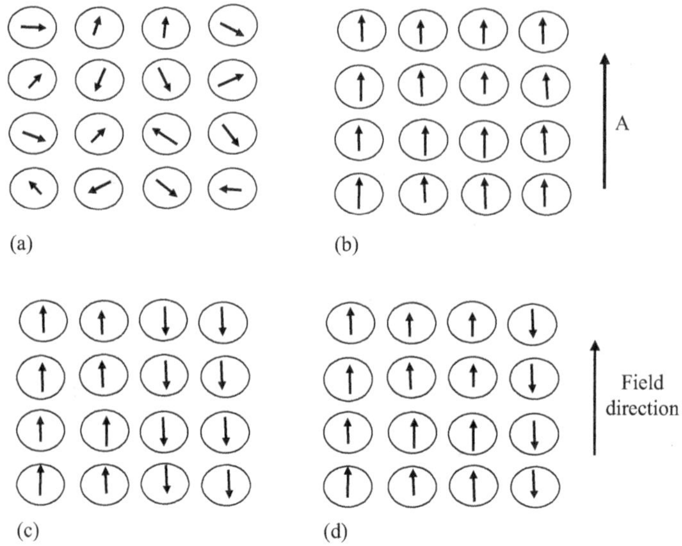

Figure 3.1 *Schematic representation of magnetic dipoles in a magnetic material: (a) randomly magnetised dipoles in a symmetrical atomic lattice, (b) dipoles all aligned in the lattice, (c) dipoles aligned in domains without an external field present and (d) influence of a small field on the dipole distribution in producing a net magnetisation in the field direction*

forces present in ferromagnetic materials which tend to make dipole moments align with each other, even when no external magnetic field is present, as shown in Figure 3.1(b). The material shown in Figure 3.1(b) is obviously in a state of magnetic saturation, all the moments adding to give self-magnetisation in direction A. The condition for dipoles always aligning in such an energetically favourable state is also satisfied in Figure 3.1(c) where it can be seen that the net magnetisation in any direction is zero so the material is in a state of demagnetisation. In this case the dipoles are arranged as two magnetic domains.

If a field is next applied in the direction shown in Figure 3.1(d), more dipoles become oriented in the field direction so the size of each domain changes leaving a net magnetisation (here equivalent to the moments of eight dipoles) along the field direction.

The process shown schematically in Figure 3.1 is very idealised and ignores the influence of other factors which affect dipole moment orientations. However, it shows the basic dipole interactions which underpin all magnetisation processes.

3.2 Weiss theory and molecular field

Weiss [5] is credited with discovering the basis of ferromagnetic domain theory with his postulation of the so-called *molecular field theory* which is still acknowledged today. At that time there were no means of testing or verifying his ideas which are taken for granted today so it is worth summarising his ground-breaking postulations as follows:

(a) Dipole moments are present all the time whether a ferromagnetic material shows signs of being magnetised or not.
(b) The dipole moments are aligned on a microscopic scale into small regions which are referred to as magnetic domains.
(c) The domains are randomly aligned in the demagnetised state and are pulled towards the direction of an applied magnetic field.
(d) The magnetisation process involves re-orientating the direction of domains to increase the volume of domains whose dipoles are oriented close to the field direction.

Weiss's theory explains how all the dipoles can be aligned at all times, although the material as a whole is not necessarily magnetically saturated along any direction. Billions of dipoles are aligned in the same direction within each magnetic domain, but there might be millions of domains oriented in different directions in the material. It was stated earlier that Figure 3.1(c) shows an oversimplified presence of two magnetic domains in a piece of magnetic material containing only a few atoms. Such simple formations do not exist in electrical steel, but the simplification helps to explain the basic structure and role of magnetic domains.

If the field applied in Figure 3.1(d) is further increased, at some point all the dipoles will have rotated into the field direction to form a single domain.[2] At this stage the material is magnetised to saturation, and, as explained in Section 2.13, at 0 K all the dipoles are perfectly aligned, the bulk magnetisation is M_0 and at any other temperature (below the Curie point) thermal agitation of the dipoles reduces the saturation to value M_s.

This simplified explanation so far ignores the existence of magnetic domain walls. Their fundamental structure and role is explained in Section 3.4.

Weiss's theory correctly postulates the existence of domains but why are they not always all aligned in the same direction in a material as in Figure 2.1(b) making all magnetic materials permanent magnets? The answer to this is that there are other factors which have major roles in determining the directions of the dipole moments, hence the orientation of the domains.

At normal operating temperatures, the molecular field overcomes thermal agitation and keeps adjacent dipole moments well aligned. In iron, a large amount of energy is needed to rotate a dipole moment out of alignment with the field. This energy is stored in the material and is called exchange energy. It is related to the way in which the energy of electrons in adjacent atoms is shared and can be thought of as a measure of the strength of coupling between the magnetic moments of the dipoles. The concept of exchange energy explains the molecular field in terms of quantum mechanical forces acting between atoms [6].

The strong positive exchange interaction in ferromagnetic materials, which acts to align atomic moments parallel to each other, is the fundamental cause of

[2]Note that the atoms themselves are fixed in position; it is only the dipole moments which change direction.

their high magnetisation[3] and results in only three ferromagnetic elements (iron, nickel and cobalt) at room temperature. The field itself is of the order of 550 MAm^{-1} in iron at room temperature [7]. To put this into perspective, it is about a million times greater than the field required to saturate non-oriented (NO) silicon–iron, so it is understandable why it is so difficult to pull dipoles out of their parallel alignment in ferromagnetic materials.[4]

3.3 Minimisation of free energy

It is necessary to introduce the concept of *energy minimisation* in order to understand the domain structures which occur in real materials. The magnetic domain structure within a material is organised so as to minimise the amount of energy stored at all times. Factors such as crystal structure, mechanical stress and the presence of an external magnetic field affect the energy balance and the domain structure adapts to keep the total stored energy as low as possible.

The material shown in Figure 3.2(a) has no exchange energy (the energy due to the exchange interaction) because all its dipoles are aligned parallel. It clearly

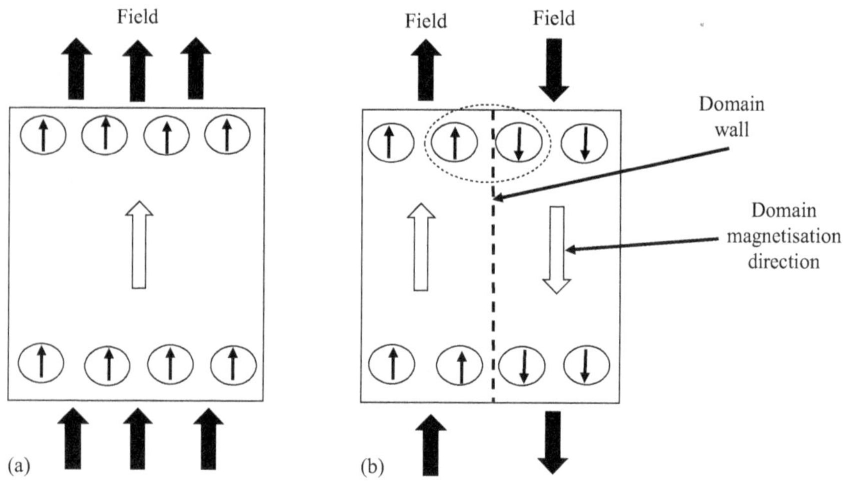

Figure 3.2 Schematic representation of dipole orientation: (a) all dipoles pointing in the same direction as a single domain and (b) lower energy arrangement of dipoles forming two domains in the same volume as in (a)

[3]Paramagnetic diamagnetic and other groups of materials are very feebly magnetic with permeabilities of the around 10^{-6} compared with values of the order of around 10^{-2} for electrical steel demonstrating a 10,000-fold difference in the size of the exchange forces between the two. Iron, cobalt and nickel become paramagnetic above their Curie temperatures.

[4]Dipole alignment occurs and magnetic domains are present in ferrimagnetic materials such as soft ferrites discussed in Chapter 17.

behaves in the same way as a permanent magnet with north and south poles creating the external magnetic field and flux paths. The energy associated with this field, the magnetostatic energy,[5] is very high in this configuration but it is reduced if two domains are formed whose dipole moments are antiparallel as shown in Figure 3.2(b). A problem with this is that some dipoles, such as the pair circled, have large exchange energy associated with them because of their antiparallel magnetic moments. However, this is considerably reduced in practice by the formation of a domain wall comprising many layers of dipoles whose magnetic moments are rotated gradually relative to each other.

3.4 Domain wall structure and motion

Figure 3.3(a) shows a simplified representation of one row of dipoles of a domain wall separating main domains A and B in a sheet of material. For clarity, the width of the wall δ is exaggerated compared to the widths of the two domains. The magnetisation direction between domains A and B is reversed within the wall by each dipole being slightly rotated normal to the $x - z$ plane as shown. In this illustration the wall is 11 dipole, or atomic layers wide. In electrical steel, such walls are 50–100 layers wide so the angle between adjacent dipoles across the wall is just 2°–3°. The wall width is determined by the requirement for it to have the lowest possible energy. If the wall was an abrupt 180° switch, such as in Figure 3.1(c), the exchange energy would be extremely large whereas, if the wall is very wide, many more dipoles are oriented away from the [100] direction so the magnetocrystalline energy[6] would be very high.

This type of domain wall is called a *180° Bloch wall* in honour of the Nobel Prize winning Swiss scientist, Felix Bloch, who discovered it. Figure 3.3(b) shows a 90° Bloch wall, present between domains whose magnetisation directions are at right angles to each other.

Other types of wall exist[7] but, as far as is known, only 180° and 90° Bloch walls are present in electrical steels.

Consider the magnetising process in the simple model illustrated in Figure 3.4(a).[8] This shows the surfaces of two domains where the arrows show their magnetising directions and the 180° domain wall is represented by the single line. The width of each domain is d and its length and thickness is 1 unit. The material is initially completely demagnetised since the vector sum of dipole moments in any direction is zero because the two domains are equal in size. Suppose a magnetic field H is applied in the direction shown in Figure 3.4(b). It does not affect the dipoles in the bulk of the two domains

[5]See Section 3.7 for an explanation of magnetostatic energy.

[6]See Section 3.6.

[7]The Bloch wall is one of the two main types of domain wall found in magnetic materials. The other is the Néel wall in which magnetisation reversal between adjacent domains occurs by dipole rotation in the plane of a very thin material whose thickness is of the same order of magnitude as the wall width itself.

[8]It will be seen in Section 3.5 that this hypothetical structure would not be expected to occur in practice, but it is useful for explaining domain wall motion in a simple way.

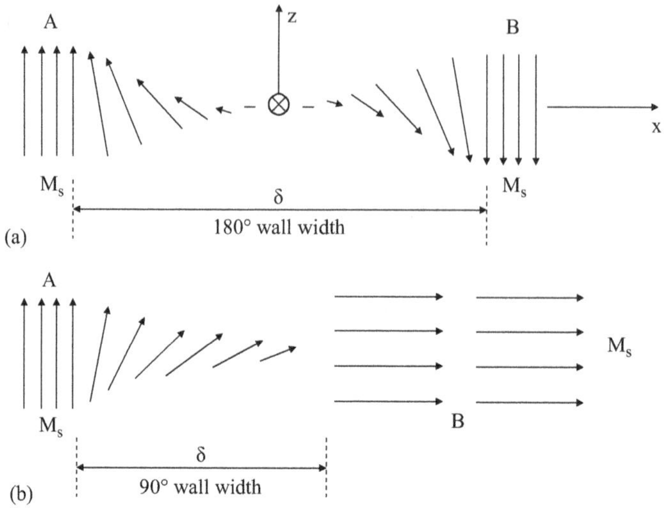

Figure 3.3 Basic rotation of dipoles in Bloch domain walls: (a) 180° wall and (b) 90° wall

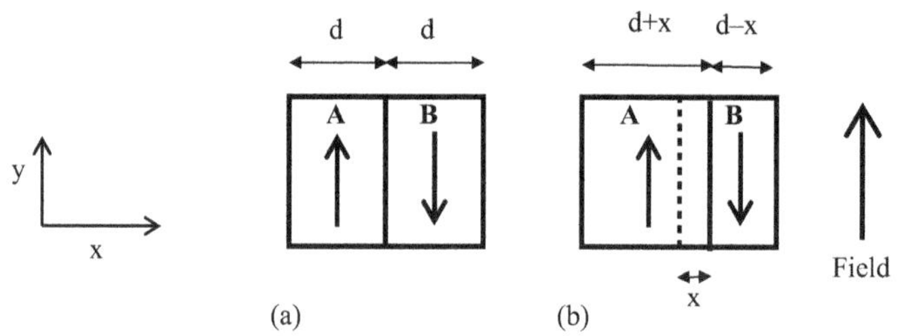

Figure 3.4 Basic motion of a 180° domain wall subject to a field applied along its longitudinal direction: (a) un-magnetised, domains of equal size and (b) magnetised along the y direction as the field forces the domain wall to move to the right

although the dipoles in domain *B* experience an increase in magnetostatic energy so the overall stored energy increases.

At the same time consider the effect of the field on the dipoles within the domain wall which has the structure shown in Figure 3.3(a). Dipoles near the edge of the wall adjoining domain *A* are oriented close to the plane of the sample so if they rotate a little their energy is reduced and at the same time the energy of the dipoles just outside the wall, at the edge of domain *B*, falls as they rotate slightly out of the plane of the material. Hence the domain wall moves linearly to the right through the material as the field is increased. The minimum energy state is maintained.

The width of the wall is unchanged during this process and of course no atoms change their positions. Suppose the wall has moved a distance x as shown due to the presence of the field. It is easy to work out the resulting magnetisation M as a percentage of the saturation magnetisation M_s. Consider the surface area of each domain which is proportional to its volume, hence the total magnetisation. The area of domain A is $(d+x)/2d$ units of the whole area and that of domain B is $(d-x)/2d$ units. The excess magnetisation in the field direction is simply the difference between these two which is $M_x x/d$. For example, if $x = d/2$ the bulk magnetisation has risen from zero to $M_s/2$. We will see later that this simple approach can be used to estimate magnetisation conditions in grain oriented (GO) electrical steel from observation of surface domain motion.

3.5 Domain changes occurring during magnetisation

The process by which a material becomes magnetised when placed in a magnetic field has been introduced in Sections 2.1.6 and 3.4. In this section, the influence of domain wall motion and domain rotation on the shape of the magnetisation curve is explained.

Consider a single crystal with the domain structure shown in Figure 3.5(a).[9] Suppose it is initially in a demagnetised state, i.e. $M = 0$ along any direction. The arrows, dots and crosses indicate the magnetisation directions of each domain. Such a structure is unlikely to be found in practice but it can be used to illustrate steps in the magnetisation process. Suppose a small field is applied from left to right along the longitudinal direction as shown in Figure 3.5(b). Domains whose magnetisations have components in the field direction grow so the material becomes partly magnetised. As the field increases a stage is reached when domains, which do not have a component of magnetisation in the field direction, suddenly switch direction until, as shown in Figure 3.5(c), all the magnetisation is directed along the easy axes closest to the magnetising direction. At this stage the crystal is effectively a single domain although, for convenience, separate domains are still shown. If the field is further increased all domains eventually become aligned along the field direction as shown in Figure 3.5(d). At this stage the crystal is saturated along the field direction, Of course, at stage (c) the material is already saturated but along the easy axis.

Although the structure shown in Figure 3.5 is hypothetical, it gives a good approximation to processes which do occur in practice.

Figure 3.6 shows portions of a typical M–H curve of a soft magnetic material. The basic processes of domain wall displacement and magnetisation rotation shown in Figures 3.4 and 3.5 are associated with three identified regions of the characteristic. Increasing the field, starting from the demagnetised state at O, the linear

[9]It will be seen in Section 3.12 that domains tend to align with certain axes in a crystalline material. In this case this idealised crystal could be iron with its surface being parallel to the (001) plane and the [100] and [010] directions being at 45° to the longitudinal direction of the strip.

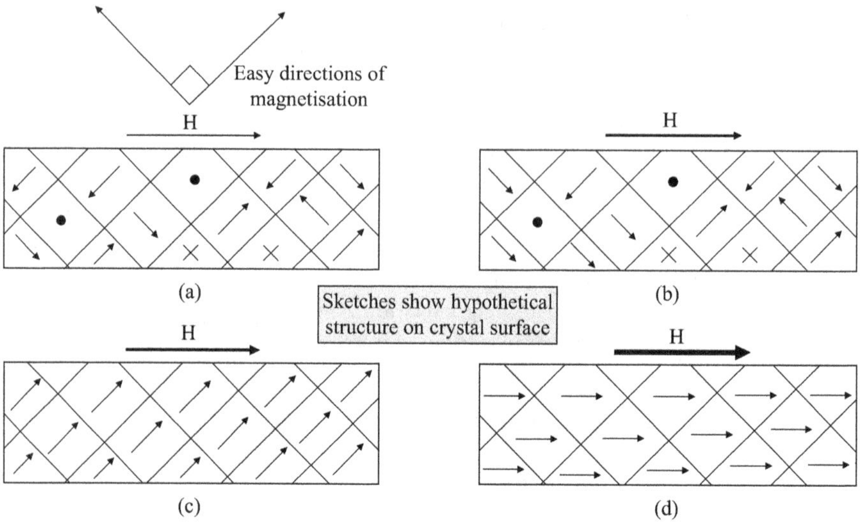

Figure 3.5 Changes in domain structure with increasing magnetisation in an idealised single cubic crystal: (a) demagnetised, random domain structure, (b) domains oriented along easy directions closest to the magnetising directing grow and others shrink when subjected to the low field, (c) near the knee of the magnetisation curve a single domain appears and (d) a single domain exists at saturation oriented along the field direction

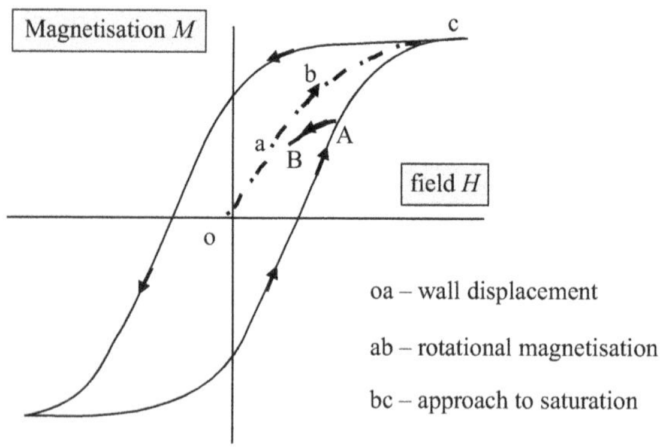

Figure 3.6 Domain restructuring in three regions of a magnetisation curve: (a) domain wall displacement region, (b) irreversible wall displacement and domain rotation region and (c) reversible zone at the approach to saturation

portion *a* of the curve is traversed. In this region almost all the magnetisation occurs by domain wall displacement of the sort schematically shown in Figure 3.4(b). As the field is increased, the knee of the magnetisation curve is reached in region b which corresponds to irreversible domain rotation as eventually all domains are switched to the condition shown in Figure 3.5(c). Region c is reached at a very high field. This is commonly referred to as the approach to saturation. During this stage, all domains rotate to line up with the field direction during the reversible rotational process.[10]

If the field is increased until the point *A* is reached and then reduced, the magnetisation follows the direction of the arrow on the curve *AB* so we say the magnetisation is not reversible because it does not have the same value when the field is increasing as it does when the field is decreasing. Note that the terms reversible and irreversible in this context also refer to whether the energy dissipated into heat is a relatively small or large proportion of the changes in stored energy occurring during the magnetisation process.

3.6 Anisotropy energy

3.6.1 Anisotropy energy in materials with cubic crystal structure (E_k)

The dependence of magnetic properties on the direction of magnetisation in a crystalline solid, such as electrical steel, is referred to as magnetocrystalline anisotropy. It was discovered in iron crystals almost 100 years ago [8]. The effect on the magnetisation curves of single crystals of iron is dramatic. As can be seen in Figure 3.7, on applying a field along the [100] direction, the intensity of magnetisation M increases far more rapidly at a much lower field than when magnetised along either of the other two principal crystal axes. For this reason <100> directions in a crystal of iron are called easy directions of magnetisation. It will be seen later that magnetisation along an easy direction causes lowest losses in an electrical steel.

Energy is stored[11] in a crystal due to the magnetocrystalline anisotropy and its magnitude depends on the direction of magnetisation. The Weiss theory concludes that a material is made up of magnetic domains always spontaneously magnetised to saturation. So it can simply be deduced that this magnetocrystalline energy is minimum when the magnetisation in any magnetic domain is directed along a <100> direction in an iron crystal. This applies to all the crystals or grains in a piece of polycrystalline iron. So, in a demagnetised state the magnetisation vectors, M_s, are distributed evenly between the six <100> directions and a state of minimum energy exists.

If a sufficiently large field is applied at an angle to the magnetisation direction within a single domain it will cause M_s to rotate out of its energetically preferred

[10]Rotational magnetisation produced by a rotating magnetic field is important in electrical machine core applications and should not be confused with the process described here. It is a completely irreversible mechanism which is discussed in detail in Chapter 8.

[11]The terms magnetic stored and free energy are synonymous throughout this book.

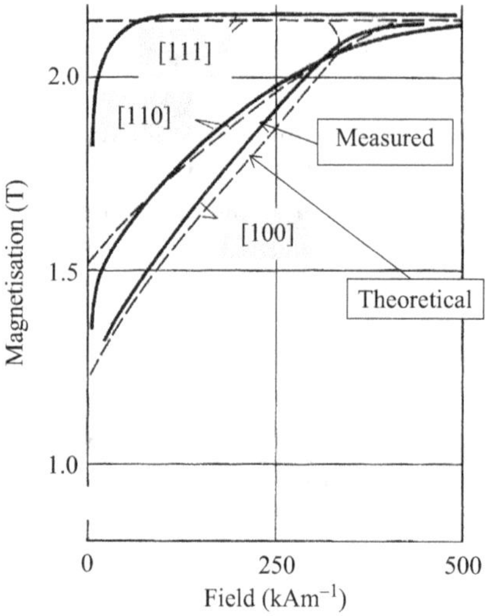

Figure 3.7 Magnetisation curves obtained on single crystals of iron magnetised along the three principal crystal directions

[100] direction towards the direction of the applied field. To do this, work must be done by the field and magnetocrystalline energy or magnetic anisotropy energy[12] is stored in the crystal. It can be deduced by symmetry [9] that the magnetocrystalline energy per unit volume E_k in a crystal with a cubic lattice structure is given the following phenomenological equation.[13]:

$$E_k = K_0 + K_1\left(l^2m^2 + m^2n^2 + n^2l^2\right) + K_2(lmn)^2\ \left(\mathrm{Jm}^{-3}\right) \tag{3.1}$$

K_0, $K_1$and K_2 are the magnetocrystalline constants (anisotropy constants) and l, m, n are direction cosines[14] of the magnetisation with respect to the crystal axes. The term K_0 is independent of direction so it is often omitted from the expression. Higher order terms such as K_3 are included in some publications but it is also argued that since they are made up of (cosine)^4 terms they are effectively included in the lower order quantities[15] [10]. It is not included in any analysis here.

The magnetocrystalline constants are structure insensitive properties which can be positive, negative or zero. For iron, at room temperature, $K_1 \approx 4.5 \times 10^4\ Jm^{-3}$

[12]This term is sometimes used but it is ambiguous since it could refer to other types of induced magnetic anisotropy so it is not used here.

[13]This is called a phenomenological equation because it is based on observed measurements rather than on any basic theory. Its accuracy (i.e. agreement with observed measurements) could probably be improved by adding more terms but it has no practical advantage.

[14]Direction (or *directional*) cosines are used in analytical geometry to define the direction of a vector in terms of the cosines of the angles it makes with the three coordinate axes.

[15]This might be the reason for the results referred to in footnote 13. In [12], positive K_3 are presented alongside the lower magnitude negative K_2 terms which might go some way towards explaining the different results in [13,14].

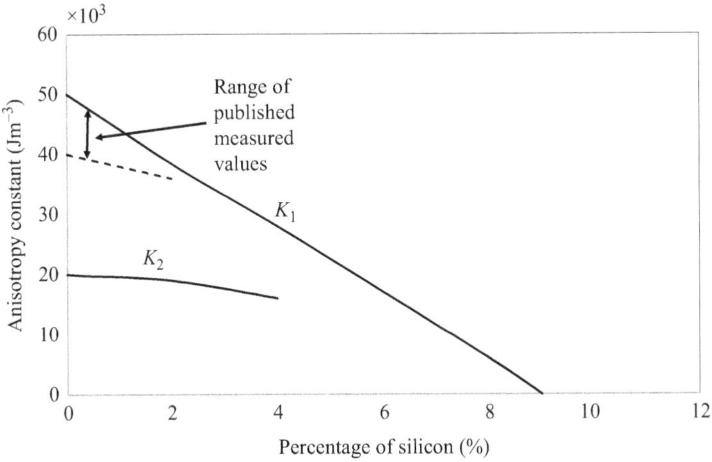

Figure 3.8 General trend of the change in magnetocrystalline anisotropy constants of silicon–iron alloys with silicon content (Averaged from various sources)

and $K_2 \approx 2.0 \times 10^4\, Jm^{-3}$. The values for nickel are $K_1 \approx -0.4 \times 10^4\, Jm^{-3}$ and $K_2 \approx 0.5 \times 10^4\, Jm^{-3}$ although measured values quoted in the literature vary widely [11]. Figure 3.8 shows the range of variation in published values of the dependence of K_1 on silicon content in iron alloys. The fall in K_1 with silicon content has a significant effect on the magnetic properties of electrical steels.

The magnetocrystalline energy stored in a crystal when magnetised along the principal axes can easily be found by substituting the relevant values of direction cosines into (3.1). Hence for the three principal directions we have

$$[100]\; l = 1, m = 0, n = 0\; E_\mathrm{k} = 0 \tag{3.2}$$

$$[110]\; l = 0, m = 1/\sqrt{2}, n = 1/\sqrt{2}\; E_\mathrm{k} = K_1/4\; \left(\mathrm{Jm}^{-3}\right) \tag{3.3}$$

$$[111]\; l = 1/\sqrt{3}, m = 1/\sqrt{3}, n = 1/\sqrt{3}\; E_\mathrm{k} = K_1/3 + K_2/27\; \left(\mathrm{Jm}^{-3}\right) \tag{3.4}$$

For a given material, we can work out which is of these is the easiest direction of magnetisation by inserting the values of K_1 and K_2 into the expressions to find which has the lowest magnetocrystalline energy. From the measured values of K_1 and K_2 shown in Figure 3.8, it can be deduced that the <100> directions are the easy directions in all the silicon–iron alloys. The lower the value of E_k in a composition, the less will be the benefit, or need, of producing a textured steel.

Since K_1 is negative and K_2 is positive in nickel, <111> directions are easy directions of magnetisation. Of course if K_1 and K_2 are both zero, magnetisation

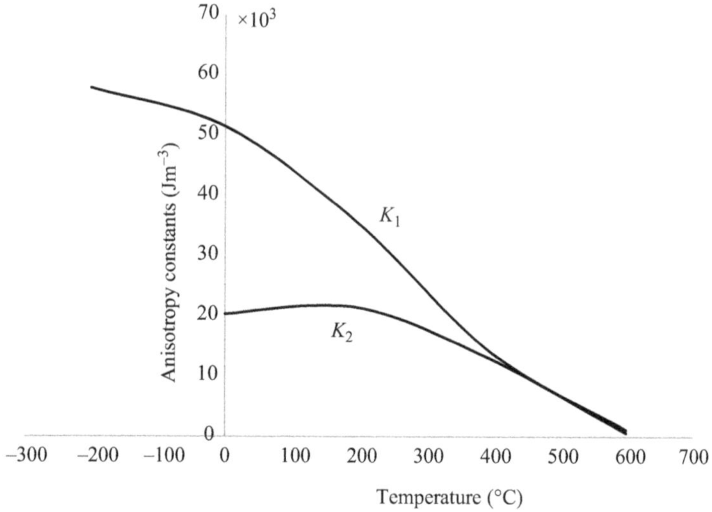

Figure 3.9 Approximate temperature variation of the magnetocrystalline anisotropy constants of iron up to 800 °C (Averaged from various sources)

can lie in any direction in a crystal without any magnetocrystalline energy being present. This is the case in 80%Ni20%Fe alloy where also the magnetostriction is approximately zero so the material responds more freely to a changing external magnetic field which gives this composition the highest permeability of all known crystalline magnetic materials.

The anisotropy constants are structure-insensitive but they can be very temperature sensitive. This contributes to the temperature sensitivity of the magnetic properties of electrical steels. Figure 3.9 shows the temperature variation of the magnetocrystalline constants of iron. Values, including the rarely used third anisotropy constant K_3, for iron over the same range are given independently in [12] and [15], the latter includes values of the anisotropy constants of silicon–iron alloys containing up to 5% Si over a temperature range from 0 °C to 300 °C.

From Figure 3.9, it can be deduced that as the temperature is reduced, the <100> directions become even easier directions of magnetisation and vice versa as the temperature is increased. This infers that domains will be less easily moved from <100> directions at very low temperatures meaning that if we wish to move the magnetisation away from one of these easy directions, for example by means of an external magnetic field or mechanical stress,[16] a higher field or stress would be needed than at room temperature. The opposite will occur at high temperatures of course.

[16]Stress sensitivity of the magnetic properties of electrical steels is practically very important and this is one of the causes of its temperature dependence.

3.6.2 Uniaxial anisotropy energy (E_u)

The simplest form of magnetic anisotropy is uniaxial anisotropy. The uniaxial anisotropy energy E_u can be expressed as

$$E_u = K_{u1} sin^2\theta + K_{u2} sin^4\theta + \dots\dots\dots (\mathrm{Jm}^{-3}) \quad (3.5)$$

where θ is the angle between the magnetisation direction and the easy direction. K_{u1} and K_{u2} are anisotropy constants. Usually K_{u2} can be neglected. As the magnetisation is rotated away from the easy direction the energy increases until it becomes greatest at 90° to the easy direction.

Uniaxial anisotropy is caused by phenomena such as magnetic annealing or the application of mechanical stress. When one of these parameters is present in iron alloys the total anisotropy is the sum of the magnetocrystalline and uniaxial anisotropies, provided no other sources are present.

Cobalt has a hexagonal crystal lattice with an easy direction of magnetisation along its <0001> axes.[17] This becomes a hard direction of magnetisation at high temperature.[18]

Suppose that only uniaxial anisotropy is present in a material and that the second term on the right hand side of (3.5) can be neglected. Suppose also that the magnetisation is very close to the easy direction, i.e. θ is very small, so $\sin\theta = \theta$. Hence

$$E_u = K_{u1}\theta^2 \quad (\mathrm{Jm}^{-3}) \quad (3.6)$$

This is equivalent to a field H_a parallel to the easy direction given by [16][19]:

$$H_a = \frac{2K_{u1}}{M_s} \quad (\mathrm{Am}^{-1}) \quad (3.7)$$

This is referred to as the *anisotropy field*. It is a useful concept since it is a measure of the strength of the anisotropy or the torque necessary to rotate the magnetisation from the easy direction. Obviously the higher the value of K_{u1}, or the lower the value of M_s, the stronger is the anisotropy.

3.7 Magnetostatic energy (E_{ms})

When a magnetic material is placed in a magnetic field, or is in a temporary or permanently magnetised state, it is likely to possess potential energy which is referred to as magnetostatic energy. This can be stored even if no external field is present (such as in a permanent magnet). The energy is a measure of the work needed

[17]For hcp cobalt, the <0001> is the lowest packed direction (perpendicular to the close-packed plane) and is the easy axis. Note extra number in indices. This is also referred to as the c direction.

[18]In cobalt, at temperatures approximately $< 50°$C and $> 400°$C, K_2 is much less than K_1 so can be neglected.

[19]For cubic anisotropy it can be shown [17] that $H_a = \frac{4}{3}\frac{K_1}{M_s}$.

to place free poles at the surfaces of a material. It is closely connected to the demagnetising factor. If $N_d = 0$ no magnetostatic energy is present. Sometimes the magnetostatic energy is called the demagnetisation energy or surface charge energy.

If a sample possessing magnetostatic energy is placed in a magnetic field it influences the magnetising process. It is closely associated with shape anisotropy which is the phenomenon whereby the magnetic properties of a sample vary with direction due to the demagnetising factor being direction dependent due to the sample shape.

The potential energy expression given in (2.3) is derived for a dipole in a magnetic field but it can easily be deduced that the same expression applies to a magnetic domain or a fully saturated specimen (which of course is also a single domain). In the case of a single domain, its saturation magnetisation is M_s and its internal field is H_e so the stored or potential energy is $-M_sH_e$ (since θ in (2.3) is now 0°). Hence the energy per unit volume of a magnetic domain is simply given by

$$E_{\mathrm{ms}} = -\int H_{\mathrm{e}}\, dM \quad \left(\mathrm{Jm}^{-3}\right) \tag{3.8}$$

where H_e is the internal field and M is the magnetisation (in Am^{-1}). If only the demagnetising field is present then $H_e = H_d = -N_dM$. Hence the magnetostatic energy of the domain can be expressed in terms of magnetisation or intensity of magnetisation J (in tesla) as

$$E_{\mathrm{ms}} = N_{\mathrm{d}}\int MdM = \frac{1}{2}N_{\mathrm{d}}M^2 = \frac{1}{2}\mu_0 N_{\mathrm{d}}J^2 \quad \left(\mathrm{Jm}^{-3}\right) \tag{3.9}$$

This equation is useful when analysing the subdivision of domains to minimise the magnetostatic energy.

3.8 Fundamentals of magnetostriction

3.8.1 Stress–strain relationship in soft magnetic materials

The basic relationship between mechanical stress and strain in a non-magnetic material within its elastic limit is needed to help explain magnetostriction. When a mechanical stress is applied to any material, a strain occurs as inter-atomic distances are changed and energy is stored. In the case of a non-magnetic material, the relationship between unidirectional[20] stress σ and strain ε within the elastic limit is simply given by

$$\sigma = \varepsilon E(\mathrm{Pa}) \tag{3.10}$$

[20]In practice stress and strain are vector quantities which might have components in any direction. The importance of including this in the analysis of electrical steels coating stresses is shown in Section 1.15 of Volume 2 of this book.

where E is Young's modulus. Single crystals of iron and silicon-iron alloys have anisotropic mechanical properties as well as magnetic properties, so E has different values along different crystal directions. It is shown in Chapter 1 of Volume 2 of this book that this can be important in electrical steels but it can be ignored for now. The elastic strain energy E_{el} stored in the materials is given by

$$E_{el} = 0.5\sigma\varepsilon = 0.5E\,\varepsilon^2 = 0.5\,\sigma^2/E \;\left(\mathrm{Jm^{-3}}\right) \tag{3.11}$$

Young's modulus can vary with stress and magnetisation in a magnetostrictive material. If we consider the strain in the material comprising a magnetostrictive component, ε_{mag} and an elastic component ε_{elast} then the simple expression for Young's modulus given in (3.10) can be rewritten as

$$E_\sigma = \frac{\sigma}{\left(\varepsilon_{elast} + \varepsilon_{mag}\right)} \quad \left(\mathrm{Jm^{-3}}\right) \tag{3.12}$$

We then quantify ΔE as

$$\frac{\Delta E}{E} = \frac{E - E_\sigma}{E_\sigma} \tag{3.13}$$

If a material is completely saturated, (i.e. it is a single domain) while it is mechanically stressed in the magnetisation direction, then no change in dimensions occurs due to magnetostriction. Hence ε_{mag} is zero so ΔE is also zero. In less common materials such as gallium–iron alloys, the ΔE effect can be more than 50%. However, it occurs to a small extent in demagnetised GO electrical steel where the initial part of the stress–strain curve is non-linear (up to a stress of about 20 MPa). It has been suggested that this could have some effect on the resonant frequencies of power transformer cores discussed in Section 5.1.3 of Volume 2 of this book [18].

A comprehensive review of the theory and measurement of the ΔE effect can be found in [19].

3.8.2 Origin of magnetostriction in soft magnetic materials

Magnetostriction is simply the process whereby the dimensions of a magnetic material change when its magnetisation changes. It was first quantified in 1842 by James Joule [20] who found experimentally that the length and width of an iron rod changed when it was magnetised. Joule could not detect any volume change. A volume change does occur but its percentage change is very small compared with the *Joule magnetostriction* [21] except at very high fields not usually experienced by electrical steels so it is not included here.

The origin of magnetostriction is easiest explained by using a simplified analysis considering a ferrous atom as having an ellipsoidal shape due to the presence of its dipole. Consider some atoms of a ferromagnetic material above its Curie temperature when their dipoles will not exist so we can assume that the atoms are spherical and, for the analysis, imagine them aligned in a lattice as shown in Figure 3.10(a).

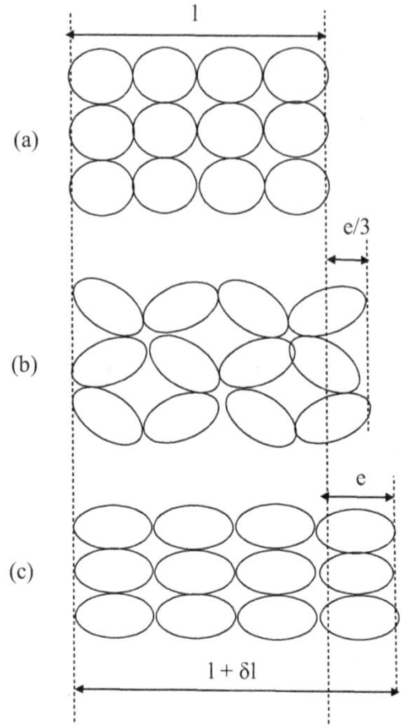

Figure 3.10 Representation of the creation of linear magnetostriction: (a) atoms above the Curie point, (b) atomic dipoles randomly oriented in unmagnetised material at ambient temperature and (c) dipoles aligned in a saturating field

When the material cools below its Curie point, dipoles are created and the Weiss molecular field makes them all align along some arbitrary direction as shown in Figure 3.10(b). These are randomly distributed, i.e. in a fully demagnetised state but the combined length of the dipoles shown is greater than that of the 'spherical' atoms they originated from. When the material is elongated along the axes of the dipoles in this way, its length is shortened in the orthogonal directions in order to maintain a constant volume. This process is called spontaneous magnetostriction.

If the length of material comprising 'non-magnetic' atoms is l and it elongates by a length δl when it becomes magnetic, the magnetostrictive strain is defined as

$$\lambda = \frac{\text{final length} - \text{initial length}}{\text{initial length}} = \frac{\delta l}{l} \tag{3.14}$$

Note that δl can be positive or negative so, to be consistent, it is always important to subtract the initial length from the final length. In this hypothetical case the longitudinal magnetostriction is positive so for the volume to remain constant the transverse magnetostriction must be negative as inferred earlier. We can easily show by equating the initial and final volume that the *transverse magnetostriction* δd of a square section bar with dimensions $l \times d \times d$, or a circular rod of length l and diameter d is given by

$$\frac{\delta d}{d} = \frac{\delta l}{2l} \tag{3.15}$$

This is an important relationship which will be used in Chapter 11 in analysing magnetostriction characteristics of GO steel magnetised in the presence of mechanical stress.

Let us now consider the *saturation magnetostriction* λ_s. This is an important practical parameter which is represented here as the proportional difference in length between the saturated condition, represented in Figure 3.10(c), and the fully demagnetised state represented in Figure 3.10(b).[21]

It can be shown [22] that the elongation of this perfectly random demagnetised structure, due to the creation of spontaneous magnetostriction, compared to the length of the 'non-magnetic' structure in Figure 3.10(a) is $\delta l/3$ where δl is defined in Figure 3.10(b).[22]

The saturation magnetostriction is given by

$$\lambda_s = (\delta l - \delta l/3)/l = \tfrac{2}{3}\delta l/l \tag{3.16}$$

If the saturation magnetostriction is measured along the [100] direction of a cubic crystal, it is defined as λ_{100}. It has a value of around 21×10^{-6} in silicon iron alloys. λ_{100} is measured in terms of metre/metre or strain. So, for iron, we can write $\lambda_{100} = 21$ micro-strain (or μ€).

It is useful to introduce a basic relationship between domain magnetisation and magnetostriction at this stage. Assume that an iron crystal in the *non-magnetic* state is a perfect unit cube with the crystal axes oriented as shown in Figure 3.11. Suppose a single domain is formed as shown in Figure 3.11(a). In this state the cube has extended by a strain λ_{100} along the [100] magnetisation direction and contracted by $-\lambda_{100}/2$ along the [010] direction (in accordance with (3.15)).

Suppose a sufficiently large field is applied to the cube along its [010] direction to make the domain magnetisation direction switch to the [010] direction as shown in Figure 3.11(b). The dimensions change as shown in order to maintain the same unit volume. The magnetostriction due to this domain, or magnetisation rotation is therefore:

$$\text{Along [100] direction } + \lambda_{100} - (-\lambda_{100}/2) = +3/2\,\lambda_{100} \tag{3.17}$$

$$\text{Along [001] direction } - \lambda_{100}/2 - \lambda_{100} = -3/2\,\lambda_{100} \tag{3.18}$$

$$\text{Along [010] direction } + \lambda_{100} - (-\lambda_{100}/2) = +3/2\,\lambda_{100} \tag{3.19}$$

[21]It could be argued that the saturation magnetostriction should be designated as the difference between the configurations in Figure 3.10(a) and (c). However, it should be remembered that these representations are entirely schematic and the only practical and meaningful structure is the fully demagnetised configuration depicted in Figure 3.10(b).

[22]Note that this hypothetical model is isotropic, so the dimensional changes depicted in Figure 3.10(b) occur along any direction. Although the dimensions change the shape will be the same.

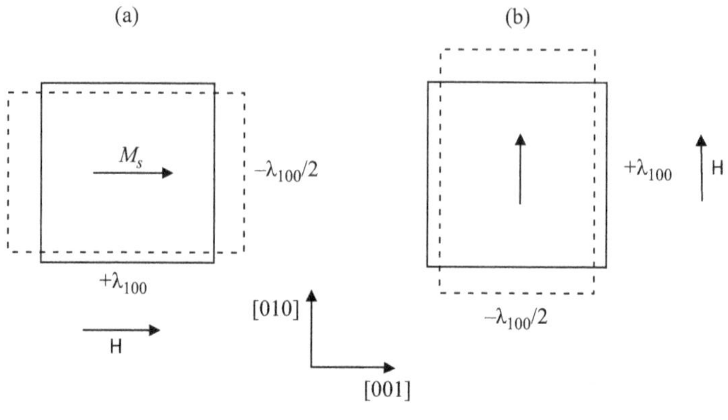

Figure 3.11 Effect of a magnetic field on the dimensions of a unit cube of a crystal with <100> directions along its axes: (a) dimensions when magnetised along the [100] direction and (b) dimensions after rotation of the domain into the [010] direction. (The solid lines depict the non-magnetic unit cube)

This approach might seem over simplistic. However, it is a useful basis for making manageable models of far more complex domain structures in electrical steels. Because the approach is based on firm physical foundations, it can result in good, quantitative approximations to what happens in real materials under static or dynamic magnetisation and also under the influence of mechanical stress.

The theoretical magnetostriction occurring when the magnetisation is switched through 90° is calculated in the simple illustration shown in Figure 3.11. It has been established that the following equation can be used to calculate the magnetostriction λ_s at any angle to the domain magnetisation direction making use of the crystalline cubic symmetry of iron:

$$\lambda_s = (3/2)\lambda_{100}\left(l^2{l_1}^2 + m^2{m_1}^2 + n^2{n_1}^2 - 1/3\right) + 3\lambda_{111}(lml_1m_1 + mnm_1n_1 + nln_1l_1) \tag{3.20}$$

where l, m, n are the direction cosines of the magnetisation direction and l_1, m_1, n_1 are the direction cosines of the magnetostriction strain with respect to the cubic axes. λ_{111} is the saturation magnetostriction measured along the [111] direction.

The expression can also be derived in terms of crystal elastic moduli and coefficients of dipole–dipole interactions [23].This is basically carried out by deriving an expression for the magnetostrictive strain due to dipole rotation and equating it to a mechanical strain to produce equilibrium and a state of minimum energy.[23] Neither

[23]In the solution λ_{100} and λ_{111} appear initially as functions of elastic moduli, again highlighting the close relationship between magnetostriction and elasticity in a crystal.

approach is exact since some other complex dipole interactions are not taken into account, but it is sufficient for engineering applications. Further insight into the more complex nature of magnetostriction analysis can be found elsewhere (e.g. [23]).

More detailed engineering aspects of magnetostriction in single cubic crystals and polycrystalline iron-based materials such as electrical steels materials are included in Section 3.16.

3.9 Magnetoelastic energy (E_{me})

One of the most important factors to take into account in the application of electrical steel is the effect of mechanical stress on its magnetic properties, commonly simply referred to as its *stress sensitivity*. Indeed, in his classic book on magnetic induction, James Ewing wrote [24]:

> No part of our subject is more interesting than that which deals with the effect of mechanical stress in altering the susceptibility, the retentiveness and other qualities of the three magnetic metals.

Elastic stress[24] introduces a strain in a material given in its simplest form by (3.10). However, magnetoelastic energy is present in a magnetostrictive material due to the exchange interaction between adjacent magnetic dipoles. Consider an array of dipoles arranged in their minimum energy state as in Figure 3.12(a). The distance between magnetic moments changes if the bond direction changes to that shown in Figure 3.12(b).

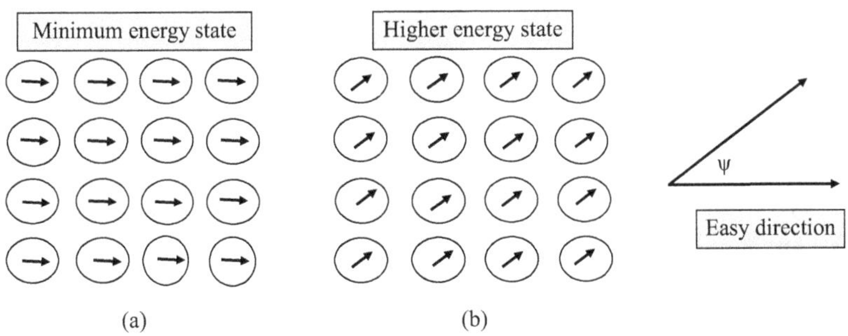

Figure 3.12 (a) Orientation of dipoles in a state of minimum stored energy and (b) dipoles, still aligned, but rotated at an angle ψ to the minimum energy direction

[24]Plastic strain occurs when a material is taken beyond its elastic limit. It is important in many applications of electrical steels as will be seen later, but the theory and analysis in this chapter applies only directly to elastic stresses and strains.

Néel's theory of magnetism can be used to show the origin of magnetostriction in a more rigorous way than that based on the simplistic model shown in Figure 3.10. It can be shown [25] that the exchange energy is given by

$$E_{ex} = l(r)\left(\left(\cos^2\psi - 1/3\right)\right)\ \left(\mathrm{Jm}^{-3}\right) \tag{3.21}$$

where $l(r)$ is a function of interatomic distance r and ψ is the bond direction indicated in Figure 3.12. When $\psi = 90°$, the energy is zero (minimum) as would be expected from the Weiss theory discussed in Section 3.2.

If the atomic lattice is distorted by an external stress the dipole direction and interatomic separation changes. The extent to which this occurs is determined by the minimisation of the sum of the energy stored in the magnetic and mechanical elastic stress systems. As shown in Chapter 11, the magnetic domain structure changes for this to happen and this in turn changes the magnetic properties of the material.

It can be seen that the effect of mechanical stress is closely linked to magnetostriction. In some text books, the effect of stress on magnetostriction is called the *inverse magnetostriction effect*. It can be either very harmful or even very beneficial depending on its direction and magnitude, the magnetisation and the material composition.[25]

It can be shown that the free energy due to magnetostriction can be written simply as

$$E_{me} = -\lambda_s\sigma\ \left(\mathrm{Jm}^{-3}\right) \tag{3.22}$$

Application of tension reduces the free energy if the magnetostriction is positive. The sign convention is used always to consider a tensile stress as being positive and compression is negative, therefore the minus sign is included in the equation.

Equation (3.22) applies to any type of material including amorphous magnetic materials which have no crystal structure. Our main focus of interest is electrical steel, which has a cubic crystal structure, so it is relevant to look at some effects of stress on a single crystal of iron which equally apply to any type of electrical steel.

Consider a single crystal of iron which comprises a single magnetic domain oriented along a direction defined by direction cosines l, m, n. Suppose a uniform tensile stress σ is applied to the crystal along a direction l_1, m_1, n_1. Then from (3.20) and (3.22) the magnetoelastic energy E_{me} stored in the crystal is given by

$$\begin{aligned} E_{me} = {} & -(3/2)\lambda_{100}\sigma\left(l^2{l_1}^2 + m^2{m_1}^2 + n^2{n_1}^2 - 1/3\right) \\ & - 3\lambda_{111}\sigma(lml_1m_1 + mnm_1n_1nln_1l_1)\ \left(\mathrm{Jm}^{-3}\right) \end{aligned} \tag{3.23}$$

We can use this equation to determine the effect of stress on the total free energy of a material. For example, in the case of iron, at room temperature,

[25] Also, temperature, to a lesser degree.

$\lambda_{100} = +21 \times 10^{-6}$ and $\lambda_{111} = -21 \times 10^{-6}$. For nickel, $\lambda_{100} \approx -25 \times 10^{-6}$ and $\lambda_{111} \approx -45 \times 10^{-6}$. If magnetisation and magnetostriction measurements are considered along the [001] direction ($l = l_1 = 0,\ m = m_1 = 0,\ n = n_1 = 1$) then $E_{me} = -\lambda_{100}\sigma$. This term is obviously negative for iron and positive for nickel. In the case of iron we can deduce the following important relationship.

Tensile stress applied along a <100> direction in an iron crystal encourages magnetisation to move towards/remain in that direction.

Conversely it can easily be deduced that

Compressive stress applied along a <100> direction in iron encourages magnetisation to move away from that direction.

The stress and magnetisation are not always in the same direction. Consider a more general case where magnetisation is along the [001] direction of a single crystal and stress is applied along an arbitrary direction l_1, m_1, n_1. Then from (3.23) the magnetoelastic energy becomes

$$E_{me} = -(3/2)\lambda_{100}\sigma\left(n_1{}^2 - 1/3\right)\left(\mathrm{Jm}^{-3}\right) \tag{3.24}$$

For the same stress magnitude and direction, the magnetoelastic energy stored in domains oriented along the other easy directions is given by

$$E_{me} = -(3/2)\lambda_{100}\sigma\left(l_1{}^2 - 1/3\right)\left(\mathrm{Jm}^{-3}\right) \tag{3.25}$$

and

$$E_{me} = -(3/2)\lambda_{100}\sigma\left(m_1{}^2 - 1/3\right)\left(\mathrm{Jm}^{-3}\right) \tag{3.26}$$

in [100] and [010] domains, respectively. When domains whose magnetisations are directed along these easy directions are present, the resulting energy differences give rise to forces on the 90° domain walls thus separating them and causing them to reposition to minimise the total free energy.

If the magnetostriction of a crystal is isotropic, then $\lambda_{100} = \lambda_{111} = \lambda$. Hence, (3.23) reduces to the following simple form

$$E_{me} = -(3/2)\lambda\sigma\cos^2\Phi\left(\mathrm{Jm}^{-3}\right) \tag{3.27}$$

where ϕ is the angle between the magnetisation and stress directions. This equation is useful because of its simplicity when analysing the general effects of stress on the orientation of magnetic domains.

Generally, the application of these equations to practical magnetic domain structures is difficult even with this approximation but it is worth considering the relative values of magnetoelastic and magnetocrystalline energy in electrical steels. As stated in Section 3.6.1, electrical steels have values of K_1 of the order of 3×10^4 Jm^{-3}. If a stress of 300 MPa is applied, and the value of λ_{100} is 20×10^{-6}, then $\lambda_{100}\sigma$ is of the order of 6,000 Jm^{-3}. At this very high stress level, the magnetoelastic energy is still only around 20% of the magnetocrystalline energy. At stress levels encountered in the application of electrical steels, normally no more than 20 MPa, it is unlikely that the domains would be pulled away from the <100> directions. However, it might make certain <100> directions more favourable, so causing a domain redistribution which, in turn, affects the magnetic properties.

3.10 Domain wall energy (E_w)

A domain wall has associated free energy. This is defined as the difference between the energy of the magnetic moments of atoms which are part of the wall and those in the domain itself. Bloch walls have two components of energy. First, the exchange energy, expressed in (3.21), is minimised by the angle ψ between adjacent layers of dipoles being as small as possible. For this to happen, the wall width, or thickness, must be as large as possible. The second component of energy is the wall's anisotropy energy. In electrical steels this is the magnetocrystalline energy which is minimum when dipoles are oriented along a <100> direction. This energy increases as the wall gets wider because more dipoles will be oriented out of a <100> direction. Hence there will be a specific wall thickness which minimises the sum of these two energy components.

It can be shown [26] that the total energy of a 180° domain wall in iron is given by[26]

$$\mathrm{E_w} = K_1\delta/_4\left(\mathrm{Jm}^{-2}\right) \tag{3.28}$$

where δ is the wall width or thickness. The wall thickness itself is given by[27]

$$\delta = \pi a M_s\left(N_w/_{2K_1}\right)^{\frac{1}{2}}(\mathrm{m}) \tag{3.29}$$

where a is the lattice constant and N_w is the Weiss molecular field constant. If the values for iron are substituted into this equation it is found that the width of a 180° domain wall is around 150×10^{-9} m. Domain wall widths in silicon iron alloys have been found experimentally, using neutron refraction methods, to be in the range 180–200 $\times 10^{-9}$ m [27].

The corresponding wall energy found by substituting the wall thickness into (3.28) is around 0.001 Jm^{-2}. It will be observed that this is the energy per unit area of domain wall whereas the magnetocrystalline and magnetoelastic energy values derived earlier are given per unit volume of magnetic domains.

[26]Several other versions of this relationship are found in the literature. We are not going to use the expression quantitatively, so all that needs to be noted is what parameters affect its value.
[27]This equation is not exact because assumptions need to be made in calculating the molecular field constant. Theoretical values of 180° wall width in iron are reported in the range 140–250 $\times 10^{-9}$ m. These are sufficient to give a reasonable estimate of the wall thickness and hence the wall energy.

3.11 Work and energy in the magnetisation process

Suppose an un-magnetised ellipsoidal shaped piece of steel with a demagnetising factor N_x is brought into an external magnetic field H_a. If both the permeability and demagnetising factor are significant, then, from (2.15), the field in the magnetised ellipsoid is $-N_dM$ so there is a linear relationship between H_a and M as shown in Figure 3.13. The slope of the curve is purely a function of sample shape, the material permeability having hardly any influence.

Suppose the sample is placed in a field which increases by a small amount dH_a then the magnetisation increases by dM and the potential energy per unit volume changes from $-N_dM$ to $-(H_a + dH_a)(M + dM)$. Hence the potential energy has decreased by an amount $H_adM + MdH_a$ represented by the shaded area in Figure 3.13(a).

The energy decrease is partly used in doing mechanical work (MdH_a) against any force restraining movement of the sample into the increasing field and partly stored as magnetostatic energy (H_adM) in the system. Hence, as the field increases from zero to its maximum value at point A, the work done, W_{ex}, against the external forces is given by

$$W_{ex} = \int_o^{H_a} MdH_a \left(\mathrm{Jm}^{-3}\right) \tag{3.30}$$

This is equivalent to the area of the triangle OAB.

Likewise the stored magnetostatic energy W_{ms} is given by

$$W_{ms} = \int_0^{M} H_adM \left(\mathrm{Jm}^{-3}\right) \tag{3.31}$$

which is equivalent to the area OAC. The total energy stored when magnetised to point A is MH Jm^{-3}.

In any practical case, particularly at medium to high magnetisation, the magnetisation curve will not be linear but more like that shown in Figure 3.13(b). If N_d is accurately known, the line OA can be drawn such that $\tan\theta = N_d$, where θ is the

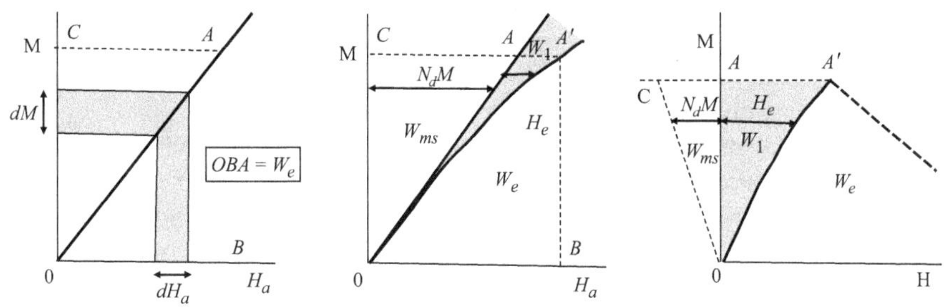

Figure 3.13 Basis relationships between M, H and stored energy: (a) mechanical and magnetic energy related to a linear M–H curve, (b) nonlinear curve showing stored energy W_1 and (c) true magnetisation curve, M–H_e

angle between OA and OC. Then, since $H_e = H_a - N_d M$, the true magnetic characteristic of the material (the $H_e - M$ curve) can be calculated and plotted as in Figure 3.13(c).

As before the mechanical work per unit volume is given by (3.5), but now the stored energy is given by

$$\int_0^M H_a dM = \int_0^M N_d M dM + \int_0^M H_e dM = W_{ms} + W_1 \left(\text{Jm}^{-3}\right) \tag{3.32}$$

W_{ms} is the magnetostatic energy, as before, and W_1 is the stored energy represented by the shaded areas in Figure 3.13(b) and (c). The magnetisation in this case is loss-free and the energy W_1 is stored in the magnetic field itself. The energy W_{ms} can be thought of as being the work done on the sample in bringing it from zero field and W_1 represents the work done by the sample.

This method of calculating the true characteristics of a sample of iron from its demagnetising factor and the $M–H_a$ curve found when placed in a field as shown is not normally accurate in practice. However, it gives further insight into the general concept of demagnetisation characteristics.

3.12 Static domain structure with minimum stored energy

Consider a single crystal plate of iron with its <100> directions parallel to the crystal edges. Suppose just one domain is present as shown in Figure 3.14(a). The magnetocrystalline energy is zero because all the dipoles are oriented along the [001] direction. There are no domain walls so no wall energy is stored, no mechanical strain is present so no magnetoelastic energy is stored. However, its magnetostatic energy is very large so it is not a preferred domain structure.

The magnetostatic energy of a domain is proportional to the demagnetising factor which, in turn, is approximately proportional to the domain width, for a fixed length. Hence the energy of the crystal will be reduced considerably by subdividing the single domain into several domains as shown in Figure 3.14(b). Of course, this introduces domain wall energy and some magnetostatic energy is still present, but the magnetocrystalline energy is unchanged so the total energy is considerably reduced.

The magnetostatic energy can be removed completely by having a closed magnetic path as in Figure 3.14(c). Each domain would have a spontaneous magnetostrictive elongation of λ_{100} in isolation but the closed structure with 90° domain walls has a large stored magnetoelastic energy.

Figure 3.14(d) illustrates the final compromise necessary to obtain the minimum energy state. Neither magnetostatic nor magnetocrystalline energy are present. This type of structure is found in actual single crystals of iron and its main alloys so it is a realistic representation. Note that 90° and 180° domain walls are present together with two common types of domains. The triangular domains at the edges are referred to as *90° closure domains* or, more commonly, just as closure domains.

The longer domains comprise the bulk of the volume of a real material; they are referred to as *180° domains*, *bar domains* or simply main domains.

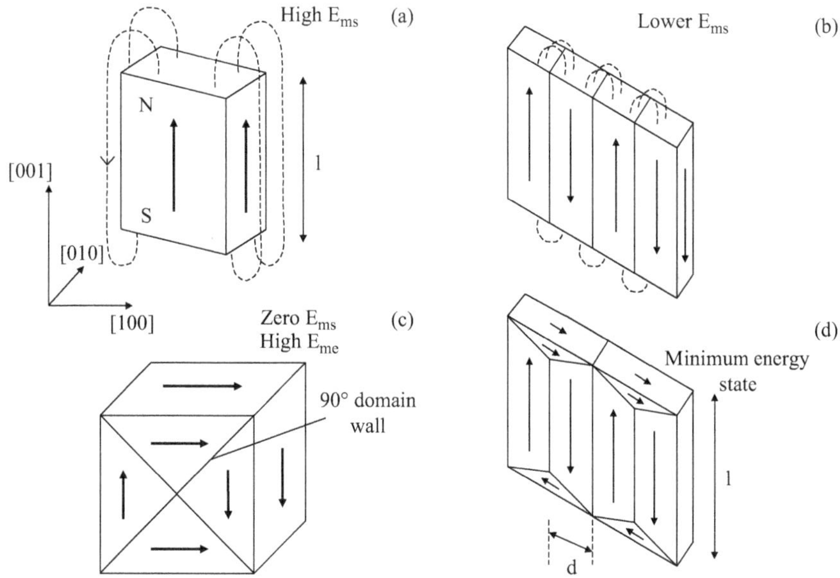

Figure 3.14 Subdivision of domains in a cubic single crystal: (a) single domain, high magnetostatic energy, (b) magnetostatic energy reduced by subdividing domains, (c) magnetostatic energy eliminated, but magnetoelastic energy stored due to closure domains and (d) minimum energy configuration

Domain structures similar to the ideal one shown in Figure 3.14(d) are commonly found in electrical steel so it is worth looking briefly at how the width d of main domains of length l can be calculated. The magnetoelastic energy of the structure can be calculated to be $E_{100}\lambda_{100}^2/2$ Jm^{-3} using (3.23) where E_{100} is Young's modulus along the [100] direction. The volume of closure domains within a unit area is simply $d/2l$ and the wall area is $l + \left(2\sqrt{2} - 1\right)d/ld$ [28]. The sum of the magnetoelastic energy and domain wall energy can be differentiated with respect to d and equated to zero to find the minimum energy and the corresponding value of d. It can then be shown that [28]:

$$d = \frac{2}{\lambda_{100}}\sqrt{\frac{E_w l}{E_{100}}}(\mathrm{m}) \tag{3.33}$$

where E_w is the wall energy per unit area. It should be noted that the domain width is proportional to the square root of its length. For a typical silicon steel, $E_w = 0.001$ Jm^{-2}, $E_{100} = 0.12 \times 10^{12}$ Jm^{-3} and $\lambda_{100} = 24 \times 10^{-6}$ hence

$$d = 7.6 \times 10^{-3}\sqrt{l}(\mathrm{m}) \tag{3.34}$$

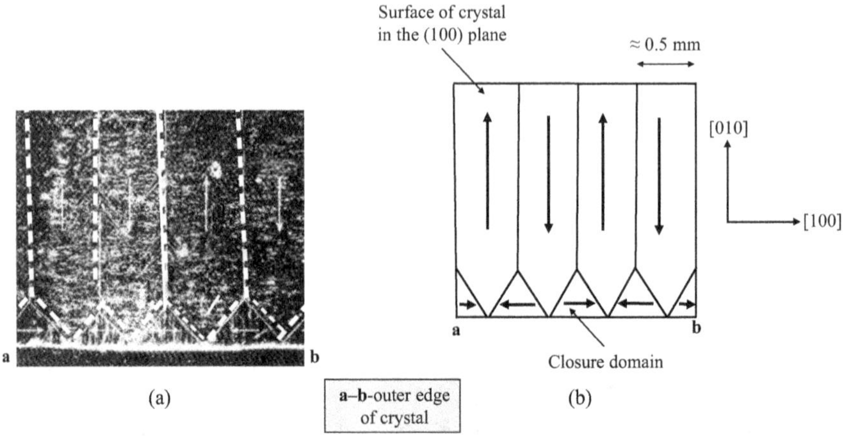

Figure 3.15 Domain pattern observed on the (100) surface of a silicon–iron single crystal: (a) observed bitter patterns shown with position of domains highlighted (Adapted from [25]) and (b) schematic diagram of the same region

The dimensions of domains observed [29] on a silicon–iron single crystal with its surface parallel with the (100) plane are shown in Figure 3.15.[28] The crystal is 2.2 mm long in the [010] direction. The measured value of d was about 0.5 mm and the value calculated from (3.33) was 0.4 mm [28]. Considering the experimental difficulties and the approximations necessary in the calculation, this was remarkable agreement at that time.

It will be seen later that the width of main domains in grain oriented silicon–iron, has a major influence on its properties. Hence, the prediction and control of domain width during the magnetisation process is important in the development of electrical steels. Equation (3.33) is useful for estimating the width of domains in a laboratory produced single crystal. However, many other factors come into play in commercial polycrystalline steels under a.c. magnetisation as can be seen in Chapters 1 and 2 of Volume 2 of this book.

3.13 Domain changes occurring during magnetisation

Magnetisation curves have been referred to several times in this chapter already. Figure 3.4 showed schematically how domains close to the magnetising direction increase in size as the field increases but stay oriented along <100> directions until the field is very high. The impact of the Weiss molecular field and high magnetocrystalline energy control this basic process.

Now consider what happens if a d.c. field is applied to the single crystal which initially has the simple demagnetised domain structure shown in position (0) in

[28]Figure 3.15 is based on an original photograph of the corresponding Bitter pattern in [29] which is too poor to reproduce clearly.

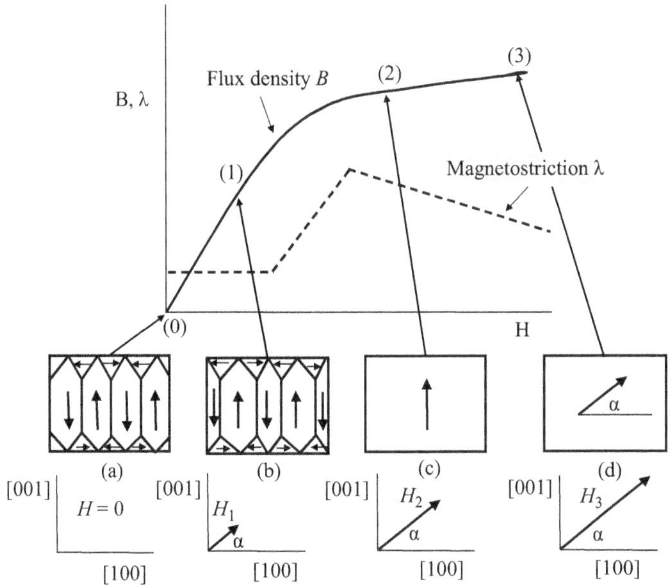

Figure 3.16 Change in idealised domain structure in an increasing d.c. field showing corresponding points on an ideal B–H curve: (a) demagnetised, (b) low external field H_1, (c) medium external field H_2 and (d) high external field H_3

Figure 3.16. Although the same fundamental domain changes occur under a.c. conditions, other processes such as wall bowing and eddy current effects make the processes more complex so they are dealt with in Chapter 1 of Volume 2 of this book. Suppose an external field is applied at angle α to the [001] direction, as shown for position (b). Initially, domains which are oriented close to the field direction grow in volume, at the expense of others, to maintain a state of minimum stored energy. The structure changes to that at (b) by 180° domain wall motion, and the component of flux density B in the field direction increases linearly with increasing field H. Also no magnetostriction change occurs since it is assumed that there is no change in volume of the closure domains during the initial stage (a) to (b).

At a much higher field (c) the closure domains are absorbed into the main domains by magnetisation rotation but at the cost of very high magnetostatic energy being stored, so although the stored energy is still as low as it can be, it is far higher than the energy stored in the demagnetised state. However, at this stage no magnetoelastic energy is present but a change of magnetostriction will have occurred. The size of the change is dependent on the volume of closure domains removed. The magnetisation process taking place between (b) and (c) is very complex. The stored energy becomes much higher and the B–H loop adopts its familiar non-linear shape.

Finally, if a very large field is applied, the dipoles in the single domain, which are naturally oriented along the easy [100] direction to minimise the magnetocrystalline energy, are rotated towards the field direction to eventually obtain full technical magnetic saturation. Extremely high fields are necessary to reach state (3) or even state (2).

The associated magnetostriction changes can be estimated considering a sample whose initial proportion of transverse domains is v per unit volume.

The initial magnetostriction along the [001] direction in the demagnetised state (a) is given by

$$\lambda_i = (1 - v)\lambda_{100} + \left(-v\lambda_{100}/_2\right) \tag{3.35}$$

The magnetostriction at point (c), at which point the material has become a single domain oriented along the [001] direction, is simply given as

$$\lambda_{\mathrm{c}} = \lambda_{100} \tag{3.36}$$

The change of magnetostriction occurring between the demagnetised state and (c) is

$$\lambda_{\mathrm{c-i}} = 3/_2 v\lambda_{100} \tag{3.37}$$

This is probably not surprising. If the magnetisation is increased to point (d) the magnetostriction can be found by putting $l = 1$, $m = 0$, $n = 0$ in (3.20) to obtain

$$\lambda_s = (3/2)\lambda_{100}\left(cos^2\theta - 1/3\right) \tag{3.38}$$

Hence the measured magnetostriction becomes

$$\lambda_{\mathrm{d-i}} = (3/2)\lambda_{100}\left(cos^2\theta - (1 - v)\right) \tag{3.39}$$

It will be noted from these expressions that as expected $\lambda_{\mathrm{d-i}} < \lambda_{\mathrm{c-i}}$ hence the magnetostriction drops as shown in Figure 3.16.

The structures shown in Figure 3.16 are very simplified. Some details are deliberately left out to avoid distraction from the basic phenomena. However, the explanations do show the general trend which help with the interpretation of actual domain structures in electrical steels.

Of course actual electrical steels are polycrystalline materials with a finite thickness. The grains vary in size and orientation so magnetisation is far more complicated but based on the simple ideas just illustrated. If a field is applied in the same way to a polycrystalline material, each grain will act individually[29] and the processes described in Figure 3.16 will occur but the initial domain structure will be different in each grain according to their size, orientation and other factors. Also magnetisation changes caused by factors such as the elimination of any closure domains occur at different field levels in each domain. The resultant magnetisation curve will be similar to an average of many grains.

[29]This is not strictly true as domain structures in the demagnetised state and during magnetisation will depend on the structure of surrounding grains. This fact is not taken into account here in order to simplify the explanation.

3.14 Energy (E_h) due to an externally applied field

Energy stored in a magnetic field by virtue of bringing a demagnetised sample into a field was introduced in Section 3.7. In this section, the more common situation of a demagnetised sample being in an increasing field is considered. It could be argued that basically there is no difference between the two magnetic situations but they are treated differently to develop the concepts of stored energy in different ways which will be brought into analyses in Chapter 11 of this book and in Chapter 1 of Volume 2. The energy stored per unit volume in a dipole oriented at an angle θ to an external magnetic field was given earlier in (2.3), repeated here as

$$W = -MH\cos\theta\ (\mathrm{Jm}^{-3}) \tag{3.40}$$

where θ is the angle between the dipole and the external field direction. It directly follows that the same expression applies to a magnetic domain in a magnetic field since it is just a collection of dipoles all oriented in the same direction. Now consider an external field applied at an angle α to a polycrystalline steel strip containing grains oriented as shown in Figure 3.17. Suppose the field causes single domains to be formed in each grain then clearly the stored energy per unit volume E_h of grain (a) is $-MH_a\cos\alpha$ and that in grain (b) is $-MH_b\cos(\alpha-\beta)$ where H_a and H_b are the resultant fields in the grains taking into account the demagnetising fields.[30] The total energy due to the external field is the volume-weighted sum of all such quantities stored in all the grains.

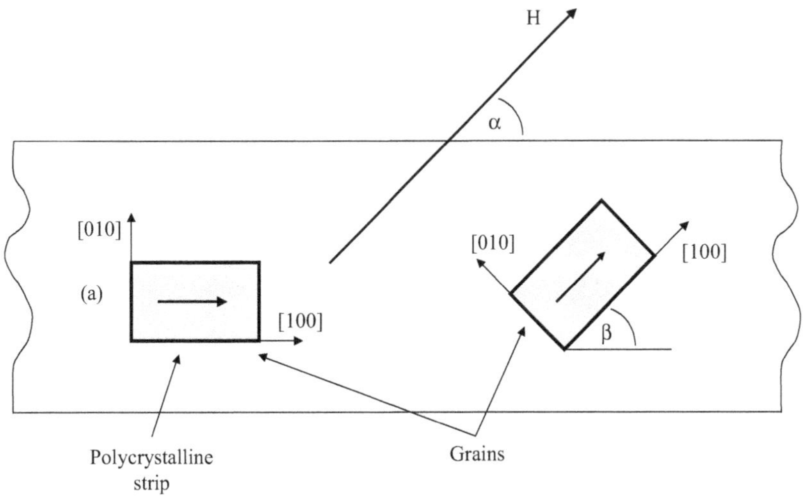

Figure 3.17 Representation of a field applied at an angle to a strip of anisotropic magnetic strip showing the positions of typical grains

[30]It is simpler, of course, to ignore the complication of the demagnetising fields in the grains and assume the field is H in each, but still their stored energies will be different.

An important point is that the true external field-induced energy in the strip is the sum of the field energies in all the grains. This is not the same as $MH \cos \alpha$ where M is the observed magnetisation along the longitudinal direction of the strip which, in practice, can be measured by means of an enwrapping B-coil.[31]

Although the external field is applied at an angle α to the magnetisation measured along the length of the strip, within each grain the field might be a different magnitude and oriented at a different angle to the longitudinal direction. It is possible to use small sensors to measure the field and flux density in individual grains of GO steel to verify this phenomenon [30]. Of course, in iron, or its magnetic alloys, the large magnetocrystalline anisotropy keeps the magnetisation along <100> directions unless the field is extremely high as at point (4) in Figure 3.16.

3.15 Effect of an applied field on a domain wall

The preceding sections of this chapter have mainly been focussed on domain wall structures and energy minimisation under static magnetisation conditions. In this section, some of the eddy current phenomena which take place when domain walls move more rapidly are considered.

The basic motion of a 180° domain wall in the presence of a field was shown in Figure 3.4. This is now considered in more detail to understand what drives the motion, particularly when eddy currents generated near moving domain walls play a major role in magnetisation processes. The domain wall movement is effectively due to a force acting on the wall making it move in the direction shown in Figure 3.4(b). This force F_w is given by [31]:

$$F_{\mathrm{w}} = -\frac{\partial E_{\mathrm{t}}}{\partial x} - \beta_x(\mathrm{N}) \tag{3.41}$$

where E_t is the total stored energy in the wall. $\partial E_{\mathrm{t}}/_{\partial}x$ is the gradient of the total energy at position x of the wall along the x-axis. β_x is a dissipative[32] force on the wall caused by a flow of eddy currents around the moving domain wall. This is said to be the most important dissipative process in electrical steels [31]:

The total energy in the wall region[33] can be written as

$$E_{\mathrm{t}} = E_{\mathrm{k}} + E_{\mathrm{ms}} + E_{\mathrm{me}} + \mathrm{E}_{\mathrm{h}} + E_{\mathrm{w}} + E_{\mathrm{ex}} \left(\mathrm{Jm}^{-3}\right) \tag{3.42}$$

All these components can be present in the domain wall region. To recap, E_{K} is the magnetocrystalline energy, E_{ms} is the magnetostatic energy, E_{me} is the magnetoelastic energy, E_h is the energy due to interaction between the external field and internal magnetisation, E_w is the domain wall energy and E_{ex} is the exchange energy.

[31]See Chapter 12.

[32]The term *dissipative* used throughout this book refers to the conversion of electromagnetic energy in a soft magnetic material magnetised by an electrical energy source, into either heat or sound energy.

[33]Of course, this applies at any point anywhere in the material.

To simplify the problem, if it is assumed that only the force on the wall is due to the presence of the applied field, then the force per unit area can be shown to be [21,25]:

$$F_w = 2J_0H = 2\mu_0 M_0 H(\mathrm{N}) \tag{3.43}$$

In practice, some of the other terms in (3.41) are usually present and disrupt this simple relationship but it is a good approximation that shows that the force on moving a domain wall is proportional to the external field.

3.16 Magnetostriction in soft magnetic materials

Equation (3.20) can be used to help understand the magnetostriction of single crystals and polycrystalline materials with a cubic crystal structure. Some relationships which are important in studies of magnetostriction, particularly in the presence of an external mechanical stress, are derived in this section.

3.16.1 Saturation magnetostriction along principal crystal axes

3.16.1.1 Saturation magnetostriction along the [100] direction

For saturation along a [100] direction, $l = 1$, $m = 0$, $n = 0$, so substituting into (3.14) we have

$$\lambda_s = (3/2)\lambda_{100}\left(l_1^2 - 1/3\right) \tag{3.44}$$

This expression can be used to obtain the magnetostriction in any direction. If we wish to know the value along the magnetising direction then $l_1 = 1$, hence $\lambda_s = \lambda_{100}$. This is, of course, expected from the definition of λ_{100}.

3.16.1.2 Saturation magnetostriction along the [111] direction

Along the [111] direction, $l = m = n = 1/\sqrt{3}$, so on substituting into (3.20) the first term becomes zero and

$$\lambda_{s111} = \lambda_{111}(lml_1m_1 + mnm_1n_1 + nln_1l_1) \tag{3.45}$$

If we put $l_1 = m_1 = n_1 = 1/\sqrt{3}$ the magnetostriction along the [111] magnetising direction can easily be shown to be λ_{111}.

3.16.1.3 Saturation magnetostriction along the [110] direction

Along the [110] direction, $l = m = 1/\sqrt{2}, n = 0$, so substituting into (3.20) gives

$$\lambda_{s110} = (3/2)\lambda_{100}\left((1/2)\left(l_1^2 + m_1^2\right) - 1/3\right) + (3/2)\lambda_{111}(l_1m_1 + m_1n_1) \tag{3.46}$$

To find the magnetostriction along the same [110] direction as the magnetisation we put $l_1 = m_1 = 1/\sqrt{2}, n_1 = 0$, to obtain

$$\lambda_{110} = (1/4)(\lambda_{100} + 3\lambda_{111}) \tag{3.47}$$

This is a neat relationship between the saturation magnetostriction values when magnetised along the three principal crystal directions.

3.16.2 The special case of (110)[001] oriented silicon–iron

The magnetostriction of GO silicon–iron is covered in detail in Chapters 1 and 2 of Volume 2 of this book. However, it is useful to introduce the basic relationship between magnetisation direction and magnetostriction measurement in the plane of a (110)[001] oriented crystal here before considering more complex practical cases later.

Suppose the [001] direction of a slightly mis-oriented grain shown in Figure 3.18 (a) lies at an angle α to the RD in the (110) plane. Consider the resulting magnetostriction at an angle θ to the [001] direction in the same plane. The magnetostriction can be calculated by putting $l = \cos 45 \cos\theta, m = \sin 45 \cos\theta, n = \cos\theta$; $l_1 = \sin 45 \cos\alpha, m_1 = \cos 45 \cos\alpha, n_1 = \cos\alpha$ into (3.20), from which

$$\lambda_s = 3/2\lambda_{100}\left(8[\sin\alpha\sin\theta]^2 + [\cos\alpha\cos\theta]^2 - 1/3\right)$$
$$+ 12\lambda_{111}\sin\alpha\sin\theta\cos(\alpha - \theta) \tag{3.48}$$

If the magnetisation is along the [001] direction, as is normally the case since the magnetocrystalline constant of silicon steel is so large, then $\alpha = 0^\circ$ so $n_1 = 1$ and the equation reduces to simply

$$\lambda_s = (3/2)\lambda_{100}(\cos^2\theta - 1/3) \tag{3.49}$$

Figure 3.18(b) shows the magnetostriction at angles to the [001] direction (in the (110) plane) obtained from (3.49). When magnetised along the [001] direction, it can be seen that the measured magnetostriction is zero at an angle of

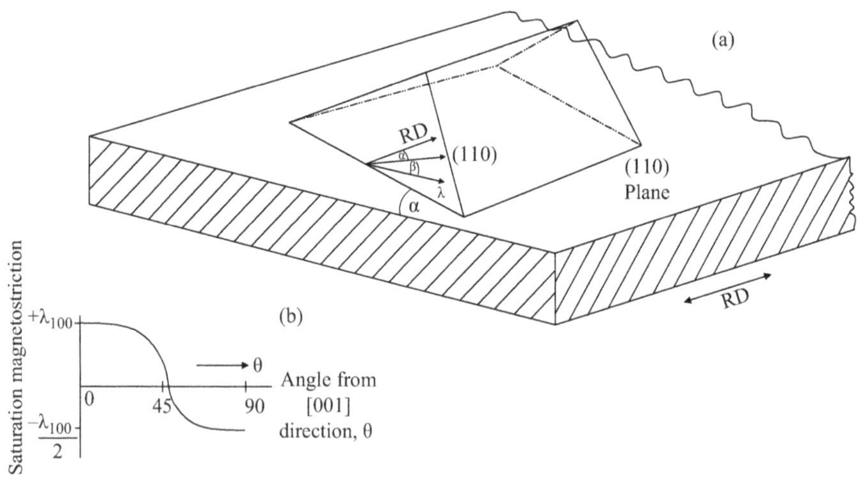

Figure 3.18 Magnetisation and magnetostriction measurement directions in a slightly mis-oriented grain in a strip of (110)[001] oriented steel: (a) representation of the mis-oriented grain and (b) saturation magnetostriction at various angles to the RD when $\alpha = 0$

approximately 54° ($\theta = \cos^{-1} 1/\sqrt{3}$) to the [001] direction before becoming negative at higher angles. It will be noticed that the sign and value of λ_{111} does not influence the size of the magnetostriction when the crystal is magnetised in this way, but it can come into play when GO silicon–iron is mechanically stressed as shown in Section 11.3.

3.16.3 Saturation magnetostriction of a polycrystalline material

Equation (3.20) is strictly only applicable to single crystals but it can be used to estimate the average magnetostriction in a polycrystalline material. Consider a crystal where the magnetisation and magnetostriction measurement directions are the same such that $l = l_1$, $m = m_1$ $n = n_1$. Equation (3.20) shows that the saturation magnetisation is then given by

$$\lambda_s = (3/2)\lambda_{100}\left(l^4 + m^4 + n^4 - 1/3\right) + 3\lambda_{111}\left(l^2m^2 + m^2n^2 + n^2l^2\right) \tag{3.50}$$

Using trigonometrical identities, (3.50) can be simplified to [8]:

$$\lambda_s = \lambda_{100} + 3(\lambda_{111} - \lambda_{100})\left(l^2m^2 + m^2n^2 + n^2l^2\right) \tag{3.51}$$

Suppose the polycrystalline material has equal numbers of crystals oriented at every possible angle in space (i.e. all possible values of l, m, n). Then the mean value of λ_s is found by carrying out a double integral of the magnetostriction over all possible angular orientations to obtain [32]:

$$\lambda_m = (2/5)\lambda_{100} + (3/5)\lambda_{111} \tag{3.52}$$

In practice, λ_m is in the range 10×10^{-6} to 20×10^{-6} in NO steel. In 3% silicon–iron $\lambda_{100} = 27 \times 10^{-6}$ and $\lambda_{111} = -5 \times 10^{-6}$. Substituting these values into (3.52) gives a theoretical value of 7.8×10^{-6}. It is not surprising that this differs significantly from experimental values for two reasons; first, the material is not isotropic therefore the averaging process to obtain (3.52) is only approximate and, second, there can be a wide range of grain sizes in a material each with a different, sometimes complicated, domain structure whereas the equation assumes each grain acts as a single domain. Despite this, the relationship expressed in (3.52) is a good guide to use in theoretical analysis and has the potential to be adapted to take some of these practical conditions more fully into account.

It is useful to know the magnetostriction of a polycrystalline material at angles to the magnetisation direction.[34] An isotropic polycrystalline material can be approximated to an isotropic single crystal in which case $\lambda_{100} = \lambda_{111} = \lambda_m$. If the angle between the magnetising direction and the magnetostriction measurement direction is θ then, from (3.20), it can be shown [32] that

$$\lambda_s = (3/2)\lambda_m\left(\cos^2\theta - 1/3\right) \tag{3.53}$$

[34]The importance of this can be seen when rotational magnetostriction is introduced in Section 8.8.

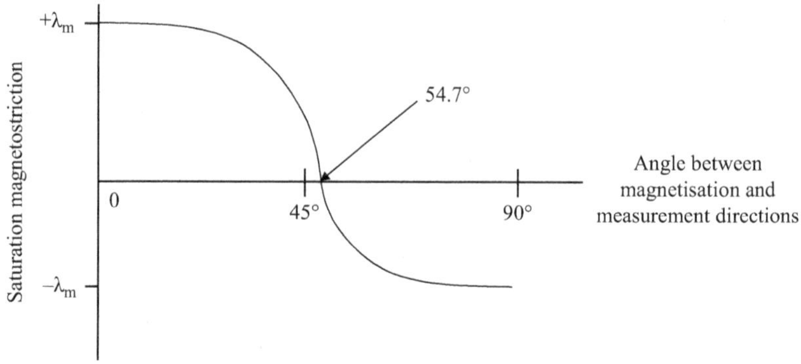

Figure 3.19 Variation of saturation magnetostriction in isotropic steel sheet with respect to the angle between the magnetising and measurement directions

This is identical in form to (3.49). Figure 3.19 shows how λ_s varies with angle between the measurement and magnetising directions in a perfectly isotropic polycrystalline material. It is not difficult to see why it is the same as the curve in Figure 3.18(b) for a single crystal, but the parameters are different.

3.16.4 Variation of magnetostriction with flux density

Section 3.16 until here has dealt mainly with magnetostriction at saturation magnetisation. However, the more important engineering and material assessment quantity is the magnetostriction at normal working flux densities. This is more difficult to quantify since it depends on the demagnetised state of the material which is difficult to determine.

It is worth emphasising two definitions of magnetostriction which are applied to electrical steels. When a material is magnetised to saturation from an isotropic demagnetised state the change in magnetostrictive strain is termed the saturation magnetostriction λ_s which is unique to a given composition and microstructure at a given temperature. If, however, the material is not fully demagnetised, or has a particular domain structure before magnetisation, the observed strain at any level of flux density is referred to as the *engineering magnetostriction* λ_e.

We can easily derive values for magnetostriction in a perfect single crystal of electrical steel ignoring the presence of closure domains. Suppose a crystal is completely demagnetised then we can imagine a hypothetical domain structure with equal numbers of domains oriented along the three easy axes as shown in Figure 3.20.[35] If a field is applied along the [001] direction, magnetisation will start simply by the 180° domain wall between domains *a* and *b* moving towards the outer edge *c* of the crystal. This process continues until 1/6 of the crystal is magnetised

[35]Of course a domain structure of this sort would not exist in practice because it would not be in its lowest energy state. However, the model does give results which agree reasonably well with magnetostriction measurements on single crystals of iron [33].

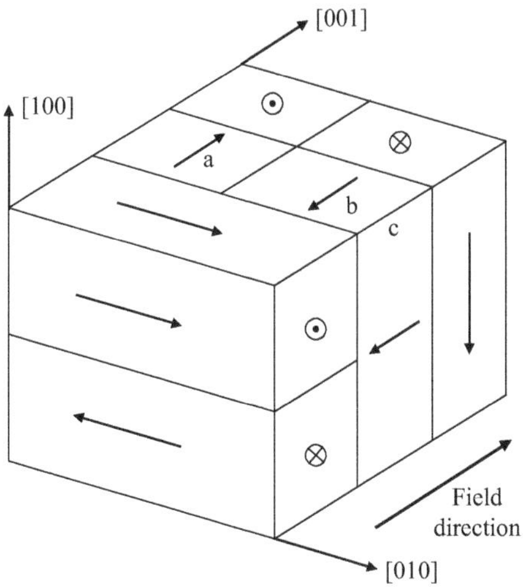

Figure 3.20 Hypothetical domain structure in a single cubic crystal used to deduce the variation of magnetostriction with flux density

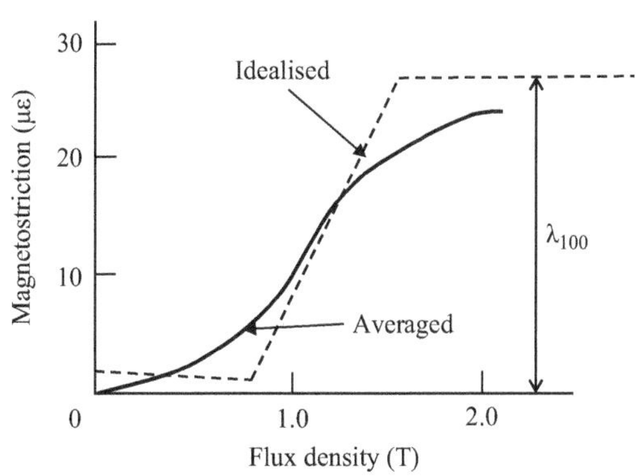

Figure 3.21 Theoretical variation of magnetostriction with flux density in a single crystal of 3% SiFe when magnetised along a <100> direction ($\lambda_{100} = 27 \times 10^{-6}$)

along the [001] direction hence the magnetisation will have risen to $M_s/3$. During this process, no change of magnetostriction occurs. Increasing the field causes the magnetisation to increase by 90° domain rotation and it can be deduced from the simple domain structure that this is accompanied by an elongation of $+3/2\lambda_{100}$ along the magnetising direction. Of course, only 2/3 of the volume changes dimensions so the total elongation is $+\lambda_{100}$ at saturation as shown in Figure 3.21. This agrees closely with measurements on single crystals [33]. Theoretical curves can be plotted for magnetostriction when magnetised along other crystal directions [33] but they are far more difficult to predict for textured NO steels.

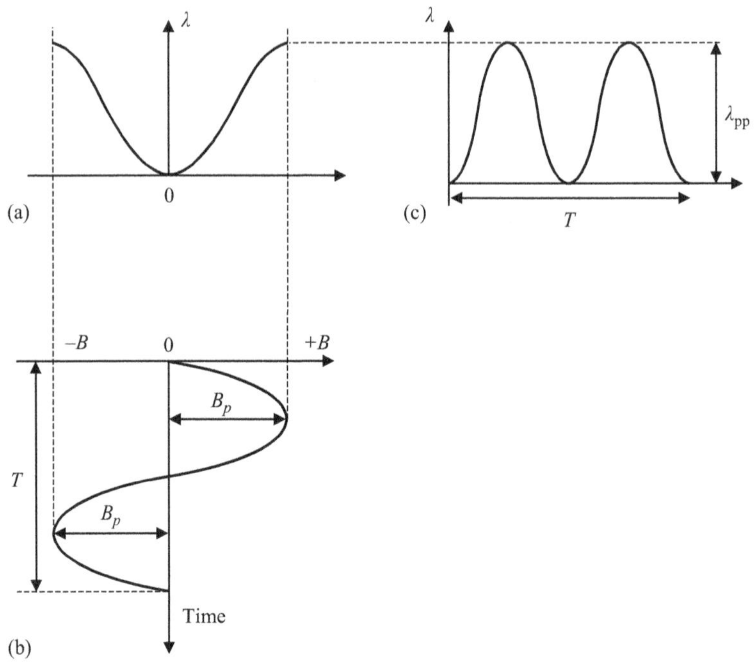

Figure 3.22 Variation of magnetostriction during one cycle of a.c. magnetisation in a perfect single crystal of SiFe magnetised along a <100> direction: (a) λ–B characteristic, (b) one cycle of B versus time waveform, (c) one cycle of λ waveform

Furthermore, magnetocrystalline and magnetoelastic energy are closely interrelated which should not be surprising, if only because of the similar forms of the phenomenological equations expressing their basic properties. In principle, the magnetocrystalline anisotropy constants (K_1 and K_2) and the magnetostriction constants (λ_{100} and λ_{111}) can be calculated theoretically from knowledge of the atomic physics of the spin-orbital coupling controlling the direction dependence of these properties in single crystals[36]. This could be extremely useful in predicting the magnetic properties of new materials to avoid tedious and expensive experimental programmes necessary to study vast numbers of un-explored possible iron alloy compositions. However, the theoretical calculations are difficult to carry out and major approximations seem necessary so no successful attempts appear to have been reported to date.

The $\lambda - B$ characteristic shown in Figure 3.22(a) is similar to the characteristic found in polycrystalline silicon–iron alloys. It can be used to show what happens when an alternating field is applied to a strip of material. It can be imagined as an

[36]Theoretical expressions for magnetostriction constants can be found in [19].

averaged value of the magnetostriction in many single crystals, each having its own particular magnetostriction characteristic similar to that given in Figure 3.21 depending on its own orientation with respect to the magnetising and magnetostriction measurement directions. It is found in practice that, over the range from zero flux density to the points $\pm a$, the magnetostriction is approximately given by $\lambda = kB_{\mathrm{p}}^2$ where B_p is the peak magnetostriction and k is a constant of proportionality.

Suppose the flux density varies sinusoidally with time as shown in Figure 3.22(b) such that $b = B_p \sin \omega t$. Over the flux density range being considered, curves (a) and (b) can be used to derive the variation of the magnetostriction through one cycle of magnetisation as shown in (c). The instantaneous value of the magnetostriction is then given by

$$\lambda = \lambda_{\mathrm{p}}(1 - \cos 2\,\omega t) \tag{3.54}$$

where λ_p is the peak value of the sinusoidal component of the magnetostriction. Important points to note are, first, that as the flux density goes through one cycle of magnetisation, the magnetostriction goes through two cycles, and, second, the magnetostriction contains a constant offset (an extension in this example) of magnitude λ_p.

In a real material, k varies with flux density and the B^2 proportionality is no longer an accurate approximation. It is shown in Chapter 10 of Volume 2 of this book that time harmonics of magnetostriction are extremely important in the generation of acoustic noise in cores assembled from magnetostrictive materials.

Magnetostriction is subject to hysteresis so its value at a particular B or H magnitude under increasing magnetisation is lower than at the corresponding value of B or H in the decreasing field. This has not been included in Figure 3.22 for the sake of clarity but some practical λ–B characteristics are shown in Figure 3.23. These characteristics are often called *butterfly loops*.

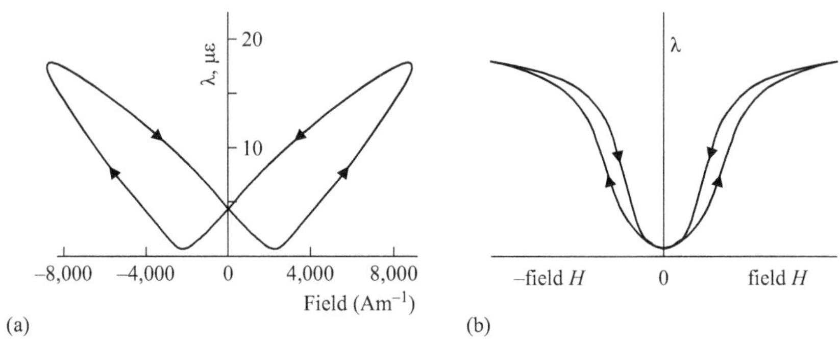

Figure 3.23 Practical examples of magnetostriction butterfly loops in electrical steels: (a) NO electrical steel and (b) grain oriented electrical steel magnetised along the RD

3.17 The Barkhausen effect

The Barkhausen effect is widely made use of in non-destructive magnetic evaluation of sub-surface inhomogeneities in a wide range of engineering materials: in particular the presence of localised stress, crack propagation and fatigue damage [34]. Several types of Barkhausen emissions occur in nature although they all have the same basic origin. These are shown in Figure 3.24. Of most interest as far as electrical steels are concerned is *magnetic Barkhausen noise* which is focussed on in this section. It has been said that this phenomenon has '*a rich structure and contains information on many aspects of the magnetisation process*' [35].

Barkhausen noise (BN) is the result of discontinuities observed during the magnetisation of any magnetic material. Figure 3.25(a) shows an example of the effect of BN on the flux density measured when a linearly, slowly increasing field is applied to a material. The *B–H* curve appears smooth but when a portion is magnified, as shown in the inset, it is seen that the flux density does not increase smoothly, but as a series of jumps. The terms *Barkhausen jumps, emissions* or *noise* are used indiscriminately when referring to this phenomenon. Figure 3.25(b) shows how the BN is largest close to the coercivity field in the *B–H* loop. This applies to both d.c. and a.c. magnetisation although the magnitude and distribution of the jumps will not be the same in each case.

The first experimental evidence of BN was reported by Professor Heinrick Barkhausen in 1919 while studying electromagnetism in an iron rod. While magnetising and demagnetising the rod, he could hear a rustling noise in a telephone ear piece connected via an amplifier to a multi-turn secondary coil which surrounded the rod [36]. In his words: 'a gradual and uniform change in MMF caused the molecular magnets to jump suddenly to their new positions so inducing non-uniform pulses in a surrounding search coil can be heard in a telephone earpiece'.

The main cause of BN is discontinuous domain wall motion. Indeed, the discovery of BN was the first indirect evidence of the existence of magnetic domains.

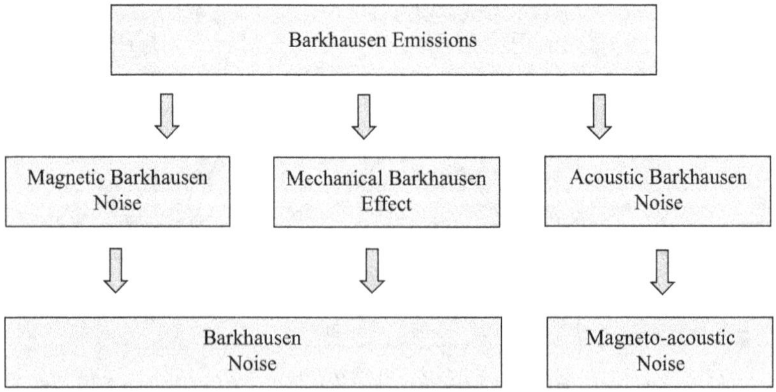

Figure 3.24 Classification of Barkhausen emissions

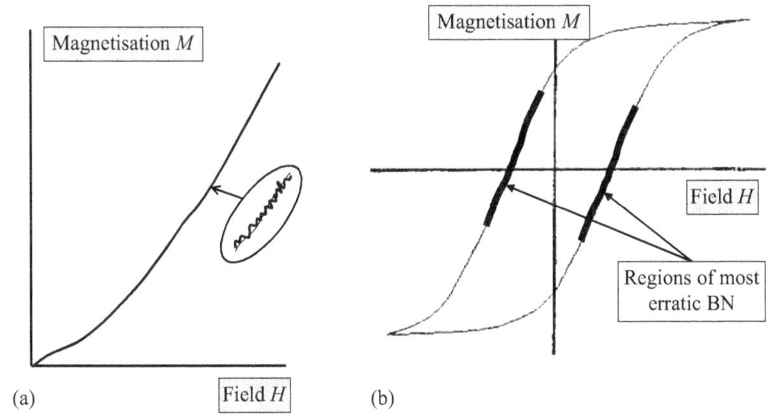

Figure 3.25 Schematic representation of Barkhausen noise present on (a) a simple B–H curve and (b) a region of a B–H loop of a magnetic material subjected to a slowly changing d.c. field

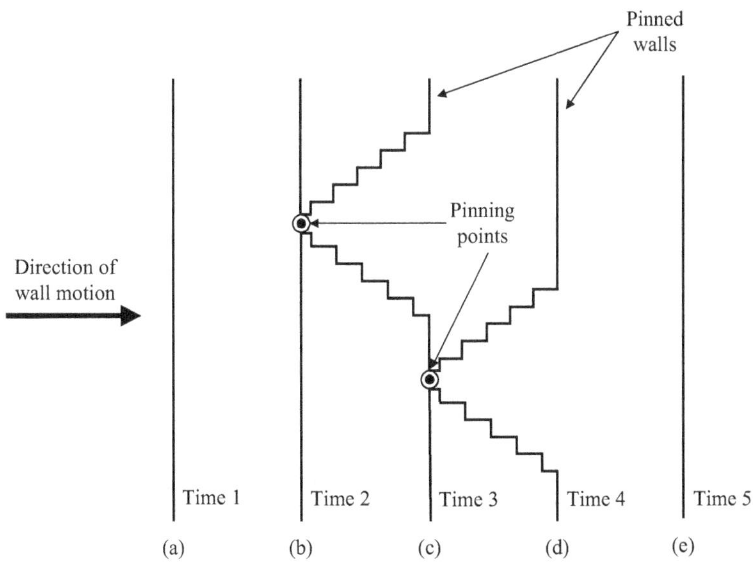

Figure 3.26 Schematic representation of process of BN emission due to pinning of a 180° domain wall in SiFe

We now know that the pulses are caused by rapid irregular motion of parts of domain walls as they travel through a magnetic material.

Consider a domain wall moving towards a small discontinuity in a material initially in position (a) in Figure 3.26. Assume that the wall is linear and moving at a constant, slow speed. At position (b), the wall has reached the discontinuity, then as the wall moves further forward it is pinned as shown in position (c). At some later stage the wall breaks away and continues its forward motion as shown in (d).

The wall pinning mechanism is far more complex than the illustration shown Figure 3.26. Small closure domains are set up near the discontinuity and later annihilated and during the process *micro-eddy currents* are set up which in turn cause losses. In this simplified description, the process occurs very rapidly and can be considered independent of the wall speed.

A real material will have huge numbers of domains encountering even greater numbers of defects, inclusions, etc. and the overall effect is the generation of heat due to the associated micro-eddy currents which is effectively the static hysteresis loss introduced in Section 7.4. Expressions can be derived for static hysteresis loss in terms of energy lost in each Barkhausen jump and by comparing with measured values of P_h it can be deduced that the duration of each jump is of the order of 10^{-8} s [35]. It has been estimated that in iron each pulse is generated by a group around 10^{14} atoms within a volume of 10^{-10} to 10^{-8} cm^3 [37].

The losses due to a single Barkhausen jump can be calculated by considering flux change and field associated with the moving wall [38,39]. Mathematical expressions for the loss can be derived but they are very complicated and far more needs to be known about the statistical distribution and nature of the jumps in real materials before they can be used in practice. BN can be measured by winding a search coil around a sample and measuring the induced e.m.f. as the flux density is changed. Figure 3.27 shows the voltage induced in one cycle of magnetisation in a piece of GO silicon–iron magnetised sinusoidally at 50 Hz. The sinusoidal *dB/dt*

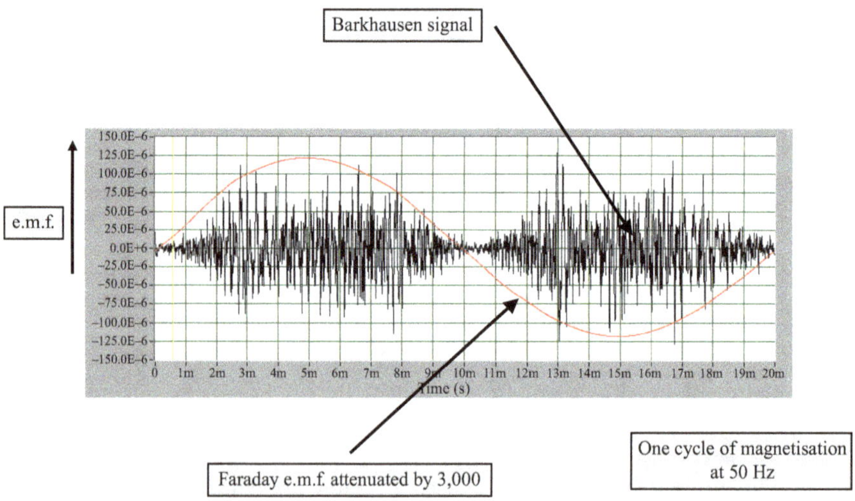

Figure 3.27 Typical BN spectrum emitted from a specimen of GO SiFe during one period of 50 Hz magnetisation

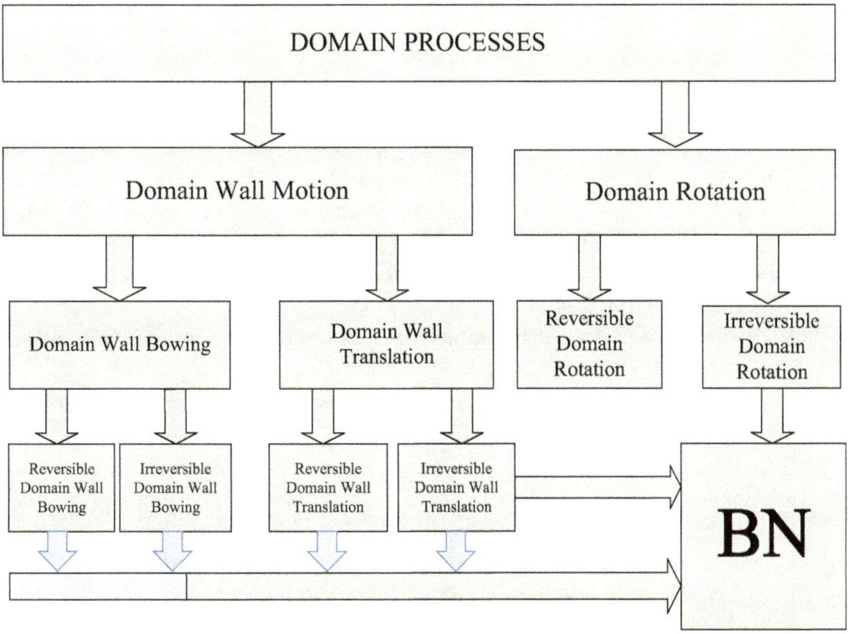

Figure 3.28 The association of BN with domain magnetisation processes in soft magnetic materials

component of the e.m.f. induced in a search coil (from which we often calculate the peak flux density) is reduced in scale by a factor of 3,000 in order to see its BN component more clearly.

The BN signal is very complex and not repeatable even in the two half cycles shown, but it can be seen that it is lowest when dB/dt is close to zero which is consistent with the sketch in Figure 3.25(b). Precautions must be made in BN measurement of the sort shown in Figure 3.27 to ensure that it is indeed all magnetic in origin since circuit noise or electromagnetic interference can often cause errors.

It can clearly be seen that BN is a stochastic process, i.e. its overall effect is unpredictable and random. Because each jump occurs so rapidly its electromagnetic effect is effectively independent of magnetising frequency and therefore directly related to the static hysteresis loss as mentioned in the previous paragraph. However, as shown in Section 1.16 of Volume 2 of this book, the total BN emission from electrical steel is frequency dependent.

Figure 3.28 shows how BN is related to magnetisation processes in soft magnetic materials. From this, it will not be surprising to see in Section 1.16 of Volume 2 of this book that the a.c. Barkhausen noise characteristics of electrical steel is related to several important properties. These include texture, domain refinement processes, precipitate and inclusion levels and coating-induced stresses.

References

[1] Hubert A., and Schäfer R. *Magnetic Domains: The Analysis of Magnetic Microstructures*. Springer, Berlin, 2008

[2] Bertotti G. *Hysteresis in Magnetism for Physicists, Material Scientists and Engineers*. Academic Press, San Diego, 1998

[3] Carey R., and Isaac E.D. *Magnetic Domains and Techniques for Their Observation*. The English University Press, London, 1966

[4] Ewing J.A. *Magnetic Induction in Iron and Other Metals*. D. Van Nostrand, London, 1900, pp. 294–354

[5] Weiss P. 'Hypothesis of the molecular field and ferromagnetic properties'. *J. Phys*. 1907, vol. 6, pp. 661–690

[6] Heisenberg W. 'On the theory of ferromagnetism'. *Z. Physik*. 1928, vol. 49, pp. 619–36

[7] Brailsford F. 'Ferromagnetism in relation to engineering magnetic materials: a review of progress'. *Proc IEE, Pt. A: Power Eng*. 1956, vol. 103(7), pp. 39–51

[8] Honda K., and Kaya S. 'On the magnetisation of single crystals of iron'. *Sci. Rep. Tohoku Univ*. 1926, vol. 15, pp. 721–752

[9] Brailsford F. *Physical Principles of Magnetism*. Van Nostrand, London, 1966, p. 121

[10] Bozorth R.M. *Ferromagnetism*. IEEE Press, Piscataway, NJ, 1993, p. 563

[11] Bozorth R.M. *Ferromagnetism*. IEEE Press, Piscataway, NJ, 1993, p. 561

[12] Gengnagel H., and Hofmann U. 'Temperature dependence of the magnetocrystalline energy constants K_1, K_2 and K_3 of iron'. *Phys. Stat. Solidi*. 1968, vol. 29, pp. 91–97

[13] Brailsford F. *Physical Principles of Magnetism*. Van Nostrand, London, 1966, p. 143

[14] Bozorth R.M. *Ferromagnetism*. IEEE Press, Piscataway, NJ, 1993, p. 570

[15] Westerstrand B., Nordblad P., and Nordborg L. 'The magnetocrystalline anisotropy constants of iron and iron-silicon alloys'. *Phys. Scr*. 1975, vol. 11, pp. 383–386

[16] Bertotti G. *Hysteresis in Magnetism for Physicists, Material Scientists and Engineers*. Academic Press, San Diego, 1998, p. 149

[17] Bertotti G. *Hysteresis in Magnetism for Physicists, Material Scientists and Engineers*. Academic Press, San Diego, 1998, p. 150

[18] Banks P.J., and Rawlinson E. 'Dynamic magnetostriction and mechanical strain in oriented 3% silicon–iron sheet subject to combined longitudinal and transverse stresses'. *Proc IEE* 1967, vol. 114(10), pp. 1537–46

[19] Daniel L., and Hubert O. 'An analytical model for the ΔE effect in magnetic materials'. *Eur. J. Appl. Phys*. 2003, vol. 45(3), pp. 1–11

[20] Joule J.P. 'On a new class of magnetic force'. *Ann. Electr. Magn. Chem*. 1842, vol. 8, pp. 219–224

[21] Bozorth R.M. *Ferromagnetism*. IEEE Press, Piscataway, NJ, 1993, p. 641

[22] Jiles D. *Introduction to Magnetism and Magnetic Materials*. Chapman and Hall, London, 1991, p. 100

[23] Chikazumi S. *Physics of Ferromagnetism*, 2nd edn. Oxford University Press, Oxford, 2007

[24] Ewing J.A. *Magnetic Induction in Iron and Other Metals*. D. Van Nostrand, London, 1900, p. 197

[25] Néel L. 'Anisotropie magnétique superficielle et surstructures d'orientation'. *J. Phys. Radium* 1954, vol. 15(4), pp. 225–239

[26] Brailsford F. *Physical Principles of Magnetism*. Van Nostrand, London, 1966. p.166

[27] Podurets K.M., and Shilstein S.S. 'Measurement of the domain wall thickness in silicon iron using the adiabatic spin-flip effect on neutron refraction'. *Physica B* 2001, vol. 1–4, pp. 263–267

[28] Bozorth R.M. *Ferromagnetism*. IEEE Press, Piscataway, NJ, 1993, p. 837

[29] Bozorth R.M. 'Magnetic domain patterns'. *J. Phys. Radium* 1951, vol. 1, pp. 308–321

[30] Mohd Ali B., and Moses A.J. 'A grain detection system for grain-oriented electrical steels'. *IEEE Trans. Magn.* 1989, vol. 256(6), pp. 4421–4426

[31] Shilling J.W., and Houze G.L. 'Magnetic properties and domain structure in grain-oriented 3 % Si-Fe'. *IEEE Trans. Magn.* 1974, vol. 10(2), pp. 195–223

[32] Brailsford F. *Physical Principles of Magnetism*. Van Nostrand, London, 1966, p. 150

[33] Webster W.L. 'Magneto-striction in iron crystals'. *Proc. R. Soc. A* 1925, vol. 109, pp. 570–584

[34] Jiles D. *Introduction to Magnetism and Magnetic Materials*. Chapman and Hall, London, 1991, pp. 366–369

[35] Bertotti G. *Hysteresis in Magnetism for Physicists, Material Scientists and Engineers*. Academic Press, San Diego, 1998, p. 285

[36] Barkhausen H. 'Zwei mit Hilfe der neuen Verstarker entdeckte Erscheinungen'. *Physik. Zeitschrift* 1919, vol. 20, pp. 401–402

[37] Tebble R.S. 'The Barkhausen effect'. *Proc. Phys. Soc. B* 1955, vol. 68, pp. 1017–1032

[38] Allia P., and Vinai F. 'Effects of Barkhausen jumps on static losses in soft magnetic materials'. *Proceedings of* 3rd *International Conference on Soft Magnetic Materials SMM-3*, paper 5–2, Bratislava, Czechoslovakia, 1977, pp. 169–177

[39] Bertotti G. *Hysteresis in Magnetism for Physicists, Material Scientists and Engineers*. Academic Press, San Diego, 1998, pp. 395–399

Chapter 4

Methods of observing magnetic domains in electrical steels

4.1 Introduction

In Chapter 3, the formation of magnetic domains was introduced and simple domain structures were used to help explain magnetisation, magnetostriction and loss processes. However, for in-depth understanding and more accurate predictions of the performance of modern soft magnetic materials, particularly under a.c. magnetisation, the more complex structures present in real materials must be taken into account. This calls for methods of directly observing and quantifying static and dynamic domain structures.

Strictly speaking, magnetic domains cannot be *observed* in the sense that they can actually be seen by the human eye. In practice their structures can be visualised by their effect on other physical quantities from which images can be produced.

Observation methods can be categorised according to the following imaging capabilities: (a) static domains directly on a prepared steel surface, (b) static surface domains beneath an insulating coating, (c) dynamic domain wall motion on a smooth steel surface, (d) dynamic domain wall motion beneath an insulating coating, (e) sub-surface static domain structure and (f) sub-surface dynamic domain wall motion.

Although surface domain observations have been carried out for over 80 years, it is only in the past decade that real progress has been made on the more challenging visualisation of domains beneath the surface of silicon–iron [1]. Sub-surface domain structure and domain wall motion are quite different to what might be observed on the surface.

To understand the main reason why surface and sub-surface domain structures can be so different, let us consider the hypothetical domain structure in a strip of non-oriented (NO) electrical steel shown in cross section in Figure 4.1. The average grain diameter could be around 50 μm in such a material. The arrowed lines represent the direction of easy magnetisation in each grain which is also the direction along which bar domains are oriented. The structure is considered to be two dimensional in order to avoid over complication.

If domains are observed on the surface of grains such as those labelled *c*, complex closure structures, will be apparent. Obviously these are unavoidable surface features which are not representative of the structure beneath the surface. A small number of grains labelled *b* will have <100> direction close to the surface

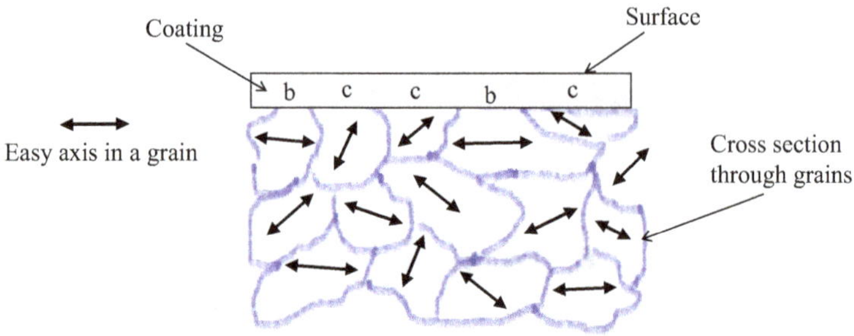

Closure domain patterns might be seen at surface regions labelled c,
180° bar domains might be seen at surface regions labelled b

Figure 4.1 Cross section through a sheet of NO steel showing a hypothetical distribution of crystal orientations

plane, so wider bar domains will be observed. The bulk of the grains are likely to contain bar domains but with complex structures near to the grain boundaries.

To some extent sub-surface grains can be deduced from surface images such as the case in grain oriented (GO) steel under applied mechanical stress, but often viewing the surface domains tells us very little about the sub-surface structures, particularly under dynamic magnetisation.

The following sections cover the most common domain observation techniques, roughly in chronological order of their first usage, which not surprisingly corresponds roughly according to their capability. They are all applicable for domain observation on any magnetic material but emphasis is placed on their relevance to electrical steels.

4.2 Powder techniques

The first recognised method of surface domain observation was the powder technique discovered independently in 1931 by two groups [2,3]. It has been said that the discovery of the powder technique was a landmark in the history of research into magnetism. At his presidential address to the British Association in 1959, the eminent UK scientist, Professor Leslie Fleetwood Bates said [4]

> Bitter's work really amounted to plotting lines of magnetic force or making visible surface magnetism with microparticles in place of the coarse iron filings which we used in our younger days and which our descendants will probably continue to use for many years to come.

More than 80 years after Bitter's discovery, powder patterns continue to be used as Bates predicted. Bitter's method was simply to place a suspension of Fe_3O_4 powder with particle sizes around 1 μm diameter on the polished surface of an iron crystal. High concentrations of particles are attracted to surface stray fields mainly emanating from domain walls as illustrated in Figure 4.2. The position of many of

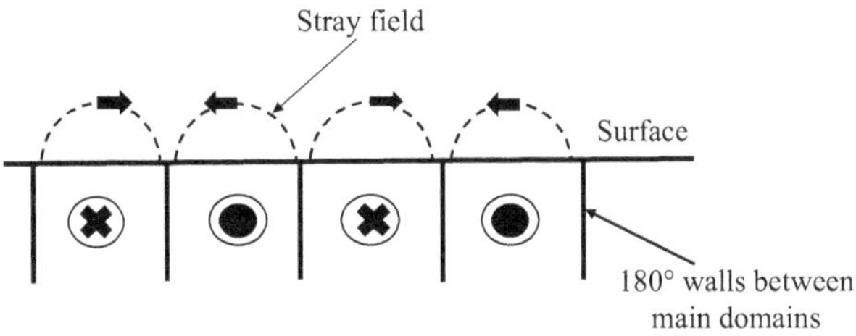

Figure 4.2 Schematic diagram showing stray fields where domain walls meet the surface of a material to produce stray external fields

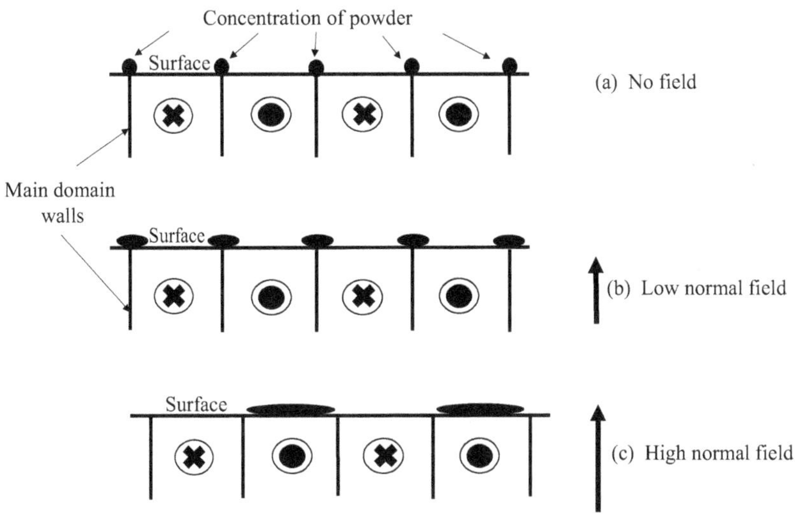

Figure 4.3 Schematic diagrams showing the idealised effect of a perpendicular enhancing field on powder concentration on a material surface: (a) no enhancing field, (b) small field and (c) large field

the domain walls in GO silicon–iron can be easily seen by eye or with a low magnification microscope. The powder domain observation method is commonly called the *Bitter technique*. However, credit should also be given to McKeehan and Elmore who first used true colloidal suspensions with Fe_2O_3 particles small enough to show Brownian motion which enables quick and clear alignment of particles with the stray fields [5].

Normally, better contrast between the domain images can be obtained by applying a d.c. field perpendicular to the sample surface by means of a planar coil to produce the powder redistribution as shown in Figure 4.3. This shows a cross section through a material at right angles to four domains separated by 180° walls. If the d.c. field is very high, as in (c), the antiparallel domains show up clearly as black and white regions. In practice, the normal field should not be too high or it

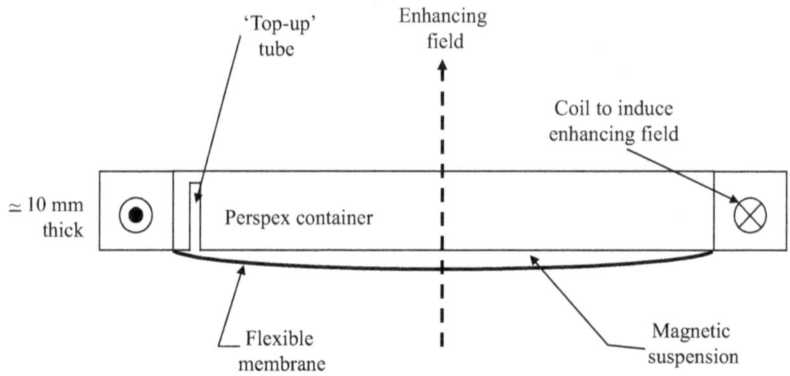

Figure 4.4 Cross section through an 'Orb domain viewer' with an observation area of approximately 9 cm diameter (not to scale)

might disturb closure structures, so the compromise in (b) clearly shows the position of the walls. However, some surface fields can be diminished because they are in the opposite direction to the enhancing field so careful consideration of the resulting images is sometimes needed.

In early powder pattern techniques, the colloid was applied directly to the sample surface. This still is best in some circumstances but it is difficult to apply the suspension uniformly and it is wasteful. In the 1970s integrated viewers such as the *3M's* domain viewer[1] were developed for observing domains in magnetic recording media without directly applying the magnetic suspension to the surface. Later the *Orb Viewer* was developed specifically for observation of domains on surfaces of GO electrical steel.

A cross sectional view of an early version of the Orb Viewer [6], produced by what now is Cogent Power Ltd, is shown in Figure 4.4. The viewing area is around 9 cm diameter and a surrounding coil can produce a variable normal field up to over 800 Am^{-1}. The viewer is simply placed on the surface to be examined with the thin membrane in close contact with the surface. After a few seconds the domain pattern can be observed by eye or its image captured for further examination. When removed from the sample surface, the powder particles are rapidly redistributed ready for the next observation. The magnetic suspension can be topped up via the trap to extend the life of the device.

Larger versions of the Orb viewer are used to observe domains across the full width of electrical steel sheet. A typical Bitter pattern observed on the surface of a strip of commercial GO silicon–iron using a domain viewer is shown in Figure 4.5.

Many types of powder and suspension media are used for the Bitter technique today. Most are capable of obtaining clear images of static surface domain patterns

[1]This was a surface domain viewer which was designed and produced by the 3Ms company in the 1960s/1970s for observation of domains on magnetic recording media. It was found to be ideal for observing domain structures on the surface of GO steel. It was probably the first magnetic domain viewer of the type and modern versions are used today.

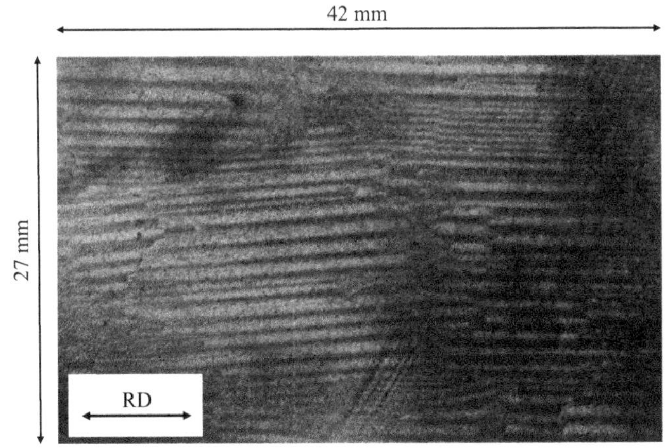

Figure 4.5 Static domain patterns observed on the surface of a strip of commercial GO silicon–iron exploiting the Bitter technique using an 'Orb domain viewer' without image enhancement

in GO steels without the need to remove the coating. Only a few seem sensitive enough to view the small domains in NO steels. The technique is mainly restricted to static images although it is possible in some cases to observe surface domain motion at magnetising frequencies of a few hertz [7].

4.3 Optical methods of surface domain observation

4.3.1 The magneto-optic effect

Optical techniques are widely used for observation of surface static and dynamic domain structures. These can be regarded as techniques for direct domain observation since they image the magnetisation vector in individual domains whereas the powder techniques are indirect since they sense stray magnetic fields.

In 1876, the Scottish pioneer of electro-optics, John Kerr, discovered that when a beam of polarised light is shone on to a magnetised surface, the plane of polarisation of the reflected beam is rotated by a small amount in one or other direction dependent on the direction of surface magnetisation [8]. This has grown into the field of physics referred to as magneto-optics where the plane and phase of a beam of polarised light is rotated when it is reflected from, or passes through, a magnetic material.

This Kerr magneto-optic (KMO) effect is extremely useful for observing static or dynamic moving domain walls on polished surfaces of ferromagnetic or ferrimagnetic materials and has been widely used in studies of electrical steels. Descriptions of the physical origin of magneto-optical effects relevant to magnetic domain observation in electrical steel can be found elsewhere (e.g. [9]).

The three basic magneto-optic effects which occur for different surface magnetisation conditions are shown in Figure 4.6. The polar effect produces the greatest

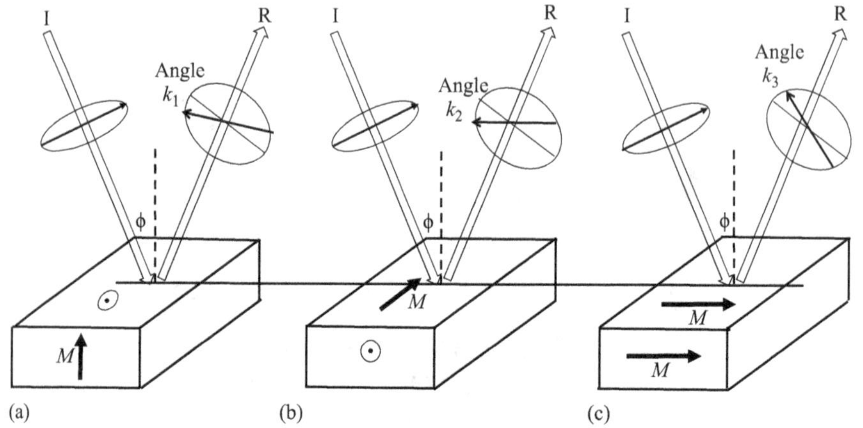

Figure 4.6 Reflection of polarised light from a surface according to the surface magnetisation conditions: (a) polar effect, (b) transverse effect and (c) longitudinal effect (Φ – angle between beam and normal direction, I – incident beam, R – reflected beam and k_1, k_2, k_3 – rotation angles)

rotation of polarisation; however, it is restricted to magnetisation out of the surface of the material which does not occur in electrical steels. In the transverse effect, magnetisation is in the plane of the sample surface and perpendicular to the plane of incidence of the beam. The longitudinal effect comes into play if the specimen is rotated through 90° relative to the light source. In the transverse mode there is no rotation of the plane of polarisation and the KMO effect though the transverse orientation occurs due to changes in the reflection coefficient of the surface of light polarised in the plane of incidence [10].

4.3.2 Domain observation using the longitudinal KMO effect

The implementation of the Kerr effect for domain studies has changed little since its early practical implementation [11]. Figure 4.7 shows how the longitudinal Kerr-effect is used to image static domains on the surface of a single crystal of GO silicon–iron. The incident beam from a high intensity light source is polarised and focused on the polished surface of the material. In the position shown, the beam falls on to regions of material magnetised in opposite directions so the plane of polarisation of the reflected beam is rotated slightly clockwise or anticlockwise according to the magnetisation direction of the two domains. The reflected beam then passes through an analysing prism which is used to extinguish one of the rotations to give the light/dark contrast between the images of the two domains as shown. The angle of incidence of the beam needs to be optimised for best contrast. Also, the angle of rotation of the polarised beam is very small so precise assembly is necessary.

Careful sample preparation is essential. Using a standard light source, a highly reflective surface is necessary to obtain clear images so in the case of GO silicon–iron the coating must be removed. The surface must be carefully polished after

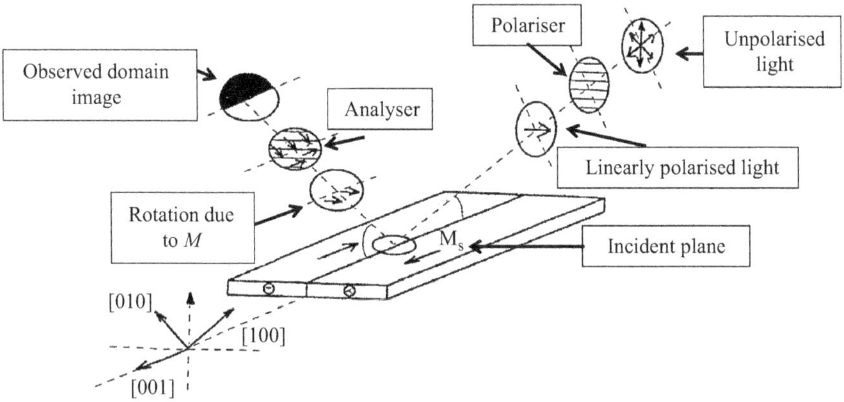

Figure 4.7 Schematic view of the use of the longitudinal Kerr effect to observe domains on the surface of a crystal of (110)[001] oriented 3% SiFe

coating removal and then the sample must be annealed to remove any stresses induced during polishing. A laser beam is often used to provide a more intense light source. The depth of sensitivity of the Kerr effect is limited to about 30 nm below the sample surface [12].

The contrast of the domain images can be enhanced by reducing the amount of regular reflected light. To do this, the sample viewing area is sometimes coated with a thin layer of dielectric material, such as zinc sulphide. This evaporation process is referred to as blooming. Improved images can be obtained but the coating thickness must be closely controlled according to the angle of incidence used, the wavelength of the light source and the refractive index of the dielectric material.

The main disadvantages of domain observation using the KMO effect is the need for careful sample preparation and the need for a skilled operator of expensive equipment, at least, compared to the powder technique. However, clearer images can be produced than by using the powder technique and its spatial resolution can be far higher.

The method can be conveniently adapted for high temperature operation or observation during the application of mechanical stress. Another advantage is that it works equally well on low anisotropy materials, such as amorphous magnetic materials. The powder pattern technique cannot work on such materials because their stray surface fields are too small to attract powder particles. This is related to the fact that the domain walls are very wide and not well defined in very low anisotropy material.[2]

Possibly the greatest advantage of the magneto-optic method is the possibility of using the technique for dynamic observations at power frequencies. This is discussed in Section 4.3.3.

[2]The wall width is inversely proportional to $K^{1/2}$ as shown in (3.29) so the anisotropy decreases as the wall width increases.

4.3.3 Observation of rapid domain wall motion using the KMO effect

Knowledge of the ways in which domain walls move and how domains are nucleated and annihilated during the magnetising process is extremely valuable in studies of losses and magnetostriction processes. As stated earlier, the powder techniques can be used up to a magnetising frequency of just a few hertz.

A major drive to observe dynamic domain motion in electrical steel was that early theories of loss separation in GO silicon steel seemed to show that the number of domain walls taking part in magnetisation processes was not constant [13]. It was confirmed later, as the result of dynamic domain studies, that this is indeed the case [14]. The number of active walls is dependent on the peak value and wave shape of flux density. The number was reported to change by a factor of around eight from close to d.c. to a magnetising frequency of 1500 Hz in a (110)[001] oriented SiFe single crystal [15].

By the late 1960s, magneto-optic systems, such as outlined in Figure 4.7, were being used in conjunction with high speed photography. A frame rate at least two orders of magnitude higher than the magnetising frequency is necessary to observe clear images of domain wall positions at a given instant in time during one magnetising cycle. For example, for 60 Hz observation a camera producing 6000 frames per second is necessary [16]. This was the highest performance obtainable at the time so a cheaper and more convenient option was developed.

A convenient way of observing domain wall motion under magnetisation frequencies up to more than 1000 Hz is to adapt the simple set-up in Figure 4.7 for use in a stroboscopic mode. The light source is synchronised with a camera and the phase of the pulse is varied to visualise domain patterns at any point in the magnetising cycle so images can be played back and examined frame by frame. For example, by flashing the light at the peak of the cycle the peak wall displacement can be obtained [17].

The stroboscopic technique has been widely used, but its accuracy suffers because domain structures in electrical steels vary a little from cycle to cycle so images are often blurred. However, it does tell us that domain structures in a demagnetised state are quite different from those at power magnetising frequencies [15].

Figure 4.8 shows a photograph of a modern, real-time, domain imaging microscope and Figure 4.9 gives a schematic view of its main components [12]. The system operates in polar or longitudinal Kerr mode and is capable of $\times 50$ magnification and a spatial resolution of 1 μm. It is capable of real time dynamic observation of domain wall motion at frequencies up to more than 50 Hz. Fuller design and performance details are given in [12] and [18].

Real-time dynamic domain observation systems are complex and difficult to use but they can give accurate information about parameters such as instantaneous domain wall position and velocity. Systems similar to that shown in Figure 4.8 have been developed for even faster imaging mainly driven by thin film and nanomaterial applications [20].

Figure 4.10 shows an example of differences in the positions of a 180° domain wall at the same time instances in successive cycles. Even at the low magnetising

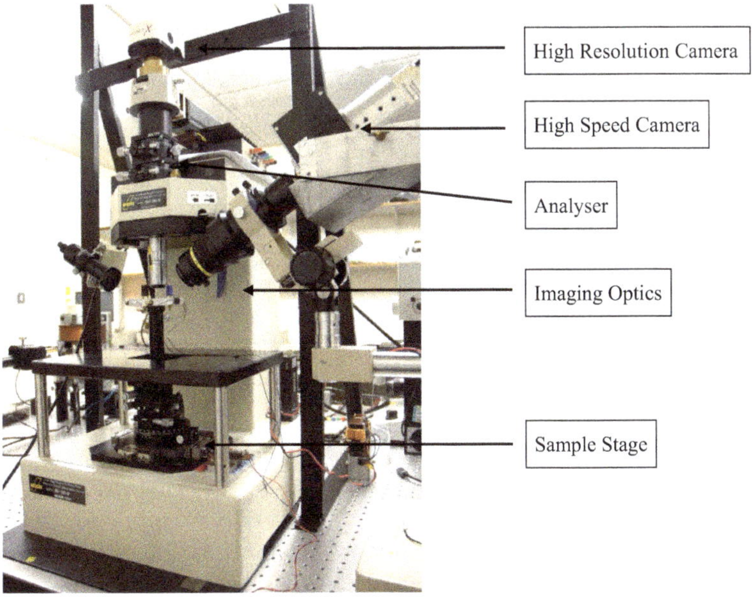

Figure 4.8 Photograph of main components of a real-time, high magnification KMO, domain imaging microscope

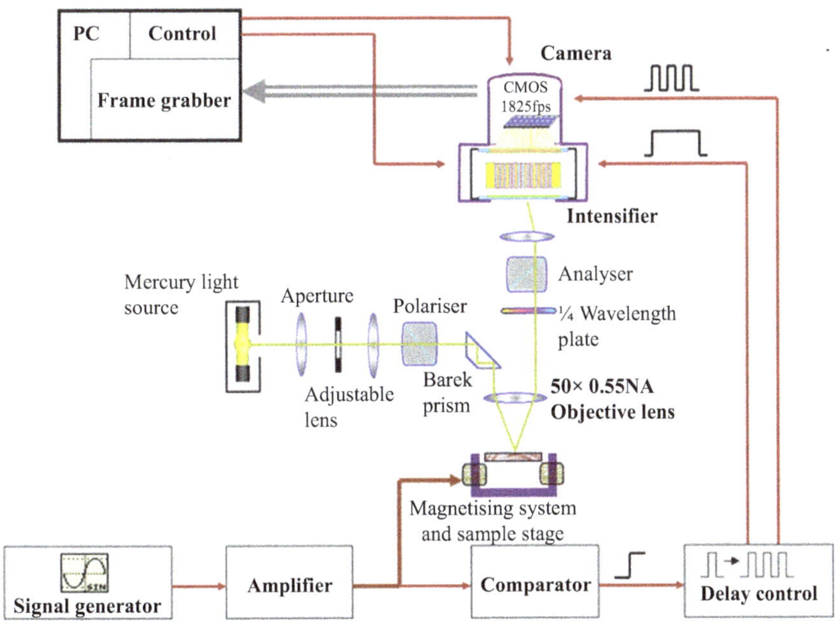

Figure 4.9 Schematic diagram of the real-time imaging KMO microscope (Adapted from figures in [19])

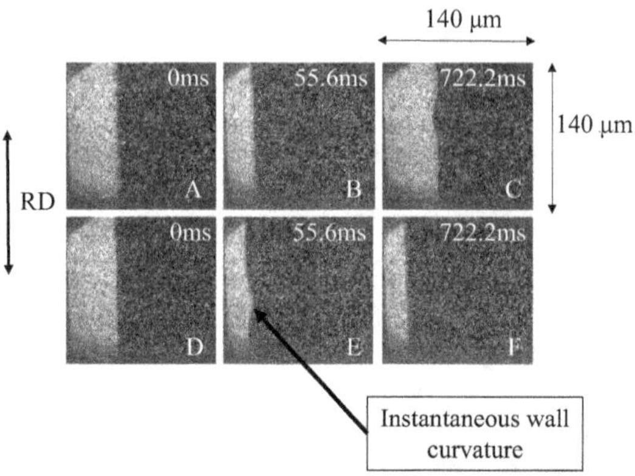

Figure 4.10 Positions of a part of the same domain wall at the surface of a 140 μm × 140 μm sample of well oriented GO silicon–iron at identical time instances in successive cycles magnetised at 1.0 Hz. (A, B, C – first cycle: D, E, F – next cycle) (Figure 4 in [21] reproduced under free licence CC-BY-4.0 and modified)

frequency used here (1 Hz), interesting[3], unexplained, wall curvature is apparent as well as a massive difference in wall position at time (c). Similar non-repeatability occurs at higher frequency.

4.4 Magnetic force microscope

Although, technically, a method of scanning the field profile on the surface of a magnetic material, the magnetic force microscope (MFM) can be used for very low-dimensional studies of magnetic domains in electrical steels. The principle of the MFM was established long before the introduction of commercial instruments. In the forerunner, domains were mapped on the surface of a strip of silicon–iron making use of the interaction of the surface stray field and the field caused by the rapid vibration of a narrow pointed strip of Permalloy [22]. A pick-up coil surrounding the probe responded to changes in magnetisation at the tip as the probe moved across the surface to produce a pattern of the domain structure on an oscilloscope.

The MFM itself was introduced in 1987 [23], soon after the more widely used atomic force microscope (AFM). The MFM measures the magnetostatic force on a ferromagnetic tip positioned just above the sample surface whereas the AFM measures the surface topology. Figure 4.11 illustrates the basic operating principle of a MFM. Its tip, at the end of a minute cantilever around 200 μm long, has a 10–150 nm thick magnetic coating so there is a magnetostatic force between the tip and the stray field which modifies its resonant behaviour.

[3]Similar non-repeatability occurs at higher frequencies [20] but the images are not clear in printed form so low frequency images are shown here.

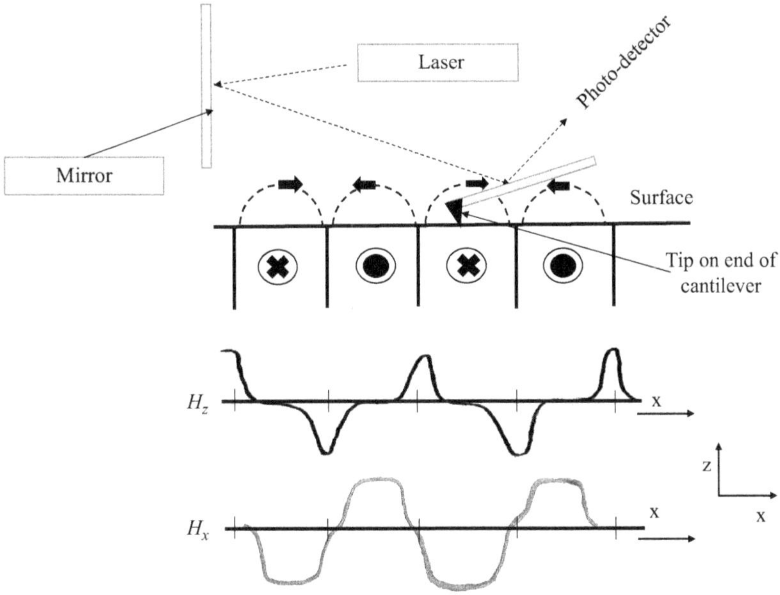

Figure 4.11 Schematic representation of stray field above the surface of a magnetic material containing a random group of domains and domain walls

The domains create stray fields whose x and z components are shown. The optical system is used to monitor the position of the tip as it is stepped along the sample. The tip is not in contact with the surface. The system detects changes in the resonant frequency of the cantilever induced by the magnetic field dependence of the tip to surface distance. The surface field pattern can be computed from the tip forces as it is scanned across a sample. An artificial domain pattern is shown in Figure 4.11 simply to illustrate the formation of a stray field.

The spatial resolution of the MFM depends on the distance from the tip to the sample surface which is normally nanometres to micrometres. If the tip is too close to the surface Van der Waals interatomic forces dominate and the magnetic component of force must be extracted from it. If the tip is too far away, the stray field will no longer be directly related to the magnetic structure of the surface.

The resolution of magnetic images obtained from an MFM can be better than 50 nm. However, it is not easy to obtain a quantitative measure of the surface field and some precautions are necessary when interpreting the images due to the magnetic tip actually disturbing the surface magnetisation. In electrical steels it is necessary to use magnetically hard tips and a wide separation between the sample and the tip to avoid this.

MFM images of magnetic structures on the surfaces of electrical steels are shown in Figure 4.12 [24]. Figure 4.12(a) shows the domain structures on the surface of a sample of uncoated GO silicon–iron. The domains are aligned along the [001]

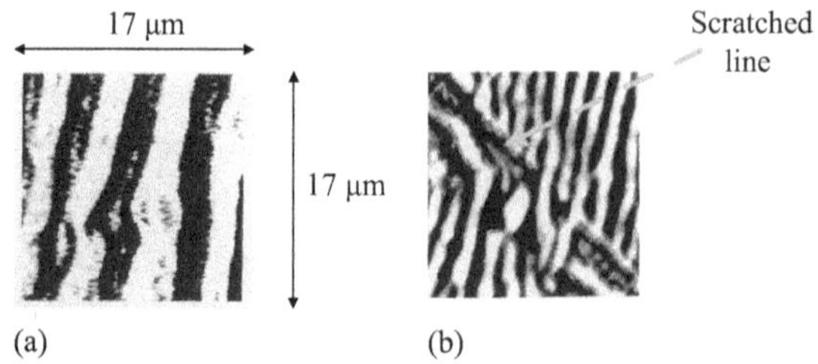

Figure 4.12 *Domain structures on the surface of electrical steels obtained using the MFM (a) [001] oriented slab domains in a 17 μm × 17 μm area of GO silicon–iron and (b) effect of a surface scratch on bar domains in a 50 μm × 50 μm region of the same steel (Figures 9 and 10 in [24] reproduced under free licence CC-BY-4.0 and modified)*

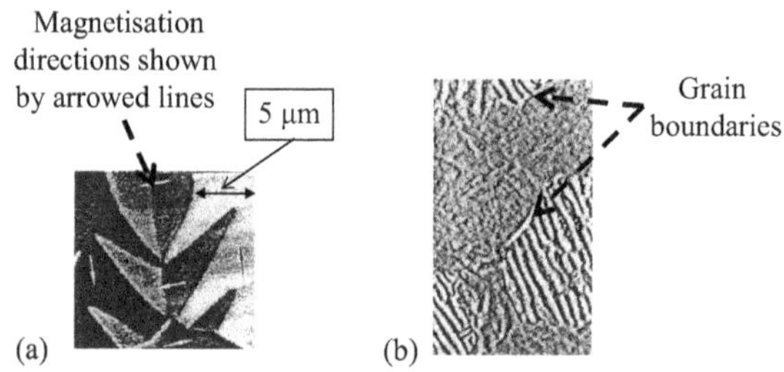

Figure 4.13 *MFM domain image on surfaces of electrical steels: (a) fine fir tree structures at a boundary between slightly misaligned 180° bar domains [25] and (b) patterns on surfaces of several grains in a NO silicon–steel (Figure 12 in [24] reproduced under free licence CC-BY-4.0)*

direction but the walls are apparently broken up in places and the domains are only around 2 μm wide which seems very small for this material. No explanation for this is given in [24] although it was said that the small kinks are possibly due to sub-surface transverse domains. Figure 4.12(b) shows the effect of a scratch forming an artificial boundary on the surface of the same sample. This seems related to the commercial domain refinement process discussed in Chapter 2 of Volume 2 of this book.

Other MFM images of domain structures on the surface of electrical steels are shown in Figure 4.13. Well known fir tree patterns are seen at a 180° domain boundary in Figure 4.13(a) [25]. The surface area of these closure domains is of the order of 20 μm^2 which is small compared to the fir tree patterns frequently reported to occur on the surface of SiFe single crystals.

Figure 4.13(b) shows the domain structure on the surface of a NO electrical steel. Grains around 4 μm diameter are present and also, presumably, [001] oriented bar domains around 0.5 μm wide. The surface patterns on some grains are difficult to see. These are possibly very small closure domains on the surface of grains without any

<100> directions close to the surface plane. The phenomenon shown in the hypothetical structure in Figure 4.1 may be present.

It is not clear whether some of the reported unexplained high resolution images are simply artefacts of the method or genuine surface patterns not possible to observe using other methods. It may be possible to adapt an MFM to make dynamic observations [26] which could generate a vast new wealth of knowledge which will reinforce or modify our existing ideas of magnetisation and loss processes.

4.5 Domain visualisation from surface field sensors

Surface domain structures can be deduced from measurement of the normal component of surface stray field using scanning field sensors. This can be measured in several ways on electrical steels. Hall effect field sensors, magneto-resistance (MR) or giant magneto-resistance (GMR) sensors, vibrating pick-up coils, optical field scanners and H-coils – all can be used. Most of these direct field sensing methods are capable of accurate field mapping without the need to remove the coating from GO steel which is a great advantage in surface domain imaging. Often, a fixed array of several sensors is scanned across a steel surface giving spatial resolution better than a few microns and a field sensitivity of around 0.1 Am^{-1} usually over an area of several square centimetres.

Figure 4.14 shows a schematic outline of a typical MR field scanner for use on a sheet of GO silicon–iron. A constant d.c. current is passed through the sensor to detect its change of resistance which is proportional to the field. A constant vertical field (not shown on the figure) is often applied to enhance the surface field, just as with the Orb domain viewer discussed in Section 4.2.

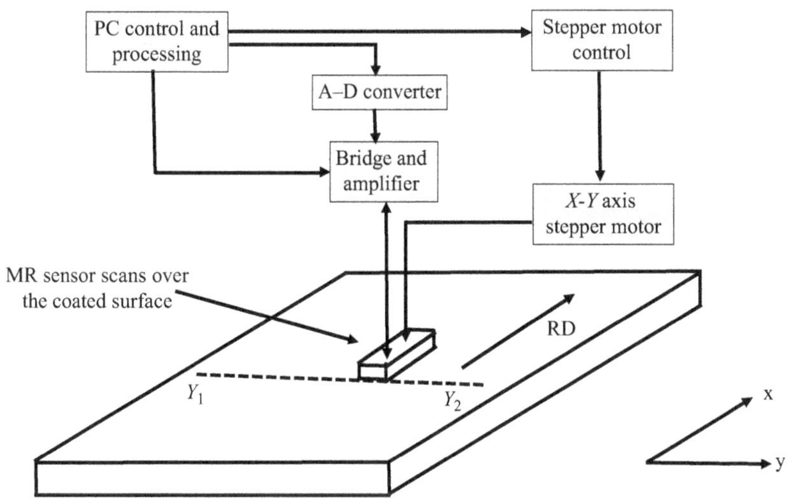

Figure 4.14 Key components and layout of a surface field scanning system for visualisation of domains on the surface of GO silicon–iron

The sensor detects the vertical component of field as it is scanned over the surface to map out a complete field contour. Three main components of stray field are likely to be present on the surface of an electrical steel. These are a field H_b above grain boundaries, a field H_w where Bloch walls reach the surface and a field H_β over the surfaces of grains mis-oriented by an angle β relative to the plane of the sheet.

The field distribution across arbitrary path $y_1 - y_2$ is similar to that in the sketch shown in Figure 4.15(a). Figure 4.15(b) shows the theoretical distribution of H_w and H_β relative to the positions of bar domains. The component H_w is effectively the same as that shown in Figure 4.2. In practice H_β is always much greater than either H_w or H_b so the resultant field is as shown in Figure 4.15(c). The actual field is affected by the presence of grain boundaries and surface closure domains and some stray air flux but these can easily be compensated for. The centres of domains are at positions such as $c_1$and c_2 and their walls are in positions on the slopes of the curve such as a_1 and a_2. A photograph of an MR sensor-based laboratory domain imaging system is shown in Figure 4.16 [27].

By relating the field intensity to a grey colour scale, and designating black and white for the highest and lowest intensity, respectively, domain images can be obtained. Figure 4.17(a) shows the field computed by scanning a 2.0 mm MR

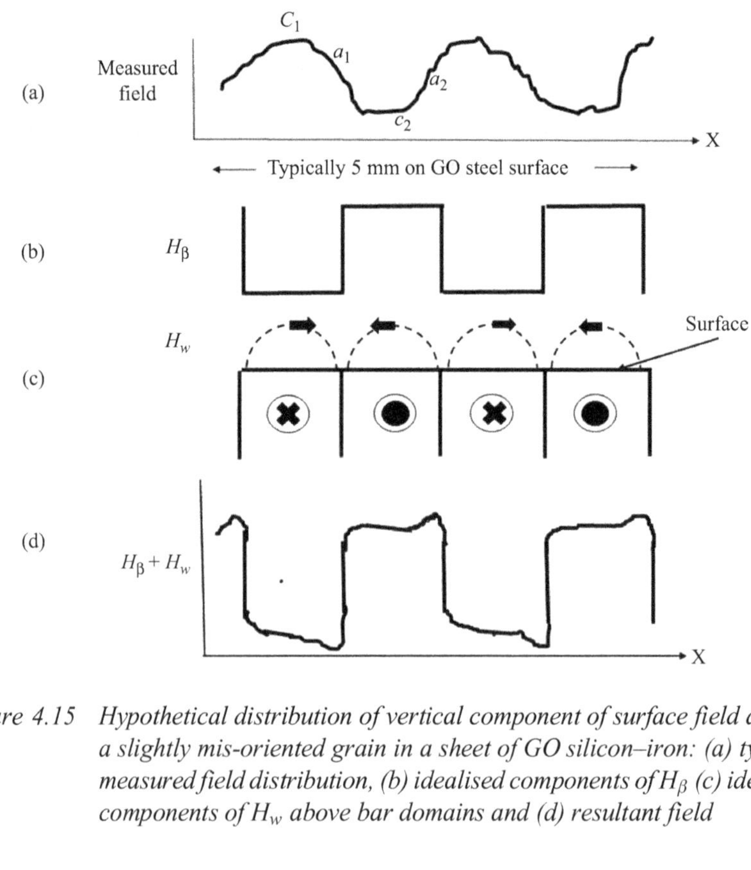

Figure 4.15 Hypothetical distribution of vertical component of surface field above a slightly mis-oriented grain in a sheet of GO silicon–iron: (a) typical measured field distribution, (b) idealised components of H_β (c) idealised components of H_w above bar domains and (d) resultant field

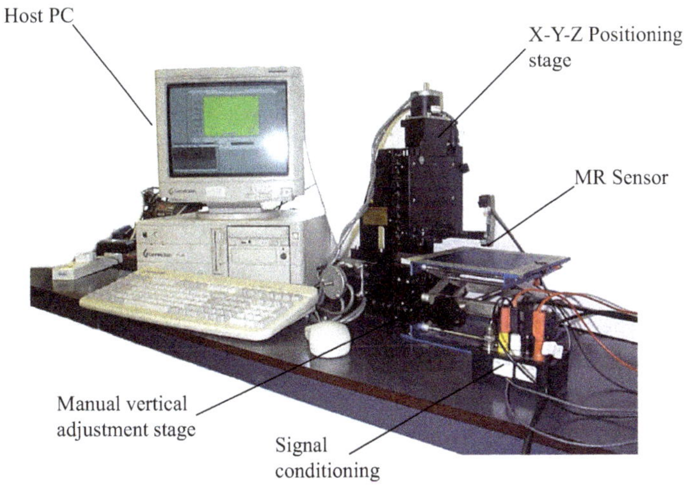

Figure 4.16 Layout of a system based on a scanning MR sensor for surface domain visualisation [27]

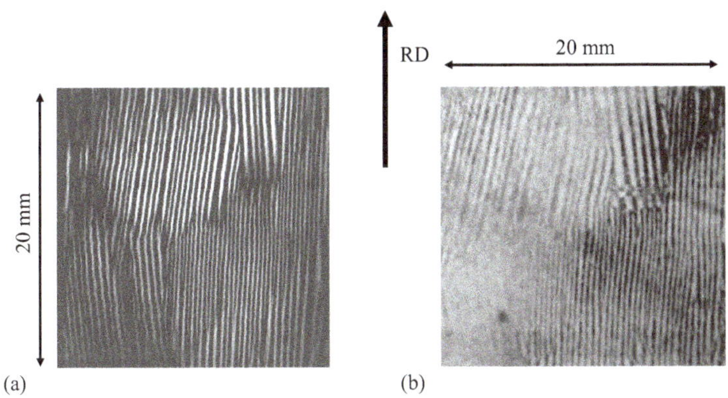

Figure 4.17 Images of sub-coating domains over a 20 mm × 20 mm area of a strip of commercial GO silicon–iron (a) computed image obtained by scanning a 2 mm MR sensor at a 100 μm step rate over the surface and (b) domain image over the same region using a commercial domain viewer based on the powder technique (832 Am^{-1} enhancing field in both cases) [27]

sensor above the coated surface of a strip of commercial GO steel. This can be compared with a view of the same region, shown in Figure 4.17(b), obtained using the powder pattern technique. The image produced by the MR scanner system is far clearer and appears to show more surface features.

In practice, a domain viewer is far cheaper and more convenient to use, so it is adequate for routine observations or trouble-shooting where sharp detail is not required.

Table 4.1 Examples of characteristics of various surface field scanning techniques

Technique	**Vibrating Permalloy probe [28]**	**Hall probe [29,30]**	**MR chevron [31]**	**MR probe [32,33]**	**Dynamic MR viewer [34,35]**
Active area/length	Few μm at tip	245 μm	2.0 mm × 1.5 mm	2.0 mm × 2.0 mm	110 μm × 2.0 mm
Effective distance from surface (μm)	1	170	25	300	35
Sampling step length (μm)	500	25	5	500	50
Image resolution	Blurred domains	Main domains	Grains	Stray field distribution	Main and supplementary domains
Image quality	Very poor	Coarse	Coarse	Good	Good
Relative imaging speed	Very slow	Slow	Slow	Medium	Fast
Ease of use	Difficult	Medium	Easy	Medium	Easy

Table 4.1 summarises typical characteristics of various field sensor techniques used for visualisation of domains on the surfaces of electrical steel. Today, most can determine domain distributions on coated GO silicon–iron, in either a demagnetised state or in a constant d.c. magnetisation. Attempts have been made to produce versions which will operate effectively under a.c. magnetisation in real time or in a stroboscopic mode. This would make the approach far more powerful than most other methods including KMO but it does not appear to have been successfully achieved yet.

4.6 Observation of sub-surface domain features

4.6.1 Electron microscope techniques

Domain structures in material less than around 0.5 μm thick can be observed using a transmission electron microscope (TEM). Lorentz forces are produced if a magnetic domain is present with its magnetisation perpendicular to the electron beam. These can be used to visualise the domains. The method can be used in stroboscopic mode up to very high frequencies.

The TEM has been used for studies of domains in electrical steels but samples have to be thinned and made very smooth. Interesting curved walls and discontinuous motion attributed to pinning sites have been observed in thinned NO electrical steel [36]. The observed domains are not representative of the bulk material [16] but appear to be useful for studying domain pinning sites.

The scanning electron microscope (SEM) is often used for domain observations in electrical steels. The use of backscattered electrons to study magnetic domains was

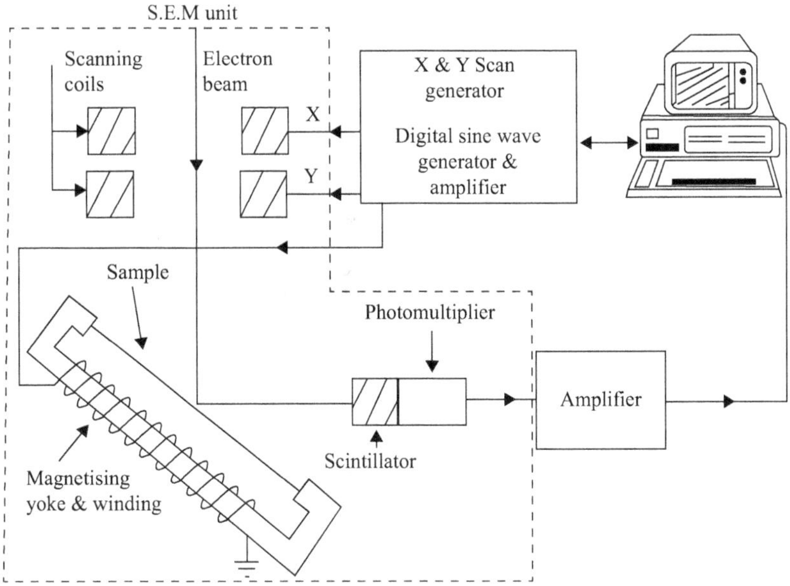

Figure 4.18 Layout of a SEM-based dynamic domain observation system (Courtesy of the late Professor P. Beckley)

first pioneered in the 1960s [37]. Domain patterns beneath coatings on GO silicon–iron can be visualised, but the surface must be polished to obtain high resolution images of small structures such as spike and lancet domains. An important advance came in 1977 when the stroboscopic mode was introduced to a HV SEM. This gave, for the first time, the possibility of revealing the true nature of domain wall motion at power frequency [38]. The restrictions of stroboscopic use were identified later.

Figure 4.18 shows the next advance which avoided stroboscopic use and enabled multiple SEM images to be taken over one cycle of 50 Hz magnetisation. The PC controls the SEM and sample magnetisation as well as signal processing and image enhancement. Figure 4.19 shows the effect of image enhancement of a raw image of domains at an instant during a 50 Hz magnetising cycle.

High voltage, stroboscopic SEM techniques are often used to observe dynamic domain structures in coated GO silicon–iron [39,40]. The depth to which wall positions can be detected depends mainly on the SEM voltage. If images are obtained at two or more voltage settings, information about domain structures between the two electron penetration depths can be obtained [41]. The method has been applied to iron-based amorphous material but a wider application of the technique has been slow to become established.

The high spatial resolution of the SEM enables researchers to delve more deeply into the domain structures of electrical steels. One study has shown complicated, unexplained, zigzag patterns of domains less than 100 μm wide close to the edges of ultra-thin NO electrical steel [42]. Results from these advanced

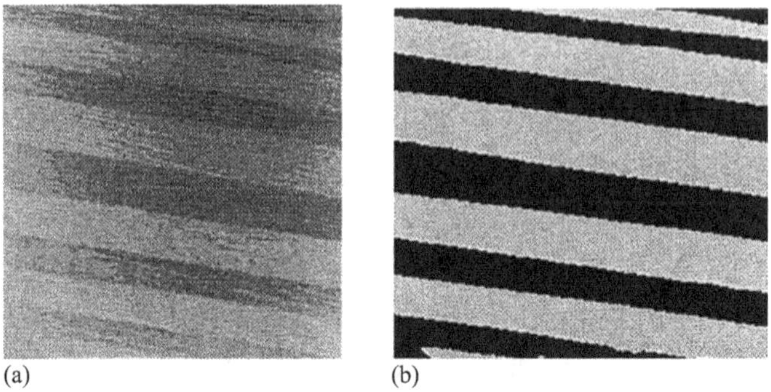

Figure 4.19 Bar domains observed under the coating of a 4.5 mm × 4.5 mm GO sample at an instant during 50 Hz magnetisation: (a) original image captured from the SEM and (b) final image after digital processing (Courtesy of the late Professor P. Beckley)

techniques are difficult to verify so it is not yet clear whether such unexplained results are either simply system artefacts or indications of fundamental phenomena which could improve our understanding of domain structures.

4.6.2 X-ray techniques

X-ray techniques have been used for many years for domain observation. X-rays undergo a change in polarisation when transmitted through a magnetic material or reflected from its surface. The distortion of a crystal lattice by domain walls can be detected, hence domains in electrical steels can be visualised. Internal domain structures in materials up to around 100 μm thick can be observed but hours of exposure time are needed just for static domain observation [16]. High energy synchrotron radiation can be used to reduce the required exposure time to enable stroboscopic observations to be made [43] but results do not seem to have been correlated with other findings or interpreted to show global sub-surface structures in electrical steels.

4.6.3 Freeze-in techniques for observing sub-surface structures

An interesting method of freezing-in domains within the volume of certain magnetic materials was discovered in the 1970s [44]. It was later applied to a silicon–iron alloy [45]. The positions of domains beneath the surface of a 1 mm thick, 12% SiFe single crystal were frozen-in as a precipitation pattern of sub-microscopic platelets after annealing at around 600 °C. The domains present at the annealing temperature can be observed using a polarisation microscope at room temperature. By removing layers of the materials and observing the surface at each stage, images of the frozen-in structure are revealed.

Figure 4.20 shows the change in domain pattern from the surface to a depth of 150 μm into the 1.0 mm thick crystal. The surface pattern is difficult to see in detail since it was unpolished but a remarkable change in structure can be seen moving into the interior.

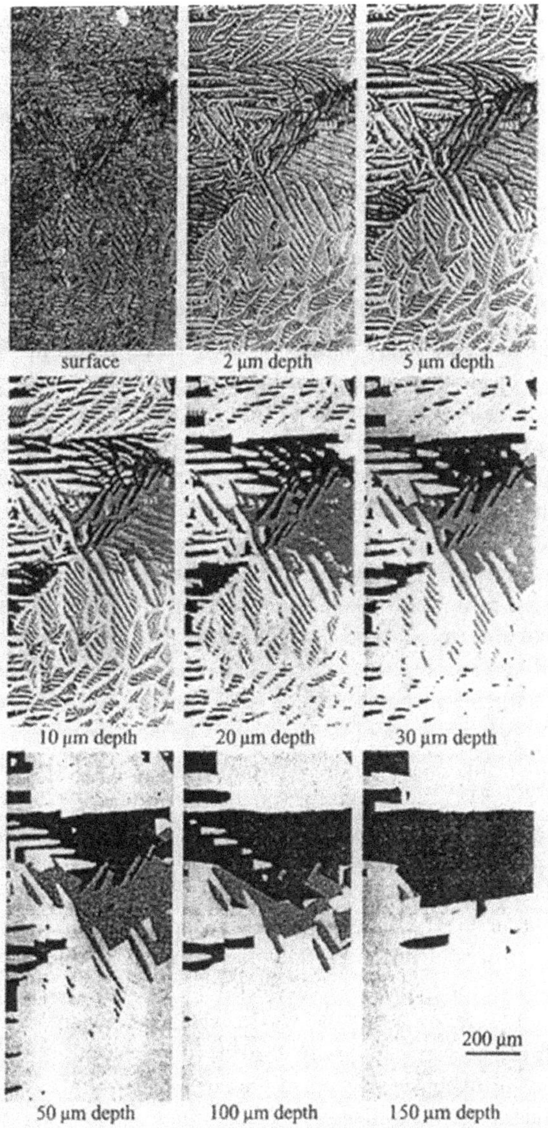

Figure 4.20 Domain patterns on the (110) surface of a 12% SiFe crystal starting with the unpolished surface and after sequential thinning to reveal structures to a depth of 150 µm beneath the original surface (Figure 8 in [45] reproduced under free licence CC-BY-4.0)

A similar investigation successfully visualised sub-surface domains in a slightly mis-oriented (110)[001] crystal of 6.5% SiFe alloy [46]. Interestingly the supplementary domain structures are different from expected according to basic domain theory but this might be due to the annealing conditions [46].

The freeze-in method is a useful step in sub-surface visualization, and could lead to better understanding of domain structures in the volume of GO steel. However, this is a destructive method requiring skilful experimentation and material processing. It can only be applied to a few alloy compositions.

The method seems to give useful insight into what structures may be present inside a material but to what extent it can ever be applied directly to electrical steels remains to be seen.

4.6.4 Visualisation of sub-surface domain structures using neutron irradiation

Potentially, neutron techniques seem to be the most powerful methods of imaging domains beneath the surface of an electrical steel. Neutrons can easily penetrate steel several millimetres thick and can interact with magnetic structure so that domain imaging is possible. Many techniques have been developed [47] and clear correlations with surface powder patterns have been reported together with complementary information about static internal structures [48].

The most promising method makes use of dark field imaging (DFI) based on the neutron grating interferometry (NGI) technique [49]. DFI allows for the in situ visualisation of the response of both bulk and supplementary domains in GO steel to the influence of an external field. Images of bulk magnetic domain structures can be obtained [48] and, also, what appears to be a correlation between internal structures in GO silicon–iron and surface KMO images has been reported [47].

Changes in main and supplementary sub-surface structures due to coating removal and externally applied mechanical stress have been observed at frequencies up to 200 Hz [50], although images are not always easy to interpret. Interesting, unexplained and unexpected frequency dependent regions of frozen-in and mobilised structures have been reported [47].

Much more work is necessary to interpret and understand static and dynamic internal domain structures visualised using DFI techniques. However, they appear to be potentially powerful research tools which can help give a far more fundamental understanding of magnetic materials than we have today [51]. Of course, there are only a few facilities in the world where NGI domain imaging can be carried out so it is necessary to continue to develop the other techniques to supplement these powerful, but restricted resources.

A comprehensive review of neutron techniques for investigation of domains and domain walls is given in [52].

4.7 Use of magnetic bacteria for domain observation

Magnetic bacteria occur widely in nature, often in damp boggy environments, and are well understood in biology [53]. Basically, they contain chains of crystals of magnetite (Fe_3O_4) or greigite (Fe_3S_4) which are sufficiently magnetic to cause the whole body of a bacterium to align with the earth's magnetic field. Live and dead bacteria align in this way.

Such bacteria placed on the surface of coated GO silicon–iron orientate themselves along stray field directions and produce patterns similar to those obtained by the conventional Bitter technique [54]. Domains less than around 60 μm wide cannot be observed in this way. The method is used in other branches of science but it seems unlikely that it will be used widely for domain observations on electrical steels.

4.8 Magneto-optical indicator films

A magnetic indicator film is a thin layer of soft magnetic material deposited on to a transparent surface. It makes use of the Faraday effect[4] to map stray surface fields. It has been applied widely to superconducting materials, hard magnetic materials, magnetic recording media and non-destructive testing. Its capability of high contrast and nanosecond time resolution suggests that the technique is capable of high frequency, real time domain observations on electrical steels.

Figure 4.21 shows the arrangement for obtaining images of surface domain patterns under the coating of GO silicon–iron [55]. Figure 4.22(a) shows an image of the structures near a grain boundary in the demagnetised state and Figure 4.22(b) shows an image in the same area at the 0.8 T peak of a cycle of 50 Hz magnetisation [56]. The images are printable but their quality is poor because the indicator was not specifically matched to electrical steel. The same author reported the presence of unexpected flux leakage at the grain boundary but this is difficult to see on the image. Also, domain wall dynamics presented on a video appeared to be different on coated and un-coated samples [57].

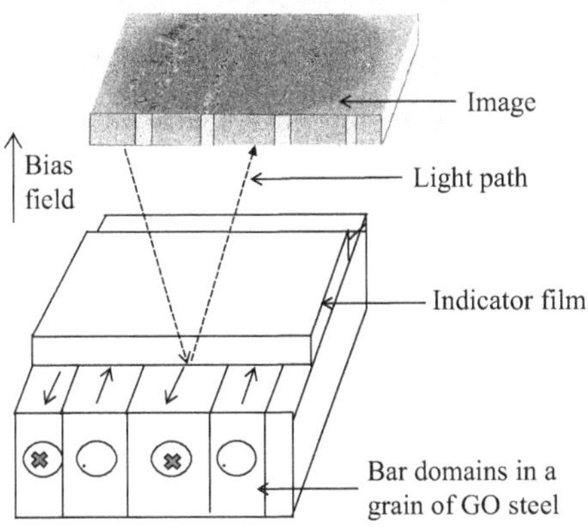

Figure 4.21 Arrangement of an indicator film system to observe surface domains on GO silicon–iron [55]

[4]The Faraday effect, discovered by Michael Faraday in 1845, is the rotation of a ray of polarised light when it is transmitted through a magnetic material. It has strong analogies with the magneto-optic Kerr effect, discovered 30 years later, which, as already seen in this chapter, is the reflection of polarised light from a magnetic surface.

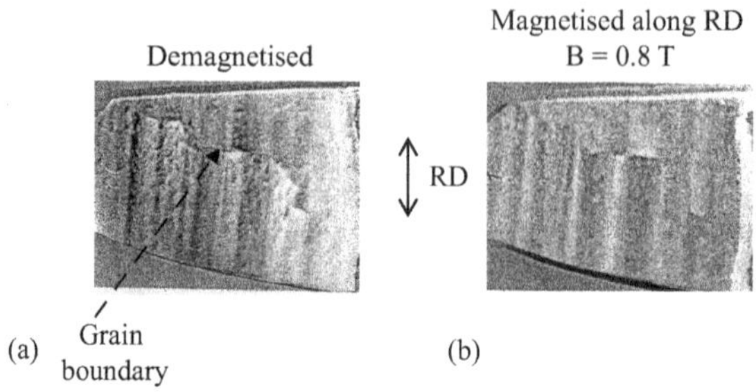

Figure 4.22 Images of domains on demagnetised and magnetised surfaces using the system outlined in Figure 4.21 [56]

The technique has been improved by other workers to obtain better domain contrast and more domain detail and can be used for static and dynamic observation through the coating of GO steel [58]. No sample preparation is required and a spatial resolution of 25 microns is achievable. Under dynamic single-shot mode, a time exposure of 1 ms is possible, showing that real time observation at frequencies up to at least tens of hertz is feasible.

It is possible that the technique can be optimised to obtain high resolution, high frequency, real time images under coatings of GO silicon–iron and similar materials to rival any other techniques.

4.9 Comparison of methods for observations on electrical steels

The methods of domain observation outlined in this chapter have been applied to electrical steels with varying degrees of success. Each has pros and cons depending very much on the requirement. If the requirement is very precisely known, the choice can be narrowed.

If low resolution surface static domain observation is required then the best choice is the Bitter technique because it is not difficult for the potential user to design and assemble a suitable domain viewer at no major expense. Inexpensive systems are sometimes commercially available.

If the requirement is more demanding, the cost of assembling or hiring a system, the need for skilled operation and interpretation must be factored in.

There are three broad categories of use. First, in industrial quality control or product development where quick results, which can easily be compared, are needed. Interpretation of the domain structure is not so important and basic surface static domain observation is usually sufficient. The second is for researches involving some aspect of electrical steel evaluation, for example the influence of mechanical stress or surface coatings on magnetic properties. In this case the interpretation of static and dynamic surface domain structures help in the understanding of their broader studies. The third category of use is for more fundamental academic research into the origins and basic understanding of magnetisation processes.

Table 4.2 General characteristics of techniques capable of observation of domains in electrical steels (H – high, M – medium, L – low capability)

Technique	Cost	Speed of preparation, observation and analysis	Spatial resolution	Surface (S) or internal (I) viewing	Static (S) or dynamic (D)	Disadvantages	Advantages
Powder pattern (Bitter)	L	H	M	S	S	Static only	Cheap/fast
Magneto-optical (Kerr)	M	M	H	S	D	Surface preparation	Dynamic
Surface field scanning	M	M	M	S	S/D	Resolution/contrast	Automatic
MFM	M	M	H	S	S	Difficult to use	High resolution
Indicator films	M	H	M	S	S		
Neutron techniques	H	L	H	I	S	Large scale facility/ very expensive/ difficult to interpret	Very high resolution/internal observations
Electron microscope/SEM	H	L	H	S/I	S	Slow/difficult to interpret	Internal observation/high resolution
Magnetic bacteria	L	H	L	S	S	Preparation	Cheap/easy to use
Freeze-in technique	M	L	M	I	S	Preparation/only possible on a few materials	Internal observation

The ultimate domain observation system might one day enable a user to observe sub-surface domain structures, at high magnetising frequency, in real time and with high spatial resolution. Only specialised researchers would make full use of such a complex facility but it would provide an opportunity to extend vastly our knowledge of the factors which really control the magnetic performance of electrical steels. However, for the less specialised user, Table 4.2 gives a guide to the features of more accessible approaches.

References

[1] Manke I., Kardjilov N., Schäfer R., *et al.* 'Three-dimensional imaging of magnetic domains'. *Nat. Commun.* 2010, vol. 1, Art. No. 125, pp.1–5

[2] Hamos L.V., and Thiessen P.A. 'Making visible the regions of different magnetic states of solid bodies'. *Z. Physik.* 1931, vol. 71, pp. 442–444

[3] Bitter F. 'On inhomogeneities in the magnetization of ferromagnetic materials'. *Phys. Rev.* 1931, vol. 38, pp. 1903–1905

[4] Bates L.F. 'The visualisation of magnetic processes'. *Address to the British Association, Section A (Mathematics and Physics)*, York, UK, 3rd September 1959

[5] Mckeehan L.W., and Elmore W.C. 'Surface magnetization in ferromagnetic crystals'. *Phys. Rev.* 1934, vol. 46, pp. 226–228

[6] Taylor R. 'A large area domain viewer'. *Anales de Fisica, Ser. B* 1990, vol. 86, pp. 126–128

[7] Beckley P. *Electrical Steels: A Handbook for Producers and Users.* European Electrical Steels, Orb Works, Newport, South Wales, 2000

[8] Kerr J. 'On rotation of the plane of polarisation by reflection from the pole of a magnet'. *Philos. Mag.* 1876, vol. 5(3), pp. 321–343

[9] Hoshtanar O. 'Dynamic Domain Observation in Grain-Oriented Electrical Steel Using Magneto-Optical Techniques'. *PhD Thesis*, Cardiff University, 2006, pp. 57–70

[10] Carey R., and Isaac E.D. *Magnetic Domains and Techniques for their Observation.* The English University Press, London, 1966

[11] Lee E.W., Callaby D.R., and Lynch A.C. 'The use of the Kerr effect for studying the magnetization of a reflecting surface'. *Proc. Phys. Soc.* 1958, vol. 72(2), pp. 233–243

[12] Hoshtanar O. 'Dynamic Domain Observation in Grain-Oriented Electrical Steel Using Magneto-Optical Techniques'. *PhD Thesis*, Cardiff University, 2006, p. 70

[13] Pry R.N., and Bean C. 'Calculation of the energy loss in magnetic sheet materials using a domain model'. *J. Appl. Phys.* 1958, vol. 29, pp. 532–533

[14] Houze G.L. 'Domain-wall motion in grain-oriented silicon steel in cyclic magnetic fields'. *J. Appl. Phys.* 1967, vol. 38(3), pp. 1089–1095

[15] Shilling J.W. 'Domain structures in 3% Si-Fe single crystals with orientation near to (110)[001]'. *IEEE Trans. Magn.* 1975, vol. 9(5), pp. 351–356

[16] Shilling J.W., and Houze G.L. 'Magnetic properties and domain structure in grain-oriented 3 % Si-Fe'. *IEEE Trans. Magnet.* 1974, vol. 10 (2) pp. 195–223
[17] Overshott K.J. 'The use of domain observation in understanding and improving the magnetic properties of transformer steels'. *IEEE Trans. Magn.* 1976, vol. 12(6), pp. 840–845
[18] Moses A.J., Williams P.I., and Hoshtanar O. 'A novel instrument for real-time dynamic domain observation in bulk and micromagnetic materials'. *IEEE Trans. Magn.* 2005, vol. 41(10), pp. 3736–3738
[19] Hoshtanar O. 'Dynamic Domain Observation in Grain-Oriented Electrical Steel Using Magneto-Optical Techniques'. *PhD Thesis*, Cardiff University, 2006, p. 86
[20] McCord J. 'Progress in magnetic domain observation by advanced magneto-optical microscopy'. *J. Phys. D: Appl. Phys.* 2015, vol. 48, Art. No. 333001
[21] Moses A.J., Williams P.I., and Hoshtanar O. 'Real time dynamic domain observation in bulk materials'. *J. Magn. Magn. Mater.* 2006, vol. 304(2), pp. 150–154
[22] Kaczer J., and Gemperie R. 'Vibrating Permalloy probe for mapping magnetic fields'. *Czech. J. Phys.* 1956, vol. 6(2), pp. 173–183
[23] Martin Y., and Wickramasinghe H.K. 'Magnetic imaging by "force microscopy" with 1000 Å resolution'. *Appl. Phys. Lett.* 1987, vol. 50(20), pp. 1455–1457
[24] Yáñez T.R., Monsalve A., Artigas A., and Houbaert Y. 'Magnetic domain observation by atomic force microscopy'. *Remetallica* 2005, vol. 25(13), pp. 35–42
[25] Belliard L., Miltat J., Thiaville L., and Tomis D. 'Magnetic force imaging of soft magnetic materials'. *7th Joint MMM-Intermag Conference*. Abstracts, San Francisco, California, USA, January 1998
[26] Magnetic Force Microscope, Wikipedia, https://en.wikipedia.org/wiki/Magnetic_force_microscope#Dynamic_(AC)_mode (Accessed June 2018)
[27] Nicholson P. I. 'Development of a Magnetic Domain Imaging System for Electrical Steel Using a Magnetoresistive Sensor Based Stray Field Scanning Technique'. *PhD Thesis*, Cardiff University, 1999
[28] Carey R., and Isaac E.D. *Magnetic Domains and Techniques for Their Observations*. The English University Press, London, 1966
[29] Matsubara K., Yamaguchi K., and Tamura R. 'Automatic measurement of magnetic domains and bloch wall distribution on 3 percent Si-Fe sheet by means of Hall sensor'. *IEEE Transl. J. Magn. Jap.* 1985, vol. 4, pp. 483–485
[30] Pfutzner H., Bengtsson C., and Leeb A. 'Domain investigation on coated unpolished Si-Fe sheets'. *IEEE Trans. Magn.* 1985, vol.21 (6), pp. 2620–2625
[31] Mohd Ali B. B., and Moses A.J. 'A grain detection system for grain oriented electrical steel'. *IEEE Trans. Magn.* 1989, vol. 25(6), pp. 4421–4426
[32] Tumanski S., and Stabrowski M. 'The magnetovision-new method of steel investigation'. *J. Magn. Magn. Mater.* 1996, vol. 160, pp. 165–166
[33] Tumanski S., and Stabrowski M. 'The magnetovision method as a tool for investigation of electrical steel quality'. *J. Meas. Sci.* 1998, vol. 9(3), pp. 488–495

[34] So M.H., Nicholson P.I., Meydan T., and Moses A.J. 'Magnetic domain imaging in coated silicon-iron using magnetoresistive sensors'. *IEEE Trans. Magn.* 1995, vol. 31(6), pp. 3370–3372

[35] Nicholson P.I., So M.H., Meydan T., and Moses A.J. 'Non-destructive surface inspection system for steel and other ferromagnetic materials using magnetoresistive sensors'. *J. Magn. Magn. Mater.* 1996, vol. 160, pp. 160–162

[36] Akase Z., Park Y-G., Shindo D., Tomida T., Yaskiki H., and Hinotani S. 'Magnetic domain structures in electrical steel sheets studied by Lorentz microscopy and electron holography'. *Mater. Trans.* 2005, vol. 46(5), pp. 974–977

[37] Philbert J., and Tixier R. 'Effects of crystal contrast in scanning electron microscopy'. *Micron* 1969, vol. 1, pp. 174–196

[38] Fukuda B., Irie T., Shimanaka H., and Yamamoto T. 'Observation through surface coatings of domain structures in 3% Si-Fe sheets by a high voltage scanning electron microscope'. *IEEE Trans. Magn.* 1977, vol. 13(5), pp. 1499–1504

[39] Mivechchi E., Beckley P., Horrocks D.H., and Porter C.H. 'Stroboscopic observation of domain motion on coated Si-Fe sheet using scanning electron microscope'. *IEEE Trans. Magn.* 1990, vol. 26(5), pp. 1975–1977

[40] Nozawa T., Mizogami M., Mogi M., and Matsuo Y. 'Magnetic properties and dynamic domain behaviour in grain-oriented 3% Si-Fe'. *IEEE Trans. Magn.* 1996, vol. 32(2), pp. 572–589

[41] Pogány L., Son D.T., Varga I., Fülöp Z., Hargitai C., and Bakonyi I. 'Measurement of internal magnetic domain structures by using backscattered electrons in a SEM'. *J. Phys. IV France* 1998, vol. 8, pp. 697–700

[42] Lee Y., Koymen A.R., Heo N.H., Na J.G., and Woo J.S. 'Surface magnetic domain observation on thin-gauged 3% Si-Fe sheets by using scanning electron microscopy with polarization analysis (SEMPA)'. *IEEE Trans. Magn.* 1998, vol. 34(4), pp. 1165–1167

[43] Hoshtanar O. 'Dynamic Domain Observation in Grain-Oriented Electrical Steel Using Magneto-Optical Techniques'. *PhD Thesis*, Cardiff University, 2006, p. 52

[44] Libovický S. 'Spatial replica of ferromagnetic domains in iron-silicon alloys'. *Phys. Stat. Sol. (A): Appl. Res.* 1972, vol. 12, pp. 539–547

[45] Schäfer R., and Schinnerling S. 'Bulk domain analysis in FeSi-crystals'. *J. Magn. Magn. Mater.* 2000, vol. 215–216, pp. 140–146

[46] Shin S., Schäfer R., and De Cooman B.C. 'Three dimensional visualization of the domain microstructure in bulk Fe-6.6 pct Si'. *Metall. Mater. Trans.* 2013, vol. A(44), pp. 4239–4243

[47] Betz B.K-J.G. Visualisation of Magnetic Domain Structures and Magnetisation Processes in Goss-Oriented High Permeability Steels Using Neutron Grating Interferometry. *DSc Thesis*, No. 7002, Federal Polytechnic of Lausanne, France, 2016

[48] Lee S.W., Kim O.Y., Kardjilov N., Dawson M., Hilger A., and Manke I. 'Observation of magnetic domains in insulation-coated electrical steels by neutron dark-field imaging'. *Appl. Phys. Exp.* 2010, vol. 3(10), Art. No. 106602

[49] Pfeiffer F., Grünzweig C., Bunk O., Frei G., Lehmann E., and David C. 'Neutron phase imaging and tomography'. *Phys. Rev. Lett.* 2006, vol. 96, Art. No. 215505
[50] Betz B., Rauscher P., Harti R.P., *et al.* 'Frequency-induced bulk magnetic domain-wall freezing visualized by neutron dark-field imaging'. *Phys. Rev. Appl.* 2016, vol. 6, Art. No. 024024
[51] Manke I., Kardjilov N., Schäfer R., *et al.* 'Three-dimensional imaging of magnetic domains'. *Nat. Commun.* 2010, vol. 1, Art. No. 125, pp. 1–5
[52] Schärpf H., and Strothmann H. 'Neutron techniques for magnetic domain and domain wall investigations'. *Phys. Scr.* 1988, vol. T24, pp. 58–70
[53] Schuler D. 'Magnetotactic bacteria'. *J. Mol. Microbiol.* 1999, vol. 1(1), pp. 76–86
[54] Harasko G., Pfutzner H., and Futschik H. 'On the effectiveness of magnetotactic bacteria for visualization of magnetic domains'. *J. Magn. Magn. Mater.* 1994, vol. 133, pp. 409–412
[55] Hoshtanar O. 'Dynamic Domain Observation in Grain-Oriented Electrical Steel Using Magneto-Optical Techniques'. *PhD Thesis*, Cardiff University, 2006, p. 182
[56] Hoshtanar O. 'Dynamic Domain Observation in Grain-Oriented Electrical Steel Using Magneto-Optical Techniques'. *PhD Thesis*, Cardiff University, 2006, p. 186
[57] Hoshtanar O. 'Dynamic Domain Observation in Grain-Oriented Electrical Steel Using Magneto-optical Techniques'. *PhD Thesis*, Cardiff University, 2006, p. 188
[58] Richert H., Schmidt H., Lindner S., Wenzel B., Holzhey R., and Schäfer R. 'Dynamic magneto-optical imaging of domains in grain-oriented electrical steel'. *Steel Res. Int.* 2016, vol. 87(2), pp. 232–240

Chapter 5
Electromagnetic induction

An electromotive force (e.m.f.) is induced in an electric circuit when magnetic flux linking with the circuit changes. This is often called *electromagnetic induction* which is very familiar to motor and transformer designers. It is very important in the characterisation and analysis of the performance and properties of soft magnetic materials under a.c. or any other form of changing magnetisation.

The three scientists who discovered the basic principle of electromagnetic induction were Michael Faraday, Heinrich Lenz and James Clerk Maxwell. Faraday published his findings, now simply referred to as Faraday's law, in 1831. Lenz published his work on the direction of the induced e.m.f. in an electric circuit just about three years later. In 1861, after more than 20 years of scientific study, Maxwell went on to unify the principles of electromagnetism by rationalising the previous works of Ampère, Gauss, Faraday and others into his classic mathematical equations. One of Maxwell's equations is effectively a more generalised mathematical form of Faraday's law.

Joseph Henry (after whom the unit of inductance is named) should be mentioned since he independently discovered the principles of electromagnetic induction at about the same time as Faraday. However, Faraday published his own work first, so Henry's work is rarely mentioned.

Faraday's law was briefly introduced in Section 2.1.4. In this section its origins and measurement relevant to electrical steels are set out in more detail.

5.1 Faraday's law

Faraday's law can be stated simply as:

> When the magnetic flux through an electric circuit changes, an e.m.f. is set up in the circuit. The magnitude of the e.m.f. at any instant in time is directly proportional to the rate at which the flux is changing at that particular time.

The change of this flux linkage with the electrical circuit can be due to either relative movement between the source of the e.m.f. and the circuit or simply a change of magnitude of the flux. It will be noted that this definition of Faraday's law does not refer to magnetic materials. It effectively refers to voltages being set up by a changing magnetic flux. This voltage, or e.m.f., can be produced by a changing electric current in passing through any conducting medium. If a magnetic material is placed in the region where the magnetic flux changes, then the flux change inside the material is amplified by an amount dependent on the relative permeability of the material.

A flux change can also be produced while a permanent magnet is being moved. In this case no electric current is apparent, but inside the magnet all the atomic dipoles are in motion which itself is effectively an electric current. Hence, although there is no direct mention of magnetic material in Faraday's law, its magnetic and electric effects are firmly intertwined in all aspects of the use of soft magnetic materials.

5.2 Lenz's law

Lenz's law defines the *direction* of an induced e.m.f. and current in a closed circuit when subjected to a change in flux linkage. It can be simply stated as:

The direction of the induced e.m.f. and resultant current is such as to oppose the change producing it.

A simple example is if a bar magnet moves towards a stationary coil of wire as shown in Figure 5.1(a), an e.m.f. is induced in the coil. If the coil forms part of a closed electric circuit a current flows in the direction shown to create a north pole. The force set up between the like poles tends to slow the magnet's motion towards the coil, hence opposing the change in magnet position which causes the e.m.f. in the first place.

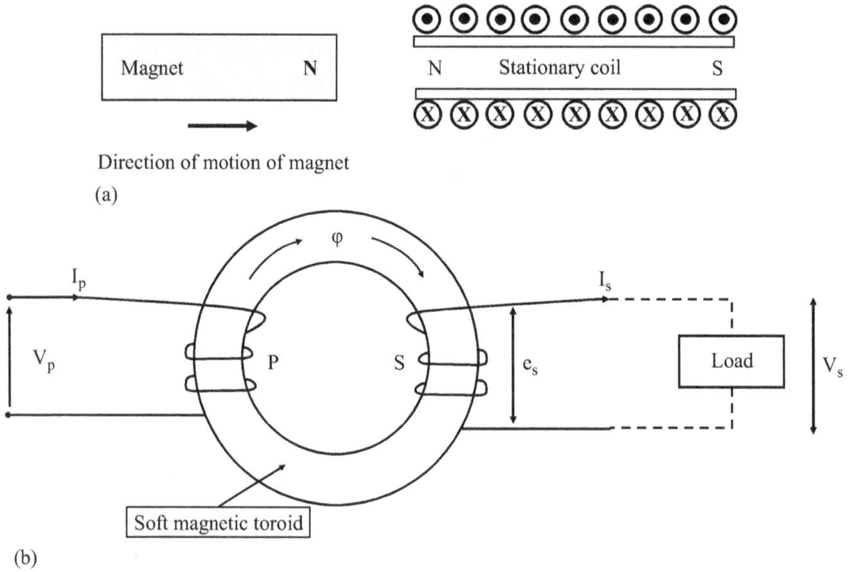

Figure 5.1 E.M.F. and resulting current induced in a coil by (a) motion of a permanent magnet and (b) a current carrying primary coil in a simple transformer

Another common example is shown in Figure 5.1(b). Here a current I_p flows through coil P to produce a clockwise field in the magnetic toroid. Suppose the current changes sinusoidally with time t at a frequency f (or ω radian.s^{-1} of the current wave). If the permeability of the toroid is constant then a sinusoidal time varying flux density $\Phi_p \sin 2\pi ft$ or $\Phi_p \sin \omega t$ is produced (ϕ_p is the peak value of the flux). This *flows* around the magnetic circuit and links with the secondary coil S which is also wound around the toroid. An e.m.f. is induced in coil S which, by Lenz's law, opposes the flux change in the toroid. If coil S is also in a closed electric circuit, a current I_s will flow in the direction shown to set up an anti-clockwise flux opposing that produced by coil P.

The use and meaning of the term e.m.f. often creates confusion so it is worth clarifying the background. The broad physical meaning of the expression is that it is the combination of electric and magnetic forces acting between charged particles. It can be attractive or repulsive and has properties associated with electromagnetic fields and light and as such it is one of the fundamental cornerstones of nature along with gravity and certain atomic forces.

An e.m.f. is not a mechanical force, it is an electric potential so it is measured as a voltage. As far as the electromagnetism associated with magnetic materials is concerned, some experts say that we should not use the expression at all since it is outdated and misleading so we should just refer to the phenomenon as an *induced voltage*. The term is most commonly used in electrical machine technology when determining, or explaining, the polarity of windings in terms of the direction or time phase of the actual voltage measured at their terminals. Both terms are used in the literature and are interchangeable in this context.

5.3 Expressions for an induced e.m.f.

The combined effect of Faraday's and Lenz's laws can be written as a simple mathematical formula. If the total flux through a circuit is ϕ at a given instant in time and it is instantaneously changing at the rate $d\Phi/dt$ then the instantaneous induced e.m.f. e is given by[1]:

$$e = -\frac{d\Phi}{dt}\,(\mathrm{v}) \tag{5.1}$$

Consider the single turn coil of conducting wire shown in Figure 5.2. Suppose it links with the changing magnetic flux. An e.m.f. given by (5.1) is induced in the coil which can be written in terms of flux density B. If the cross sectional area of the coil is A then, since $\Phi = BA$, the e.m.f. can be expressed simply as

$$e = -A\frac{dB}{dt}\,(\mathrm{v}) \tag{5.2}$$

[1]The minus sign is usually included in (5.1) simply as a sign convention basically saying that if the circuit is complete the e.m.f. simply causes a current to flow in a direction to set up flux to oppose the original flux. This is, of course, Lenz's law.

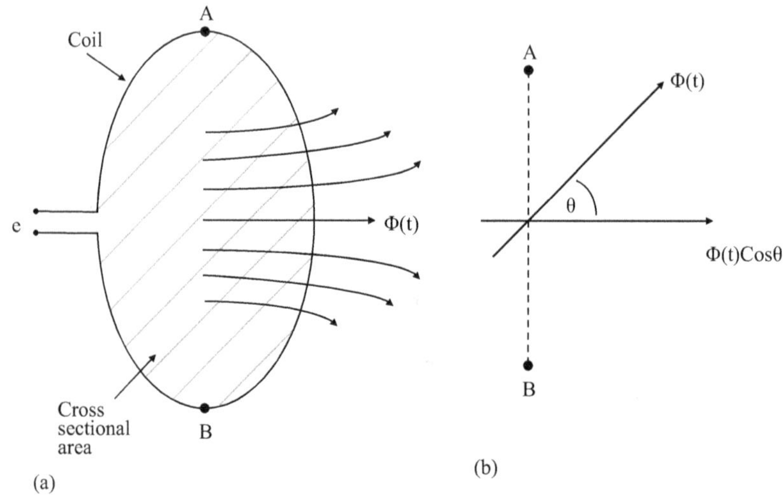

Figure 5.2 E.M.F. induced in a stationary coil due to change in flux linkage: (a) time varying flux linking the coil and (b) vector representation of flux at an angle to plane AB of the coil

If we have a multi-turn coil with N identical turns, each embracing the same flux and each with the same cross sectional area A, then the e.m.f. will simply be N times the value given in (5.2).

It should be noted that when applying (5.1) or (5.2) to a coil, such as the one shown in Figure 5.2 that the calculated values of flux and flux density are their components along the axis of the coil. If a flux $\phi(t)$ is passing through the coil at an angle θ to this axis, then the induced e.m.f. is proportional to $\frac{d(\phi(t)\cos\theta)}{dt}$. This is particularly important in the measurement of flux distributions within electrical machine cores.

Suppose the coils P and S in Figure 5.1(b) are wound from N_p and N_s turns, respectively. If the flux is varying sinusoidally, the induced e.m.f. e_s in coil S is given by

$$e_s = -N_s A \frac{d(B_p \sin \omega t)}{dt} = -N_s A B_p \, \omega \cos \omega t(\text{v}) \tag{5.3}$$

where B_p is the peak, or maximum, value of the sinusoidally varying flux density. The varying flux in the toroid also sets up an e.m.f. in coil P which is given by

$$e_p = -N_p A B_p \, \omega \cos \omega t(\text{v}) \tag{5.4}$$

Since the flux density is constant all around the toroid, and the cross sectional area of the toroid is constant, we can see from (5.3) and (5.4) that

$$\frac{e_p}{e_s} = \frac{N_p}{N_s} \tag{5.5}$$

Furthermore, we can apply Ampère's law[2] to the closed magnetic circuit of the toroid and obtain

$$N_p I_p = N_s I_s \tag{5.6}$$

Hence

$$\frac{e_p}{e_s} = \frac{N_p}{N_s} = \frac{I_s}{I_p} \tag{5.7}$$

Equations (5.5)–(5.7) and the basic geometry of the two windings on the toroid may be recognised as relating to an ideal single-phase transformer core with its primary circuit *P* and secondary circuit *S* where the primary and secondary voltages will be equal to e_p and e_s, respectively. Flux leakage, winding resistance and reactance are not considered in this ideal case.[3] As well as showing the basic concept of flux linkage in an ideal transformer, the same magnetic circuit forms the basis of many systems used for magnetising and measuring the magnetic properties of electrical steels in various forms ranging from single strips to laboratory scale stacked cores.

Of course, e_s in (5.3) is the instantaneous value of induced e.m.f. but because it is varying sinusoidally we can express the peak value E_p, r.m.s. (root mean square) value E_{rms} or average value E_{av} of e.m.f. as

$$E_p = N_s A B_p \omega, E_{rms} = \frac{1}{\sqrt{2}} N_s A B_p \, \omega, E_{av} = \frac{2}{\pi} N_s A B_p \omega (\mathrm{v}) \tag{5.8}$$

When the flux density varies sinusoidally, any of these expressions can be used to calculate the peak flux density but it is shown in Chapter 6 that if the flux density is not sinusoidal then the first two expressions in (5.8) are no longer applicable[4] and there are limitations on the use of the average e.m.f.

In electric circuit analysis, r.m.s. values of currents and voltages are commonly used. In the application of electrical steels, the peak flux density is a far more important characteristic than its r.m.s. value. It is simply because the power conversion capacity of an electrical machine is proportional to the peak value of flux density in its magnetic core. For example, in a rotating electrical machine, such as an alternator, the total output power is proportional to the product of its speed, the armature ampère-turns and the peak flux density in the air gap [1].

Reference

[1] Goodlet B.L. *Basic Electrotechnics*, 3rd edn. Edward Arnold Ltd, London, 1962, p. 118

[2]This is explained in Chapter 10.

[3]These are discussed in Chapter 5 of Volume 2 of this book.

[4]It is shown in Chapter 6 that there are also limitations on the use of average e.m.f. to determine peak flux density.

Chapter 6
Fundamentals of a.c. signals

Engineers and physicists will be familiar with many aspects of waveform analysis. It is widely applied to measurement and computational analysis of *B–H* characteristics, magnetostriction and losses of electrical steels and other soft magnetic materials. The fundamentals of waveform analysis relevant to these applications are presented in this chapter for the benefit of readers who might be less familiar with the topic.

The chapter gives a brief introduction to waveform analysis before focusing on the occurrence and common applications related to magnetic parameters particularly important in the application of soft magnetic materials in power devices.

Fundamentally, harmonic analysis is a mathematical tool developed to help deal with complex, time varying waveforms. The French scientist Joseph Fourier was the first to demonstrate how such periodic mathematical functions can be defined, either exactly or approximately, by a series of trigonometrical sine and cosine functions.

The normal way of carrying out waveform analysis in most areas of science and engineering is to apply Fourier theory, usually simply by inputting, or manipulating data, using powerful software tools. These are easy to use but give little indication of the physical nature of the analysis of magnetic parameters.

In this chapter, the importance of a physical understanding of the analysis is highlighted keeping mathematical analysis to a minimum. The development and engineering application of Fourier analysis is covered in standard mathematics textbooks such as [1] or [2]. Readers can refer to these to obtain a broader background, understanding and knowledge of the range of ways in which waveform analysis can be applied in more challenging applications.

6.1 Waveform terminology

Voltage, current and magnetic flux encountered in electrical engineering and magnetics technology often vary with time in a complex manner. Sometimes they vary in a repetitive way such as the typical magnetic waveforms shown in Figure 6.1. In each example one full cycle of the repetitive waveform is shown over one period of magnetisation. The simplest waveform shown is the flux density whose magnitude varies in a sinusoidal fashion with time as shown in Figure 6.1(a).[1]

[1]Of course other wave shapes such as square, triangular, etc. are also simple in this respect; some of the advantages of sinusoidal, periodic waveforms will be seen later.

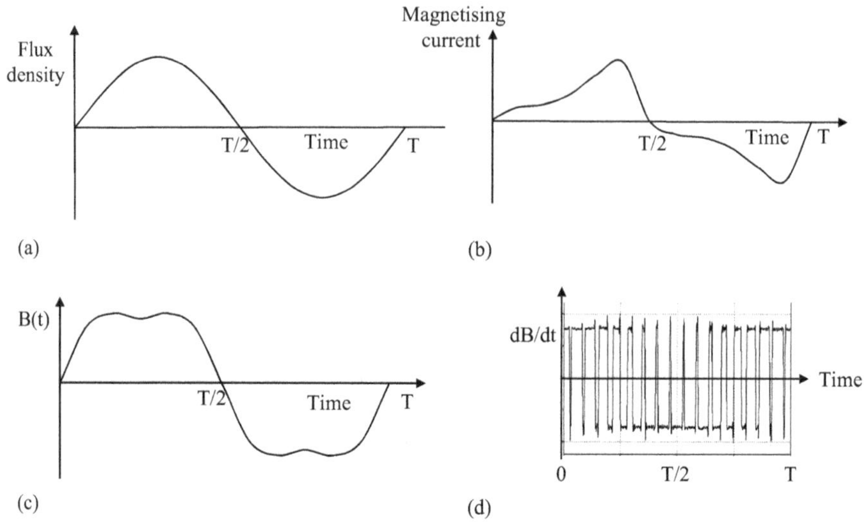

Figure 6.1 Typical periodic magnetic waveforms: (a) ideal flux density in the core of a power transformer, (b) transformer magnetising current, (c) localised flux density in a part of a lamination in a transformer joint and (d) rate of change of flux density dB/dt in a transformer core magnetised by a pulse width modulation (PWM) voltage source

The period T of this waveform is the time taken for the flux density to complete its reversal. Since this waveform is periodically repeated, we can define its frequency f as the number of times the flux density waveform repeats itself every second. The well-known relationship between the two is simply written as

$$T = \frac{1}{f}(\mathrm{s}) \tag{6.1}$$

The same sinusoidal repetitive waveform has an angular frequency ω measured in radians per second. Now 2π radians is equal to an angular rotation of 360° which takes place during one complete cycle during the period of the wave, so we can write the angular velocity of the wave as

$$\omega = \frac{2\pi}{T} = 2\pi f\left(\mathrm{rads}^{-1}\right) \tag{6.2}$$

Such expressions will be very familiar to many readers but they are included as a reminder of their basic meanings since they are constantly used, directly and indirectly, in waveform analysis.

The other wave shapes in Figure 6.1 are also symmetrical but are clearly more complicated than the sine wave.

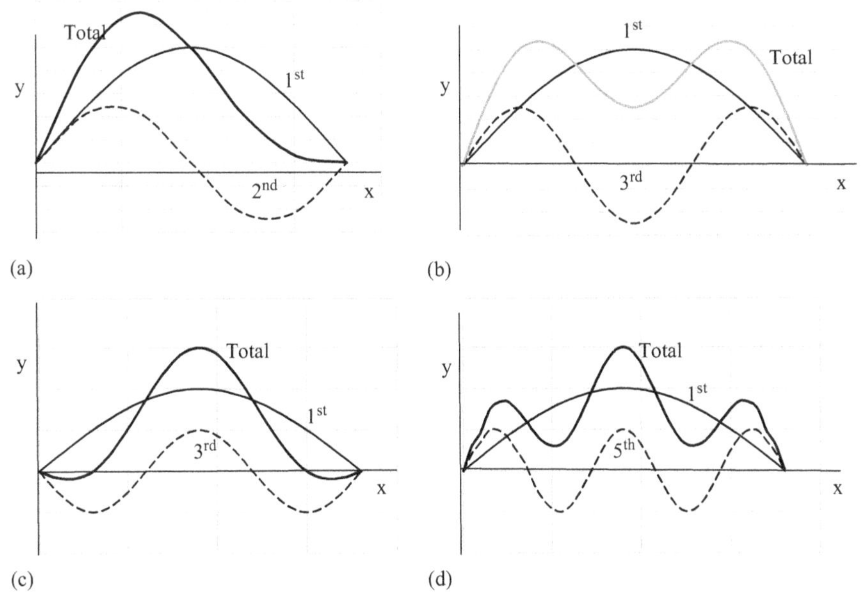

Figure 6.2 Graphical analysis of simple distorted waveforms: (a) second harmonic in phase with the fundamental, (b) third harmonic in phase with the fundamental, (c) third harmonic antiphase to the fundamental and (d) fifth harmonic in phase with the fundamental

Figure 6.2 shows how some simple, non-sinusoidal, wave shapes can be graphically built up from sinusoidal waves with different time periods.[2] For example, the waveform shown in Figure 6.2(a). It can be seen that it is equivalent to the sum of two sinusoidal waves, shown dotted, of different magnitudes. In this case one has the same period as the full wave shape and the other has a lower peak value but whose period is one half that of the others. Hence the instantaneous value of the full curve at any instant in time is equal to the sum of the instantaneous values of the two dotted sinusoidal waves at the same instant. The waveform can be expressed mathematically as

$$y(t) = Y_1 \sin \omega t + Y_2 \sin 2\omega t \tag{6.3}$$

This waveform comprises two components which are referred to as the first harmonic, or fundamental, component and the second harmonic component which varies at twice the frequency of the fundamental component. In this case, the second harmonic component is said to be in phase with the fundamental component since, after the instant in time when both their magnitudes are zero, they both start increasing in the same (negative or positive) direction.

[2]The independent variable here is time t. Of course, similar wave shapes can occur with many different forms of independent variables which can be analysed in the same way as illustrated in this chapter.

If the same analysis is carried out for the waveform in Figure 6.2(b), a third harmonic component which is in phase with the fundamental component has to be added in order to obtain $y(t)$, i.e.

$$y(t) = Y_1 \sin \omega t + Y_3 \sin 3\,\omega t \tag{6.4}$$

The waveform shown in Figure 6.2(c) comprises two components, but the third harmonic component is 180° out of phase (i.e. antiphase) with the fundamental waveform. In this case

$$y(t) = Y_1 \sin \omega t - Y_3 \sin 3\omega t \; or \; y(t) = Y_1 \sin \omega t + Y_3 \sin(3\omega t + 180) \tag{6.5}$$

Figure 6.2(d) shows a waveform comprising a fifth harmonic component in phase with the fundamental component. Any unknown waveform can be analysed graphically in this way to identically replace it by a series of sinusoidal harmonic components. This is difficult to do accurately particularly for a complex wave shape which might be composed of a large number of harmonics all with different peak values and also different phases with respect to the fundamental wave.

If the wave shapes containing the second harmonic component with either of the wave shapes containing just the third harmonic component are compared, an important fact becomes evident. The wave repeats itself exactly every half cycle of the fundamental component when only the third harmonic is present. However, if the waveform only contains the second harmonic component it does not repeat itself in this way and it can be clearly seen that in the first half cycle $y(t)$ increases rapidly with time but in the second half cycle it rises slowly. The more wide ranging principle demonstrated here is the important fact that waveforms containing only odd harmonic components repeat themselves every half cycle whereas if any even harmonic components are present (with or without any odd harmonics) the waveform only repeats itself every full cycle.

Even harmonic components of flux density are rarely encountered. Their presence implies non-symmetrical magnetisation or magnetic poles of unequal strength in electrical machine cores [3]. In practice, normally only low frequency harmonics (third, fifth and seventh) occur but under complex magnetisation conditions, such as in rotating machine stator cores energised by PWM voltages, much higher harmonics, at least up to the 51st, may be encountered. These can cause adverse effects in electrical and magnetic systems.

The magnitude of each harmonic component in a waveform is normally expressed as a percentage of the amplitude of the fundamental component although sometimes it is stated as a percentage of the peak value of the overall waveform.

The peak value of a distorted flux density is an important parameter in magnetisation analysis, but the r.m.s. value is important for current or voltage analysis. Suppose a non-sinusoidal waveform can be expressed as

$$y = Y_1 \sin \omega t + Y_2 \sin(\omega t + \varphi_2) + Y_3 \sin(\omega t + \varphi_3) + \cdots + Y_n \sin(\omega t + \varphi_n) \tag{6.6}$$

where φ_n is the phase angle of the nth harmonic with respect to the fundamental component. The r.m.s. value $Y_{\text{r.m.s.}}$ of such a distorted waveform can be shown simply to be the square root of the sum of the squares of the peak values Y_n of the individual harmonics in the waveform [3]. Mathematically, this is expressed as

$$Y_{\text{r.m.s.}} = \sqrt{\frac{Y_1^2 + Y_2^2 + Y_3^2 + \ldots\ldots Y_n^2}{2}} \tag{6.7}$$

When a harmonic analysis of magnetic quantities such as B or H is carried out, a basic superposition technique is being used. This assumes that the effect of the distorted waveform on a parameter is the same as the sum of the effect of each harmonic occurring in isolation. Superposition is commonly used to study complex circuit problems in electrical engineering where the final solution is the result of several influences acting simultaneously. The main constraint is that there must be a linear relationship between the cause and effect. So, for superposition theory to be valid in a magnetic circuit in which H is the input and B is the output, they must be directly proportional to each other. It is well known that this rarely occurs in practical situations so, strictly speaking, superposition does not work. However, if used carefully, it is useful for finding approximate solutions in magnetic applications.

6.2 Distortion factor

The amount of distortion of a wave can be quantified in terms of the distortion factor (DF) or the total harmonic distortion (THD). In fact these terms are synonymous and the THD term is mostly used in magnetic waveform analysis.

There is no recognised standard definition of THD. It is defined in one of two ways. It can be defined as the ratio of the r.m.s. value of the harmonic components to the r.m.s. value of the total waveform as

$$\text{THD}_R = \frac{\sqrt{Y_2^2 + Y_3^2 + \ldots\ldots Y_n^2}}{\sqrt{Y_1^2 + Y_2^2 + Y_3^2 + \ldots\ldots Y_n^2}} \tag{6.8}$$

Alternatively, it can be defined as the ratio of the r.m.s. value of the harmonic components to the r.m.s. value of the fundamental component and written as[3]:

$$\text{THD}_F = \frac{\sqrt{Y_2^2 + Y_3^2 + \ldots\ldots Y_n^2}}{Y_1} \tag{6.9}$$

At low distortion levels (say less than 0.2) the difference between results using the two approaches is negligible. At levels of over approximately unity, the difference can be significant. In many publications the used approach is not quoted, so results can be misrepresented. For example, in the case of a pure square wave, $\text{THD}_F = 0.483$ and $\text{THD}_R = 0.435$ [4]. If no harmonics are present the DF is zero.

[3]In fact the peak values of each component are included in (6.8) and (6.9). Each is $\sqrt{2}$ times the corresponding r.m.s. value so if the equations are solved using peak or r.m.s. the same result is obtained.

6.3 Distorted voltages on power systems

Normally, magnetic cores for power applications are energised by voltages which have close to perfect sinusoidal time varying waveforms. In growing numbers of applications electronically produced, non-sinusoidal voltage waveforms are used. Sinusoidal voltages happen to be relatively convenient to generate, and their mathematical analysis is straight forward but the main reason for their use is that the sine wave is the only wave shape whose differential curve is the same shape.

Take the case of the ideal single phase transformer shown in Figure 5.1(b). It can easily be deduced that, if the primary voltage varies sinusoidally with time, the secondary voltage will also be sinusoidal (from (5.3) and (5.4)). Suppose that the primary voltage V_p[4] is non-sinusoidal and, for simplicity, contains just one harmonic component so it can be written as

$$V_p = V_1 \sin \omega t + V_n \sin n\omega t (\mathrm{v}) \tag{6.10}$$

where V_1 and V_n are the peak values of the fundamental and nth harmonic components, respectively. The secondary voltage $V_s = dV_p/dt$ hence

$$V_s = \omega(V_1 \cos \omega t + nV_n \cos n\omega t)(\mathrm{v}) \tag{6.11}$$

Hence the nth harmonic component of the secondary voltage waveform is $n\omega$ times that of that in the primary voltage waveform, so the wave shape is different. When more harmonic components are present the effect is more complex but can be analysed in the same way.

The output voltage of a system transformer energised by a distorted flux density due to a voltage waveform, such as that in (6.11), is more distorted so if this, in turn, energises a second transformer in a distribution system the core loss of the second transformer would be increased due to the increased harmonic content. More importantly, the wave shape of its secondary voltage would be further distorted. Such a condition could lead to system malfunction.

Figure 6.3 shows an extreme situation where a square wave e.m.f. is applied to an ideal transformer. The flux waveform is triangular as shown in (b) since it is proportional to the time integral of the e.m.f. In a real machine, the waveform is more like that shown in (c) due to the presence of winding resistance and magnetising current. The output voltage, proportional to the differential of the flux density, is then very distorted, as in (d), which is nothing like the input voltage. From this result, it is easy to deduce that the presence of just a small harmonic component in the primary voltage causes a further distortion of the secondary voltage waveform.

[4]It is possible that confusion over conventionally used symbols for peak values of some quantities in electrical transformer technology and basic magnetic terminology is possible. For example, B_p invariably means the peak value of a cyclic flux density waveform. However, V_p might mean the peak value of a cyclic voltage waveform or the voltage across the primary winding of a transformer without specifying whether it is r.m.s., average, peak or instantaneous. Fortunately, if the context is carefully noted, no confusion should arise in well written literature.

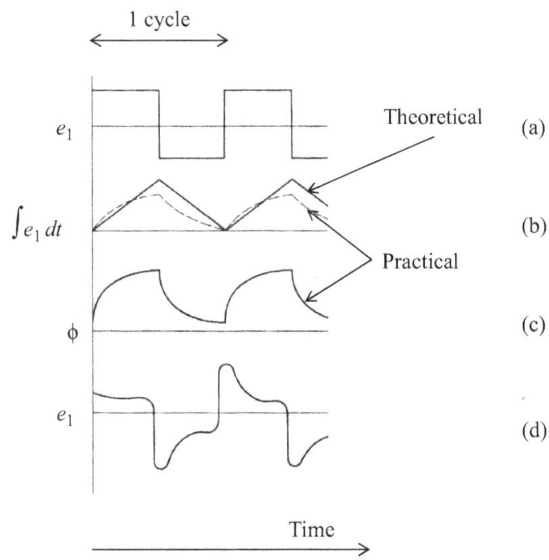

Figure 6.3 Change in the shape of an e.m.f. waveform containing high order harmonics as it passes from primary to secondary winding of a transformer: (a) square wave primary e.m.f, (b) ideal flux waveform, (c) possible practical flux waveform and (d) secondary voltage waveform

6.4 Distorted *B* or *H* waveforms due to non-linear magnetisation curves

Under a.c. magnetisation there are many reasons why it is preferable for the flux density and magnetic field in an electrical steel to vary sinusoidally with time as will be seen particularly in Chapter 7 of Volume 2 of this book. Apart from at very low flux density, this is not possible because of its non-linear *B–H* characteristic.

A common way in which harmonic components of *B* or *H* are generated in a magnetic material subject to cyclic magnetisation is illustrated in Figure 6.4. For convenience only one cycle of magnetisation is shown and hysteresis is ignored. The *B–H* curve shown in (a) is a function of the material, its shape determined as the domains in the material redistribute according to the influence of the applied field. Suppose the applied field *H* varies sinusoidally in time as in (b), at any time t_1, when the magnitude of the instantaneous field is h_1, the corresponding value of flux density is b_1. It can clearly be seen that when the *H* waveform is sinusoidal the *B* waveform is non-sinusoidal, i.e. it is due to the non-linear *B–H* characteristic. Conversely if *B* is sinusoidal then *H* will vary non-sinusoidally with time. In general, both *B* and *H* waveforms might contain harmonic components in which case there is an infinite range of possible wave shapes.

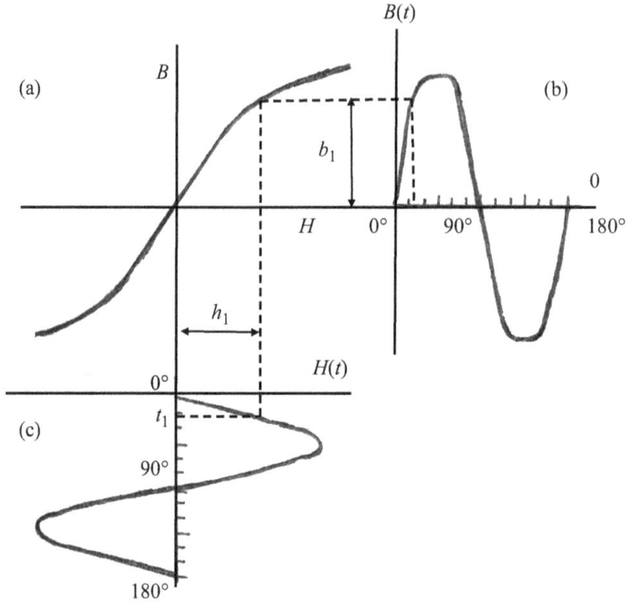

Figure 6.4 Creation of distorted flux density waveform in a material with a non-linear B–H characteristic magnetised by a magnetic field varying sinusoidally with time: (a) basic B–H characteristic of the material, (b) magnetising waveform and (c) resultant flux density waveform

6.5 Effect of the electric circuit on waveform distortion

Consider the toroidal iron core illustrated in Figure 5.1(b). Suppose the primary coil carries an instantaneous current i_p which produces the field to magnetise the core with the secondary winding on open-circuit ($i = 0$). From Ampère's law,[5] $Hl = N_p I_p$ where N_p is the number of magnetising turns of the primary winding, l is the mean circumference of the toroid and H is the mean field. This obviously shows that the field in the toroid is directly proportional to the primary current which, in turn, means that their waveforms must be identical.

If the primary coil has an electrical resistance R_p then an alternating primary voltage whose instantaneous value is v_p can be regarded as being the sum of two components, one of which is the voltage drop across R_p and the other is a

[5]See Chapter 10 for the background to Ampère's law.

component opposing the voltage induced in the coil by the change of flux linkage. So we can write

$$v_p = R_p i_p - \left(-\frac{d\Phi}{dt}\right) = R_p i_p + \frac{d\Phi}{dt}\,(\text{v}) \tag{6.12}$$

We can replace i_p, the instantaneous primary current, by Hl/N_p. Also, we can replace $d\phi/dt$ by $N_p AdB/dt$ where A is the cross sectional area of the toroid to obtain

$$v_p = \frac{R_p lH}{N_p} + \frac{N_p AdB}{dt}\,(\text{v}) \tag{6.13}$$

This now expresses v_p in terms of its field and flux density dependence. Because of the non-linear relationship between H and B, neither H nor dB/dt can both vary in time in the same way as v_p. In particular, if v_p varies sinusoidally with time neither H nor dB/dt can be sinusoidal.

6.6 General relationship between harmonics in *B* and *H* waveforms

Consider again a material in the form of a ring or toroid magnetised in the way shown in Figure 5.1(b) and suppose its B–H characteristic is non-linear. If the input voltage is sinusoidal, i.e. $v_p = V_p \sin \omega t$, harmonics must appear in the H and dB/dt waveforms as shown in Section 6.5. The following general expression can be written as

$$H = H_0 \sum_{n=1} h_n \sin(n\omega t + \phi_n)\ \left(\text{Am}^{-1}\right) \tag{6.14}$$

$$\frac{dB}{dt} = \omega B_0 \sum_{n=1} b_n \sin(n\omega t + \varphi_n)\ \left(\text{Ts}^{-1}\right) \tag{6.15}$$

H_0 and B_0 are the peak values of the distorted H and B waveforms and the angles ϕ_n and φ_n are the phase angles between the respective n^{th} harmonics and an arbitrary common reference. h_n and b_n are the percentages of each harmonic with respect to H_0 and B_0. The products $H_0 h_n$ and $B_0 b_n$ give the magnitude of the nth harmonic component of H and B, respectively.

If these expressions are put into (6.13), it can be solved to give some interesting relationships between harmonics dependent on magnetic circuit design [5]. If n is odd then solving the equation gives:

$$\Phi_n - \varphi_n = 2\pi\text{n} \tag{6.16}$$

$$\frac{h_n}{b_n} = -Q \tag{6.17}$$

where Q is the 'Q factor'[6] which is simply the ratio of the reactance of the coil at a specific frequency to its resistance. For the geometry in Figure 5.1(b) it is given by[7]:

$$Q = \frac{\omega L}{R_p} = \frac{N_p{}^2 A\omega\mu_0\mu_r}{R_p l} \tag{6.18}$$

where L is the self-inductance of the primary coil.

Consider the practical meaning of (6.16) and (6.17). First, from (6.16) we can deduce that there is always a 180° phase difference between harmonics in current and voltage (or H and dB/dt). Second, from (6.17) we can see that the percentage of any harmonic in the current or H waveform is Q times higher than the percentage of the corresponding harmonic in the dB/dt waveform. The practical consequence of the second observation is that a value of Q between five and ten should be aimed for in measurements at power frequencies and inductions in order to have reasonably sinusoidal flux without the need for additional control [5].

A special case in which the shape of the magnetic circuit has a large influence on the magnitude of the harmonics is the three-phase, three-limb transformer core. In this case there is no complete steel path for triplen[8] harmonics to flow, so this suppresses the presence of third harmonic flux in particular. This is dealt with in more detail in Section 5.14 of Volume 2 of this book.

6.7 Calculation of flux density under distorted magnetisation conditions

Expressions for the peak flux density B_p under sinusoidal flux density are given in Section 5.3, but, in practice, both in measurements[9] and in applications of soft magnetic materials, distorted flux waveforms are common so it is necessary to see how the peak flux density can be quantified in these cases.

[6]The Q (Quality) factor of a coil is the ratio of its reactance (ωL) to its resistance (R). It is effectively a measure of the efficiency of a coil. The term is commonly used in high frequency electronic circuit applications where it is desirable for a coil to have low resistance (to restrict its i^2R loss) and high reactance to enable it to store a large amount of energy. It is not commonly used in low frequency transformer analysis but it is a useful term to apply in analysing the effect of the magnetising circuit parameters on the harmonics in B and H in electrical steels.

[7]The expression for the inductance of the coil given by $L = N^2\mu A/l$ has been used to obtain the final term in this equation.

[8]Triplen harmonics are odd multiples of three of the fundamental frequency. For example, if the fundamental frequency is 50 Hz the triplen harmonics are 150 Hz, 450 Hz, 750 Hz, etc. Their special feature is that they are in phase when present in three-phase transformer windings.

[9]Although standard measurements on electrical steels, and built up cores, are carried out under overall sinusoidal flux density, the flux in individual laminations in a stack being measured or at localised regions of magnetic cores will be distorted. This applies even to NO SiFe tested in the Epstein frame where, fortunately, the least overall distortion in individual strips coincides with the IEC recommended stacking method [6].

Suppose that the non-sinusoidal flux density in a material is given by

$$B(t) = \sum_{h=a} a_h \sin(h\omega t + \varphi_h)(\mathrm{T}) \tag{6.19}$$

where a_h is the peak value of the hth harmonic component of B. This reaches its peak value B_p at some time τ during the magnetising cycle so

$$B(\tau) = \sum_{h=1} a_h \sin(h\omega\tau + \varphi_h) = B_p(\mathrm{T}) \tag{6.20}$$

At a time π/ω, half a period later the flux density becomes

$$B(\tau + \pi/\omega) = \sum_{h=a} a_h \sin(h\pi + h\omega\tau + \varphi_h)(\mathrm{T}) \tag{6.21}$$

If the waveform only contains odd harmonics then

$$B(\tau + \pi/\omega) = -B(\tau) \tag{6.22}$$

Now

$$\frac{dB}{dt} = \sum_{h=a} a_h\, \omega \cos(h\omega t + \varphi_h)(\mathrm{Ts}^{-1}) \tag{6.23}$$

This is zero at times $(\tau + n\omega/\pi)$ where n is an integer. Hence the average value of *db/dt* over one half period of the fundamental component is

$$\begin{aligned}\left(\frac{dB}{dt}\right)_{av} &= \frac{\omega}{\pi}\int_{\tau}^{\tau+\frac{\pi}{\omega}} \sum_{h=1} a_h\, h\omega\cos(h\omega t + \varphi_h)\\ &= \frac{\omega}{\pi}\sum_{h=1} a_h[[\sin(h\omega\tau + \varphi_h) - \sin(h\pi + h\omega\tau + \varphi_h)]](\mathrm{Ts}^{-1})\end{aligned} \tag{6.24}$$

If n is odd, the expression reduces to

$$\left(\frac{dB}{dt}\right)_{av} = \frac{2\omega}{\pi}\sum_{h=1} a_h \sin(h\omega\tau + \varphi_h)(\mathrm{Ts}^{-1}) \tag{6.25}$$

However, the summation on the right hand side of (6.24) is equal to B_p so we can write

$$B_p = \frac{\pi}{2\omega}\left(\frac{dB}{dt}\right)_{av}(\mathrm{T}) \tag{6.26}$$

Hence B_p can be obtained from the average value of the e.m.f. induced in a search coil provided the B waveform only contains odd harmonics. It can easily be shown that (6.26) reduces to that given in Section 5.3 when the waveform is sinusoidal.

A minor loop is formed in a *B–H* characteristic when harmonics in the flux density waveform cause more than one positive and negative peaks in each cycle resulting in the flux density not continually increasing or decreasing between peak values. Apart from forming minor loops, which add to the losses, this causes expressions such as (6.26) to become invalid.

Minor loop formation depends on the magnitude and phase of the harmonics. The form of a minor loop cannot be determined by simple analysis [7]. If just the third harmonic is present, minor loops form when the magnitude of the harmonic is greater than 10% to 30% of the fundamental value (dependent on the relative phase of the harmonic). In the case of the fifth harmonic this value is around 4% to 14%.

If a second harmonic component is present it can be shown that the fractional error in calculating B_p from (6.26) is of the order of $\frac{a_2}{a_1}\sin(\varphi_2 - \varphi_1)$ where the subscripts 1 and 2 refer to the fundamental and second harmonic [8]. So the presence of a 5% second harmonic would cause an error of up to 5 % in the calculated flux density.

Of course, it is a relatively trivial procedure to carry out an exact analysis of an experimentally obtained e.m.f. waveform using a fast Fourier transform (FFT) and software to calculate the harmonic components of any cyclic flux density waveform. This can quickly and simply be applied for most measurement needs.[10] However, it is still useful to understand some of the physical aspects of waveform distortion covered in this chapter.

References

[1] Lowry H.V., and Hayden H.A. *Advanced Mathematics for Technical Students: Part Two.* Longmans, Green and Co, London, 1962

[2] Croft A., and Davison R. *Mathematics for Engineers: A Modern Interactive Approach.* Prentice Hall, London, 2004

[3] Goodlet B.L. *Basic Electrotechnics.* Edward Arnold, London, 1962, p. 190

[4] http://en.wikipedia.org/wiki/Total_harmonic_distortion (Accessed June 2018)

[5] Astbury N.F. *Industrial Magnetic Testing.* The Institute of Physics, London, 1952, pp. 28–29

[6] Banks P.J., and Rawlinson E. 'Measurement of the total loss of cold-rolled nonoriented 3% silicon iron using an Epstein square'. *Proc. IEE* 1965, vol. 112(1), pp. 179–182

[7] Nakata T., Ishihara Y., and Nakano M. 'Iron losses of silicon steel core produced by distorted flux'. *Elect. Eng. Jap.* 1970, vol. 90(1), pp. 10–20

[8] Astbury N.F. *Industrial Magnetic Testing.* The Institute of Physics, London, 1952, p. 37

[10] This is covered in Chapter 12.

Chapter 7

Losses and eddy currents in soft magnetic materials

An important property of a soft magnetic material is its loss when subjected to a time varying magnetic field. This is effectively the conversion of useful magnetic energy into some other form, such as thermal or acoustic energy. Energy conversion of this form is often loosely referred to as energy loss or power loss because, normally, in practical applications, it is a source of inefficiency. Such energy transfer in magnetic material is often referred to as iron loss and when the same steel forms part of a magnetic core, the same energy conversion is referred to as core loss. This common terminology will be used throughout this book.

In this chapter general aspects of losses are introduced. This is built on in Chapters 1 and 2 of Volume 2 of this book where more detail of losses in electrical steels is presented. In such cases, physical properties, such as strip thickness, and material properties, such as grain size and texture, become controlling factors.

7.1 Physical and engineering approaches to magnetic losses

Methods of analysing and interpreting losses in soft magnetic materials can broadly be separated into physical and engineering concepts illustrated in Figure 7.1 [1]. The physical approach uses the Poynting theorem to analyse the power flow P into and out of a region through a surface area a at a time t. This involves integrating the Poynting vector over one period T of an alternating magnetisation process [2] using

$$P = \frac{1}{T}\int_0^T \oiint_a (\boldsymbol{E} \times \boldsymbol{H})\,da\,dt\ \left(\mathrm{Wm}^{-3}\right) \tag{7.1}$$

where $\boldsymbol{E}$ and $\boldsymbol{H}$ are the vector electric and magnetic fields, respectively. This equation can be transformed [3] into the widely used expression for magnetic loss in a volume v as

$$P = \frac{v}{T}\int_o^T H\frac{d\boldsymbol{B}}{dt}dt\ (\mathrm{W}) \tag{7.2}$$

This physical approach is illustrated in Figure 7.1(a) and is explained more in Section 7.3.

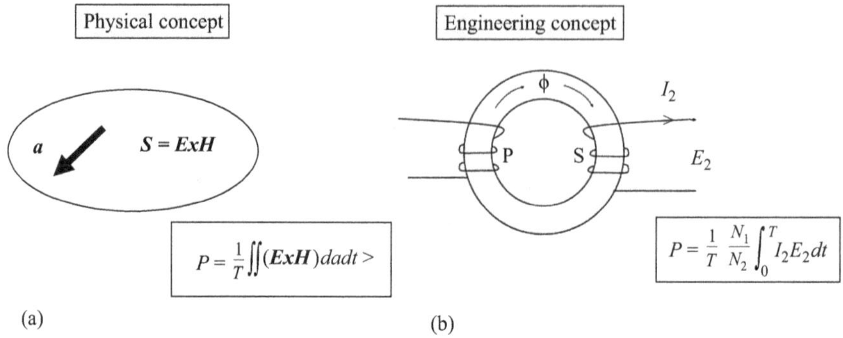

Figure 7.1 Equivalent approaches to magnetic losses in a material: (a) physical concept and (b) engineering concept

The engineering approach is built on the energy fed into a magnetic circuit shown in Figure 7.1(b). This is the same as the basic single-phase transformer and the Epstein Square covered in Section 12.2.1 where it is shown that the magnetic loss can be written as

$$P = \frac{v}{T}\int_0^T I_p V_s \, dt(\mathrm{W}) \tag{7.3}$$

In an ideal magnetic circuit I_p is proportional to the magnetic field strength (Ampère's law gives $Hl = I_p N_p$) and the secondary winding voltage is proportional to *dB/dt* (according to Faraday's law). It can be seen from the equations that the physical and engineering approaches are equivalent.[1] It is more convenient to use one or other concept for different applications.

7.2 Energy dissipation derived from the area enclosed by a *B–H* loop

The concept of energy being stored in a material during the magnetisation process was introduced in Chapter 3 and the energy components related to single value[2] *B–H* or *M–H* curves are depicted in Figure 3.13. Magnetostatic energy is stored in a material when magnetised and then later returned to the source in such a loss-free system. The stored energy involved is shown as being proportional to the shaded area in Figure 3.13(c).

[1]It will be noted that (7.2) and (7.3) give the loss per unit volume. In engineering applications losses are usually quoted in loss per unit mass which is simply obtained by replacing v by m/D, where D is the density.

[2]Single value here simply means that the flux density is the same at a given field level whether the field is increasing or decreasing, i.e. no hysteresis is present in the *B–H* loop.

Let us return to the *M–H* loop of a practical material which, from experience, we know that magnetic energy is converted into thermal energy during the magnetisation process simply from the observation that the material may become hot. The size of this loss can be deduced from the *M–H* curve shown in Figure 7.2.

Consider the portion of the characteristic when the field is increased from zero to its maximum value along path ABC. The energy provided from the field source and stored in the material, as discussed in Section 3.11, is represented by the area ABCDA. Although we say that the energy is stored in this way, it is not strictly true since it will be shown later in this section that some of this *stored* energy is not recoverable. As the field is reduced to zero along the path CD energy proportional to the area CDD' is returned to the source. An identical process is followed when the field increases in the opposite direction for the magnetisation to follow the path DEF, before it is reduced again to complete the cycle and arrive back at point A.

Following this process it can immediately be deduced that the energy not returned to the source on completion of the magnetisation cycle is proportional to the area enclosed by the *M–H* curve. This area, the product *MH* has units of $\mathrm{TAm^{-1}}$ which is equivalent to $\mathrm{Jm^{-3}}$.

The loop can be considered as being made up of an infinite number of strips each of area $H\delta M$ so the total loop area is given by

$$\text{Loop area} = \oint H dM \quad (\mathrm{Jm^{-3}}) \tag{7.4}$$

This area can also be expressed as

$$\text{Loop area} = 2\int_A^D H dM = 2\int_A^C H dM - 2\int_C^D H dM \ (\mathrm{Jm^{-3}}) \tag{7.5}$$

As mentioned earlier, if the *M–H* characteristic is linear, the loop area is zero, so no loss occurs. There is sometimes confusion over whether the loss of a given material is

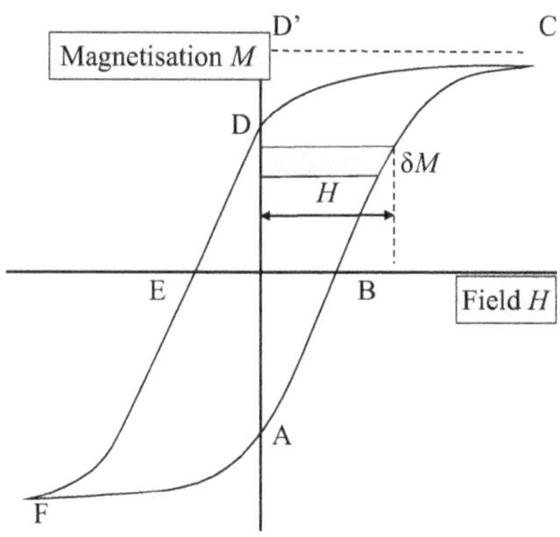

Figure 7.2 Typical M–H loop of a soft magnetic material showing areas proportional to stored and dissipated energy

proportional to the area enclosed by the *M–H* loop or the *B–H* loop. This is easily clarified as follows. Since $B = \mu_0 H + M$ then the change in flux density caused by a small change in field is given by $\delta B = \mu_0 \delta H + \delta M$, so (7.4) can be written as

$$\text{Loop area} = \oint H dM = \oint H dB - \oint H dH \left(\text{Jm}^{-3}\right) \tag{7.6}$$

But over a complete cycle of magnetisation $\oint HdH$ is zero hence

$$\oint H dM = \oint H dB \; \left(\text{Jm}^{-3}\right) \tag{7.7}$$

Equation (7.7) is only strictly true for the loss measured over a complete cycle, but since, apart from at very high fields, $B \approx \mu_0 H$ in electrical steels either expression can be used. In the case of more complex magnetisation loops containing minor loops, the expressions have to be modified to take account of the additional minor loop areas.

Equation (7.7) is consistent with both the physical and engineering concepts of losses discussed in Section 7.1.

So far in this chapter it has been shown that the loss that occurs when a material is taken through one cycle of magnetisation is proportional to the area enclosed by the *M–H* loop. It was pointed out in Section 2.1.6 that the *M–H* or *B–H* loop can be referred to as a hysteresis loop if the magnetisation is taken through the cycle very slowly. This leads to the nature of the loss occurring during magnetisation. If the loop is traversed very slowly, typically less than a few cycles per second, we can say that the area enclosed is indeed proportional to the hysteresis which is assumed to be the only loss mechanism occurring at such a low rate of change of magnetisation. This is labelled low frequency loop in Figure 7.3(a).

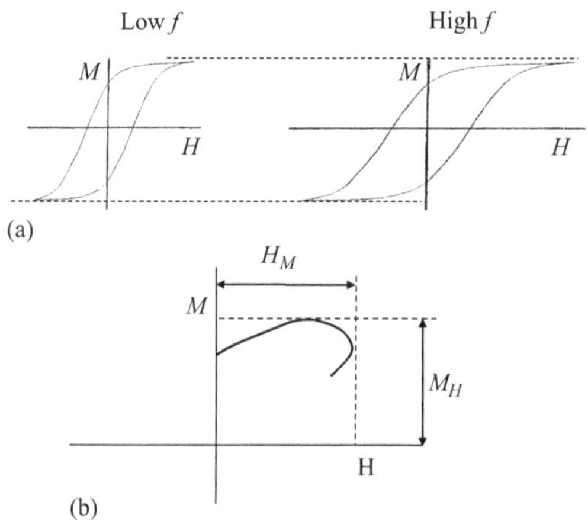

Figure 7.3 (a) Typical shapes of low and high frequency M–H curves in a soft magnetic material and (b) rounding of tip of a M–H loop caused by M and H not reaching their peak values at the same time

If the loop is traversed rapidly at constant frequency the broader high frequency loop shown in Figure 7.3(b) applies. The increased area is proportional to eddy current losses which become more prominent as the magnetising frequency increases. It should be mentioned that, of course, separating portions of the a.c. loop in this way is a convenient way of measuring the magnitude of loss components but it does not indicate the physical regions of the loop where hysteresis and eddy current losses dominate separately.

A point often not realised, when investigating a.c. *M–H* or *B–H* loops, is that B_p and H_p need not occur at the same time. Suppose a magnetisation loop is produced by varying H at a very slow rate. When the field reaches its highest value the corresponding flux density will also be reached.[3] This is particularly apparent at low flux density. Under a.c. magnetisation B or M lag behind H in time since domains cannot reconfigure instantaneously in response to an instantaneous change in field. This results in a rounding of the loop near its apexes as shown in Figure 7.3(b) where H_M and B_H are the values of field and flux density at the instants in time when the flux density and field are maximum.

7.3 Derivation of the dependence of loss on *B* and *H* using the Poynting vector theorem

The Poynting vector, named after its originator, John Henry Poynting, is a mathematical representation of energy flow in electric and magnetic fields based on Maxwell's equations. Poynting worked under James Maxwell at the Cavendish Laboratory in Cambridge in his early life, later moving to Birmingham University in 1880 where he published his ground breaking energy theory in 1884. Like so many pioneering scientists of the time working on electricity and magnetism, he made significant breakthroughs in very different disciplines; in fact, he is more famed as being the first person to determine the weight of the earth.

The most prominent application today of the Poynting theorem is in vector analysis of high frequency electromagnetic wave propagation. Although it is rarely directly applied to soft magnetic materials operating at power frequencies, it is instructive to look at its implications by reference to its underpinning mathematical analysis but without any rigorous use of the complex vector analysis necessary to fully follow the detail of its development. The brief overview is given here to demonstrate the dependence of magnetic losses in a material on the magnetic field and flux density in a more rigorous manner without the need to make deductions from an experimentally observed *B–H* loop, as in Section 7.2.

Of course, the two approaches lead to the conclusion that the area of the *B–H* loop under a.c. magnetisation is proportional to the total loss but the Poynting vector approach gives a deeper insight. This is useful in studies of losses under

[3] The labelling of B_p and H_p on Figure 2.7 presume that they *do occur* at the same time in this case, of course.

rotational or other types of complex magnetisation which are more difficult to visualise as a loop area.

The Poynting vector is basically a theoretical energy balance equation. It can be expressed in words as follows [4]:

$$\begin{pmatrix}\text{Power}\\ \text{entering}\\ \text{a volume}\end{pmatrix} = \begin{pmatrix}\text{Rate of increase}\\ \text{of stored}\\ \text{electrical energy}\end{pmatrix} + \begin{pmatrix}\text{Rate of increase}\\ \text{of stored}\\ \text{magnetic energy}\end{pmatrix} + \begin{pmatrix}\text{Power}\\ \text{converted}\\ \text{to heat}\end{pmatrix} \tag{7.8}$$

The total power input comprises three components, the first is the rate of increase of electric energy inside the volume in question, the second term represents the rate of energy stored in the magnetic field and the third is the energy converted into heat.[4] The relationship can be applied to the total flow δp into, or out of, a small volume δv to be stored or dissipated. This can be written as [5]:

$$\delta p = \left\{ \boldsymbol{E}\left(J + \frac{\partial D}{\partial t}\right) + \boldsymbol{H}\frac{\partial B}{\partial t} \right\} \delta v \ (\mathrm{Js^{-1}m^{-3}}) \tag{7.9}$$

where J is the local value of current density and D is the electric flux density. In a metal $\partial D/\partial t$ is negligible so it is omitted from here on. From Poynting's theorem, the total stored energy density u in the electromagnetic field in the volume V can be written as [6]:

$$u = \int_v \frac{1}{2}(\varepsilon \boldsymbol{E}^2 + \mu \boldsymbol{H}^2)dv \ (\mathrm{Jm^{-3}}) \tag{7.10}$$

The instantaneous power p related to this energy is

$$p = -\frac{\partial u}{\partial t} = -\int_v \left(\varepsilon \boldsymbol{E}\frac{\partial \boldsymbol{E}}{\partial t} + \mu \boldsymbol{H}\frac{\partial \boldsymbol{H}}{\partial t} \right) dv \ (\mathrm{Js^{-1}m^{-3}}) \tag{7.11}$$

This can be transformed to express the power as

$$p = \int_v \sigma \boldsymbol{E}^2 dv + \oint_A (\boldsymbol{ExH})dA \ (\mathrm{Js^{-1}m^{-3}}) \tag{7.12}$$

where σ is the conductivity of the material. The vector product $\boldsymbol{ExH}$ is referred to as the Poynting vector. Its integral over the closed surface gives the total outward flow of energy per unit time [7]. The first term on the right hand side of (7.12) is the electric power dissipated and the second term is electromagnetic energy flowing into the region across area, A.[5]

[4]Strictly speaking, conversion into mechanical and acoustic energy can take place. This is normally small compared with the thermal energy conversion so it is not included here.

[5]More comprehensive background reading on electromagnetic energy in magnetic materials can be found in many text books such as [4] and [5] devoted to electromagnetic theory.

The definition of energy flow in terms of the integral of the Poynting vector can be used to express (7.1) mathematically as

$$-\int_s (\boldsymbol{E}x\boldsymbol{H}).ds = \frac{\partial}{dt}\int_v \frac{1}{2}(\varepsilon \boldsymbol{E}^2 + \mu \boldsymbol{H}^2)dv + \int_v EJdv \tag{7.13}$$

Again, the first term on the right represents stored energy. The second term is the energy converted into heat, i.e. the total magnetic losses. The minus sign on the left of the equation infers that the power flow is out of the volume being considered.

Either (7.12) or (7.13) form the basis for the analysis of losses of a sheet of ferromagnetic material. In the general case of two-dimensional magnetisation of a sheet, the instantaneous power p streaming into the sheet through its two surfaces of area $2A$ is [7]:

$$2(\boldsymbol{E}x\boldsymbol{H}).A = \begin{bmatrix} E_x & H_x & 0 \\ E_y & H_y & 0 \\ 0 & 0 & 2A \end{bmatrix} = 2A(E_xH_y - E_yH_x) \tag{7.14}$$

According to Faraday's law, the electric field $\boldsymbol{E}$ is related to $-dB/dt$[6] from which it can be shown that the total energy loss p_t of the material is given by [8]:

$$p_t = \frac{1}{DT}\int_0^T \left(H_x\frac{dB_x}{dt} + H_y\frac{dB_y}{dt}\right)dt \; (\mathrm{Jkg}^{-1}) \tag{7.15}$$

It should be noted that this is only truly valid if $\boldsymbol{B}$ and $\boldsymbol{H}$ are always co-linear [9]. This is not the case under commonly occurring rotational magnetisation of electrical steel discussed in Chapter 8. Furthermore, under a.c. magnetisation both B and H can vary widely through the thickness of a strip of material such as electrical steel, so it is not clear in the derivations given so far in this section what specific values of instantaneous H and dB/dt should be used in (7.15).

It is shown in Section 7.9 that the losses under a.c. magnetisation depend on the average value of dB/dt throughout the cross sectional area of a sheet and on the tangential component of H at the sheet surface.

The Poynting energy flow theorem is not directly applied in this book. One reason is that, although the Poynting vector represents the power flow, it needs to be interpreted with care in practical materials [10].

7.4 Hysteresis loss

The magnetic loss which occurs when magnetisation changes very slowly, such as in a low frequency cyclic state, is referred to as hysteresis loss. This is proportional to the area enclosed by the d.c., or static, magnetisation loop. It is caused by the generation of micro-eddy currents associated with domain wall motion and pinning and produces i^2R heating in a material. Note that all magnetic losses originate from

[6]Faraday's law can be written in differential form as $Curl.\boldsymbol{E} = -dB/dt$ and used to derive (6.15).

the presence of eddy currents; there is no fundamental difference between eddy current and hysteresis loss as will become clearer in Section 7.8.

Hysteresis loss occurs in a randomly localised manner where domain walls are pinned at impurities, grain boundaries and other defects. Whilst it is dependent on material structure, it cannot be directly calculated so it has to be deduced from measurements. Steinmetz found that, for a wide range of flux densities in iron samples of varying purities, the hysteresis loss per cycle satisfies a simple relationship given by [11]:

$$W_h = k_h B_p^n \ (\mathrm{Jm^{-3}}) \tag{7.16}$$

where k_h is the Steinmetz coefficient and n is a constant which lies between around 1.5 and 2.5. B_p is the peak value of flux density which occurs at the extremes of the B–H loop. Although referred to as a constant, k_h varies widely in a given material. For instance, even when subjected to d.c. magnetisation, the value of n changes significantly near the knee point of the magnetisation curve [12]. Generally speaking, materials with high magnetocrystalline anisotropy have high hysteresis [13].

If a material is magnetised at f cycles per second, the hysteresis loss per second is simply $W_h f$. This is of course the power loss P_h which is usually written as

$$P_h = k_h B_p^n f \ \left(\mathrm{Wm^{-3}}\right) \tag{7.17}$$

This is the hysteresis loss which occurs during magnetisation at any value of f. However, there is no reason why the hysteresis loss occurring during each cycle of magnetisation, at say 50 Hz, should be the same as that occurring in one cycle when the magnetisation is changing very slowly, e.g. under d.c. magnetisation. Magnetisation conditions vary with magnetising frequency so the hysteresis loss per cycle is not constant, although it is assumed to be so in many loss models. This is discussed further in Chapter 4 in Volume 2 of this book.

7.5 Eddy current generation in a rod of conducting material

Macro-eddy currents are produced around closed paths of least electrical resistance linked with the flux whenever it changes inside a conducting material. The eddy currents produce heat (i^2R loss) and the magnetic field, which they themselves produce, opposes the flux change.

The fundamental creation of eddy currents in a conducting material is shown in Figure 7.4(a). A single turn carries a current i instantaneously flowing in the direction shown. This sets up a magnetic field H along the axis of the single turn as shown previously for a simple current loop in Figure 2.1. The rod inside the coil is magnetised to a flux density B. Suppose the current is increasing, then the flux density will also be increasing. By Faraday's law, it will set up e.m.f.s along concentric paths in the rod which, because it is a conductor, will carry eddy currents as shown. Note that the eddy currents are flowing in the opposite direction to the driving current so they set up a so-called eddy current field h_e which directly

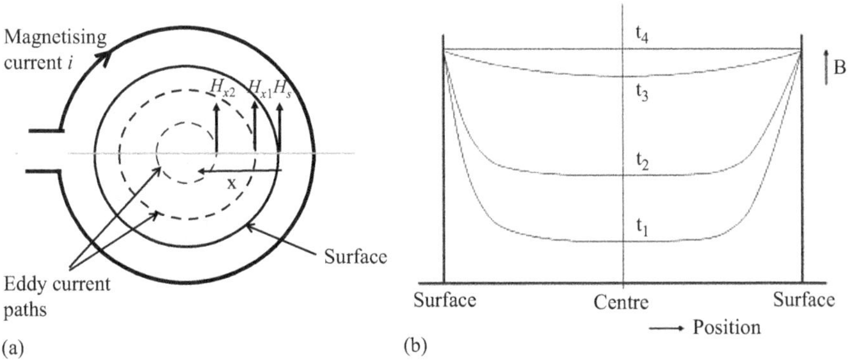

Figure 7.4 (a) Eddy current generation inside a bar caused by an enwrapping current carrying coil and (b) flux density across a diameter of the bar at successive times t_1 to t_4 after the field is applied

opposes the main field. Hence the material is more difficult to magnetise when the eddy currents are present because the effective field is reduced.

Now suppose the resultant field at a distance x from the surface of the rod is H_x as shown in the inset on Figure 7.4(a) where H_s is the tangential component of the field at the surface. If Ampère's law is applied to the path 1-2-3-4-1 and j is the density of the encircled eddy currents then

$$H_x = H - \int_0^x j dx \left(\mathrm{Am^{-1}}\right) \tag{7.18}$$

If the magnetising field is suddenly applied, initially the field on the axis of the rod must be zero so the total eddy current density per unit length must be equal to the ampere-turns per unit length of the coil. As this happens, the eddy currents in the centre of the rod gradually die away due to the dissipative resistance of the material and the flux density can build up as shown in Figure 7.4(b).

It is clear that the magnetisation takes time to penetrate into the rod. If an alternating magnetic field is applied, the flux has to reach its uniform state after one quarter of a magnetising cycle so at high frequency the flux in the centre of the rod never has time to build up to its surface value. This incomplete magnetisation is referred to as the *magnetic skin effect.*

7.6 Eddy currents in a thin sheet

The concept of incomplete flux penetration into a material was introduced in the previous section. In this section, an approximate analysis of eddy currents and their associated losses in a thin sheet, such as a lamination of electrical steel, is presented. The simplified analysis that follows makes some major assumptions which, it must be noted, are not always justified in applications of electrical steels.

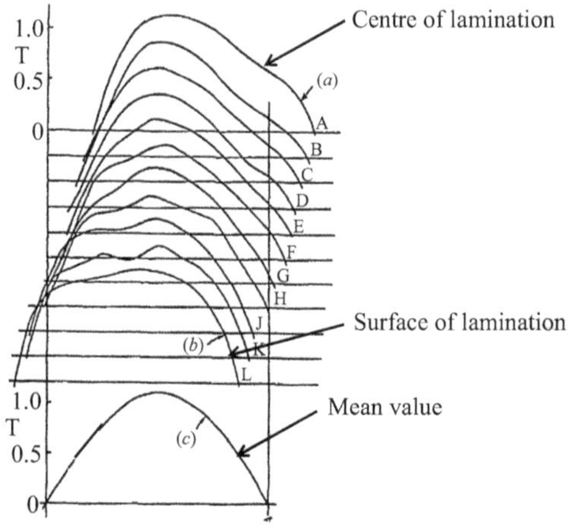

Figure 7.5 Flux waveforms at different depths in a lamination of NO silicon iron. Waveform (a) is at the centre, (b) is at a surface and (c) mean flux density (Figure 2 in [14] reproduced under free licence CC-BY-4.0 and updated)

The major difficulty in the analysis is the very non-linear relationship between *B* and *H*. If the flux density within a lamination can be imagined to vary during part of a magnetising cycle in a similar manner to that shown in Figure 7.4(b), then the variation of internal magnetisation would be a series of continuously varying *B–H* loops going deeper into the material. This means that even if the total flux in a lamination varies sinusoidally with time, at any depth beneath the surface it is distorted.

The type of internal flux waveform distortion to be expected can be demonstrated using an experimental resistance analogue which simulates the eddy current distribution through the thickness of a single lamination. Figure 7.5 shows the calculated flux density variation during a simulated half cycle of magnetisation at 50 Hz of NO steel whose average peak flux density is 1.12 T [14]. The average flux density through the thickness is almost sinusoidal but the flux density at each of the 11 equi-spaced layers is non-sinusoidal, it being flat topped at the surface and triangular at the centre. The average third and fifth harmonic distortion of the flux density waveform is 14% and 5% of the overall value, respectively. These are not sufficient to account for the excess loss [14]. The simulation assumes constant permeability and does not take account of the presence of domains. However, flux distortion in real laminations might not be very different to that calculated here [14].

A similar distortion appears in individual laminations in a stack whose overall flux density is sinusoidal.

Attempts at mathematical analysis of such a complex internal magnetic distribution are advancing but are difficult to verify because the internal magnetisation depends on other factors, not least the magnetic domain structure. The main assumption made in the following analysis is that the *B–H* loop is linear. This, and other

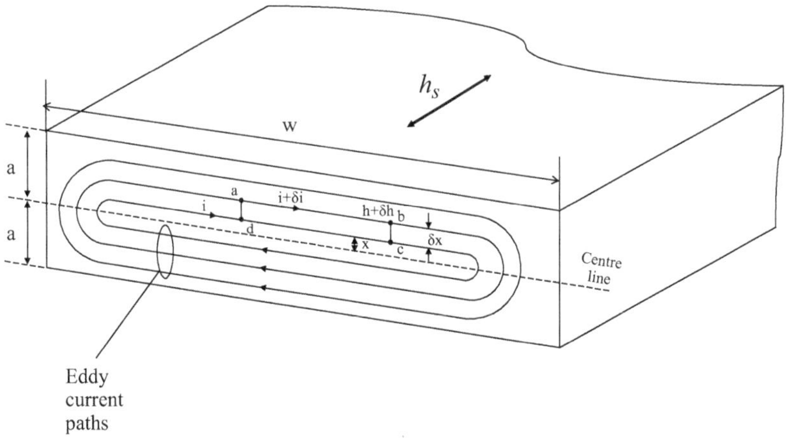

Figure 7.6 Schematic representation of eddy current paths viewed through a cross section of a thin strip of steel magnetised along its longitudinal direction

simplifications and approximations, made in the following analysis eliminates the complexity just mentioned to give a useful insight of the phenomenon, but numerical results based on this analysis should be treated with caution.

Suppose an alternating magnetic field $h_s = H_s \cos \omega t$, where h_s and H_s are the surface values of instantaneous and peak field respectively, is applied as shown in Figure 7.6 to a thin strip of steel of thickness $2a$.[7] The field sets eddy currents circulating around the paths shown. Consider a layer of material of thickness δx at a distance x from the centre of the strip. Let the instantaneous values of flux density, magnetic field and eddy current, which are functions of time, be b, h and i respectively. Assume that the permeability μ is constant. The lamination thickness is very small compared to its width and the effect of eddy currents in modifying the flux density distribution may be neglected. The flux density b_s at the surface of the lamination is equal to $B_s \cos \omega t$.

The flux in the element of width δx is $b\delta x$ so, from Ohm's law, we can write

$$\rho(i + \delta i) - \rho i = \frac{\partial(b\partial x)}{\partial t} = \partial x \frac{\partial b}{\partial t} \tag{7.19}$$

Hence

$$\rho \frac{\partial i}{\partial t} = \frac{\partial b}{\partial t} \tag{7.20}$$

[7]In the analysis of eddy currents and losses in electrical steels the thickness is often written, as here, as $2a$. This is simply to make the algebra neater than when using a term for the full thickness. In other treatment of electrical steels a term, normally t or d, is used to represent the thickness. This mixed usage is adopted in this book but it should not lead to any confusion.

By Ampère's law, the magnetomotive force (MMF) around the circuit abcd is given by

$$i\delta x = (h + \delta h) - h \tag{7.21}$$

Therefore

$$i = \frac{\partial h}{\partial x} = \frac{1}{\mu}\frac{\partial b}{\partial x}(\mathrm{A}) \tag{7.22}$$

From (7.19) and (7.20), we can show that

$$\frac{\partial^2 b}{\partial x^2} = \frac{\mu}{\rho}\frac{\partial b}{\partial t} \tag{7.23}$$

This equation can be solved to give the instantaneous flux density at any position inside the lamination at any instant during the magnetising cycle. Solving the equation gives the flux density at any point as [15]:

$$b = B_s\left(\frac{\cosh 2px + \cos 2px}{\cosh 2pa + \cos 2pa}\right)^{\frac{1}{2}} \cos(\omega t - \gamma)(\mathrm{T}) \tag{7.24}$$

where γ is a complex trigonometrical function of p, x and a and

$$p = \left(\frac{\pi\mu f}{\rho}\right)^{\frac{1}{2}} \tag{7.25}$$

From (7.24) it can be deduced that if B_x is the amplitude of the flux density at the distance x then

$$\frac{B_x}{B_s} = \left(\frac{\cosh 2px + \cos 2px}{\cosh 2pa + \cos 2pa}\right)^{\frac{1}{2}} \tag{7.26}$$

The phase of the flux density varies with depth into the lamination as indicated in (7.24). When this is taken into account, the magnitude of the flux density B_m averaged over the complete cross sectional area of the lamination is given by [16]:

$$\frac{B_m}{B_s} = \left(\frac{\cosh 2pa - \cos 2pa}{\cosh 2pa + \cos 2pa}\right)^{\frac{1}{2}} \Big/ \left(\sqrt{2}pa\right) \tag{7.27}$$

The curve shown in Figure 7.7(a) is derived from (7.26) to show the dependence of the variation of flux density with depth on the value of $2pa$. If its value is low (< 1.0) then the flux is uniform throughout the thickness and the magnetic properties of the material are used to the full extent. To achieve this, the product of the thickness ($2a$), the permeability (μ) and magnetising frequency (f) must be as

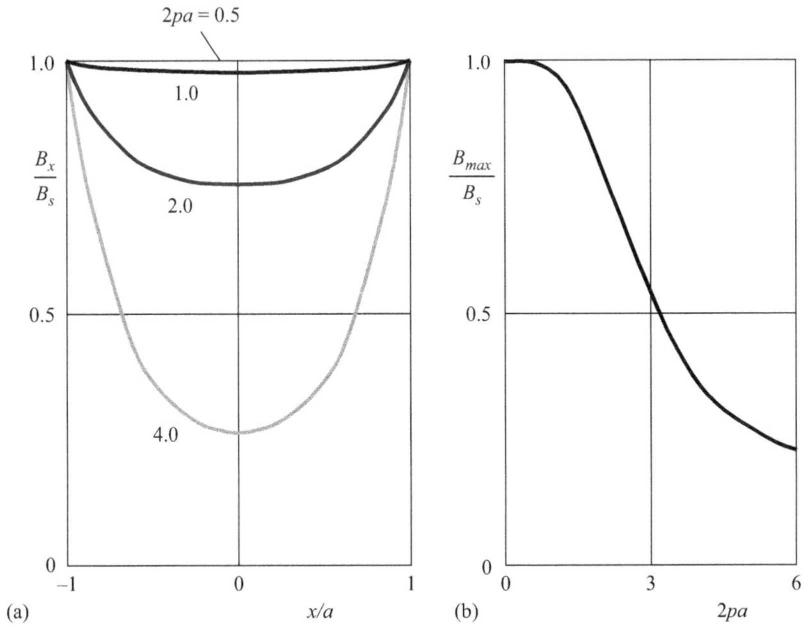

Figure 7.7 Flux penetration into a strip of thickness 2a (a) relative to depth in sheet and (b) relative to magnitude of 2pa (Figure 7.2 in [16] reproduced under free licence CC-BY-4.0)

low as possible and the resistivity (ρ) as high as possible. Less and less flux penetrates to the centre of the lamination as the value of pa rises. The distribution shown here should not be confused with that shown in Figure 7.7(b). They are, of course, caused by the same eddy current phenomenon but Figure 7.7(b) illustrates how the flux density builds up throughout a material subjected to a constant magnetic field over a short time after it is applied, whereas Figure 7.7(a) shows how the peak flux density under sinusoidal magnetisation varies with depth.

Figure 7.7(b) shows how B_m drops rapidly as $2pa$ increases. This shows how the flux carrying effectiveness of a lamination falls rapidly as we change thickness, permeability, frequency or resistivity.

7.6.1 Skin depth and equivalent depth of uniform magnetisation

The term skin effect was introduced in Section 7.5. We can now define it with respect to the results of the eddy current analysis just presented. The *equivalent depth of uniform magnetisation d* can be defined as the thickness of each of two layers, one on either side of a lamination of thickness 2a, which, if uniformly magnetised, would carry the same amount of flux as that carried throughout the whole sheet under a.c. magnetisation. The depth of equivalent magnetisation is sometimes called the *skin depth*. This can be a little confusing because the term skin depth is often defined differently as shown later in this section. The relationship between d and a is simply given as

$$dB_s = aB_m \tag{7.28}$$

Hence from (7.27)

$$d = \left(\frac{\cosh 2pa - \cos 2pa}{\cosh 2pa + \cos 2pa}\right)^{\frac{1}{2}} \Big/ \sqrt{2}p\ (m) \tag{7.29}$$

Figure 7.7(a) [17] shows how d decreases with increasing magnetising frequency for different values of relative permeability in a 0.33 mm thick sheet. This was derived for a GO steel with a resistivity of 48×10^{-8} Ωm. In this case the skin effect is negligible up to around 100 Hz provided the relative permeability is less than 10,000. It should be noted that as the permeability increases, as is often the goal in electrical steel development, the upper frequency at which the skin effect is negligible decreases.

The skin depth s for conductors in general is defined differently as the distance into the lamination at which point the magnetisation has dropped to $1/e$ times the surface value, or around 36% of B_s. For engineering applications d is usually a more useful parameter than s, one reason being that B_s is not a practically accessible quantity. It can be shown [18] that the skin depth is given approximately as

$$s = \sqrt{\frac{2\rho}{\omega\mu}}\ (\mathrm{m}) \tag{7.30}$$

For GO silicon–iron, at 1,000 Hz magnetisation and $\mu_r = 10{,}000$, this becomes

$$s = \sqrt{\frac{2 \times 48 \times 10^{-8}}{2\pi \times 1{,}000 \times 4\pi \times 10^{-7} \times 10{,}000}} = 0.11\ \mathrm{mm} \tag{7.31}$$

The sketch in Figure 7.8(b) shows the value of d obtained from Figure 7.8(a) relative to s for the same material. Obviously, care is needed in making this comparison since, at first sight, it appears that d is showing us that the material is being fully utilised whereas s infers partial magnetisation.

Finally, it is interesting to compare the skin depth of electrical steel with that of copper. The skin depth in copper is important in electrical machine windings where Litz wire[8] is sometimes used to reduce the effect. The resistivity of copper is around 25 times less than that of 3% silicon steel whereas its relative permeability is, of course, unity. This means that for a relative permeability of 10,000 the skin depth in the steel is approximately 5% of that in copper.

[8]Litz wire is a multistrand conductor designed to reduce the high frequency skin effect and proximity effect losses.

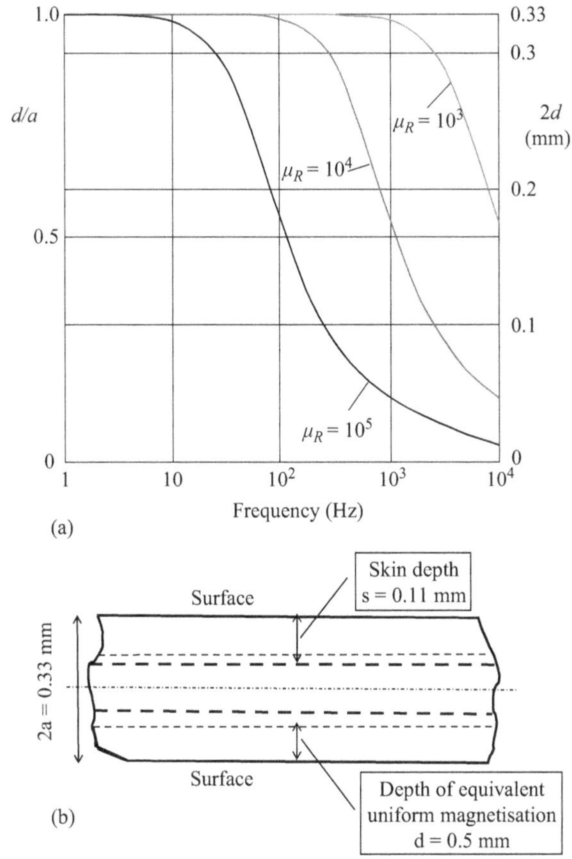

Figure 7.8 (a) Equivalent depth of uniform magnetisation in relation to frequency for a SiFe sheet for several values of relative permeability (Figure 10.8 of [17] reproduced under free licence CC-BY-4.0) and (b) comparison of skin depth s and depth of equivalent magnetisation d using typical parameters of SiFe at 1,000 Hz magnetisation frequency ($\mu_r = 10{,}000$, $\rho = 48 \times 10^{-8}\ \Omega m$)

7.7 Classical eddy current loss

The eddy currents produced by the mechanism depicted in Figure 7.6 cause localised ohmnic heating referred to simply as eddy current loss or, more exactly, as classical eddy current loss since its magnitude is calculated based on the eddy current distribution derived from Maxwell's equations. The instantaneous eddy current loss at any point in a strip is i^2R (Wm^{-3}) where i is the eddy current density (Am^{-2}) at that point and the average loss at this point over one cycle of magnetisation is obtained by integrating over the period from zero to $2\pi/\omega$.

If this is integrated over the thickness of the strip, the mean eddy current loss per unit volume of the whole lamination in one cycle of magnetisation, W_{clas}, is given by

$$W_{clas} = \int_{-a}^{+a} \int_{0}^{\frac{2\pi}{\omega}} i^2 \rho \, dt \, . \, dx \left(\mathrm{Jm}^{-3}\right) \tag{7.32}$$

This equation might not look too difficult to solve but it will be noted that (7.27) also must be solved to obtain i, and all the practical assumptions made in its own derivation apply. The solution of (7.32) for a flat sheet is given in [19].

However, if (7.32) is solved we obtain

$$W_{clas} = \rho p B_s^2 \Big/ 2a\mu^2 \left(\frac{\sinh 2pa - \sin 2pa}{\cosh 2pa + \cos 2pa} \right) \left(\mathrm{Jm}^{-3}\right) \tag{7.33}$$

For a thin sheet, when pa is small, the quantity in the bracket in (7.33) approaches unity and if we also write the equation in terms of the more practical term B_p instead of B_s it can be shown [17] that

$$P_{clas} = \frac{\pi^2}{\rho k_{clas}} \left(2afB_p\right)^2 \left(\mathrm{Wm}^{-3}\right) \tag{7.34}$$

This expression is almost universally used to calculate the classical eddy current loss of a thin sheet *under a sinusoidal flux density* for which $k_{clas} = 6$. If the B waveform is triangular then k_{clas} becomes $3\pi^2/4$.[9] It must be remembered that approximations are made in its derivation. Unfortunately, it is used indiscriminately in some publications.

It is interesting to look at the effect of the assumption that pa is small. If this assumption is not made the more complete version of (7.34) becomes [21]:

$$P_{clas} = \frac{\pi a f p B_p^2}{\mu} \left(\frac{\sinh 2pa - \sin 2pa}{\cosh 2pa - \cos 2pa} \right) \left(\mathrm{Wm}^{-3}\right) \tag{7.35}$$

When typical values of μ, ρ and a are inserted into this equation, it is found that the calculated eddy current loss does not differ by more than a few percent from that given by the approximate equation up to around 500 Hz [21]. However, the possible inaccuracies due to the other assumptions made in the eddy current analysis already mentioned still apply. A far more rigorous analysis of eddy currents in thin sheets is given in [22].

7.7.1 Reduction of eddy current loss by use of laminations

Eddy current loss is clearly proportional to the square of the thickness of a sheet of magnetic material as shown, for example, in (7.34). The effect of laminating a thick

[9] k_{clas} also depends on the shape of the sample, here assumed a thin strip. $k_{clas} = 20$ for a sphere and $k_{clas} = 16$ for a cylinder where a becomes the radius in each case [20].

plate can be determined using another approach to calculate the losses, again supposing that the flux density is sinusoidal and that no skin effect is present, i.e. the peak value, B_p, is uniform throughout the slab. Consider the strip shown in Figure 7.6 as a thick slab. The r.m.s. flux enclosed by the current path at a distance x from either side of the centre of the slab is given by

$$\phi_{rms} = \frac{B_p}{\sqrt{2}} 2wx(\mathrm{Wb}) \tag{7.36}$$

where w is its width and edge effects are neglected. The induced e.m.f. driving the eddy currents around a circuit of width δx is given by

$$e = -\frac{d\Phi}{dt} = \sqrt{2}\Phi\omega \cos \omega t(\mathrm{v}) \tag{7.37}$$

Hence the r.m.s. value of the e.m.f. is $\phi\omega$. The resistance of the elemental eddy current path is $2\rho l/\delta x$ so the eddy current loss in the element is given by

$$\delta P = \frac{\Phi^2 \omega^2 \delta x}{2w\rho}(\mathrm{W}) \tag{7.38}$$

The loss in the whole slab is therefore

$$P = \oint_0^a \frac{B_p^2 4\pi^2 f^2 wx^2}{\rho} dx(\mathrm{W}) \tag{7.39}$$

This can be written in terms of loss per unit surface area, P_{area} as

$$P_{\mathrm{area}} = \frac{B_p^2 \ 4\pi^2 f^2 \ a^3}{3\rho} \ (\mathrm{Wm}^{-2}) \tag{7.40}$$

Suppose the thickness of the slab is $2a$ and it is split into N laminations each of thickness ${}^{2a}/_{N}$ the loss changes to

$$P_{\mathrm{area}} = N\left(\frac{B_p^2 \ 4\pi^2 f^2 \ \left({}^{a}/_{N}\right)^3}{3\rho}\right) \ (\mathrm{Wm}^{-2}) \tag{7.41}$$

This shows that the loss in a stack of N laminations is $1/N^2$ times the loss of the un-laminated slab. To demonstrate the effect, consider a laminated core 100 mm high assembled from 0.35 mm or 0.27 mm thick laminations. It is quick to apply this relationship to show that the eddy current loss of the stack assembled from the thinner laminations is around $(286/370)^2$ or 60% that of the stack assembled from the thicker laminations provided that everything else remains the same. In practice, it is not surprising to find many other factors vary with lamination thickness so the loss difference in practice is far lower than given here.

7.8 Separation of losses into eddy current and hysteresis components

7.8.1 Hysteresis loss components

Hysteresis can be separated into two components predominately occurring at high and low fields in electrical steel. The low-induction component of hysteresis, introduced in [23], increases monotonically with increasing magnetising angle from the RD, whereas the high induction component reaches a peak at about 50° to the RD.

This induction dependent hysteresis loss separation is discussed in detail in [24]. The boundary in electrical steel is around 0.8 T. The low-induction component increases linearly up to about 1.2 T, while the high-induction component is zero up to about 0.7 T, and then increases according to a power law. Grain size has a significant effect on low-induction losses whereas texture is more influential on the high-induction component.

Figure 7.9 gives a schematic breakdown of the components and shows how they vary with flux density in a typical NO steel. The two components are probably controlled by different energy dissipation mechanisms and should be modelled separately. However, it has been said that ‘there is no quantitative figure of the effect of grain-size and magnetic texture and their interplay with magnetizing behaviour and the specific magnetic losses, even for electrical steels’ [25].

7.8.2 Separation of total loss into two or three components

The concept of loss separation is rather artificial since there is no fundamental difference between eddy current and hysteresis loss. Loss separation techniques are useful for predicting losses in magnetic cores and also in basic studies of electrical steel but it is becoming widely recognised that it should be used with caution when applied to thin steels or at extremes of magnetisation. The basic background is given here and it is discussed further in Section 4.4 in Volume 2 of this book.

The fact that hysteresis loss is frequency dependent has been raised many times, going back to the 1960s [26]. Methods have been developed to investigate the rate of increase of magnetisation dependent hysteresis which increases with magnetising frequency in GO SiFe and Co-based amorphous ribbon due to the larger number of mobile domain walls being present at higher frequencies in each case [27].

Common expressions for the static hysteresis loss and the classical eddy current loss are given in (7.17) and (7.34), respectively. Their sum per cycle, which equates to energy loss, can be written as

$$\frac{P_{\mathrm{total}}}{f} = k_h B_m^n + k B_m^2 f \ \left(\mathrm{Jm}^{-3}\right) \tag{7.42}$$

This relationship has been used by electrical machine designers and material developers from early in the twentieth century for accounting for losses in electrical steel. At a constant flux density, this can be written more simply as

$$\frac{\mathrm{P}_{\mathrm{total}}}{f} = k_1 + k_2 f \ \left(\mathrm{W\,kg}^{-1}\right) \tag{7.43}$$

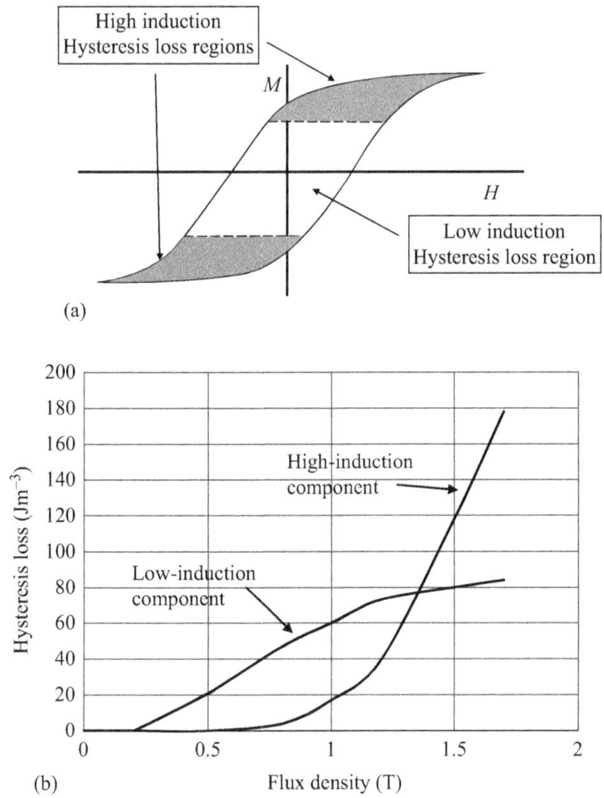

Figure 7.9 (a) Breakdown of hysteresis loss into high-field and low-field components and (b) trends in the components with increasing flux density in NO steel

where k_1 is the hysteresis loss per cycle and k_2 is a material dependent eddy current constant at a given flux density. Equation (7.43) infers a linear relationship between loss per cycle and frequency. The way in which this relationship was (and occasionally still is) used is illustrated in Figure 7.10(a). The loss per cycle is measured at a reference flux density at two convenient frequencies, f_1 and f_2. Assuming the linear relationship, a straight line can be extrapolated to 'zero frequency' to determine k_1 and k_2 as the slope of the straight line. Hence the hysteresis and classical eddy current losses can be predicted at any flux density.

In the early 1900s this method was shown only to be reliable for frequencies lower than around 100 Hz and relative permeability less than 5,000 [28]. Soon afterwards, it became widely recognised in the emerging telecommunications industry that the linear law did not apply to thin nickel–iron alloys [29].[10] Impurities

[10]The discrepancy was accredited to some sort of magnetic after-effect. An after-effect is simply a change in magnetic properties with time occurring after any change of external magnetic field has ceased.

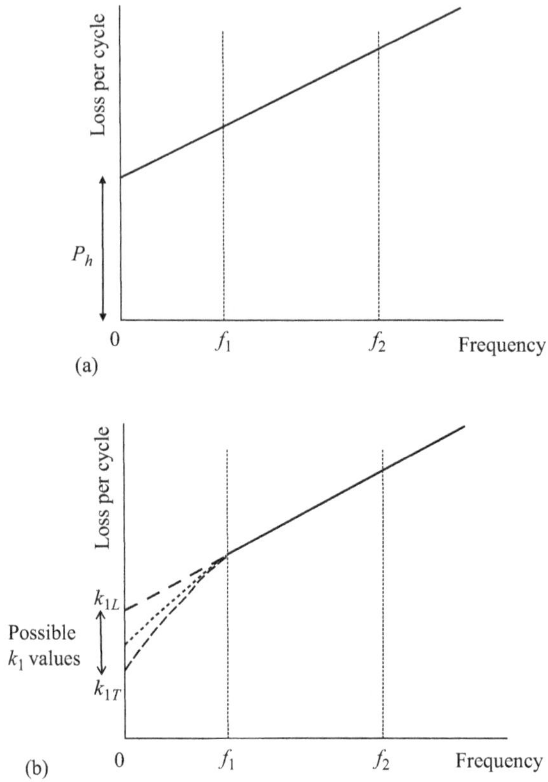

Figure 7.10 Traditional separation of loss into classical eddy current and hysteresis components: (a) loss per cycle versus magnetising frequency and (b) low frequency deviation from linear extrapolation of the curve

and defects effectively cause small distortions in the lattice of a crystalline material which hinder domain wall motion and hence increase a.c. losses. Today, the more commonly accepted explanation that the additional component of loss is caused by domain wall pinning means that the losses are always higher than that given by the equation that was proposed in the 1930s [30]. Many other explanations have been proposed over the years. One proposal, based on results of a comprehensive experimental study of losses in electrical steels, was that the extra loss was caused by 'the slowness of domain wall movements, microscopic eddy currents or by some other form of friction' [31]. This is not far from our present understanding.

The weakness of the two component analysis of the loss became more evident from results of lower frequency experimental measurements carried out in the 1960s. Previous to this, it was difficult to carry out low frequency measurements very accurately so measurements tended to be carried out over a linear range between f_1 and f_2 shown on the hypothetical characteristic shown in Figure 7.10(b). From this, the static hysteresis would be assumed to be k_{1L} and theories were

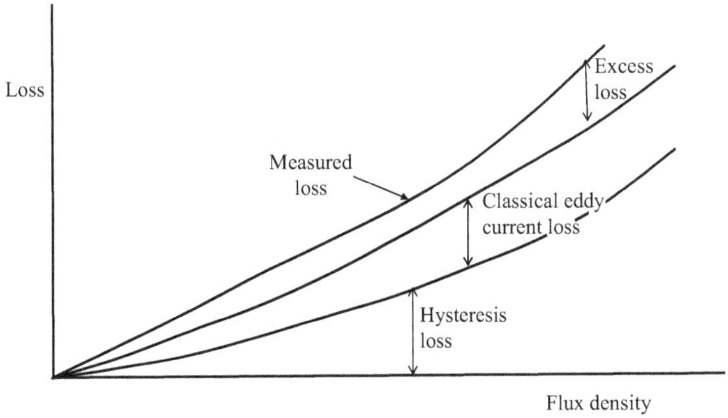

Figure 7.11 Variation of loss components with flux density at constant frequency showing the presence of excess loss

developed to be consistent with this value found from linear extrapolation. If the full characteristic is measured it might show the hysteresis loss to be k_{1T}. Intermediate values of k_{1I} would be obtained depending on the lowest measurement frequency. Of course, the static hysteresis loss can be measured by cycling the magnetisation at a very low frequency, such as in a torque magnetometer.[11] Ideally this gives the same value k_{1T}.

Early researchers attributed the difference between the values of hysteresis measured by linear extrapolation and from quasi-static measurements as being due to experimental errors in one or both of the methods.

As the accuracy of experimental measurements of loss improved it became more apparent that not only was the linear loss versus frequency characteristic incorrect but also the measured value of loss per cycle at any frequency is always higher than the sum of the measured static hysteresis loss and the eddy current loss calculated from (7.34). This difference is illustrated in Figure 7.11. Figure 7.11 shows how the loss components generally vary with flux density at a given magnetising frequency. The figure is not to scale; the relative size of each loss component can vary significantly depending on the magnetising frequency, flux density waveform, material composition and microstructure.

When this phenomenon was first investigated in the 1960s, it was widely believed that that the difference between measured and calculated loss could not be attributed to inaccuracies in applying the eddy current theory or in calculating the hysteresis loss so the extra loss was considered to be due to a third component of power loss. It was named *anomalous loss* because it could not be definitely explained so it was simply quantified as being the difference between the measured

[11]The torque magnetometer is rarely used today. A mechanical torque exerted on a specimen, usually disc shaped, is measured as an external field is slowly rotated around the specimen. Any magnetic anisotropy makes the specimen tend to line up with the field and measurement of the restraining force opposing this physical alignment gives rapid information about the anisotropy of the specimen.

loss of a material and the sum of the hysteresis and classical eddy current components.

The anomalous loss can be quantified in terms of the anomaly factor η defined as

$$\eta = \frac{W_{clas} + W_{ex}}{W_{clas}} \tag{7.44}$$

where W_{ex} is the excess loss.

The anomaly factor is unity when no excess loss is present. In a given material it varies with B_p, f, magnetising waveform and direction of magnetisation in the plane of the sheet. In electrical steel it has been found to reach values greater than eight [31] but in modern steels it is normally far lower.

When anomalous loss was first recognised, its origins in GO steel were listed as follows [32]:

- non-sinusoidal, non-uniform flux density,
- non-repetitive variation of flux density,
- lack of flux penetration and domain wall bowing,
- non-sinusoidal flux density and localised variation of flux density,
- mechanisms dependent on grain size, grain orientation and specimen thickness and
- nucleation and annihilation of domain walls.

As understanding of the phenomenon has grown, the 'additional loss' became known as excess eddy current loss, later in time as excess hysteresis loss or dynamic hysteresis loss. Today, because it is still not fully understood, it is most commonly referred to just as excess loss.

The total loss P_{total} can simply be written as the sum of the three traditional components of power loss, P_h, P_{clas} and P_{ex} as

$$P_{\text{total}} = P_h + P_{clas} + P_{ex}\ \left(\text{Jm}^{-3}\right) \tag{7.45}$$

Equation (7.45) can be expressed in terms of loss per cycle of magnetisation as

$$\frac{P_{\text{total}}}{f} = k_h B_p^n + \frac{\pi^2 4a^2 B_p^2 f}{6\rho} + \frac{P_{ex}}{f}\ \left(\text{Jm}^{-3}\ \text{cycle}^{-1}\right) \tag{7.46}$$

This is shown graphically in Figure 7.12 for a constant peak flux density B_p. It can be seen that the classical eddy current loss per cycle is directly proportional to magnetising frequency. In practice, the hysteresis loss is still often obtained by extrapolation as in Figure 6.8(b). The classical eddy current loss is calculated directly, the total loss per cycle over the full magnetising frequency range is measured from which the excess loss at any frequency is obtained. In such a simple analysis the hysteresis loss per cycle is assumed constant, i.e. independent of magnetising frequency.[12]

[12]As already hinted, this is not always the case because the influence of magnetic domains in the magnetisation process is not taken into account in this basic analysis. The influence of domains on modelling of losses in electrical steels is discussed in Chapter 4 of Volume 2 of this book.

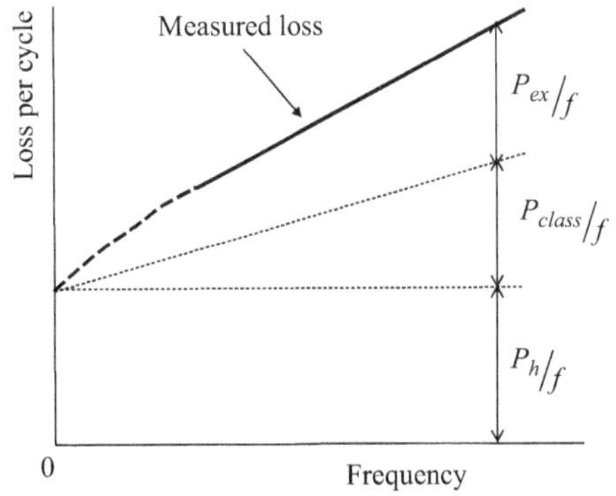

Figure 7.12 Separation of loss per cycle of magnetisation versus magnetising frequency into three traditional components at a constant flux density (not to scale)

Many attempts have been made to model or predict the excess loss in a thin sheet, the most commonly quoted being [33]:

$$P_{ex} = 8\sqrt{\sigma G A V_0}(B_m f)^{3/2}\left(\mathrm{Jm}^{-3}\right) \tag{7.47}$$

where G is a constant related to energy dissipation at a moving domain wall, A is the cross sectional area of the sheet and V_0 is a complex parameter dependent on the material microstructure.

This simple approach is a useful guide but it must be constantly noted that the analysis into these components applied in this way has little or no physical foundation.

Hence the justification for separating losses into two, or even three, components is questionable. It was said many years ago that adding a third component to loss models is not fruitful if the models of the other two are incorrect [34]. Steel users need accurate ways of predicting losses under magnetising conditions outside the narrow limits within which the method of loss separation discussed in this section are at least partly valid.

7.9 Total loss within a sheet

It is useful to recap on the basic relationship between losses and the B–H loop in a strip of steel uniformly magnetised in its longitudinal direction. Under slowly changing magnetisation through one cycle it can be shown that the area of the loop is equal to the total loss dissipated during the cycle. This is hysteresis loss measured in joules/m^3/cycle. Flux density and field vary with depth into a strip under a.c. magnetisation so it is necessary to identify what actual components of these quantities are directly related to the loss.

B and H are often vaguely referred to as if they are both accessible quantities which we can quantify experimentally at any point inside a strip of material. Of course, this is not the case and furthermore, only changes in B can be measured.

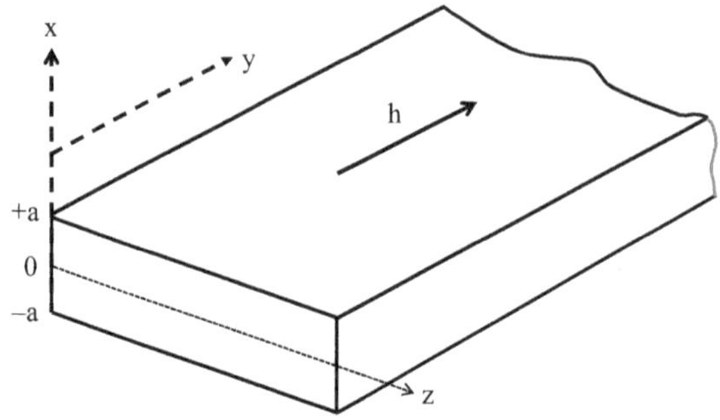

Figure 7.13 Coordinates of a strip magnetised along its longitudinal y-axis

This can be done by enwrapping the strip with a suitable search coil, measuring dB/dt from which the space-average value of B inside the material can be obtained.

Unless microscopic holes can somehow be made in a material to insert suitable field sensors, without disturbing the internal field and flux density, the internal field cannot be measured. Only the surface tangential component is accessible. The tangential component in the air on the sample surface can be measured with a field detector. This component of field is equal to the value just inside the surface of the strip so the tangential component of the surface field is accessible in this way.[13]

Fortunately, it is possible to find an expression for the total loss inside a sheet in terms of these experimentally accessible quantities. The following mathematical analysis is complex, but leads to the conclusion that losses can be calculated from knowledge of the space-average flux density and surface value of field.

Consider a sheet of magnetic material of thickness $2a$ magnetised in the y direction as shown in Figure 7.13. Assume that the magnetisation is confined to the y direction and that the eddy currents are confined to the z direction (as depicted in Figure 7.6). Equations (7.20) and (7.22) can be rearranged as

$$\frac{\partial i}{\partial x} = \frac{\partial^2 H}{\partial x^2} = \frac{1}{\rho}\frac{\partial b}{\partial t}\ (\mathrm{Am^{-1}}) \tag{7.48}$$

$$i = \frac{\partial H}{\partial x} = \frac{1}{\mu}\frac{\partial b}{\partial x}(\mathrm{A}) \tag{7.49}$$

The following identity can be written for magnetisation components in any yz plane

$$\frac{\partial}{\partial x}\left(H\frac{\partial H}{\partial x}\right) = H\frac{\partial^2 H}{\partial x^2} + \left(\frac{\partial H}{\partial x}\right)^2 \tag{7.50}$$

[13] An explanation of change in field and flux density direction on either side of a steel-air interface is given in Section 10.6.

Substituting values from (7.48) and (7.49) into this expression gives

$$\frac{\partial}{\partial x}\left(H\frac{\partial H}{\partial x}\right) = \frac{H}{\rho}\frac{dB}{dt} + i^2 = \frac{1}{\rho}\left(H\frac{\partial B}{\partial t} + i^2\rho\right) \tag{7.51}$$

Since H and B in this equation are the localised values at layer x in the material, the first term in the bracket on the right-hand side is equal to the localised energy loss due to hysteresis δP_h and the second term is the eddy current loss δP_{clas}, at the same spot, both expressed as per unit volume per cycle. Hence the sum of the eddy current and hysteresis loss averaged over the full cross section of the material can be written as

$$E_h + E_{clas} = \frac{\rho}{2a}\int_{-a}^{+a}\frac{\partial}{\partial x}H\left(\frac{\partial H}{\partial x}\right)dx = \frac{\rho}{a}H_a\left(\frac{\partial H}{\partial x}\right)_a \ (\mathrm{Jm}^{-3}) \tag{7.52}$$

where E_h and E_{clas} are the instantaneous hysteresis and classical eddy current losses, respectively. The integrated expression in (7.52) is obtained making use of the fact that (by symmetry) $+H_a = -H_a$, where subscript a depicts these as being specific values of the quantities at the sheet surfaces. The equation shows two important factors which underpin many measurement and analytical techniques performed on electrical steels, and, indeed, any soft magnetic material in thin plate, or strip, form. First, the total loss is proportional to the experimentally accessible quantity H_a. The second important point becomes clear when the output of a search coil wound around the full cross sectional area of the sample is considered. The instantaneous induced e.m.f. of the coil can be written as the integral of the localised e.m.f.s induced throughout the cross section of the material from one surface to the other and expressed as

$$e = \frac{k_c}{2a}\int_{-a}^{+a}\frac{dB}{dt}dx(\mathrm{v}) \tag{7.53}$$

where k_c is a constant relating to the search coil design.[14] A solution to this equation can be found by making use of the relationship between B and H given in (7.52) leading to

$$e = \frac{k_c\rho}{2a}\int_{-a}^{+a}\left(\frac{\partial^2 H}{\partial x^2}\right)dx = \frac{k_c\rho}{a}\left(\frac{\partial H}{\partial x}\right)_a \ (\mathrm{v}) \tag{7.54}$$

This shows that the quantity $\left(\frac{\partial H}{\partial x}\right)_a$, which seems quite inaccessible is, in fact, directly proportional to the e.m.f. induced in a search coil enwrapping the sample. The physical significance of this surprising finding is beyond the scope of this book so it will not be discussed any further but simply made use of in the following chapters.

[14]In fact k_c is equal to the product of the number of turns in the coil and its effective cross sectional area.

The result shown in (7.54) can be used to rewrite (7.52) as

$$E_{\text{total}} = E_h + E_{clas} = \frac{1}{k_c} H_a e \; (\text{Jm}^{-3}) \tag{7.55}$$

where E_{total} is the total *instantaneous energy loss*. H_a is the tangential component of surface field parallel to the magnetising direction which is co-linear with B. This expression now will be developed further in the next section to shown how it can be used to obtain the *power loss* under practical a.c. conditions.

The approach starts with the same concept of internal flux and field distribution in a sheet developed in Section 7.6 from which the classical component of eddy current loss was calculated in Section 7.7 based on the assumptions given earlier. The loss calculated in this section is claimed to include all loss producing mechanisms in a sheet, although only two components, hysteresis and classical eddy current loss appear in the equations. An interesting point is that (7.55) works in practice to measure losses and permeability of electrical steels as is seen in Chapter 12, so the excess loss must be effectively included in one of these two terms.

In Section 7.8, a question was raised over the validity of the traditional separation of losses into either two or three independent components. An assuring point about the implication of (7.55) is that, although one component is termed the hysteresis component and the other is the eddy current component, they are not calculated independently as was done in Section 7.6 where the classical eddy current loss was calculated.

It is interesting to note that some of the relationships, such as in (7.51), derived in this section are analogous to those derived in Section 7.2 based on the Poynting theorem. This is not surprising since they both are based on versions of Maxwell's equations followed by suitable mathematical manipulation to obtain solutions which appear to be related to widespread experimental findings. It is questionable whether these approaches are fully compatible with our knowledge of domain dynamics in these materials [35]. This throws doubt on the general implementation of Maxwell's equations in the basis analysis of electrical steel, but not their fundamental validity.

Measurements of losses in NO silicon iron laminations using a fieldmetric method (using components of B and H as discussed in Chapter 12) have been shown to agree well with a method of obtaining losses directly from measurement of the initial rate of rise of temperature [36]. However, no comprehensive direct experimental comparisons have been reported under more complex conditions, nor in highly anisotropic material, such as GO silicon–iron, where appreciable differences might be anticipated.

7.10 Total power loss of a strip expressed in terms of *B* and *H*

Suppose a sheet of soft magnetic material is subjected to an a.c. periodic flux density expressed as

$$B(t) = B_1 \sin \omega t + B_3 \sin(\omega t + \varphi_{b3}) + B_5 \sin(\omega t + \varphi_{b5}) + \cdots B_{2n-1} \sin(\omega t + \varphi b_{2n-1})(\text{T}) \tag{7.56}$$

and the driving magnetic field is

$$H_a(t) = H_{a1}\sin(\omega t + \varphi_{h1}) + H_{a3}\sin(\omega t + \varphi_{h3}) + h_{a5}\sin(\omega t + \varphi h_5) + \cdots H_{a(2n-1)}\sin(\omega t + \varphi_{h(2n-1)})\ (\text{Am}^{-1}) \tag{7.57}$$

where $B_1 - B_n$ and $H_1 - H_n$ are the peak values of the harmonic components of $B(t)$ and $H_a(t)$, respectively and φ_{nb} and φ_{nh} are the phase angles of the nth harmonic components of B and H respectively with respect to the fundamental component of the flux density waveform.

Equation (7.55) gives the total instantaneous energy loss in a sheet in terms of the surface field H_a and the e.m.f., dB/dt, induced in a search coil enwrapping the sample. This can be rewritten and integrated over one cycle of magnetisation to give the total energy loss per cycle as

$$E_{\text{total}} = \frac{\omega}{2\pi}\int_0^{\frac{2\pi}{\omega}} H_a \frac{dB}{dt}\,dt\ (\text{Jm}^{-3}) \tag{7.58}$$

If the harmonic series for H_a and dB/dt (obtained by differentiating (7.56)) are inserted into this and the function integrated,[15] then the total power, *provided only odd harmonics of flux density are present,* can be shown to be given by

$$P_{\text{total}} = 2\pi f \sum_{n=0} H_{an}B_{an}\sin(\varphi_{nb} - \varphi_{nh})\ (\text{Wm}^{-3}) \tag{7.59}$$

If no harmonics are present then

$$P_{\text{sine}} = 2\pi f H_{a1}B_1 \sin\varphi_1\ (\text{Wm}^{-3}) \tag{7.60}$$

where φ_1 is the phase angle between the fundamental components of B and H. If H_a and B are in phase, then $\varphi_1 = 0$, hence the loss is zero.[16]

Since $B = \mu H$, it might appear that we can also express the energy loss in terms of B and dH/dt, as

$$E_{\text{total}} = \frac{\omega}{2\pi}\int_0^{\frac{2\pi}{\omega}} B \frac{dH}{dt}\,dt\ (\text{Jm}^{-3}) \tag{7.61}$$

This is correct in concept, but it must be remembered that in this equation B is the surface value and dH/dt is its average value across the thickness of the material; both are inaccessible experimental quantities.

As pointed out in Section 6.4, normally, when an a.c. field is applied to a material, the B waveform, and the H waveform, cannot both be sinusoidal. So, for

[15]The integration steps are not included because they are rather tedious and can easily be carried out with a little knowledge of calculus, remembering that the integral of products of unequal harmonic numbers (cross products) are zero over one cycle of magnetisation.

[16]Tracing the *B–H* characteristic from such B and H waveforms on a *B–H-time* projection clearly reveals that the *B–H* characteristic is linear, hence no losses occur.

example, if the B waveform is forced to be sinusoidal then the H waveform contains odd harmonics. However, the loss is simply the product of the differential of the fundamental component of $B(t)$ and the fundamental component of $H(t)$ since all the other harmonics in the H waveform given in (7.57) do not contribute to the losses. The same applies if H is sinusoidal.

Equation (7.59) is developed for a single lamination or strip. It can be shown that the same expression can be applied to a stack of laminations provided the tangential field components are equal at each pair of adjacent surfaces in the stack [37]. This is valid for a perfectly homogenous stack of identical laminations symmetrical about the centre line of the stack and with laminations separated by negligible air gaps. The validity of these assumptions in practice does not seem to have been reported.

References

[1] Sievert J. 'The measurement of magnetic properties of electrical sheet steel – survey on methods and situation of standards'. *J. Magn. Magn. Mater.* 2000, vol. 215–216, pp. 647–651

[2] Tumanski S. *Handbook of Magnetic Measurements.* CRS Press, Boca Raton, FL, 2011, Equation (2.39)

[3] Bettotti G. *Hysteresis in Magnetism for Physicists, Material Scientists and Engineers.* Academic Press, San Diego, 1998 p. 99

[4] Schwarz S.E. *Electromagnetics for Engineers.* Sanders College Publishing, Orlando, 1990, p. 258

[5] Carter G.W. *The Electromagnetic Field in Its Engineering Aspects.* Longmans, London, 1962, p. 315

[6] Tumanski S. *Handbook of Magnetic Measurements.* CRS Press, Boca Raton, FL, 2011, Equation (2.123), p.46

[7] Pfützner H. 'Rotational losses of grain oriented silicon steel sheets: fundamental aspects and theory'. *IEEE Trans. Magn.* 1994, vol. 30(5), pp. 2802–2807

[8] Tumanski S. *Handbook of Magnetic Measurements.* CRS Press, Boca Raton, FL, 2011

[9] Carter G.W. *The Electromagnetic Field in Its Engineering Aspects.* Longmans, London, 1962, p. 199

[10] Schwarz S.E. *Electromagnetics for Engineers.* Sanders College Publishing, Orlando, 1990, p. 259

[11] Steinmetz C.P. 'Note on the law of hysteresis'. *Electrician* 1890, vol. 10, p. 137

[12] Brailsford F. 'Alternating hysteresis loss in electrical sheet steels'. *J. IEE* 1939, vol. 84, pp. 399–407

[13] Jiles D. *Introduction to Magnetism and Magnetic Materials.* Chapman and Hall, London, 1991, p. 92

[14] Brailsford F., and Burgess J.M. 'Internal waveform distortion in silicon-iron laminations for magnetisation at 50 c/s'. *Proc. IEE* 1961, vol. 108(C), pp. 458–462

[15] Brailsford F. *Physical Principles of Magnetism*. Van Nostrand, London, 1966, p. 234
[16] Brailsford F. *Physical Principles of Magnetism*. Van Nostrand, London, 1966, p. 235
[17] Brailsford F. *Physical Principles of Magnetism*. Van Nostrand, London, 1966, p. 236
[18] Chikazumi S. *Physics of Magnetism*. R E Krieger, New York, 1978, p. 322
[19] Russell A. *Alternating Currents*. Cambridge University Press, Cambridge, 1914
[20] Jiles D. *Introduction to Magnetism and Magnetic Materials*. Chapman and Hall, London, 1991, p. 271
[21] Brailsford F. *Physical Principles of Magnetism*. Van Nostrand, London, 1966, p. 237
[22] Bertotti G. *Hysteresis in Magnetism for Physicists, Material Scientists and Engineers*. Academic Press, San Diego, 1998, pp. 391–430
[23] Landgraf F.J.G., Emura M., Teixeira J.C., Campos M.F., and Muranaka C.S. 'Anisotropy of the magnetic losses components in semi-processed electrical steels'. *J. Magn. Magn. Mater.* 1999, vol. 196–197, pp. 380–381
[24] Landgraf F.J.G., de Campos M.F., and Leicht J. 'Hysteresis loss subdivision'. *J. Magn. Magn. Mater.* 2008, vol. 320, pp. 2494–2498
[25] Houbaert Y., and Schneider J. 'Effect of texture and grain size on magnetic losses and permeability'. Proceedings of the 2nd International *Workshop on Magnetism and Metallurgy*, Freiberg, Germany, 21st–23rd June, 2006, pp. 90–106
[26] Overshott K.J. 'Hysteresis and eddy current loss in grain-oriented silicon-iron'. *Proceedings of the 3rd International Conference on Soft Magnetic Materials, SMM-3,* Bratislava, Czechoslovakia, paper 9.1, 1977, pp. 265–268
[27] Winner H., Naber W., Sander R., and Grosse-Nobis W. 'Dynamic and hysteresis losses'. *Phys. Scr.* 1989, vol. 39, pp. 52–55
[28] Campbell A., Booth H.C., and Dye D.W. 'Report on five samples of magnetic sheet material tested for total loss and hysteresis at the Physikalisch-Technische Reichsanstalt, the Bureau of Standards and the National Physical Laboratory'. *J. IEE* 1912, vol. 48(211), pp. 269–280
[29] Snoek J.L. 'Time effects in magnetisation'. *Physica* 1938, vol. 5(8), pp. 663–688
[30] Becker R., and Doring W. *Ferromagnetismus*. Julius Springer, Berlin, 1939
[31] Stewart K.H. 'Losses in electrical sheet steel'. *Proc. IEE: Part II Power Eng.* 1950, vol. 97(56), pp. 121–125
[32] Overshott K.J., and Hill S. 'An assessment of the origin of losses in grain-oriented 3% silicon iron'. Proceedings of Soft Magnetic Materials 2, European Physical Society, Cardiff Conference, April 1975, pp. 115–120
[33] Bertotti G. *Hysteresis in Magnetism for Physicists, Material Scientists and Engineers*. Academic Press, San Diego, 1998, p. 428
[34] Becker J.J. 'Magnetization changes and losses in conducting ferromagnetic materials'. *J. Appl. Phys.* 1963, vol. 34(4), pp. 1327–1332

[35] Moses A.J. 'Energy efficient electrical steels: magnetic performance prediction and optimization'. *Scr. Mater.* 2012, vol. 67(6), pp. 560–565
[36] Ragusa C., Appino C., and Fiorillo F. 'Comprehensive investigation of alternating and rotational losses in non-oriented steel sheets'. *Przeglad Elektrotechniczny (Electric. Rev.)* 2009, vol. 85, pp. 7–10
[37] Pegler S. M. 'The Effect of Bonding upon the Power Loss & the Magnetostriction of Assemblies of Transformer Laminations'. *PhD Thesis*, University of Wales, 1975

Chapter 8
Rotational magnetisation and losses

It is shown in Chapters 5 and 6 of Volume 2 of this book that rotational magnetisation is a very important phenomenon in rotating electrical machines and in transformers because it can produce high losses and complex magnetostriction. This chapter introduces fundamental aspects of the rotational magnetisation process and its general effect on losses and magnetostriction in both isotropic and highly anisotropic material, such as GO electrical steel.

The localised magnetisation within a core assembled from electrical steel, other thin magnetic material, or indeed bulk soft powder composites, can be very complicated. The magnetic field and flux density probably will vary in magnitude from point to point within the material. Also, under a.c. magnetisation its time varying waveform might be non-sinusoidal. Furthermore, the direction of B and H at a given point within the core material might vary with time. These effects are loosely referred to as being rotational magnetisation phenomena.

In electrical steels, this rotation mainly occurs in the plane of the sheet which makes analysis much simpler than when the magnetisation has three degrees of freedom in a magnetic material.[1] If the flux density has a fixed magnitude and rotates at constant velocity in the plane of the sheet it is referred to as a pure rotational flux density or circular flux density. Likewise, a pure rotational, or circular, magnetic field is one which has constant magnitude and rotates at constant velocity.

An argument could be made that a better generic term for rotational magnetisation is simply *two-dimensional magnetisation*. Common abbreviations for this are simply 2D or 2-D both of which are used throughout this book together with 1D and 3D when referring to unidirectional and three-dimensional variation of magnetisation, respectively.

Strictly speaking, almost all magnetisation conditions within magnetic cores involves B and H changes in all three directions, but the terminology referred to here is intended to help to focus on the most relevant, or dominant, magnetic processes associated with any particular analysis, characterisation or application of the materials.

[1]In transformer core joints, flux transfers between laminations in adjacent layers so this is strictly magnetisation in three directions. This is discussed in Chapter 5 of Volume 2.

A more general form of rotational magnetisation occurs when its directional locus takes the form of an ellipse. In fact, such elliptical magnetisation is far more common in electrical steels than circular magnetisation.[2]

Another form of complex magnetisation of growing importance, particularly in power and distribution transformer cores, is the simultaneous presence of a.c. and d.c. magnetic fields which are superimposed to create a distorted *B–H* loop. In transformer parlance, this condition is commonly referred to as a *d.c. offset*. If the a.c. and d.c. components are not co-linear they set up a complex rotational magnetisation condition. Characterisation of electrical steels magnetised under d.c. offset conditions is covered in Section 12.3.5.

If a magnetic field is steadily increased along a fixed direction at some angle to an easy axis in an anisotropic magnetic material, the corresponding flux density will be found to be at an angle to the field direction. This is a form of rotational magnetisation but, normally, the flux density only deviates by a few degrees from the field direction as the field rotates during the magnetisation cycle. During what we normally call the rotational magnetisation process both field and flux density rotate through 360° during one cycle of magnetisation. This partial rotational process is discussed in Chapter 9.

Historic rotational magnetisation terminology is sometimes misused even today. Many publications refer to a.c. components of rotational flux, or field at power frequencies. This is completely valid in mathematical analyses of rotational fields but, occasionally, authors treat these a.c. components as if they are physically creating losses in their own right, rather than just being a convenient analytical process.

Another misconception sometimes appears in technical publications related to methods of analysing rotational magnetisation processes in strip or sheet. In such cases, analysis is built on the assumption that magnetic properties under rotational conditions can be directly inferred from measurements made under a.c. fields applied separately along orthogonal directions in the plane of the material to set up the same flux densities as the orthogonal components of the rotational locus. However, although basic mathematical analysis of losses shows that this is theoretically valid for a homogenous material, magnetic domain studies show that magnetisation conditions under rotational and a.c. conditions can be very different, so the analysis is not consistent with what happens in real materials.[3] Furthermore, it is still not clear how well the mathematical treatment of rotating fields and flux density truly account for the domain dynamics which occur during the process, and the extent of the associated errors [1].

Poynting's theorem was introduced in Chapter 7 as a means of determining losses under uniaxial magnetisation but it can be applied to any form of magnetisation provided the vector time and spatial relationships between field and flux density are fully accounted for. It is not appropriate to go into detail of the mathematical relationships here but a useful review is given in [2]. A more detailed account of rotational magnetisation and losses in soft magnetic materials is given in [3].

[2]In fact such magnetisation is even more complex but can be split into time harmonic components of a series of ellipses.

[3]See Section 8.5.

8.1 Vector representation of a pure rotating magnetic field

Suppose a magnetic field of constant magnitude H rotates in a clockwise direction through 360° in space in the $x-y$ plane of a sheet of soft magnetic material. The changing field during the cycle of rotational magnetisation is depicted in Figure 8.1(a). The field is initially directed along the x-axis of the sheet in position 1 and rotates by 30° in equal time periods as shown until it returns to the x-axis, position 1 after one cycle of magnetisation.

This is the locus of a *circular rotational field* since it has constant magnitude and constant angular velocity.

A pure rotational d.c. field could be set up by sequentially setting a constant magnitude field at angles to the reference x-axis through 360° to produce a locus which can be represented in the same way but without having the time function included.[4] Physically, such a pure rotational field could be created by rotating a magnet around a disc of magnetic material as shown schematically in Figure 8.1(b). This is the basis of the torque magnetometer in which the variation of mechanical torque is measured as the magnet is slowly rotated around a disc in order to measure properties such as its rotational hysteresis or texture as discussed in Section 8.3.

Suppose orthogonally positioned magnetising coils carrying a.c. currents, which are 90° out of phase, are used to magnetise the disc as shown in Figure 8.2(a). Let the currents in the coils produce fields $h_x = H_x \sin \omega t$ and $h_y = H_y \sin(\omega t + 90°)$ in the directions shown. It can be clearly seen from the vector diagram shown in Figure 7.2(b)

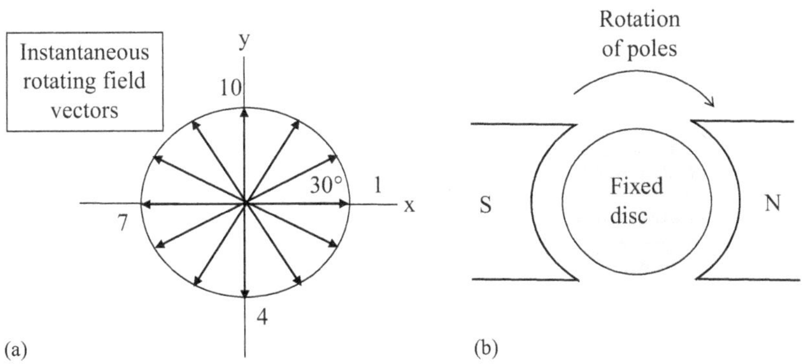

Figure 8.1 (a) Representation of one cycle of a pure rotating field as a circular locus and (b) creation of a d.c. circular rotational field by a two pole magnet

[4]A d.c. rotational field of this sort is usually set up in the laboratory by slowly rotating a source of constant d.c. field such as a permanent magnet or electromagnet normally at a speed of less than 1 revolution per minute.

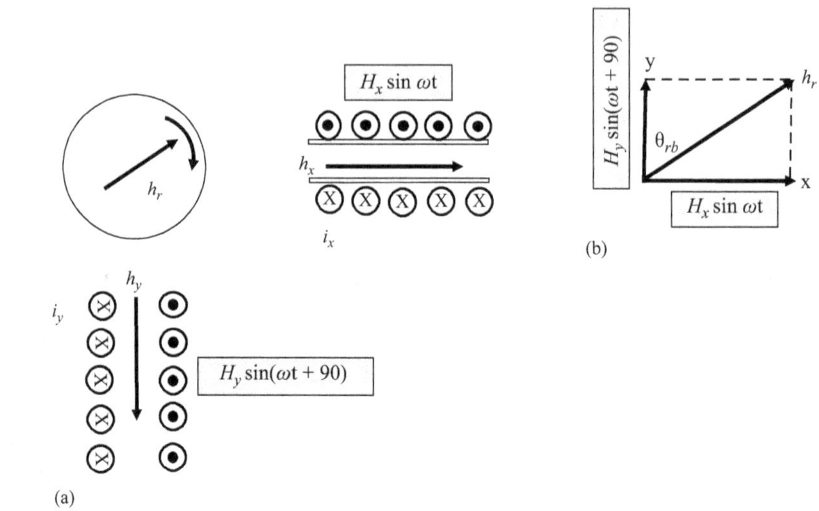

Figure 8.2 (a) Production of a rotating field using orthogonally positioned magnetising coils carrying a.c. currents and (b) vector representation of the rotating field

that at any instant in time the magnitude h_r of the resultant field and its direction θ_{rh} with respect to the y-axis is given by

$$h_r = \sqrt{h_x^2 + h_y^2}\ \left(\mathrm{Am}^{-1}\right) \tag{8.1}$$

and

$$\tan\theta_{rh} = \frac{h_y}{h_x} = \frac{H_y \sin \omega t}{H_x \sin(\omega t + 90^\circ)} \tag{8.2}$$

If the field is to be purely rotational of magnitude H_{rot}, then $H_{rot} = H_x = H_y$. Hence we can write

$$h_r = H_{rot}\ \left(\mathrm{Am}^{-1}\right) \tag{8.3}$$

and

$$\theta_{rh} = \omega t \tag{8.4}$$

Equations (8.3) and (8.4) show that the two coils, carrying equal magnitude a.c. currents which are 90° out of phase, produce a pure rotating field. In this case it has a magnitude H_{rot} and is rotating at the same angular frequency as the a.c. current,[5] i.e. if the frequency of the a.c. current is 50 Hz, the field rotates 50 times every second.

[5]This is equivalent to the way in which pure rotating fields are produced by two or three phase windings in rotating machine cores where the coil axes are displaced 90° and 120° apart, respectively.

If the frequency of the currents in the orthogonal magnetising coils is very low, the rotational field which it produces is identical to that produced by the rotating magnet. It will be noted that the rotational field is approximately constant as the electromagnet is rotated. It will be seen in Section 12.4 that rotational loss measurements are usually carried out under constant flux density conditions today, but there is on-going debate over which is most appropriate for the assessment of electrical steels under magnetising conditions which are most prominent in electrical machine cores.

8.2 Rotational flux density

A rotational field applied to a soft magnetic material produces a rotational flux density. As stated earlier, we define a pure rotational flux density as being of constant magnitude and rotating at constant angular velocity in the plane of the magnetisation. We can represent it as a vector in the same way as the rotating magnetic field.

If a pure rotational field is applied to an anisotropic material, generally, the resulting flux density will not be circular. There is a simple analogy with the case of the magnetisation of any material in one direction under a.c. excitation where both B and H cannot vary sinusoidally at the same time when its B–H relationship is non-linear as discussed in Section 6.4. This is to be expected, and can be explained if we make the gross assumption that the flux density at any instant in time is the resultant of two components produced independently by two orthogonal a.c. components of field. If the material is anisotropic and its relative permeability under a.c. magnetisation along the x and y directions is μ_x and μ_y then the instantaneous rotational flux density can be written as

$$b_r = \sqrt{b_x^2 + b_y^2}\,(\mathrm{T}) \tag{8.5}$$

from which

$$b_r = \sqrt{(\mu_x H_{rx} \sin \omega t)^2 + (\mu_y H_{ry} \sin(\omega t + 90))^2}\,(\mathrm{T}) \tag{8.6}$$

where H_{rx} and H_{ry} are the components of the rotational field in the x and y directions.

It is possible to solve this equation using basic trigonometrical identities. However, this is very difficult because of the need to insert the appropriate values of permeability at different time instances and the gross assumption that it is not necessary to use a tensor permeability factor to account for the non-linear way in which the relative permeability changes in magnitude as the material is magnetised at angles between the x and y directions. Equation (8.6) is introduced just to illustrate the complexity of rotational magnetisation; it is not developed further here.

Rotational flux in a specific region of a steel sheet can be measured using orthogonally positioned search coils such as the pair illustrated in Figure 8.3(a).[6] Suppose that, at any instant in time $\omega t = \delta$, the magnitude b_r and direction θ_{rb}

[6]Practical techniques for measuring flux density using search coils are explained in Section 12.3.1.

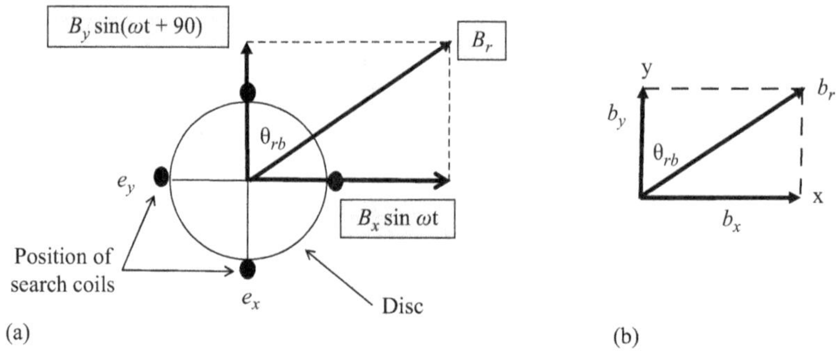

Figure 8.3 (a) Components of flux density detected by orthogonal search coils and (b) summation of instantaneous components of flux density to obtain b_r

(relative to the y-axis) are as shown in Figure 8.3(b). If the instantaneous flux density is replaced by components b_x and b_y, which vary periodically with time, they will have the general forms:

$$b_x = \sum_{n=1} B_{nx} \sin(n\omega t + \varphi_{nx})(\mathrm{T}) \tag{8.7}$$

and

$$b_y = \sum_{n=1} B_{ny} \sin(n\omega t + \varphi_{ny})(\mathrm{T}) \tag{8.8}$$

The corresponding electromotive forces (e.m.f.s) induced in the search coils are proportional to the rates of change of flux density at right angles to their axes. Their average values e_{xav} and e_{yav} are proportional to B_x and B_y according to (5.8). Hence

$$B_x = \frac{\pi e_{xav}}{2NA\omega}, \quad B_y = \frac{\pi e_{yav}}{2NA\omega}(\mathrm{T}) \tag{8.9}$$

where A and N respectively are the cross sectional area and number of turns of each search coil.

It is more convenient to treat the harmonic components individually and to superimpose the values to obtain the magnitude and position of b_r at any instant in time during the magnetisation cycle. The orthogonal components of flux density can be expressed simply as

$$b_x = B_x \sin \omega t(\mathrm{T}) \tag{8.10}$$

$$b_y = B_y \sin(\omega t + \Phi)(\mathrm{T}) \tag{8.11}$$

The corresponding induced e.m.f.s are

$$e_x = -NAB_x\omega \cos \omega t(\mathrm{T}) \tag{8.12}$$

and

$$e_y = -NAB_y\,\omega \cos \omega t(\mathrm{T}). \tag{8.13}$$

The magnitude and direction of the resultant flux density at an instant in time, when $\omega t = \delta$, are given by

$$b_r = \sqrt{b_x^2 + b_y^2} = \frac{\pi}{2NA\omega}\sqrt{(e_{xav})^2 \sin^2\delta + (e_{yav})^2 \sin^2(\delta + \Phi)}(\mathrm{T}) \tag{8.14}$$

$$\theta_{rb} = \tan^{-1}\frac{e_{yav}\sin(\delta + \Phi)}{e_{xav}\sin\delta} \tag{8.15}$$

The expressions given in (8.14) and (8.15) are derived assuming only sinusoidal components of b_x and b_y are present. When harmonics are present, Fourier analysis can be applied to the e.m.f.s and the harmonic components of flux density can be added (in vector form) to obtain the instantaneous total values, assuming that this superposition of the waveforms is valid.

In practice the B waveforms will contain harmonic components which can be analysed individually using (8.14) and (8.15) to obtain the complete locus.[7] Figure 8.4 [4] shows a typical flux density locus measured in the joint of a three-phase transformer where rotational flux is present. In this case fundamental and third harmonic elliptical components rotating in different directions are present.[8] The numbers on the locus represent equal time intervals, so it can be deduced that the angular velocity of the total rotational magnetisation is not constant.

The locus of the fundamental component is elliptical, as is widely found in practise. The strong magnetocrystalline anisotropy of this material is the cause of the direction of rotation of b_r reversing for short times during the cycle.

8.3 Torque curves and stored magnetocrystalline energy

Consider a single crystal disc of an anisotropic material which has an easy direction of magnetisation along the x direction as shown in Figure 8.5. Suppose a field H, applied along the x direction, is sufficiently large to form a single domain oriented along the same direction. Since the field and the domain are oriented along the same direction, the magnetic free energy stored in the disc is a minimum. If the

[7] More direct modern methods of such analysis on electrical steels are presented in Chapter 12.

[8] Of course counter rotating harmonics exist only in the mathematical representation. It should not be confused with the physical processes which occur under any component of magnetisation. A similar issue occurs in the mathematical treatment of rotational loss where, as will be seen in Section 8.7, negative components of loss are sometimes measured but the vector sum of the losses is always positive. This mathematical treatment of magnetisation and losses is very convenient in measurements and magnetisation modelling but attention should always be paid to the actual magnetic processes discussed throughout this chapter.

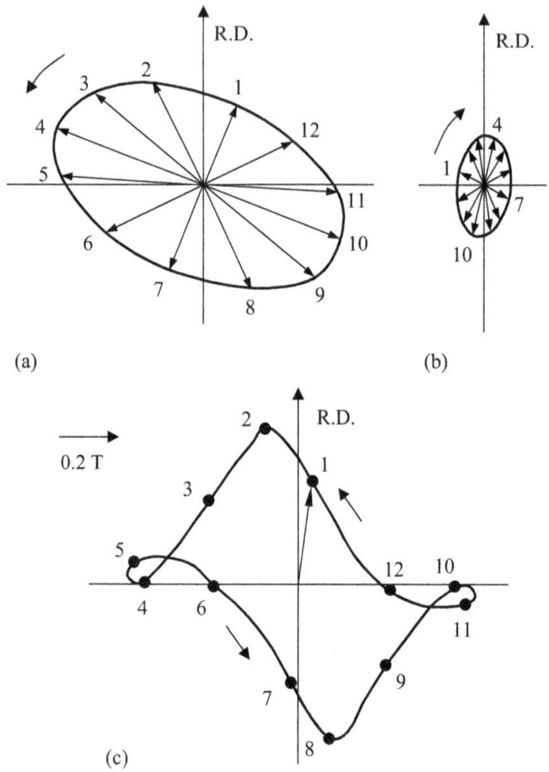

Figure 8.4 Magnitude and direction of flux density occurring in a joint of a three-phase transformer core assembled from GO SiFe laminations: (a) fundamental component at 30° intervals in time, (b) third harmonic component at 10° intervals in time and (c) total flux density at 30° intervals in time (Figure 4 in [4] reproduced under free licence CC-BY-4.0)

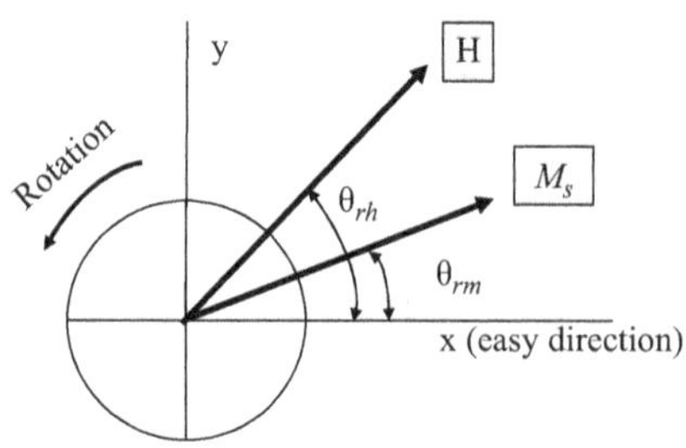

Figure 8.5 Angle between field and saturation magnetisation in a sample in a torque magnetometer

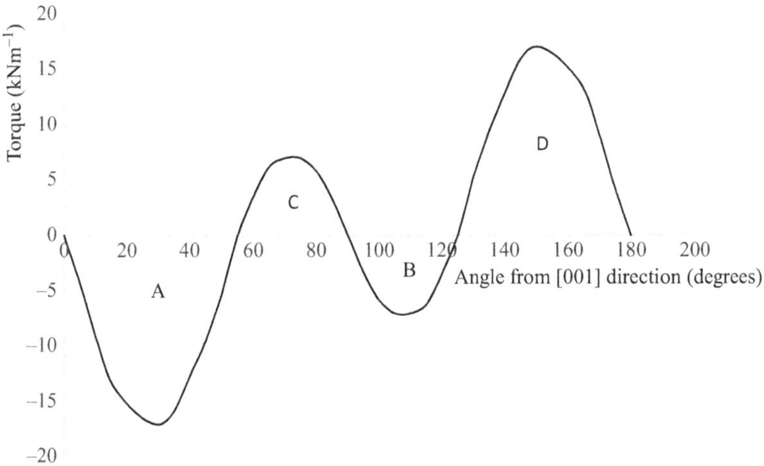

Figure 8.6 Theoretical torque-angle curve of a single crystal of (110)[001] oriented silicon–iron ($K_1 \approx 28{,}700\ Jm^{-3}$, $K_2 \approx 10{,}000\ Jm^{-3}$)

field is slowly rotated to an angle θ_{rh} from the x-axis the magnetisation will be rotated to the position shown in Figure 8.5 where the sum of the field energy (proportional to $\sin \alpha$) and anisotropy energy (proportional to θ_{rm}) is a minimum. At the same time, the disc experiences a mechanical torque tending to align the field and flux density.[9] The torque T is given by

$$T = -HM_s \sin \alpha (\text{Nm}) \tag{8.16}$$

The negative sign appears because the torque opposes the clockwise motion of H_r. Since the disc comprises a single domain, the flux density is approximately equal to its saturation magnetisation M_s. If the disc rotates by a small angle $d\theta$ the work done, in a lossless situation, amounts to the stored energy. If it is assumed that the only contribution to the change in stored energy comes from the change in magnetocrystalline energy E_k, we can write

$$T = -\frac{dE_k}{d\theta} \left(\text{Jm}^{-3} \text{ or Nm/m}^3\right) \tag{8.17}$$

If the field is rotated through 360° and the torque on the disc is continuously measured, a torque-angle curve for that material under the particular field conditions can be plotted. This is the principle of a torque magnetometer. Although it is rarely used today because the material information can be obtained far more conveniently using modern techniques, it still has advantages for high field measurements. Figure 8.6 shows the general shape of the torque curve of a single crystal

[9]This is analogous to the force on the magnet in Figure 2.1(c) tending to align the magnet with the external field.

disc of (110)[001] SiFe. The characteristic is very similar to that of a specimen of well oriented GO electrical steel.

If the direction cosines relative to the crystal magnetisation are substituted into the theoretical expression for E_k, given in (3.1), and the expression differentiated with respect to θ, the theoretical torque curve for a (100)[001] crystal can be expressed as [5]:

$$T = -\left({}^{K_1}/_4 + {}^{K_2}/_{64}\right)\sin 2\alpha - \left({}^{3K_1}/_8 + {}^{K_2}/_{16}\right)\sin 4\alpha + \left({}^{3K_2}/_{64}\right)\sin\, 6\alpha \mathrm{Nm/m^3} \tag{8.18}$$

This is the torque-angle expression obtained when the material is saturated ($M = M_s$). The shape of the curve in Figure 8.6 contains information about the magnetisation. Over region A, the torque is negative, effectively trying to twist the disc so the field lies along the [001] direction. In region C the preferred minimum energy condition is actually achieved when the field lines up with the [110] direction. No torque is exerted on the disc when the angle between the field and the [001] direction is about 55°, the exact angle depends on the relative magnitudes of the anisotropy constant K_1 and K_2. These can be found from the curve although, in practice, it is not particularly accurate, partly because of the difficulty in ensuring that the whole of the sample is fully magnetically saturated at all times.

The variation of magnetocrystalline energy of a (110)[001] oriented crystal with magnetisation direction in its (110) plane is shown in Figure 8.7. This characteristic is obtained by inserting $l = \cos\theta, m = (1/\sqrt{2})\cos\theta, n = (1/\sqrt{2})\sin\theta$ into (3.1) to give

$$E_k = \frac{K_1}{2}\left(\cos^2\theta\left(1 + \frac{\sin^2\theta}{2}\right)\right) + K_2\left(\cos^4\theta\sin^2\theta/4\right)\left(\mathrm{Jm^{-3}}\right) \tag{8.19}$$

The plots shown in Figure 8.7 are simply the trigonometrical functions in the brackets in (8.19) independent of the numerical values of K_1 and K_2. Stable conditions when the energy is minimum and these correspond with magnetisation along the [001] and [110] directions when the torque is zero. It can be seen that the contribution of the second term is lower than that of the first. In iron, and silicon–iron, the magnitude of K_2 is around one third that of K_1 and it can be ignored in most studies. The torque experienced by the material is also zero when magnetised at 54.7° to the [100] direction which corresponds to the [111] direction. The relative values of K_1 and K_2 determine the ratio of the peak magnitudes of the torque but do not affect the zero torque positions.

Figure 8.7 clearly shows that the favoured direction of magnetisation in a (110) [001] oriented single crystal is the [001] direction since the stored energy is lowest. The fact that the torque on a sample magnetised along the [001] direction is zero indicates that it is independent of the numerical values of K_1 and K_2. The large energy stored when the magnetisation is rotated only a small angle from the [001] direction is evidence of the reason why domains in these materials only move out of

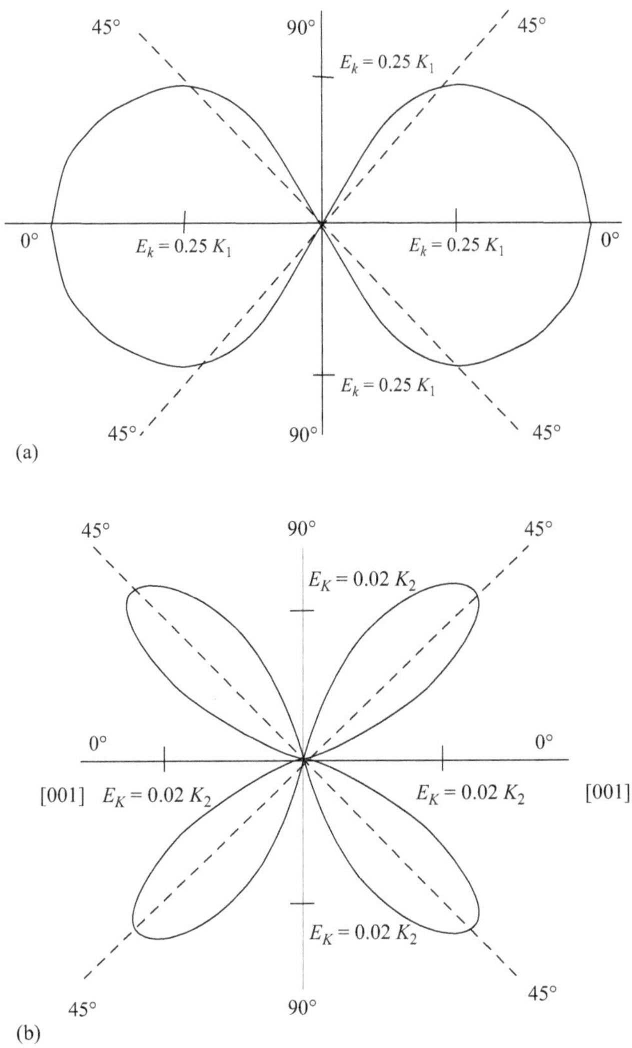

Figure 8.7 Angular variation of stored magnetocrystalline energy in a (110)[001] oriented SiFe crystal: (a) contribution from the K_1 term and (b) contribution from the K_2 term

the {100} directions if a far higher energy contributor, such as an extremely large stress or magnetic field, is present.[10]

[10]Equation (8.19) can be differentiated and equated to zero to find angles at which the energy is maximum or minimum. If the first term is differentiated a maximum at 45° will be found and if the second term is differentiated a maximum appears when $\theta = \tan^{-1}(1/\sqrt{2})$ or 35.26° (90°–54.74°). To find the true minimum, the relative values of the anisotropy constants must be included and this will give an intermediate angle showing that K_2 does play a role in determining the anisotropic properties of GO silicon–iron.

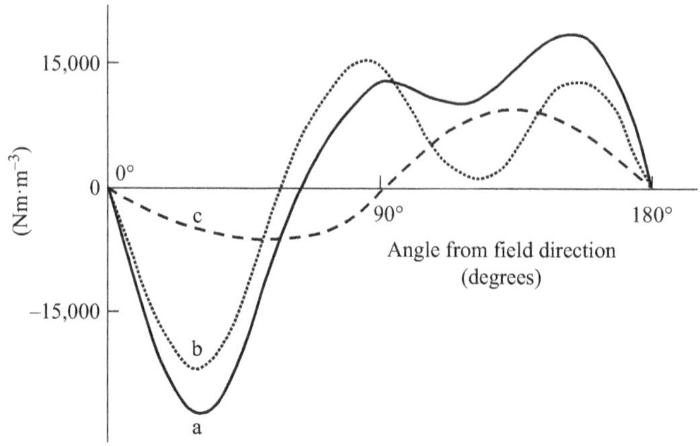

Figure 8.8 Effect of magnetic anneal on the torque curves of a sample of NO 2.7 % silicon–iron (a) before magnetic anneal, (b) after magnetic anneal and (c) change due to magnetic anneal (Figure 8 in [6] reproduced under free licence CC-BY-4.0)

The shape of a practical torque curve, measured in a saturating field, is temperature dependent but it also depends on the total magnetic anisotropy of a sample. The temperature sensitivity occurs simply because the magnitudes of K_1 and K_2 are temperature dependent as shown in Section 18.1.3.

The shape of a torque-angle curve of a polycrystalline material is dependent on its texture and other factors such as the effect of magnetic or stress annealing. Figure 8.8 [6] shows the effect of magnetic annealing on the torque curve of a NO steel. Before annealing, the maximum torque is only around 30,000 dyne.cm/cm^3 (3,000 Nm/m^3)[11] and the irregular curve[12] is related to the crystal texture of the steel. After magnetic annealing the torque curve is still irregular but the difference between the curves, shown as curve (c), is the effect of an additional torque of the form $T_u = K_u \sin 2\theta$ where θ is the angle between the field direction and the measurement direction. This is the characteristic torque curve caused by any form of uniaxial anisotropy in a material which is briefly discussed in Section 3.6.2.

8.4 Rotational hysteresis loss

The torque and energy characteristics given in Section 8.3 are based on the material in question acting as a single magnetic domain in a completely saturated state. In such a case the areas enclosed by the positive and negative portions of the torque-angle curves are equal when summed over one cycle of rotational field.

[11]The peak torque in grain oriented steel is of the order of ten times this value giving an indication of the effect of its high crystal anisotropy.

[12]The shape is dependent on the crystal texture. If the material had a fully random (isotropic) structure the torque would be zero.

Suppose a magnetic field of any magnitude is applied to a fixed disc of magnetic material and slowly rotated through 360°. The energy flowing into the system is given by

$$P_{rin} = -\int_{\theta_{rh}=0}^{\theta_{rh}=2\pi} T\,d\theta\left(\mathrm{Jm}^{-3}\right) \tag{8.20}$$

If no losses occur the energy E_{rhys} is stored in the material. The net work done during the cycle is zero if H and M are constant.[13]

At a lower value of field, when the disc no longer comprises a single domain, the rotational magnetisation process becomes more complicated; both domain wall displacement and, perhaps, domain rotation, occurring simultaneously. In this situation all of the energy stored is not recovered during the 360° rotation but is converted into heat which we refer to as *rotational hysteresis loss*. We can combine (8.16) and (8.17) to obtain an expression for the rotational hysteresis loss per cycle as

$$W_{rhys} = \int_{0}^{2\pi} B_r H_r \sin\alpha\,d\theta\,\left(\mathrm{Jm}^{-3}\right) \tag{8.21}$$

where B_r and H_r are the corresponding instantaneous values of rotational flux density and field, respectively. The origin of this loss is the same as that of the uniaxial d.c. hysteresis loss introduced in Section 7.2.

When rotational hysteresis loss occurs the sum of the areas A and B on Figure 8.6 are not identical to the sum of areas C and D and the energy fed into the material is greater than that returned to the source. This is because rotational loss has occurred. Rotational loss is not generated at a constant rate throughout a 360° rotation of field which would cause the equivalent of a d.c. offset to the complete curve but in practice it manifests itself as a distortion in the shape of the curve.

A further point to note is that the Steinmetz equation (7.17) applies to the area of a B–H loop obtained under a.c. unidirectional magnetisation so it should not be expected to fit very accurately any rotational loss characteristics [7].

The quantification of the rotational hysteresis from torque-angle curves measured at increasing values of applied field was established more than 120 years ago by Bailey [8]. He was probably the first[14] to carry out a quantitative investigation of rotational magnetisation of soft magnetic materials [9]. His results for 'soft iron' were reproduced in Sir James Ewing's classic work and it is reproduced in Figure 8.9 [10]. It is interesting that Ewing pointed out that Steinmetz's formula had no application at high rotational magnetisation. At the same time, the fact that the hysteresis loss should vanish at high rotational field seemed so improbable that it was put forward as a criticism of Ewing's molecular theory of magnetism which

[13]Simply substitute for T from (8.12) and integrate (8.13) to find it equal to zero.

[14]This is not a historical review, so we only refer to limited early research into rotational magnetisation. However, it may be of interest to say that around ten years before Bailey's publication, Ferraris was possibly the first to investigate hysteresis loss in a rotating field, when, in 1888, he deduced that a laminated iron core could be made to rotate in a magnetic field by virtue of its hysteresis.

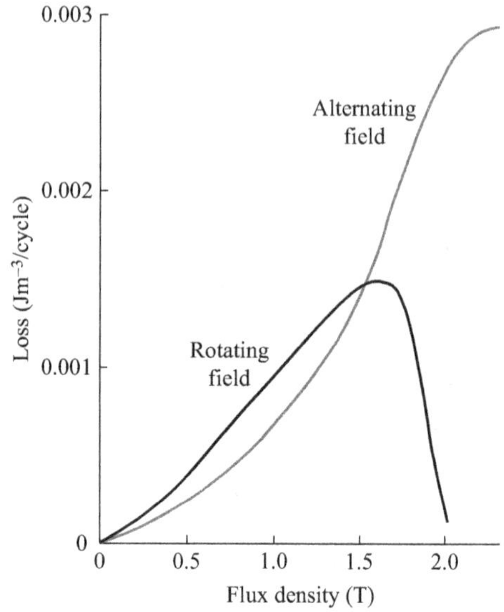

Figure 8.9 Early comparison between rotational and a.c. (hysteresis) losses in soft iron (Figure 151 in [10] reproduced under free licence CC-BY-4.0 and updated)

he had developed from the original concept forwarded by Weber around 50 years earlier.

From a historical view point, perhaps it is worth a reminder that when explanations of the reduction of rotational hysteresis at high magnetisation were first being proposed magnetic domains were unheard of, all losses in a magnetic core were considered to arise from hysteresis processes and the role of eddy currents was not yet appreciated. (Indeed the term *eddy current* is not mentioned in Ewing's classical work.)

It is interesting to look at the numerical result in Figure 8.9. The alternating loss shown (no frequency specified but probably less than 50 Hz) is apparently measured up to 23,000 gauss (2.3 T) which seems remarkably high but the peak of the rotational loss occurs at around 1.6 T which is not far from what is found in modern materials. However, its peak value of 15,000 ergs/cc (presumably per cycle), which is equivalent to 1,500 J/m^3/cycle, is around five times higher than is found in today's standard grades of NO electrical steels.

Figure 8.10(a) shows the rotational hysteresis characteristic of a GO steel sample compared with the hysteresis characteristics of the same material under a.c. magnetisation [11]. The rotational hysteresis loss drops to close to zero as expected at high flux density and at the same time the a.c. hysteresis is lower than the rotational loss at low flux density. It should be noted that by a.c. hysteresis in this context the author [11] means the static hysteresis loss given by Steinmetz's equation at the given flux density.

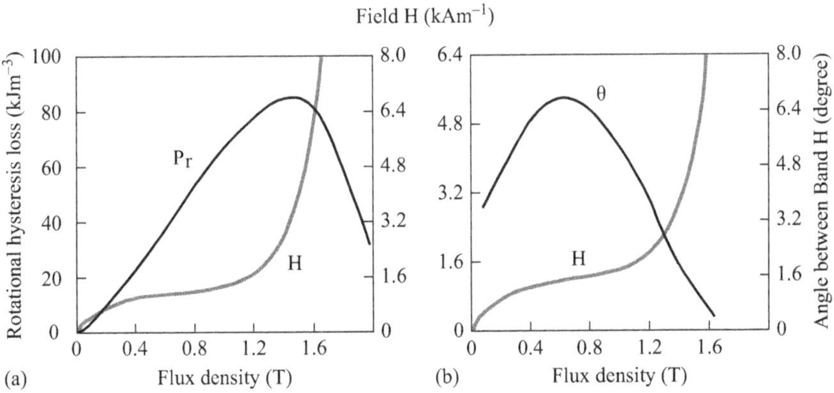

Figure 8.10 (a) Variation of the magnitude of rotational and a.c. hysteresis with flux density in a GO steel sample (Figure 11.30 in [11] reproduced under free licence CC-BY-4.0 and updated) and (b) variation of phase angle between B_r and H_r and the corresponding variation of H_r (Figure 11.29 in [12] reproduced under free licence CC-BY-4.0 and updated)

The phase angle between rotational *B* and *H* drops to close to zero as the flux density nears saturation as shown for a different material in Figure 8.10(b) [12]. The flux density at which the rotational hysteresis is maximum does not necessarily correspond with the highest phase difference between B_r and H_r. This might not appear consistent with the fact that loss is dependent on the sine of the angle, but it should be remembered that the field required to magnetise the sample is not directly proportional to the loss. The peak is more closely related to the domain structures present under rotational magnetisation discussed in Section 8.5.

The ratio of the rotational hysteresis loss to the a.c. hysteresis loss depends on many factors and can be much higher than shown in Figure 8.9. It has been found to increase rapidly with reducing thickness, doubling in value from a maximum of around 5 to 10 as thickness is reduced from 0.013 inches (0.33 mm) to 0.002 inches (0.05 mm) in GO steel [13].

At low fields, the demagnetising factor of the disc must be taken into account when calculating the true field inside the sample. Likewise, accurate measurement of the uniformity of the d.c. flux density within a normally tested thin disc sample is often questionable. Because of these points the accuracy of rotational loss characteristics obtained from torque curves is often poor.[16]

It can be concluded that, in general, the measurement of rotational hysteresis loss using the torque magnetometer is useful as a qualitative test but satisfactory

[16]It should be remembered that Figure 8.10 shows d.c. characteristics which do not include the effect of eddy current loss. Characteristics at power frequency are discussed in Section 8.7.

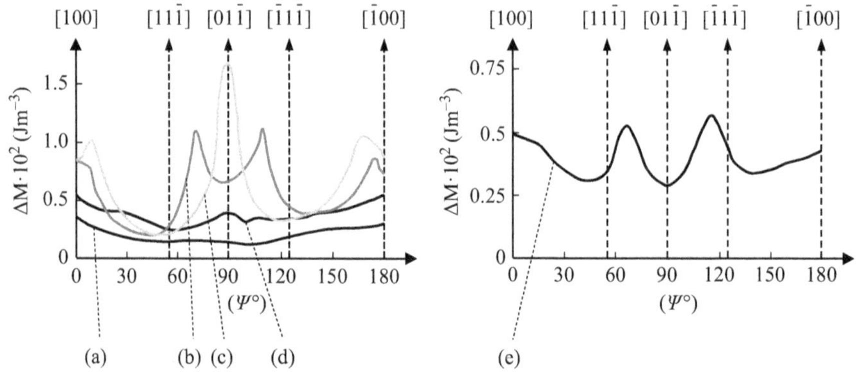

Figure 8.11 Variation of rotational hysteresis loss with angular position of the rotational field in a (110)[001] single crystal of silicon iron relative to the [001] direction at magnetisation levels M/M_s of (a) 0.40, (b) 0.61, (c) 0.76, (d) 0.87 and (e) 0.95. (Note (e) is shown on separate graph to improve the clarity (Figures 2 and 3 in [15] reproduced under free licence CC-BY-4.0 and combined)

quantitative results are difficult to achieve [14]. This may be true, but interesting material characteristics can be obtained with an accurate magnetometer which might still be difficult to carry out using any modern approach.

One use of the torque magnetometer is to determine the instantaneous variation of rotational hysteresis loss throughout a cycle. This can be obtained as the difference between measurements taken when rotating the field in opposite directions [15]. Figure 8.11 shows how the loss reaches peak values at different positions relative to the [001] direction as the flux density in a (110)[001] oriented single crystal of 3.25% silicon–iron is increased from around 40% to 95% saturation. At a low flux density, the loss is virtually independent of the angle of magnetisation, at intermediate field weak maxima appear when the field is along the [100] direction and, as the field is increased further, maxima occur when the field is along [011] and [111] directions. At a very high field the peaks occur just when magnetised close to the [111] direction at around 54.7° to the [100] direction. Close correlation exists between the loss variation with position of the field and the variation of static domain structure under rotational field as shown in Section 8.5.

8.5 Magnetic domain structures under rotational magnetisation

When a rotating magnetic field is applied to a magnetic material, magnetisation occurs by domain wall displacement and domain rotation processes described in Section 3.5. Domains are distributed when the field is oriented along any direction during slow or rapid rotational magnetisation to produce a minimum energy state. The domain structure therefore depends on factors such as temperature, magnitude

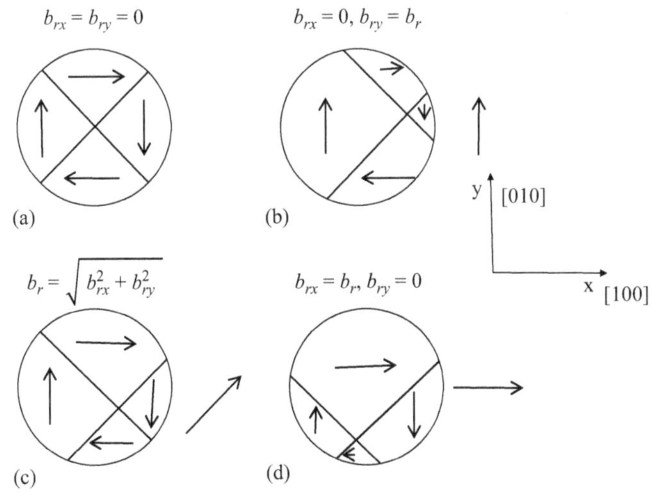

Figure 8.12 Schematic domain structure illustrating the basic wall displacement during a cycle of magnetisation rotation in an idealised (100)[100] iron crystal

and speed of rotation of the field, crystal orientation, or texture, and the presence of mechanical stress. If the magnetocrystalline anisotropy constants are high, as they are for electrical steel, then magnetisation will take place predominantly by domain wall displacement.

Consider a single crystal of silicon–iron with [100] and [010] directions as shown in Figure 8.12. The [001] direction is perpendicular to the plane and no domains are oriented out of the plane. Figure 8.12(a) shows its idealised demagnetised structure. Suppose the crystal is subjected to a slowly rotating clockwise field H. Figure 8.12(b) shows the domain structure when the field is oriented along the [010] direction. This is the most basic domain structure which could exist in the crystal to satisfy the minimisation of energy requirement. In this position the greatest domain volume is oriented in the y direction giving a net flux density b_{ry} as shown at this instant. When H rotates so that it is at 45° to the easy axes, the flux density, b_r, has rotated likewise but the domains themselves are still oriented along easy directions. When the field has rotated to the [100] direction, as in Figure 8.12(d), the domain structure becomes identical to that in (b), but rotated through 90°. The two components, b_{rx} and b_{ry} are physically present in the material. If the field is constant in magnitude the instantaneous values of flux density vary with angular position.

Important points to note are:

- the domain pattern is at no instant the same as the demagnetised structure shown in Figure 8.12(a);
- at no time does the magnetisation move out of {100} directions, even when the field is not directed along an easy direction of magnetisation because of the high anisotropy in this case;

- at some instances during the rotation, flux density components are present which are not oriented along the field direction. This does not generally occur under uniaxial magnetisation along an easy direction of magnetisation such as shown in Figure 3.15 and
- the simplified magnetisation process described above is restricted to domain wall displacement. In practice, it dominates in electrical steels because of the high magnetocrystalline anisotropy tending to make the magnetisation remain in {100} directions, apart from at very high fields.

Let us briefly consider the case when domain magnetisation rotation is the only process which occurs when a rotating field is applied to a material. Suppose a rotational field applied to a disc specimen of isotropic material is large enough to ensure that a single domain exists while the field is slowly rotated through one revolution. As the field is rotated, the domain stays perfectly aligned with the field. No domains are formed, or annihilated, and no domain wall motion occurs, so no losses occur.[17] If the same field is rapidly rotated in the isotropic crystal then the flux density will lag the field by an angle θ_{bh}.

If the material is anisotropic, with an easy axis in the plane of rotation, this angle varies through the cycle as illustrated in Figure 8.13. The angle between the domain orientation and the field increases as the field moves away from the easy axis as dipoles tend to move less freely from that direction. As the field approaches the easy axis the opposite occurs and the domain becomes oriented closer to the easy axis than the field provided that the crystal anisotropy is high. In each case, a single domain is present. In (a) dipole moments rotate very slowly so no losses occur but in (b) and (c) eddy current loss occurs as dipoles rotate rapidly.

It can be concluded that, at a very high field, when all the magnetisation is due to domain rotation, no rotational hysteresis loss occurs, although eddy current loss is still likely to occur.

Figure 8.14 shows schematic images of actual domain patterns observed on the (001) surface of a single crystal of 3.25% Si Fe [16]. When the field is applied along the [100] direction the domains initially oriented in that direction grow at the expense of those oriented along the [$\bar{1}$00], [001] and [00$\bar{1}$] directions. If the field is sufficiently large, a single domain parallel to the field direction is formed as expected. When the field is applied along the [110] direction the process in (b) occurs. Initially, domains in the [100] and [001] easy directions closest to the field direction become larger. As the field increases, eventually they occupy the whole sample and finally they are rotated out of these easy directions. Similar structures on the surface of a (100)[100] oriented single crystal of 3% silicon–iron were reported later [17].

Figure 8.15 shows the domain structure on the surface of a sheet of commercial (110)[001] grain oriented steel under rotational magnetisation. The surface domain structure at the point when the field has rotated to produce a flux density of 0.7 T at

[17]If the field is rotated rapidly, eddy currents are produced due to the rotating dipoles but this is not regarded as rotational hysteresis loss here.

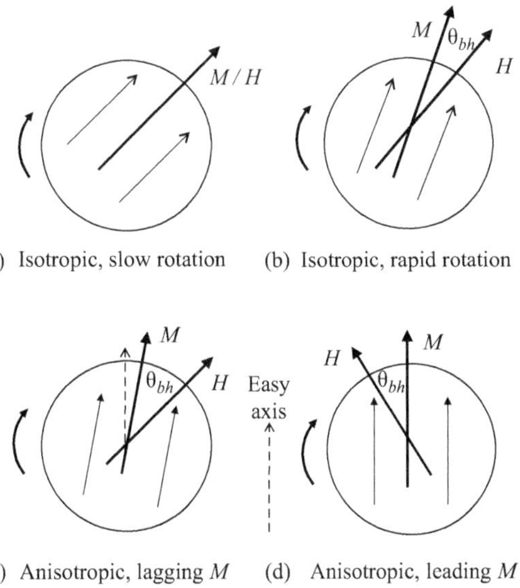

Figure 8.13 Alignment of a hypothetical single domain in a disc (a) isotropic, slowly rotating field, (b) isotropic rapid rotation of field, (c) anisotropic, slow or rapid rotation, H rotating from an easy axis and (d) anisotropic, slow or rapid rotation, H approaching an easy axis

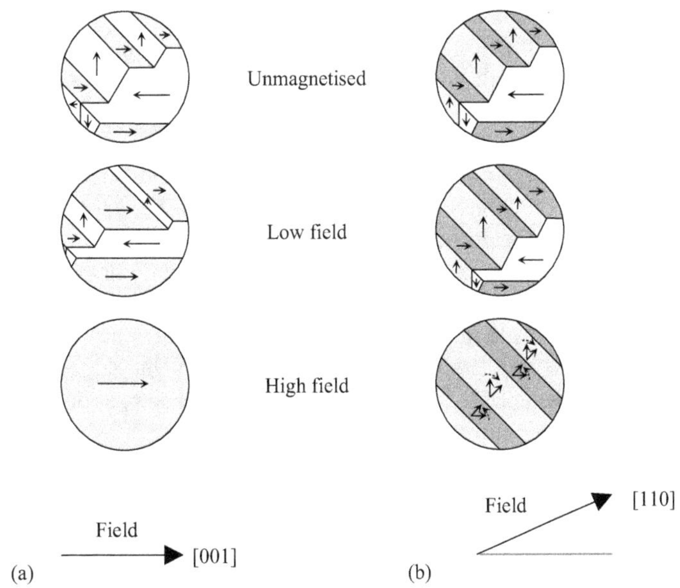

Figure 8.14 Domain wall displacement observed on the surface of a [100)(001) oriented crystal of silicon–iron during rotational magnetisation: (a) field along the [001] direction, (b) field along the [110] direction (field of view ≈ 0.5 mm diameter) (Figure 6.19 in [16] reproduced under free licence CC-BY-4.0)

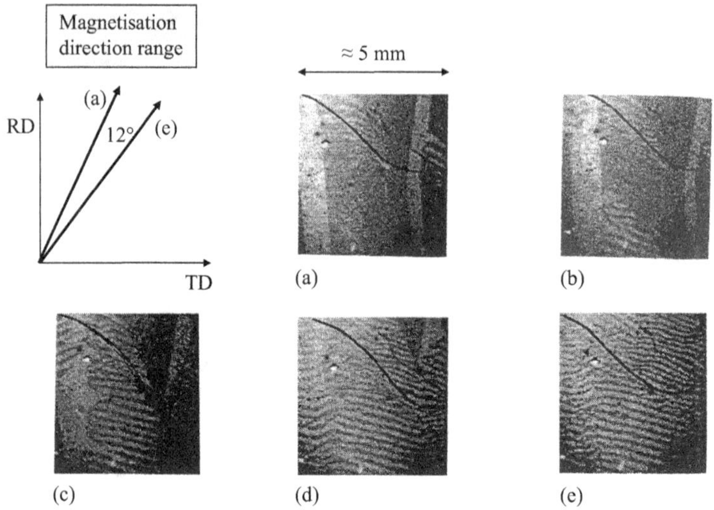

Figure 8.15 Domain observations on the surface of a (110)[001] oriented electrical steel as a d.c. field rotates from 25° to 37° to the RD: (a) 25°, (b) 28°, (c) 31°, (d) 34° and (e) 37° (Figure 7 in [18] reproduced under free licence CC-BY-4.0 and modified)

25° to the RD (rolling direction), is shown in (a). The wide domain (dark region) in the centre of the photograph is oriented along the [100] directions of each grain giving a net component of magnetisation parallel to the field direction. When the flux density has rotated to 28° to the RD (image (c)) the same domain breaks up more as indicated by surface transverse domains. As the field is rotated further, the volume of transverse domains increases until, when the flux density has rotated to 37° to the RD, they cover the entire surface.

The transverse structure is, in fact, made up of closure domains on the surface of the steel oriented along the [100] and [$\bar{1}$00] directions. This structure is very similar to that caused by compressive stress applied along the RD of GO steel discussed in Section 11.3. The domain structure beneath the surface is probably complex and can be deduced to some extent from the surface patterns but it appears that most activity occurs during an interval as the field rotates by only around 10° in space. It is expected that the instantaneous power loss is far higher than average during this part of the rotational cycle.

The theory developed in Section 8.2 shows that if the spatial locus of flux density is circular then the components of B_r in the orthogonal directions will be sinusoidal as stated.

However, it does not prove the converse, i.e. if we have sinusoidal orthogonal components then the locus of B_r must be circular although this is widely assumed and accepted.

Figure 8.16 shows the positions of single-turn search coils wound through holes in a specimen of GO steel subjected to rotational magnetisation. Magnetisation was

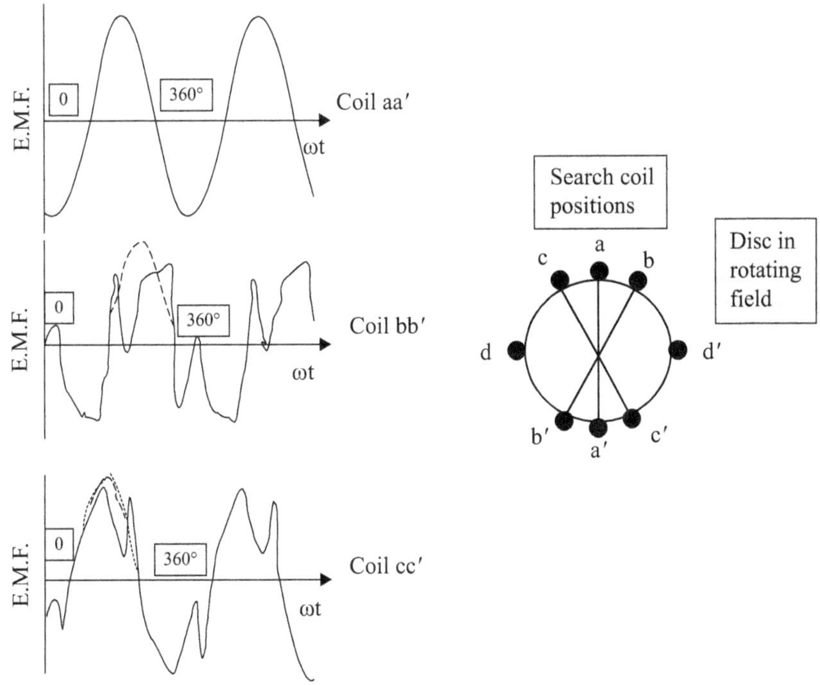

Figure 8.16 The e.m.f. waveforms induced in search coils mounted on a sample of GO SiFe subject to controlled rotational flux density of magnitude B showing the controlled sinusoidal e.m.f. waveform induced in coil aa' and waveforms of induced e.m.f.s induced in coils bb' and cc' spaced at ± 30° to coil aa'. The dotted curves show the e.m.f. waveform induced in half a cycle of magnetisation in aa' superimposed on those induced in the other coils to illustrate similarities and differences (Results from personal records held by Professor A.J. Moses)

controlled such that sinusoidally varying e.m.f.s, 90° out of phase and with equal peak values were set up in coils *aa'* and *dd'*. This implies that a circular rotational flux density is set up in which case the induced voltages in coils *bb'* and *cc'* should also be sinusoidal and displaced ± 30° with respect to the e.m.f. induced in *aa'*. It can be seen that this is not the case and the induced voltage due to the flux density in the uncontrolled regions is far from sinusoidal and each is around 55° out of phase with the e.m.f. induced in *aa'*.

The validity of the results in Figure 8.16 has not been confirmed. The waveforms were photographed from an oscilloscope screen and no experimental details exist.[18] We are not aware of such measurements being repeated in a more

[18]These measurements were made during the development of a rotational loss tester at the Wolfson Centre for Magnetics at Cardiff University in the mid-1970s. Accuracy and reproducibility are not known.

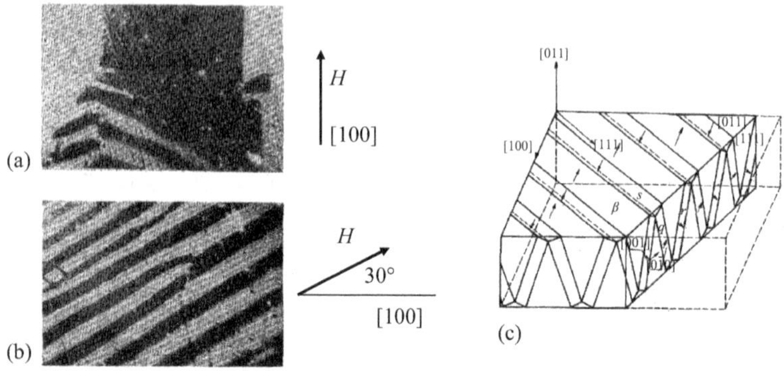

Figure 8.17 Domain patterns on the surface of a (110)[001] oriented single crystal of SiFe under rotational magnetisation: (a) when the field (15,840 Am^{-1}) is directed along the [100] direction, (b) when the field directed at 30° to the [001]direction and (c) proposed configuration of the structure in (b) (Figures 7 and 11 in [15] reproduced under free licence CC-BY-4.0 and modified)

systematic way or if similar findings have been reported. There are many reasons why such distorted waveforms might arise but if the results are valid, there might be a connection with the rapid domain switching over a narrow spatial angle observed in the domain observations shown in Figure 8.15, but this is only speculation.

Figure 8.17(a) shows how the domain structure observed on the surface of the (110)[001] single crystal, whose rotational hysteresis loss is shown in Figure 8.11, changes with field direction [14]. The dynamics of the narrow domains oriented at around 30° to the [100] direction are believed to be a part of the reason for peaks in instantaneous loss shown in Figure 8.11.

At first sight, the structures shown in Figure 8.17(b) may appear to differ significantly from those in Figure 8.15 observed on the surfaces of a well oriented (110)[001] grain in polycrystalline material . However, on closer inspection it does appear that transverse structures observed in [16] are actually the [111] oriented domains referred to in [16]. In earlier work on d.c. rotational magnetisation in a (110)[001] single crystal of silicon–iron, the two types of domains were observed at different flux densities when the magnetisation was along a [110] direction. In that case, at low flux density, the surface domains oriented along the [110] direction dominated but as the flux density is increased the domains at an angle to the [110] direction become more established [19]. The inclined domains are not oriented along an easy direction and the way in which they are reorganised is the cause of the loss peak. It is intended here to illustrate the complexity of these structures, not to try to explain them.

It should be noted that the images shown in Figures 8.14, 8.16 and 8.17 were obtained under d.c. magnetisation, i.e. a field is applied in a set direction, measurements are made and repeated after the field has been rotated to a new position. So, although the authors refer to this as rotational magnetisation, in fact, it is magnetisation at fixed angles to an easy direction in an anisotropic material. If the field were rotating continuously (which we define as a true rotating field) then the domain structures depend on the rotational speed in a similar way to the dependence of domain structures on magnetising frequency under unidirectional a.c. magnetisation. Under such conditions the structures are more complex [20] and difficult to interpret. However, an adequate understanding of the general phenomenon can be obtained from the d.c. images shown here.

8.6 Combined alternating, rotational and d.c. offset magnetisation

In a magnetic material subject to d.c. magnetisation, the flux density and magnetic field at any point within the material are, of course, constant and the condition is relatively simple to visualise and analyse. The analysis is a little more difficult if the magnetisation varies sinusoidally with time, under what is often referred to as sinusoidal a.c. conditions. If, under a.c. conditions, both B and H vary non-sinusoidally with time the condition becomes more complex again. The next stage of complexity comes when the magnetisation varies in direction but either B or H remain constant which is referred to as pure rotational magnetisation. The highest state of complexity is when the magnetisation contains d.c., alternating and rotational components.

Under such complex magnetisation conditions, domain structures, internal magnetisation and loss mechanisms become difficult to measure, analyse or predict. This is of growing practical interest as users and producers of soft magnetic materials become more aware that such conditions are common in many applications. Aspects of such complex magnetisation are briefly introduced in Sections 8.6.1 and 8.6.2.

8.6.1 Combined alternating and rotational magnetisation

The flux density in localised regions of materials used in many applications contains both alternating and rotational components. Some combinations are depicted in Figure 8.18. In (a), the angular frequency of the rotational field and the frequency of the a.c. field are equal, neither contain any harmonic components and the magnitudes of the peak value of the a.c. field and the magnitude of the rotation field are equal. The resultant is simply the vector sum of H_{xac} and H_{xr} in the x direction and the value of H_{yr} in the y direction at any instant in time. This is simply given by

$$h_{xr} = H \sin \omega t + H \cos \omega t \left(\mathrm{Am}^{-1}\right) \tag{8.22}$$

$$h_{yr} = H \sin \omega t \left(\mathrm{Am}^{-1}\right) \tag{8.23}$$

The resultant is the elliptical shaped field locus already shown in Figure 8.4. This type of elliptical field produces a far more complex flux density locus because of the non-linear relationship. The representation of other combinations of a.c. and d.c. fields are shown in Figure 8.18(b) and (c).

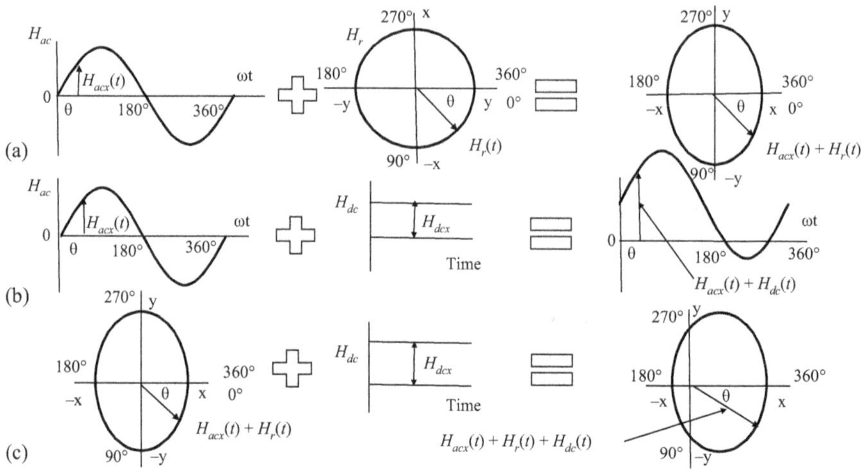

Figure 8.18 *Waveforms of combined fields over a cycle of a.c. magnetisation: (a) superimposed circular rotational field and a.c. component of the same period to give an elliptical spatial locus, (b) superimposed a.c. and d.c. offset field and (c) superimposed circular rotational field and d.c. offset field along its the principal axis to give an offset elliptical locus*

Losses under such complex magnetisation conditions can be measured using methods discussed in Chapter 12. There are obviously infinite combinations of magnitude, phase and frequency of the components of alternating and rotational magnetisation so it is not surprising that no systematic characterisation of materials has been reported. However, one comprehensive study [21] resulted in an interesting finding that the rotational loss can be lower under elliptical magnetisation with axes inclined at an angle to the RD, than when the principal axis is along the RD of GO steel. This does not seem to have been fully quantified or verified by other workers.

Figure 8.19 shows what is possibly the first reported variation of loss with flux density under combinations of rotational and alternating field ranging from pure alternating (0%) to pure rotating (100%) in NO and GO steels [22]. The main trends to note are that, over the measured flux density range, the rotational loss in both materials is always greater than under a.c. magnetisation for the same peak flux density, and that the relative ratio of rotational loss to a.c. loss in the GO material is higher than in the NO steel. Similar findings were reported by other authors [23,24]. Similar characteristics are found in modern steels.

The trend seems to be that the ratio of rotational loss to a.c. loss at corresponding flux densities and magnetising frequencies increase with the degree of material anisotropy as suggested from early work on a limited range of NO electrical steels whose loss anisotropy factor varied from 1.3 to 1.9 and the corresponding ratios of rotational loss to a.c. loss magnetised along the RD varied from

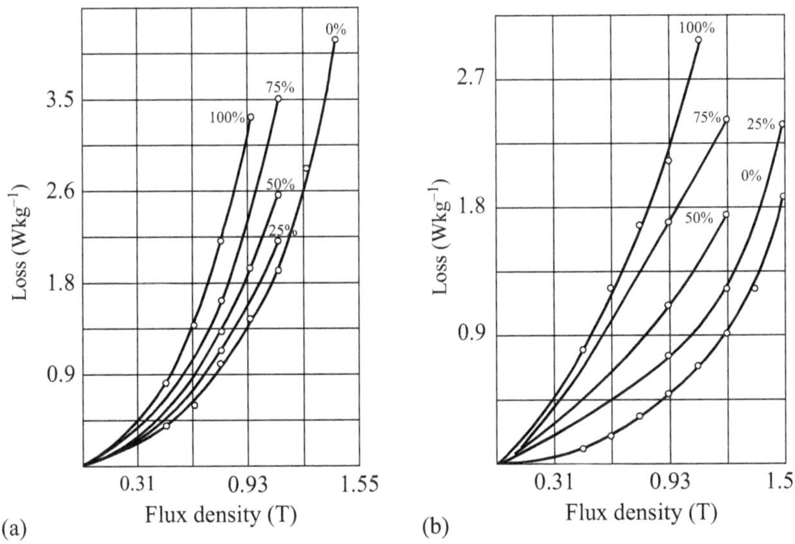

Figure 8.19 Variation of loss with flux density under combinations of alternating and rotating fields: (a) NO steel (b) GO steel (%= component of B in TD/component of B in RD) × 100 = 2.2 W kg^{-1} (Figure 1 from [22] reproduced under free licence CC-BY-4.0 and updated to SI units)

2.7 to 3.3 [25]. In laboratory produced (100)[001] material[19] the ratio of rotational loss to a.c. loss at the same flux density has been found to be far lower than in GO silicon–iron showing that it might be a useful material in transformer cores whose corner joints are subjected to large rotational flux [23].

Table 8.1 shows experimental measurements of the ratio of rotational loss to a.c. loss in various materials. Because the ratio varies significantly with B, the values presented correspond with the flux densities at which the ratios are highest (if possible) or a few are chosen to show the trend. It is not surprising that there is a wide spread of values for nominally the same materials and magnetisation conditions because of the difficult experimentation. However, the one trend which is very apparent is the increase in the ratio with increasing anisotropy expected from domain studies and the differences in the rotational B–H characteristics, particularly the variation of the spatial angle between B_r and H_r during a magnetising cycle.

It was proposed in [24] that the loss under an elliptical applied field can be calculated from the sum of the losses caused by major axis and minor axis fluxes existing separately, i.e.

$$P_{rot} = \frac{\pi^2 f (2a)^2}{6\rho} \left(B_x^2 + B_y^2\right) \left(\mathrm{Wm}^{-3}\right) \tag{8.24}$$

[19]See Section 2.2.2 of Volume 2 of this book for a description of (100)[001] oriented electrical steel.

Table 8.1 Ratios of rotational loss to alternating loss measured in various materials at 50 Hz magnetisation

	Ratio of rotational loss to unidirectional AC loss						**Comments**	**Reference**
Flux density	**< 0.4 T**	**0.9 T**	**1.0 T**	**1.3 T**	**1.5 T**	**2.0 T**		
Fe-based amorphous ribbon	5.0	2.0	2.0	2.0	–	–	A.C. magnetisation along ribbon axis	[26]
(100)[001] oriented SiFe			2.0	1.9	2.0	–	Localised loss measurements	[23]
GO 6.5% SiFe			2.0		1.2			[27]
Co-Fe-V strip			2.4		1.8	1.6	Localised loss measurements	[28]
28% Co-Fe			1.7		1.25			[29]
Fe49Co49V2			1.3		1.3	0.75	0.2 mm commercial, Fieldmetric measurements[a]	[30]
48% NiFe			1.9	2.3			Cubic texture	[29]
GO SiFe			6				Anisotropy factor = 5.3. Commercial, grade not specified	[31]
GO steel		4.5					M7 grade	[22]
HGO steel			5.7		5.2		Commercial, high permeability GO steel	[27]
Uncoated HGO			2.0		2.0		surface polished	[27]
NO steel		2.5					3.75% Si	[22]
NO steel			1.85		1.60		0.2 mm 3% SiFe	[30]
NO steel			2.0		1.25		Anisotropy factor = 1.5. Commercial, grade not specified	[31]
NO steel		2.1					Grade M330-35	[29]
NO steels			2.73.3				Range of steel grades	[25]

[a]This is the term used for determining magnetic properties of strip or sheet material derived from measurements of spatial components of flux density and field using sensors. These are discussed in Chapter 12.

where B_x and B_y are the flux densities along the major and minor axes. This has been backed up analytically for a circular rotating field [32]. This approach is sometimes used by machine core designers today, but it should be noted that it is reasonably accurate only over a narrow range of magnetisation conditions. There is no fundamental reason why this analysis should be completely valid since domain studies show that quite different magnetisation processes occur under rotational and alternating fields.

8.6.2 Alternating magnetisation combined with d.c. offset fields

A d.c. biased field, commonly referred to as a d.c. offset field, is becoming more common in power devices, normally superimposed on an alternating field. It has long been known that the presence of a d.c. offset field in a magnetic core distorts the *B–H* loop and consequently, usually, results in extra losses [33]. Material characterisation under offset fields was developed in the early years of electrical steels and the difficulty of systematic and unambiguous measures have long been recognised [34]. The resulting loss increase as well as the introduction of harmonics are practically important. In an early study it was found that the d.c. magnetisation offset caused by a d.c. current of just 1.5% of the rated primary current of a 200 kVA distribution transformer increased the losses by the maximum allowed amount of around 10%. It also introduced harmonics which were deemed to have a negligible effect on the system [35].

The effect of a d.c. offset on the *B–H* characteristics is shown in Figure 8.18(b) for the simplest case where the offset field is oriented along the same direction as the a.c. magnetisation. The resultant shown in the figure is simply given by

$$h_{xr} = H_{dc} + H_{xac} \sin \omega t \ \left(\mathrm{Am}^{-1}\right) \tag{8.25}$$

In practice, it is usually found that $H_{ac} > H_{dc}$. However, the resulting flux density may be highly distorted and reach saturation during alternate half cycles of magnetisation.

The combination of an elliptical field and a d.c. offset field shown in Figure 8.18(c) can be present in practice in a transformer core but its occurrence or analysis is not widely reported. The influence this combination has on loss depends on the relative magnitudes of the components of field compared to the characteristic loss versus *B* or loss versus *H* characteristics of a given material at a given magnetising frequency.

Characterisation of materials under controlled d.c. bias is difficult to carry out because of the need to measure the d.c. component of flux density, so innovative approaches are needed. [36–38]. These are discussed further in Chapter 12.

Some investigators have produced formulae for calculating losses under d.c. bias conditions based on combining hysteresis, eddy current and excess loss into a Steinmetz expression of the form [39]:

$$P_{\text{bias}} = k_{dc} f^{\iota} (B_{ac})^{\kappa} (\text{arbitrary units}) \tag{8.26}$$

where k_{dc}, ι and κ are parameters related to physical material characteristics and the magnetisation level. The results shown in Figure 8.20 [39] shows that this model can give close agreement with measured values, the large increase in loss under d.c. excitation is apparent.[20]

Today, the term d.c. offset is always taken to mean that the d.c. bias occurs along the magnetising direction in the plane of the lamination; hence, it distorts the

[20]The authors tested this approach on one material only, a 0.5 mm thick NO steel whose a.c. loss at 1.5 T was 4.7 W kg^{-1}.

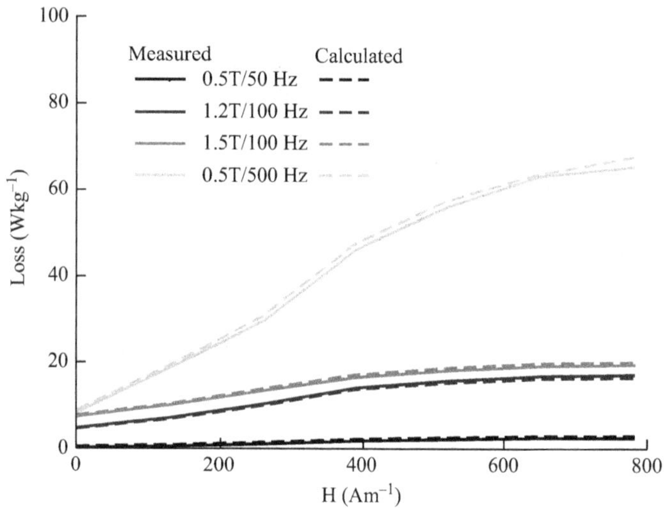

Figure 8.20 Measured and calculated loss of a NO steel under various d.c. bias conditions (Figure 8 in [39] reproduced under free licence CC-BY-4.0 and modified)

B–H loop in a harmful way and increases the a.c. losses. However, a d.c. field sometimes can be present normal to the plane of laminations magnetised by alternating or rotating magnetic fields. In this case, the d.c. field will tend to encourage domains to take up positions closer to the plane of a lamination hence making in-plane magnetisation easier and the losses will be lower. A d.c. bias field of 4 kAm^{-1} has been shown to reduce rotational losses in NO and GO silicon–iron by around 5% [40]. This might not have any practical benefit in magnetic cores but it illustrates that every form of d.c. offset need not be harmful.

A d.c. field applied perpendicular to the plane of a lamination subjected to an in-plane rotating field can cause magnetisation changes in any direction. This is sometimes referred to as three-dimensional (3D) magnetisation and is discussed further in Section 8.9.

8.7 Rotational loss at power frequency

8.7.1 Distinction from rotational hysteresis loss

It should be pointed out again that rotational hysteresis loss discussed in Section 8.6 is the loss occurring in a *slowly rotating field*, analogous to the static hysteresis loss produced under a.c. conditions discussed in Section 8.4.

The rapidly rotating field which occurs in transformer and motor cores produces flux which, not only varies in magnitude and rotational speed, but also produces eddy currents, which in turn, produce losses. In this case the losses are normally simply referred to as rotational losses or even a.c. rotational losses. This terminology is not strictly correct since, in this context, 'a.c.' is meaningless. By definition a pure rotational flux or magnetic field has a constant magnitude and does not alternate in the same sense as a flux, current or voltage which change magnitude, perhaps in a cyclic manner, but are unidirectional. The terminology has come into common usage because of the way in which rotational flux density and field is analysed into a.c. components in the way described in Sections 8.1 and 8.2.

In loss separation studies under a.c. magnetisation, it is generally accepted that there is a finite upper magnetising frequency limit which can be used to measure H_{rh}. The limit is around 5 Hz, depending on the material being tested; below this, eddy current loss and excess loss can be neglected. Similar limitations apply for separating rotational loss into static and dynamic components. The limit is then around five revolutions per second loss speed [41]. However, another report shows that, even at 2 Hz, in NO steel the peak measured rotational loss can differ from the assumed H_r by more than 10% [42].

8.7.2 Total rotational loss in terms of B and H

Equation (8.17) was derived to obtain the rotational hysteresis loss in terms of instantaneous magnitudes of B and H and the spatial angle between them during one slow period of field rotation. In the case of rapid rotation, the equation can be rewritten to give the power loss per unit volume per rotational cycle of magnetisation as

$$W_r = \frac{1}{T}\int_0^T B_r H_r \sin \alpha dt \left(\mathrm{Wm}^{-3}\right) \tag{8.27}$$

Although (8.21) and (8.27) might appear very similar, there are important differences in their application. In (8.17), the angle α is simply related to the torque and the energy stored in the materials when both B and H are constant in magnitude throughout the thickness of the material. However, (8.27) is related to the Poynting vector where H_r is the rapidly rotating tangential component of the surface field. B_r is the component of flux density averaged through the full cross sectional area of the material, as derived for the uniaxial a.c. case in Section 8.3.

Now consider a sheet of electrical steel subjected to in-plane magnetisation of any form. The instantaneous H and B vectors can be represented as shown in Figure 8.21. At this moment in time, they make angles θ_{rh} and θ_{rb} to the arbitrary x-axis and can be split into components H_x, H_y, B_x and B_y, as shown. Hence we can rewrite (8.27) as

$$W_r = \frac{\omega}{T}\int_0^T \left(H_x B_x + H_y B_y\right) dt \left(\mathrm{W\,m}^{-3}\right) \tag{8.28}$$

The angle between them is given by [43]

$$\alpha = \tan^{-1}\left(\frac{H_y B_x - H_x B_y}{H_x B_x + H_y B_y}\right) (\mathrm{radian}) \tag{8.29}$$

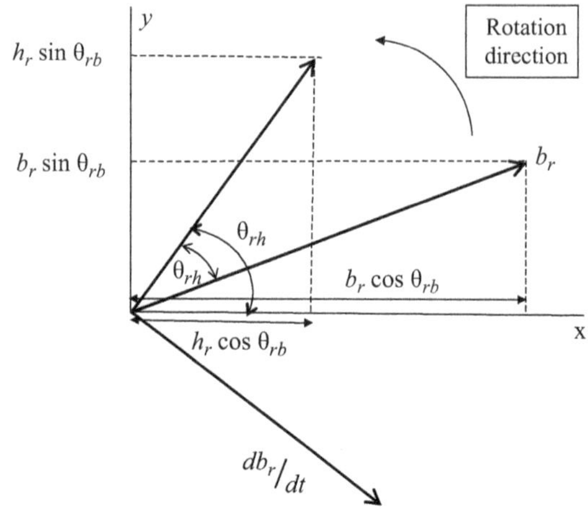

Figure 8.21 Vector representation of instantaneous magnitude and direction of rotating field and flux density as components in arbitrary x and y directions

If either B or H are purely rotational then an expression for power loss can be written in terms of experimentally obtainable quantities as [44]:

$$W = \frac{1}{TD}\int_0^T \left(\frac{dB_x}{dt}H_x + \frac{dB_y}{dt}H_y\right)\mathrm{dt}\ (\mathrm{W\,kg^{-1}}) \tag{8.30}$$

The material density, D, is introduced to convert to the more common format of loss per unit mass.

During magnetisation rotation at power frequency the spatial angle $(\theta_{rh} - \theta_{rb})$ between the field and flux density varies substantially in an anisotropic material. Figure 8.22 shows how the angle between B and H varies over a cycle of magnetisation when the flux density is forced to remain purely rotational, i.e. under the condition when the voltages induced in orthogonal search coils are equal in magnitude, sinusoidal and 90° out of phase. In the NO steel H_r always leads B_r by between 20° and 70°. This relatively small change in phase angle is caused by the presence of some texture in the NO steel causing domains not to be completely randomly oriented.

In the typical GO material the angle by which h_r leads b_r changes rapidly from around −10° to +75° as the flux density rotates by an angle of approximately 15° in space. This, of course, infers that if the rotational velocity of h_r is constant, then the instantaneous rotational velocity of b_r varies during the period of rotational magnetisation. If this is the case, the rapid changes in instantaneous b_r during a cycle leads to extra losses.

This phenomenon has been mentioned in several publications, although its occurrence in electrical steels has not been quantified. Since it is neglected in models of rotational processes applied to material and machine core analysis its

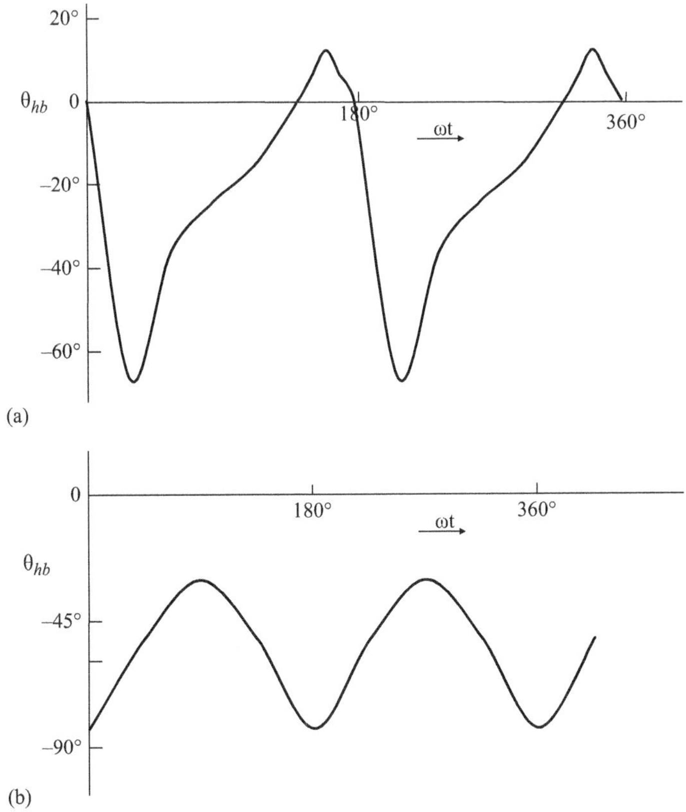

Figure 8.22 *Typical variation of angle $\theta_{\rm hb}$ between b_r and h_r in one revolution of magnetisation in (a) NO steel and (b) GO steel ($\theta_{\rm hb}$ is positive when b_r leads h_r)*

possible significance in such applications needs to be established. The phase angle between h_r and b_r tends to increase with the anisotropy of the steel and is not greatly affected by magnetising frequency [30].

The angular relationship shown in Figure 8.22 is calculated from B and H sensors which detect the spatial mean, instantaneous value of B_r and the instantaneous value of the tangential component of magnetic field. The angular difference between them, effectively the angle α in (8.27), is needed in equations used to calculate the losses, although its physical meaning is a little difficult to understand.

It was shown in Section 7.6 that the magnitude and phase of a unidirectional a.c. flux density varies with depth in a thin lamination due to the classical eddy current shielding. Figure 8.23 [45] shows a computed result of the same phenomenon under rotational magnetisation. The magnitude and phase of the instantaneous local rotating flux density varies from the surface to the centre of a lamination. This particular result, like most others, is calculated from Maxwell's equations making

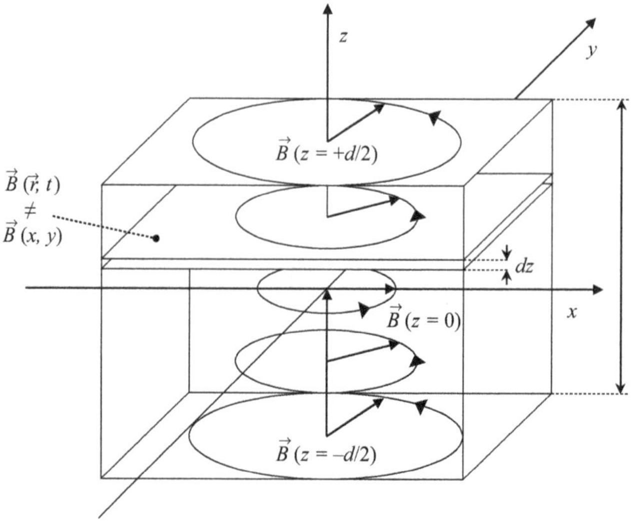

Figure 8.23 Computed flux density in x − y planes inside a sheet of iron (thickness t in the z direction) at an instant in time under circular magnetisation in the x − y plane (un-numbered figure in [45] reproduced under free licence CC-BY-4.0)

all the assumptions and approximations made in similar a.c. solutions presented in Section 7.6. Although it is based on an established theoretical basis, the quantitative values are difficult to relate to the magnetic domain structures within a sheet. Matching theory of this sort with magnetic domain observations is still very vague. How far this can be advanced and the benefits which can arise from better correlation between models based on Maxwell's equations and the complex domain structures found in electrical steels remains to be seen.

The rotational loss under circular magnetisation can be calculated by describing the magnetisation in terms of a tensor permeability relating the anisotropy of a steel to the rotational field throughout a cycle of magnetisation. It can be shown that [46]:

$$W_r = \frac{B^2 \pi}{\rho T} \left(\frac{1}{\mu_x} + \frac{1}{\mu_y} \right) \sin \theta_{bh} \left(\text{W m}^{-3} \right) \quad (8.31)$$

This is restricted to constant values of B_r and rotational velocity. The approach can be used in numerical field calculations in magnetic circuits but its accuracy and general applicability is not quantified.

8.7.3 Loss separation under rotational magnetisation

Decomposition of rotational loss per cycle into static hysteresis, classical eddy current and excess loss components for a material under unidirectional magnetisation was discussed in Section 7.8. The same loss separation process can be applied for rotational magnetisation [47,48] and we can simply write

$$P_r = P_{rh} + P_{rclas} + P_{rex} \left(\mathrm{J\,m^{-3}}\right) \tag{8.32}$$

where P_{rclas} is the 2D version of the classical eddy current loss equation and P_{rex} is the rotational excess loss. This is analogous to (7.45). Equation (8.28) can be justified since the same loss mechanisms occur under rotational magnetisation, although perhaps in different proportions. However, it must be remembered that the formula has no physical basis, and domain wall activity is not taken into account [9]. Measurement of rotational hysteresis P_{rh} using a torque magnetometer was discussed in Section 8.3. It is more likely to be measured today at low frequency 2D magnetisation as is discussed in detail in Section 12.4.

Methods of predicting its magnitude in electrical steel, based on the Stoner–Wohfarth model and other empirical formulae, have been attempted but they have not yet proved sufficiently accurate for engineering applications [49].

If the locus of the rotational flux density is circular, the rotational classical loss can be written as [44]:

$$P_{rclas} = 2C_{clas}(fB_r)^2 \left(\mathrm{J\,m^{-3}}\right) \tag{8.33}$$

where $C_{clas} = \pi^2 t^2/6\rho$ and t is the sheet thickness. Of course, this value is simply double the classical loss under a.c. magnetisation at the same frequency and flux density.

An expression for the excess rotational loss, by analogy with the excess loss under a.c. magnetisation, can be written as

$$P_{rex} = C_{ex}(fB_r)^{1.5} \left(\mathrm{J\,m^{-3}}\right) \tag{8.34}$$

where C_{ex} is a constant which varies with f and with B_r. If it is assumed that the excess loss under rotational magnetisation can be analysed in a similar way to that under a.c. magnetisation, Equation (8.30) can be written as [50]:

$$P_{rex} = 8.76\sqrt{\sigma G A V_0}\left(B_p f\right)^{3/2} \left(\mathrm{Jm^{-3}}\right) \tag{8.35}$$

This general form of variation of excess loss with B and f agree well with one of the earliest experimental studies of excess loss components under very low frequency, pure rotational magnetisation, which showed that it seems to follow the same trends as the d.c. hysteresis component [51].

Loss separation under rotational magnetisation can be carried out in the same way as under a.c. magnetisation by plotting the measured loss per cycle against frequency and extrapolating the curve to zero frequency to obtain P_{rh}. Few reports of the relative values of these components have been published, but for NO electrical steel at 50 Hz magnetisation, the ratio $P_{rh} : P_{rclas} : P_{rex}$ is of the order of 1.0:0.2:0.3 compared with around 1.0:0.6:0.4 for a.c. magnetisation. For a GO steel the ratio, at 1.4 T, 50 Hz, rotation has been reported as 1.0:0.25:0.9 [52] the higher proportion of P_{rex} being in line with a typical ratio of $P_h : P_{clas} : P_{ex}$ of around

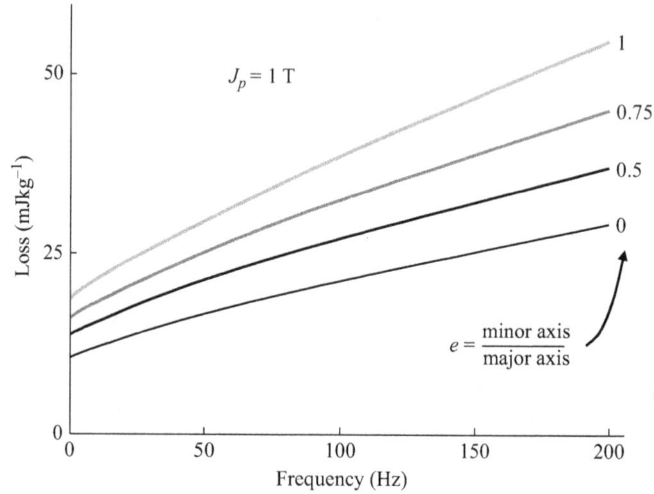

Figure 8.24 Loss separation under rotational ($e = 1$) elliptical ($e = 0.5, 0.75$) and alternating magnetisation ($e = 0$) in NO electrical steel ($B_p = 1.0$ T) (Figure 2 in [54] reproduced under free licence CC-BY-4.0)

1:0.8:2.0 in GO steel at 1.6 T 50 Hz [53] under unidirectional magnetisation. It will be noted that the proportion of excess component in the GO steel is around three times higher than in NO steel under a.c. and rotational magnetisation.

These values are only indicative since in practice they vary widely with material thickness, material microstructure, flux density and magnetising frequency. However, in general they do seem to indicate that hysteresis is more prominent under rotational magnetisation which might be due to the far more complex domain activity.

The ratio of the loss components present under rotational and a.c. magnetisation is also interesting to look at. Some data for NO steel from [54] indicate that P_{rh}/P_h drops from around 2.0 to 1.0 as the flux density is increased from close to zero to near to saturation. The ratio P_{rclas}/P_{clas} is, not surprisingly, constant with a value of 2.0 as expected from (8.29) and P_{rex}/P_{ex} varies in a similar manner to the hysteresis ratio.

Figure 8.24 [54] shows a typical set of loss per cycle versus magnetising frequency characteristics of NO steel at 1.0 T. It can be seen here that the slope of the curve is greater for the pure rotational magnetisation condition.

Equation (8.33) applies to purely circular flux density. It can be slightly modified to give the loss under the more common elliptical magnetisation as [50]:

$$P_{clas}^{2d} = P_{clas}^{ac}\left(1 + e^2\right)\ \left(\text{J m}^{-3}\right) \tag{8.36}$$

where e is the ellipticity factor B_y/B_x, which is zero for unidirectional a.c. magnetisation and unity for pure rotational flux. It is interesting that using this loss separation approach results in the ratios of rotational hysteresis to a.c. hysteresis, classical eddy current loss to a.c. eddy current loss and the rotational excess loss to a.c. excess loss at a given flux density are independent of frequency.

It is still not clear what physical meaning such loss analysis has. The assumptions and approximations made in a.c. analysis are significant and they probably cause even greater discrepancies under rotational magnetisation. It is also interesting that there is apparently good correlation between the loss derived from components of rotational and a.c. magnetisation despite their quite different magnetisation processes as demonstrated by the different changes in domain structures occurring in the two magnetisation processes.

The losses under elliptical magnetisation are even more difficult to model or predict and relate to a.c. losses in anisotropic materials. Rotational hysteresis loss has been correlated with a.c. values up to the knee of the magnetisation curve from torque characteristics using the average value of the permeability in the plane of an anisotropic material, but it is subject to limitations and assumptions, even over this narrow range [47].

The total loss under elliptical magnetisation can be estimated using the values of the major and minor axis flux densities in functions to modify the a.c. hysteresis loss P_h and classical eddy current loss P_{clas} using the following [55]:

$$P_{rh} = P_{hx}\left[\frac{B_x}{B_m} - \left(\frac{B_x}{B_m}\right)^2\right] + P_{hy}\left[\frac{B_y}{B_m} - \left(\frac{B_y}{B_m}\right)^2\right] \ (\mathrm{Jm^{-3}}) \tag{8.37}$$

$$P_{rclas} = P_{clas}\left[\left(\frac{B_x}{B_m}\right)^2 + \left(\frac{B_y}{B_m}\right)^2\right] \ (\mathrm{Jm^{-3}}) \tag{8.38}$$

Equation (8.38) reduces to the expression in (8.36) if the flux density is circular, i.e. $B_x = B_y$. The model is not based on any physical basis and is dependent on the values of a.c. hysteresis and eddy current losses being reliable. This is not normally the case and also the approach ignores the presence of excess losses.

Another approach claimed to be successful for estimating losses of GO steel under elliptical magnetisation is to treat the P_{rh} as having rotational and alternating components such that the total value varies linearly as e varies from 0 to 1 [56]. Combining this with the classical eddy current loss of a form similar to that in (8.32) leads to good agreement between measured and calculated rotational loss under elliptical magnetisation, but over a limited peak flux density range up to around 1.1 T in GO steel and 1.3 T in NO steel.

The rotational loss is often estimated from the sum of components of the losses measured separately under a.c. unidirectional magnetisation along orthogonal directions. In electrical steel these would normally be the RD and the TD (transverse direction of a strip). It has been claimed [57] that the choice of a more appropriate set of reference axes gives more meaningful correlation between the orthogonal components of B–H data with material performance, but this has not been confirmed. It has been shown that calculation of eddy current loss in this way is valid, subject to some approximations such as assuming that the magnitude of B is constant throughout the cross sectional area of a lamination for anisotropic materials

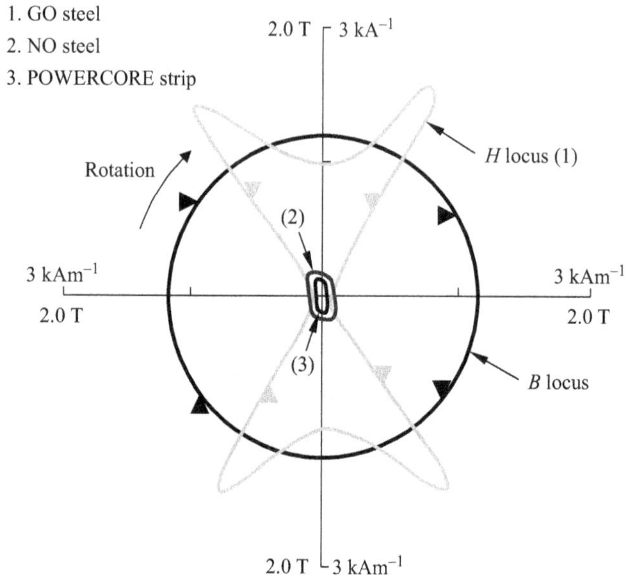

Figure 8.25 Typical B and H loci under rotational magnetisation of three contrasting types of material: (1) GO steel, (2) NO steel and (3) Iron-based amorphous material (part of Figure 5 in [63] reproduced under free licence CC-BY-4.0 and modified)

with non-linear B–H curves [58]. This gives a reasonable approximation for the eddy current component of the total loss despite the fact that it does not take account of the variation of B_r through the thickness [46] or magnetic domain dependent processes, such as hysteresis loss, which can dominate in many materials at power frequencies.

Another rigorous treatment using Poynting vector analysis shows that rotational loss based on separation into P_x and P_y loss components for elliptical magnetisation cannot be assumed to be consistent with domain theory in anisotropic material [59]. However, it has also been claimed [60] that analysis of rotational power loss into P_x and P_y components gives useful information about the properties of electrical steel. Furthermore, it is still not clear whether analysing in terms of the orthogonal components guarantees that the total loss is included. [61,62].

Figure 8.25 [63] shows the locus of field and flux density in three materials during one cycle of magnetisation (at 50 Hz) when the flux density is forced to be purely rotational at 1.2 T. The required fields of the NO steel and the POWERCORE[21] strip have elliptical loci with their major axes along easy directions, indicating a small degree of anisotropy in both materials. The field required to

[21]POWERCORE is a bonded stack of iron-based amorphous ribbon. It is discussed more fully in Section 15.1.7.

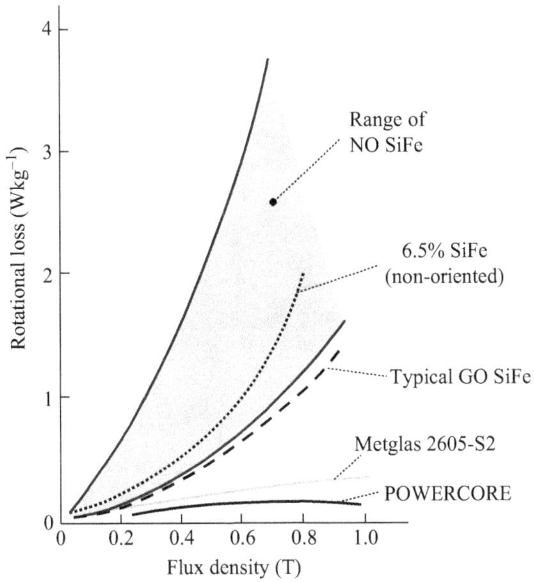

Figure 8.26 Variation of rotational loss with flux density in typical soft magnetic materials at 50 Hz (Figure 7 in [63] reproduced under free licence CC-BY-4.0)

magnetise the GO steel is around 20 times larger when the flux is at approximately 57° to the RD than when it is along the RD itself. Under a.c. conditions the ratio typically has a value of around ten, a consequence of the far different magnetisation occurring in the material under rotational flux compared to that under unidirectional conditions.

The corresponding rotational losses for such materials are shown in Figure 8.26 [63].

Most of this chapter has focused on rotational magnetisation in electrical steel because the majority of the publications in this area have been devoted to this family of alloys. Many of the characteristics such as the representation of B and H, loss modelling and decomposition, domain dynamics, etc. can be applied to other soft magnetic materials taking care to consider their fundamental characteristic domain structure and anisotropy.

One of the few direct comparisons of rotational loss of electrical steel with other materials is shown in Figure 8.27 [29]. The CoFe alloy has low crystalline anisotropy so the three a.c. characteristics are similar whereas the NO SiFe has a typical a.c. loss variation with angle to the RD. The rotational loss of the NiFe alloy reaches a peak usually observed in SiFe alloys at a higher flux density. The rapidly increasing loss of the NiFe material magnetised at 45° to its RD is due to its almost cubic crystalline texture.

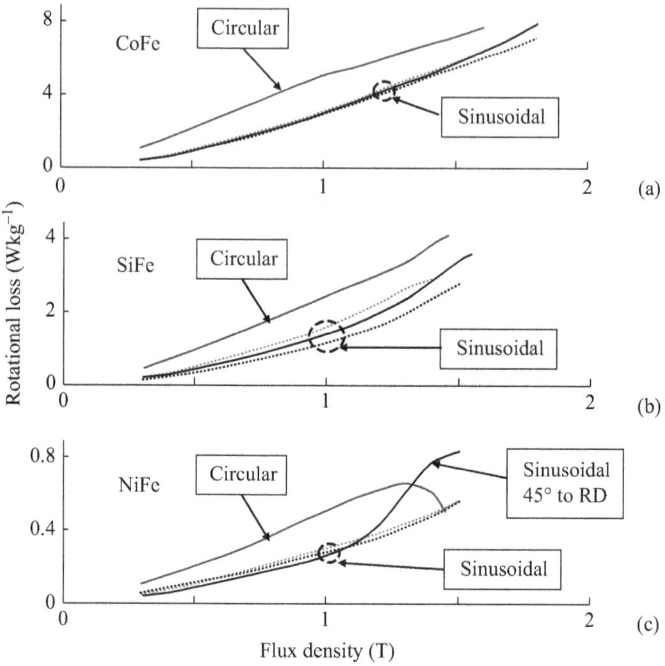

Figure 8.27 Variation of loss with a.c. and circular flux density (50 Hz) in CoFe, NiFe and NO SiFe (Figure 1 in [29] reproduced under free licence CC-BY-4.0 and modified)

8.8 Magnetostriction under rotational magnetisation

8.8.1 Multidirectional magnetostriction

Normally, magnetostriction of soft magnetic materials is simply measured along the same direction as the magnetising direction in a piece of the material. This is sufficient for many purposes, but one of the main uses of magnetostriction measurement data is for studying material deformation in electrical machine cores in order to predict their vibration characteristics. If magnetostriction along one direction is known from such measurements, it might be possible to calculate its value along other directions for more accurate core vibration assessment but this is not always straight forward, particularly under complex magnetisation conditions.

Magnetostriction is effectively always *multidirectional*, so if a unidirectional magnetic field is applied to a material, dimensional changes in the field direction will be accompanied by dimensional changes in all other directions as illustrated in Figure 3.11. Before discussing the more complicated variations of magnetostriction under rotational fields it is useful to return briefly to the characteristics of magnetostriction in a time varying unidirectional field which was introduced in Section 3.16.

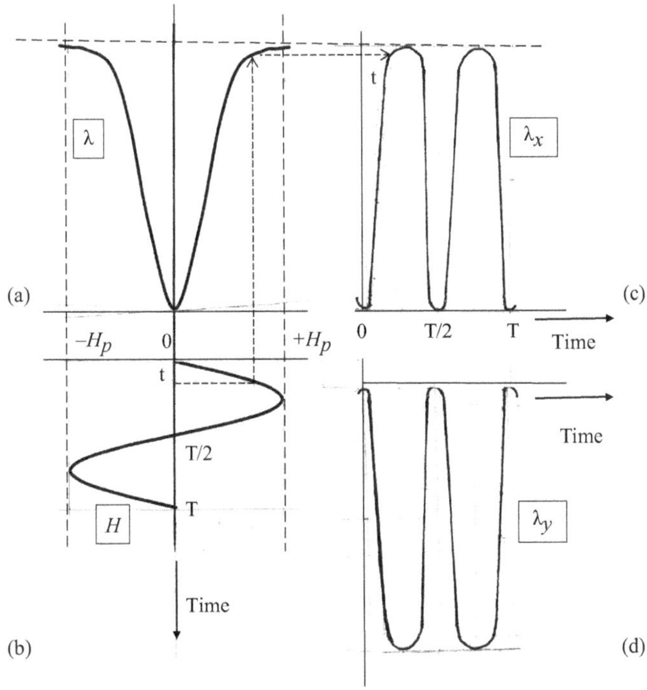

Figure 8.28 Simple representation of the evolution of a.c. magnetostriction-time waveforms: (a) non-linear magnetostriction versus field characteristic, (b) sinusoidal applied field waveform, (c) time variation of component of magnetostriction λ_x along the x direction and (d) time variation of component of magnetostriction λ_y along the y direction

Figure 8.28(a) shows the shape of the magnetostriction versus H characteristic of a soft magnetic material during one full cycle of magnetisation. This is closer to reality than the idealised curve shown in Figure 3.22, but it is simplified by omitting the hysteresis of magnetostriction to focus on the main effects related to multidirectional magnetostriction. The shape of the curve is material dependent and determined by the internal domain structure which, of course, in turn is magnetising frequency dependent, but this is not of concern here.

If a unidirectional sinusoidally time varying a.c. field is applied along the x direction then it is obvious that the corresponding magnetostriction during one cycle of magnetisation will be similar to that shown in Figure 8.28(c). The actual wave shape of this curve depends on the shape of the λ-H curve but the points to note are that the λ-*time* curve is not sinusoidal and λ is never negative. As will be seen later, it is common practice to represent this non-sinusoidal λ-*time* characteristic as a series of harmonic components. A similar characteristic could be produced if the flux density was forced to be sinusoidal.

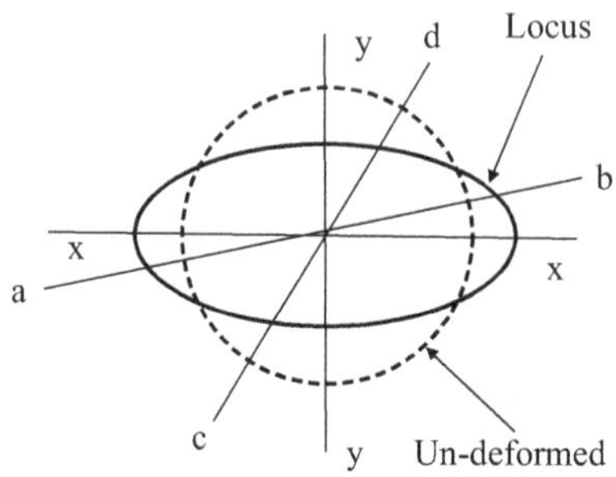

Figure 8.29 Locus of in-plane deformation at the instant in time when the field is maximum ($t = T/2$)

Suppose the sample being magnetised is an infinitely thin sheet, then to maintain constant volume the displacement in the y direction at any instant in time must be negative and the same magnitude as the value in the x direction so its waveform is as shown in Figure 8.28(d). In this case, the peak to peak components of the magnetostriction along the two directions are identical in magnitude but the instantaneous values of λ are always positive in the one case and negative in the other.

If the sheet has a finite thickness the peak to peak value of λ_y drops to half the value of λ_x to maintain constant volume.

Finally, suppose the infinitely thin sample is circular and still subject to a uniform sinusoidal magnetic field along the x direction. We can show the instantaneous deformation of the disc at any instant in time by referring to the components in Figure 8.28. Figure 8.29 shows the locus of the deformation at time T/4 in the a.c. cycle. The dotted circle represents the un-deformed disc. At this instant, magnetostriction is positive along some arbitrary direction *ab* and at the same time along *cd* it is negative.

This observation might not be surprising but it is introduced because it forms the basis of the more complex magnetisation characteristics under rotational magnetisation discussed in Section 8.8.2. The quantified magnetostriction versus angle characteristic of a uniaxially magnetised disc of real material could be deduced approximately in a similar way as for the infinitely thin sheet of non-hysteretic, isotropic ideal material represented in Figure 8.29. However, some thought would be necessary to decide on the best way of dealing with any anisotropy or hysteresis in the $\lambda-H$ curve.

8.8.2 Simulation of rotational magnetostriction

Magnetostriction under rotational field conditions is far more complex than under unidirectional conditions [64–66]. In the case of a thin sheet, the problem can be simplified by analysing the rotational flux density in terms of orthogonal components in the plane of the sheet as is usually done for other rotational magnetisation problems. The magnetostriction of an isotropic sheet under 2D magnetisation can be written as [67]:

$$\tau_m \frac{d}{dt}\begin{bmatrix}\lambda_x \\ \lambda_y \\ \Upsilon_{xy}\end{bmatrix} + \begin{bmatrix}\lambda_x \\ \lambda_y \\ \Upsilon_{xy}\end{bmatrix} = \begin{bmatrix} {}^{1}/_{M_m} & -{}^{\xi}/_{M_m} & 0 \\ -{}^{\xi}/_{M_m} & {}^{1}/_{M_m} & 0 \\ 0 & 0 & 2(1+\xi)/_{M_m}\end{bmatrix} \frac{1}{2\mu_0}\begin{bmatrix} b_x^2 \\ b_y^2 \\ b_x b_y\end{bmatrix} \tag{8.39}$$

where b_x and b_y are orthogonal components of the instantaneous flux density, λ_x and λ_y are the magnetostriction components along the corresponding directions, Υ_{xy} is the shear magnetostriction between the two directions, M_m is defined as the magnetic modulus and ξ is the magnetic Poisson's ratio.[22] This expression is derived based on analogy with the basic 2D mechanical elastic stress–strain relationship [68] and the history dependent time constant τ_m which controls the width of the magnetostriction versus flux density loop.[23]

The expression in (8.39) was derived for an isotropic material but it can conveniently be extended to model anisotropic steels by introducing components of the magnetic moduli and magnetic Poisson's ratio along orthogonal directions to give

$$\tau_m \frac{d}{dt}\begin{bmatrix}\lambda_x \\ \lambda_y \\ \Upsilon_{xy}\end{bmatrix} + \begin{bmatrix}\lambda_x \\ \lambda_y \\ \Upsilon_{xy}\end{bmatrix} = \begin{bmatrix} {}^{1}/_{M_{mx}} & -{}^{\xi_y}/_{M_{my}} & 0 \\ -{}^{\xi_x}/_{M_{mx}} & {}^{1}/_{M_{my}} & 0 \\ 0 & 0 & {}^{1}/_{D_{xy}}\end{bmatrix} \frac{1}{2\mu_0}\begin{bmatrix} b_x^2 \\ b_y^2 \\ b_x b_y\end{bmatrix} \tag{8.40}$$

where D_{xy} is the shear magnetic modulus. Figure 8.30 [67] shows the simulated 2D magnetostriction characteristics of the butterfly loops of isotropic and highly anisotropic materials such as NO and GO electrical steels, respectively. Table 8.2 shows the parameters used to obtain the solutions. The pure rotational flux density in each case was set at 1.3 T ($b_x = b_y = b$) rotating at 50 Hz. In the isotropic case the $\lambda_x - b$ and $\lambda_y - b$ loops are identical hence they overlay each other on the figure.

Loci of in-plane magnetostriction of NO steels under circular magnetisation of 1.3 T and 1.9 T are shown in Figure 8.31 when the instantaneous flux density is at 0°, 45° and 90° to the x direction [67]. The overall trends are similar and only small influences of flux density or material thickness on the general shape of the loci are observed. More detailed experimental and simulated loci for electrical steels are presented in Section 2.4 of Volume 2 of this book.

The magnetostrictive deformation under 2D magnetisation can be far higher than under a.c. magnetisation with the same peak flux density [69] just as rotational power loss can be higher than the equivalent a.c. value. This might have implications in

[22]Poisson's ratio is the ratio of the proportional decrease in lateral measurement to the proportional increase in length of a sample when stretched. In a perfect material it has a value of 0.5. By the magnetic Poisson's ratio we mean that the ratio includes magnetostrictive and elastic lateral strain components.
[23]This is simply the hysteresis found in the λ–H or λ–B loops caused by the same magnetic domain phenomena that forms normal B–H loops.

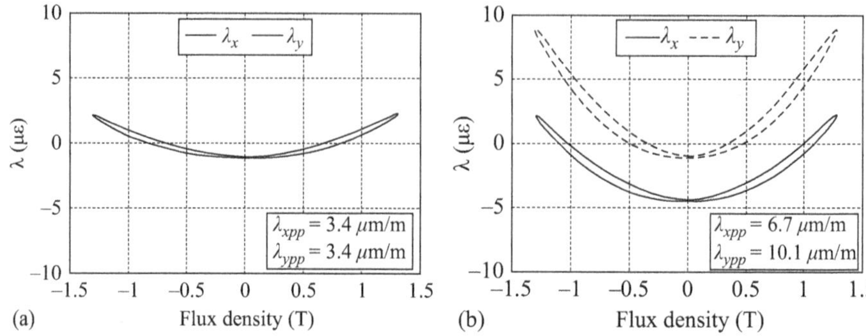

Figure 8.30 Simulated magnetostriction in orthogonal x and y directions in the plane of a sheet under pure rotational magnetisation (1.3 T, 50 Hz): (a) purely isotropic material (λ_x and λ_y are identical) and (b) highly anisotropic material (λ_x lower than λ_y) [67]

Table 8.2 Parameters used for simulated magnetostriction under 2D magnetisation

	M_{mx} (GPa)	ξ_x	M_{my} (GPa)	ξ_y	D_{xy} (GPa)	τ_m (ms)
Isotropic	300	0.5	300	0.5	$M_{mx}/3$	0.2
Highly anisotropic	300	0.5	$M_{mx}/4$	0.5	40	0.2

noise and vibration characteristics of transformer or rotating machine cores as seen in Chapter 10 of Volume 2 of this book. A further complication in measurement and analysis of 2D magnetostriction is that it is highly shape dependent. This is particularly noticeable when magnetisation is carried out at angles to the RD of electrical steel. This is discussed further in Section 2.6 of Volume 2.

The 2D magnetostriction characteristics of NO steel are closely linked to the magnetic Poisson's ratio[24] and the magnetostrictive anisotropy of the steel. For isotropic material, 2D magnetostriction is 3/2 times greater than a.c. magnetostriction [70]. The magnetic Poisson's ratios of NO steel can be close to unity due to random crystal orientation. It is claimed to be possible for this to cause large radial deformation at the tooth root and back iron regions of stator cores. A simple magnetic domain model has been developed to explain the influence of the texture on the 2D magnetostriction and the possible need to include the effect of the magnetic Poisson's ratios in computational models of electrical machine core deformation and vibration in order to obtain more accurate predictions [70].

[24]This is simply a modification to the normally used value of Poisson's ratio to take account of the magnetostrictive strain.

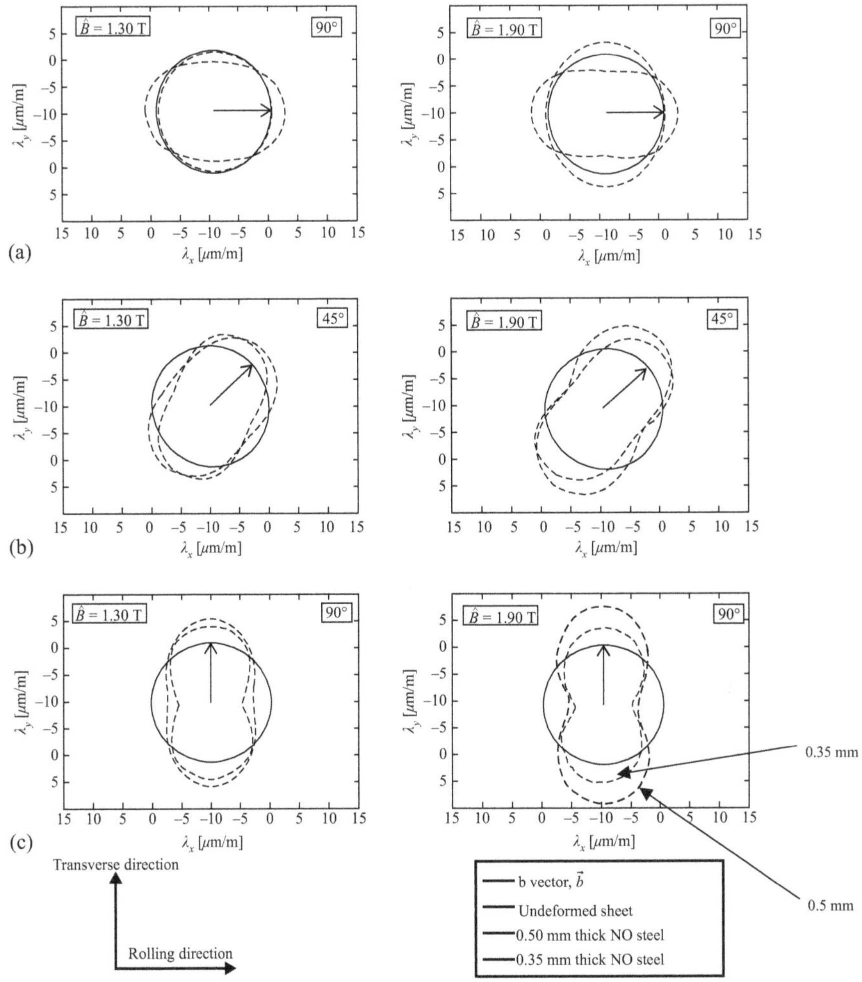

Figure 8.31 Loci of magnetostriction in the x – y plane of 0.35 mm and 0.5 mm thick NO steels at 1.3 T and 1.9 T at instances in time when a pure rotational flux density is oriented at (a) 0°, (b) 45° and (c) 90° to the x direction [67]

8.9 Three-dimensional magnetisation

Laboratory measurements under 3D magnetisation were briefly mentioned in Section 8.6. Parts of laminations in power transformer cores and in electrical machine cores are subjected to 3D magnetisation. This occurs in transformer joints where alternating or rotational flux can be present in the same regions as inter-laminar flux. It can occur in regions of transformer and rotating machine cores where laminations are subjected to flux components perpendicular to their planes.

This can occur in the end regions of stator cores of a large generator subjected to stray flux from nearby electrical windings or in large transformer cores due to stray flux from the windings, particularly when on load. These phenomena can lead to overheating if not controlled.

Small electrical machine cores make use of soft magnetic composite (SMC) components designed to allow flux to travel in 3D paths without the restriction of being confined to one plane as in laminated cores.

Material characterisation under 1D or 2D magnetisation can be carried out under well-established reference conditions. The Epstein square is, of course, a well-known, standardised system for 1D measurements and, although there is not yet any agreed methodology for testing and characterising under 2D magnetising conditions, users tend to choose similar reference magnetising conditions which helps when exchanging material characteristics. However, 3D magnetic testing is still in its infancy and no methods of measuring or analysing magnetic properties under such conditions are yet established, as will be seen from the rest of this section.

Components of 3D flux density can range from very low to near to saturation in any direction in SMC cores. In laminated cores, the localised flux density normal to the plane of the laminations rarely exceeds 0.1 T.[25]

In the laboratory, although not a simple procedure, cuboid samples, either a solid piece or a stack of square laminations can be magnetised under wide ranges of unidirectional, circular or elliptical magnetisation conditions. Ideally, spherically shaped samples would be used to avoid having to take demagnetisation factors into account, but this is a costly and time consuming exercise.

Analysis of magnetisation processes and loss generation under 3D excitation is normally based on Poynting vector theory and all the assumptions referred to earlier for 1D or 2D magnetisation. The loss in a thin lamination can be written as the sum of the three components as follows:

$$P_{3D} = \frac{1}{T}\int_0^T \left(H_x\frac{dB_x}{dt} + H_y\frac{dB_y}{dt} + H_z\frac{dB_z}{dt}\right)dt \ \left(\mathrm{Jm^{-3}}\right) \tag{8.41}$$

This is used indiscriminately in measurements and calculations although the validity and accuracy of the approach has not yet been verified. Its derivation for 1D magnetisation was explained in Chapter 7, limitations in its extension to 2D magnetisation were mentioned in Section 8.7.2. To what extent the even more complex domain activity under 3D excitation, which is ignored in (8.41), affects its practical accuracy is an open question.

Characterisation of the losses or other magnetic properties of laminated soft materials subject to such magnetisation is still only in its infancy and there is no strong evidence that it would be of general use in material quality assessment. Of course, for composite materials 3D magnetic isotropy can be of importance in applications, so interest in this area might grow.

[25]In the overlap regions of corner joints, it can far exceed this value, but this occurs over a very small volume of a core.

The loss of a cube specimen of soft magnetic composites can be measured under controlled 3D excitation making use of a suitably placed array of B and H sensors and applying (8.41) [71]. However, the calculated core losses of a motor deduced from such 3D measurements can differ by 20% from the measured motor losses [72]. This large difference is attributed to errors in the way in which losses due to harmonic components of flux in the core are accounted for, rather than any possible deficiency in the theory of loss analysis.

Measurement of losses under 3D magnetisation of a lamination or a stack of laminations using (8.41) is regarded as impracticable because of the difficulty in measuring the field H_z perpendicular to the plane. However, the loss can be measured using a thermal method[26] without the need for H measurement. Using such a technique, it has been found that the loss of a lamination magnetised in its plane increases substantially when subject to a superimposed normal component of a.c. flux. Furthermore, the loss is very dependent on the lamination surface area due to in-plane eddy currents rather than the quality of the steel itself [73]. Surface area dependency is not included in (8.41). This introduces more doubt over its practical significance in material testing or localised loss calculation in laminated machine cores.

The challenges in separating losses into static and dynamic components for 2D magnetisation were raised in Section 8.7.3. The skin effect and flux harmonics must be taken into account in order to model excess eddy current loss in SMC iron powder cores under 3D magnetisation [74]. So-called variable coefficient and indirect orthogonal decomposition models give similar predictions showing the excess loss to be around double that calculated under 2D excitation. It is further claimed that traditional methods of loss separation have restricted value under 3D magnetisation [70]. However, the physical significance and engineering value of loss separation under 3D (or even 2D) magnetisation remains to be seen.

3D magnetostriction can be deduced from an understanding of the basic theory of magnetostriction discussed in Chapter 3. It is of little practical interest in laminations since the planar dimensions are usually very large compared to the thickness, so, even if the magnetostriction $\delta l/l$ in the z direction is relatively large compared to that in the other directions, the actual dimensional change δz is small. It can be deduced by measuring the much larger changes in the in-plane x and y dimensions and assuming no volume change to calculate δz from the simple relationship

$$\frac{\delta z}{z} = \frac{x\delta y + y\delta x}{(x + \delta x)(y + \delta y)} \tag{8.42}$$

where, y and z are the dimensions of the sheet or strip.

[26]Thermal methods of measuring localised and overall core losses were used prior to the 1970s and since that time laboratory measurement of losses in soft magnetic materials mainly make use of *db/dt* and H sensors based on Poynting's law. Both have merits and disadvantages which are discussed in Section 11.5 of Volume 2.

In a solid block subjected to any form of magnetisation, the dimensional changes can be similar in magnitude along any direction so magnetostriction can become a more important parameter. Practically, it can be measured using any of the techniques discussed in Section 12.4. In theory, it can be calculated using the $3D$ version of the $2D$ strain (8.39) or (8.40).

References

[1] Moses A.J. 'Possible future trends and research challenges related to 1 & 2D magnetic properties of soft magnetic materials'. *Przeglad Elektrotechniczny (Electr. Rev.)* 2011, vol. 87(9), Part II, pp. 11–16

[2] Pfützner H. 'Fundamental aspects of measuring field and losses in anisotropic soft magnetic materials'. In *Two-Dimensional Magnetic Measurements and Its Properties* (Ed. Enokizono M.). JSAEM Studies in Applied Electromagnetics, 1992, pp. 57–69

[3] Zurek S. *Characterisation of Soft Magnetic Materials under Rotational Magnetisation*. CRC Press, London, 2018

[4] Moses A.J., and Thomas B. 'Measurement of rotating flux in silicon iron laminations'. *IEEE Trans. Magn.* 1973, vol. 9(4), pp. 651–654

[5] Brailsford F. *Physical Principles of Magnetism.* D Van Nostrand Co Ltd, London, 1966, p. 141

[6] Moses A.J., and Thompson J.E. 'The effect of heat treatment in a magnetic field or under an applied stress on the magnetic properties of silicon-iron'. *J. Mater. Sci.* 1970, vol. 5, pp. 463–469

[7] Carter G. W. *The Electromagnetic Field in its Engineering Aspects.* Longmans Green & Co Ltd, London, 1962 p. 200

[8] Bailey F.G. 'The hysteresis of iron and steel in a rotating magnetic field'. *Philos. Trans. R. Soc.* 1896, vol. 187(A), pp. 715–746

[9] Guo Y., Zhu J.G., Zhong J., Lu H., and Jin X.J. 'Measurement and modelling of rotational core losses of soft magnetic materials used in electrical machines: a review'. *IEEE Trans. Magn.* 2008, vol. 44(2), pp. 279–291

[10] Ewing J.A. *Magnetic Induction in Iron and Other Metals.* D. Van Nostrand, London, 1900, p. 328

[11] Bozorth R.M. *Ferromagnetism.* IEEE Press, *Piscataway*, NJ, 1993, p. 516

[12] Bozorth R.M. *Ferromagnetism.* IEEE Press, *Piscataway*, NJ, 1993, p. 515

[13] Archenhold W.F., Sandham H.F., Thompson J.E. 'Rotational hysteresis loss in grain-oriented silicon-iron'. *J. Appl. Phys.* 1960, vol. 11, pp. 46–49.

[14] Brailsford F. 'Rotational hysteresis loss in electrical sheet steels'. *J. IEE* 1938, vol. 83, pp. 566–575

[15] Zbroszczyk J., Drabecki J., and Wysłocki B. 'Angular distribution of rotational hysteresis losses in Fe-3.25%Si single crystals with orientations (001) and (011)'. *IEEE Trans. Magn.* 1981, vol. 17(3), pp. 1275–1282

[16] Chikazumi S. *Physics of Magnetism.* R E Krieger, New York, 1978, p. 125

[17] Boon C.R., and Thompson J.E. 'Rotational hysteresis loss in single-crystal silicon-iron'. *Proc. IEE* 1964, vol. 111(3), pp. 605–609

[18] McQuade F., Moses A.J., and Meydan T. 'Domain observations under DC representation of rotational magnetisation'. *Studies in Applied Electromagnetics and Mechanics 10, Non-Linear Electromagnetic Systems*. IOS Press, Amsterdam, 1996, pp. 430–433

[19] Narita K., and Yamaguchi T. 'Rotational and alternating hysteresis losses in 4% silicon-iron single crystal with the {110} surface'. *IEEE Trans. Magn.* 1975, vol. 11(6), pp.1661–1666.

[20] Tiunov V.E., Korzunin G.S., and Inisheva L.A. 'Dynamic behaviour of the domain structure and magnetic losses of Fe-3% Si crystals in rotating magnetic field'. *Phys. Met. Metall.* 1989, vol. 68(4), pp. 58–67

[21] Ishihara Y., and Takahashi N. 'Iron losses of silicon steel due to rotating fluxes (inclined rotating flux and distorted one)'. *Memoirs of the School of Engineering, Okayama University Japan* 1979, vol. 14(1), pp. 15–28

[22] Kaplan A. 'Magnetic core losses resulting from a rotating flux'. *J. Appl. Phys.* 1961, vol. 32(3), pp. 370S–371S

[23] Boon C.R., and Thompson J.E. 'Alternating and rotational power loss at 50 c/s in 3% silicon-iron sheets'. *Proc. IEE* 1965, vol. 112(11), pp. 2147–2151

[24] Kapoor A.K. 'Measurement and separation of total iron losses under combined alternating and rotating fields' *J. Inst. Eng. (India)* 1971, vol. 51(6), pp. 123—133

[25] Arabi M.A., and Moses A.J. 'An improved method for measuring rotational power loss in electrical steels'. *J. Magn. Magn. Mater.* 1984, vol. 41, pp. 227–229

[26] Bottoni G., Candolfo D., Cecchetti A., and Masoli F. 'Rotational and alternating loss in amorphous ferromagnet'. *J. Magn. Magn. Mater.* 1982, vol. 26, pp. 130–132.

[27] Yamaguchi H., Pfützner H., and Krismanic G. 'Consequences of novel technological measures on the multidirectional magnetization performance of SiFe steel'. *Przeglad Elektrotechniczny (Electr. Rev.)* 2007, vol. 83(4), pp. 35–37

[28] Thomas B. 'A quantitative assessment of the change in magnetic properties of Co-Fe-V alloy when subjected to flux and stress conditions similar to those encountered in an aircraft generator'. *IEEE Trans. Magn.* 1980, vol. 16 (5), pp. 1299–1309

[29] Spornic S.A., Kedous-Lebouc A., and Cornut B. 'Anisotropy and texture influence on 2D magnetic behaviour for silicon, cobalt and nickel iron alloys'. *J. Magn. Magn. Mater.* 2000, vol. 215–216, pp. 614–616

[30] Appino P., de la Barrière O., Beatrice C., Fiorillo F., and Ragusa C. 'Rotational magnetic losses in nonoriented Fe-Si and Fe-Co laminations up to the kilohertz range'. *IEEE Trans. Magn.* 2104, vol. 50(11), Art. No. 2007104

[31] Nencib N., Spornic S., Kedous-Lebouc A., and Cornut B. 'Macroscopic anisotropy characterization of SiFe using a rotational single sheet tester' *IEEE Trans. Magn.* 1995, vol. 31(6), pp. 4047–4049

[32] Zhu J.G., Ramsden V.S., and Zhong J.J., 'Rotational core losses with circular H and/or B'. *Proceedings of International Symposium for Electromachining,* 1997, Braunschweig, Germany

[33] Edgar R.F. 'Loss characteristics of silicon steel at 60 cycles with D-C excitation'. *Trans. Am. Inst. Electr. Eng.* 1933, vol. 52(3), pp. 721–725

[34] Ball J.D. 'The unsymmetrical hysteresis loop'. *Trans. Am. Inst. Electr. Eng.* 1915, vol. XXXIV(2), pp. 2693–2720

[35] Kiiskinen E. 'The effect of the DC component on the primary current and iron losses of distribution transformers'. *SAHKO Electricity in Finland* 1972, vol. 45, pp. 329–332

[36] Miyagi D., Yoshida T., Nakano M., and Takahashi N. 'Development of measuring equipment of DC-biased magnetic properties using open-type single-sheet tester'. *IEEE Trans. Magn.* 2006, vol. 42(10), pp. 2846–2848

[37] Marketos P., Moses A.J., and Hall J. P. 'Effect of DC voltage on AC magnetisation of transformer core steel'. *J. Electr. Eng.* 2010, vol. 61(7), pp. 123–125

[38] Zhang Y., Wang J., Sun X., Bai B., and Xie D. 'Measurement and modelling of anisotropic magnetostriction characteristic of grain-oriented silicon steel sheet under DC bias'. *IEEE Trans. Magn.* 2014, vol. 50(3), Art. No. 7008804

[39] Chen W., Huang X., Cao S., Ma J., and Fang Y. 'Predicting iron loss in soft magnetic materials under DC bias conditions based on Steinmetz premagnetization graph'. *IEEE Trans. Magn.* 2016, vol. 52(3), Art. No. 6301404

[40] Hadoud S., and Meydan T. 'Characterisation of electrical steels under three-dimensional excitation'. *Proceedings of 5th 2 DM Workshop*, Grenoble, France, 1975, pp. 127–135

[41] Zurek S. 'Static and dynamic rotational losses in non-oriented electrical steel'. *Przeglad Elektrotechniczny (Electr. Rev.)* 2009, vol. 85(1), pp. 89–92

[42] Fiorillo F., and Rietto A.M. 'The measurement of rotational power losses at I. E.N: Use of thermometric methods'. Proc*eedings of* 1st International Workshop on Magnetic Properties of Electrical Sheet Steel Under 2-D Excitation, Physikalisch-Technische Bundesanstalt, Braunschweig (PTB), Germany, September 1991, pp. 162–172

[43] Brix W., Hempel K.A., and Schroeder W. 'Method for the measurement of rotational power loss and related properties in electrical steel sheets'. *IEEE Trans. Magn.* 1982, vol. 18(6), pp. 1469–1471.

[44] Moses A.J. 'Importance of rotational losses in rotating machines and transformer'. *J. Mater. Eng. Perform.* 1992, vol. 1(2), pp. 235–244

[45] Hempel K.A., and Birkfeld M. 'A phenomenological description of the anisotropic and non linear properties of electrical sheet under general two-dimensional magnetic excitation by means of reluctance tensor'. *Proceedings of 5th 2 DM Workshop*, Grenoble, France, 1975, pp. 93–99

[46] Brix W., and Hempel K.S. 'Tensorial description of the rotational magnetization process in anisotropic silicon steel'. *J. Magn. Magn. Mater.* 1984, vol. 41, pp. 279–281.

[47] Kapoor A.K. 'Analytical correlation between rotational and alternating hysteresis loss in silicon-iron laminations'. *J. Inst. Eng. (India)* 1971, vol. 51, pp. 351–359

[48] Fiorillo F. 'A phenomenological approach to rotational power losses in soft magnetic laminations'. Proceedings of the 1st International *Workshop on Magnetic Properties of Electrical Sheet Steel Under 2-D Excitation. Physikalisch-Technische Bundesanstalt, Braunschweig (PTB)*, Germany, September 1991, pp. 11–24

[49] Zhu J.G., and Ramsden V.S. 'Improved formulation for rotational core losses in rotating electrical machines'. *IEEE Trans. Magn.* 1998, vol. 34(4), pp. 2234–2242

[50] Appino C., Fiorillo F., and Ragusa C., 'Loss decomposition under two-dimensional flux loci in non-oriented steel sheets'. *Przeglad Elekrotechniczny* 2007, vol. 83(4), pp. 25–30

[51] Brailsford F., and Fogg. R. 'Anomalous iron loss in cold reduced grain-oriented transformer sheet at very low frequencies'. *Proc. IEE* 1966, vol. 113(9), pp. 1562–1564

[52] Dupré L.R., Fiorillo F., Appino C., Rietto A.M., and Melkebeek J. 'Rotational loss separation in grain-oriented Fe-Si'. *J. Appl. Phys.* 2000, vol. 87(9), pp. 6511–6513

[53] Brailsford F. *Physical Principles of Magnetism.* D Van Nostrand Co Ltd, London, 1966, p. 239

[54] Appino C., Fiorillo F., and Rietto A.M. 'The energy loss components under alternating, elliptical and circular flux in non-oriented alloys'. *Proceedings of 5th 2 DM Workshop*, Grenoble, France, 1997, pp. 55–61

[55] Strattant R.D., and Young F.J. 'Iron losses in elliptically rotating fields'. *J. Appl. Phys.* 1962, vol. 33(3), pp. 1285–1286

[56] Salz W., and Hempel K.A. 'Power loss in electrical steel under elliptically rotating flux conditions'. *IEEE Trans. Magn.* 1996, vol. 32(2), pp. 567–571

[57] Kedous-Lebouc A. 'A new hysteresis loop for rotating losses'. *J. Magn. Magn. Mater.* 1996, vol. 1609, pp. 45–46

[58] Mayergoyz I., and Serpico C. 'Rotational losses in anisotropic media subjected to rotating magnetic field'. *J. Appl. Phys.* 1999, vol. 85(8), pp. 5199–5201

[59] Pfutzner H. 'Rotational magnetization and rotational losses of grain oriented silicon steel sheets: fundamental aspects and theory'. *IEEE Trans. Magn.* 1994, vol. 30(5), pp. 2802–2807.

[60] Zurek S. 'Quantitative analysis of P-x and P-y components of rotational power loss'. *IEEE Trans. Magn.* 2014, vol. 50(4), Art. No. 6300914

[61] Gorican V., Jesenik M., Hamler A., Stumberger B., and Triep M. 'Performance of rotational single sheet tester (RRSST) at higher flux densities in the case of GO materials'. 7th In*ternational Workshop on 1 & 2 Dimensional Measurement and Testing, Ludenscheid 2002, PTB-E81 Report*, 2003, pp. 143–149

[62] Moses A.J. 'The case for characterisation of rotational loss under pure rotational field conditions'. *Przeglad Elektrotechniczny* 2005, vol. 81(12), pp. 1–4
[63] Moses A.J. *'Eddy Current Losses'. Wiley Encyclopaedia of Electrical and Electronic Engineering* (Ed. John G. Webster). John Wiley & Sons, Inc. New York, 1999, vol. 6, p. 143
[64] Enokizono M., Suzuki T., and Sievert J.D. 'Measurement of dynamic magnetostriction under rotating magnetic field'. *IEEE Trans. Magn.* 1990, vol. 26(5), pp. 2067–2069
[65] Hasenzagi B. 'Magnetostriction of 3 % SiFe for 2-d magnetization patterns'. *J. Magn. Magn. Mater.* 1996, vol. 160, pp. 55–56
[66] Pfützner H., and Hasenzagl A. 'Fundamental aspects of rotational magnetostriction'. In *Studies in Applied Electromagnetics and Mechanics. 10, Nonlinear Electromagnetic Systems* (Eds. Moses A.J., Basak A.). IOS Press, Amsterdam, The Netherlands, 1996, pp. 374–379.
[67] Somkun S. '*Magnetostriction and Magnetic Anisotropy in Non-Oriented Electrical Steels and Stator Core Laminations*'. PhD Thesis, Cardiff University, UK, 2010, p. 54.
[68] Lundgren A. '*On Measurement and Modelling of 2D Magnetostriction of SiFe Sheets*'. PhD Thesis, Royal Institute of Technology, Sweden, 1999
[69] Somkun S., Moses A.J., Zurek S., and Anderson P.I., 'Development of an induction motor core model for measuring rotational magnetostriction under PWM magnetisation'. *Przeglad Elekrotechniczny (Electr. Rev.)* 2009, vol. 85 (1), pp. 103–107
[70] Moses A.J., Anderson P.I., and Somkun S. 'Modelling 2-D magnetostriction in nonoriented electrical steels using a simple magnetic domain model'. *IEEE Trans. Magn.* 2015, vol. 51(5), Art. No. 6000407
[71] Zhi W., L., Zhu J.G., Guo Y.G., Zhong J.J., and Lu H.W. 'B and H sensors for 3-D magnetic property testing'. *Int. J. Appl. Electromagn. Mech.* 2007, vol. 25, pp. 517–520.
[72] Zhu J.G., Lin Z.W., Guo Y.G., and Huang Y. '3D measurement and modelling of magnetic properties of soft magnetic composites'. *Przeglad Elektrotechiczny (Electr. Rev.)* 2009, vol. 85(1), pp. 11–16
[73] Basak A. and Moses A.J. 'Effects of normal flux on power loss in grain oriented silicon-iron laminations'. *IEE Conference Proceeding*, vol. 142, London, UK, 1976, pp. 33–36
[74] Li J., Yang Q., Li Y., Zhang C., and Qu B. 'Measurement and modeling of 3-D rotating anomalous loss considering harmonics and skin effect of soft magnetic materials'. *IEEE Trans. Magn.* 2017, vol. 53(6). Art. No. 6100404

Chapter 9

Anisotropy of iron and its alloys

The property which makes GO electrical steel almost unique as a soft magnetic material is the anisotropy of its magnetic properties. Depending on the circumstances, this can be beneficial or harmful. The basic concept of the dominating role which magnetocrystalline energy plays in determining the magnetic domain structure was explained in Chapter 3. The most significant point in the case of electrical steel is that the stored magnetocrystalline energy becomes extremely high if the magnetisation in a single grain moves away from a preferred {100} direction of magnetisation.

Magnetisation within grains in a sheet of electrical steel tends to remain oriented along preferred directions unless an extremely high external field or mechanical stress is applied to change the energy balance.[1] This leads to an important fact, touched on often in this book, that in-plane magnetisation in electrical steel is likely to occur by magnetisation along preferred crystal directions within individual grains.

The deviation between B and H in a sheet varies according to the type of material anisotropy, the magnetising frequency, peak flux density and the influence of microstructure and magnetic domain structure. In this chapter some practical effects of the phenomenon particularly relevant to the measurement or characterisation of anisotropic materials are introduced. One purpose is to demonstrate the influence of magnetic anisotropy on flux and field direction in single specimens and in assemblies. It is useful to understand the phenomenon in order to interpret effects of microstructure on structure-dependent magnetic properties of electrical steels.

A comprehensive engineering approach to the anisotropy of iron-based soft magnetic materials including Fe-based amorphous ribbon is given in [1].

[1]There is a subtle difference between the terms *preferred* and *easy* directions of magnetisation. As stated above, a preferred direction refers to an axis in a single crystal which can be identified from the known anisotropy constants of a material. An easy direction, such as the RD in GO steel, defines a direction in a polycrystalline material along which magnetisation leads to higher permeability and lower losses than when magnetised along any other direction. Strictly speaking, of course, a material can have more than one easy direction where some might be "easier" than others. There is a close connection between preferred and easy directions of magnetisation but other factors come into play when creating an easy direction of magnetisation. In practice, the terms are often interchanged but this rarely affects understanding or meaning.

It is most convenient to deal with the measurement or analysis of the properties of soft magnetic materials under unidirectional magnetisation if the magnetic flux density is induced along the same direction as the applied magnetising field. It is easy to assume that when a piece of soft magnetic material is magnetised, the resulting flux density will naturally be along the field direction. In some circumstances this is not the case. For example, when a demagnetising field within a single grain of iron causes a difference between the direction of the local B and H, or when a mechanical stress causes B to move towards, or away from, the principal stress direction, or when it is energetically more favourable for B to take up a direction closer to an easy direction of magnetisation than the direction of the magnetising field.

9.1 Magnetisation at an angle to a preferred crystal direction

Consider an isotropic disc shown in Figure 9.1(a). Suppose it has an infinite radius and is initially demagnetised such that it has equal volume of domains oriented in all planar directions. Suppose an external field is applied such that the uniform field H on the surface of the disc is directed as shown. Depending on the size of the field, a certain volume of domains will be reoriented to set up a net flux density B_{AB} along the field direction. It will be remembered that B and H are vector quantities so both have components B_θ and H_θ along any direction θ in the plane of the disc.

Suppose the same field is applied to a single crystal with just one preferred direction at an angle α to the field direction as shown in Figure 9.1(b). Before magnetisation, the domains will be in the preferred direction. Also, assuming the disc is demagnetised, equal numbers of domains will be oriented along oe and oe' in equal volumes. When the field is applied it encourages the volume of domains along direction oe to grow at the expense those along oe' so a resultant flux density B_a is directed at the angle α to the field direction. A B–H curve derived from components of field and flux density along ab is effectively a curve of $B_{\alpha\alpha}$ against

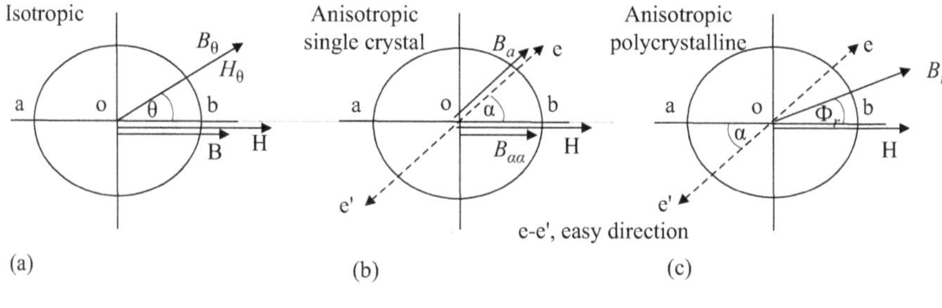

Figure 9.1 Schematic representation of B and H components in the plane of a disc with infinite diameter: (a) fully isotropic case, (b) single crystal with field applied at a fixed angle to its single preferred axis of magnetisation and (c) polycrystalline material with field applied at a fixed angle to its single easy axis, stress-free disc of infinite diameter

H where B_{aa} is simply $B_a \cos \alpha$. If α is known, and is fixed, then the measured B–H curve is a function of α provided domains do not move from the aa' direction. However, if the field is very large some domains rotate out of the preferred direction so the resultant flux density is at an angle of less than α to the field direction. When the field becomes very large, all the domains are oriented along the field direction and the material is saturated.

The above process occurring in a single crystal means that the shape of a B–H curve, plotted from the demagnetised state to saturation, can be distorted at different stages of magnetisation, initially dependent only on $\cos \alpha$ and, later, also by the magnetocrystalline anisotropy constants of the crystal.

In many situations, as will be seen later, α varies during the magnetisation process in which case the characteristic obtained from the components B and H along the direction ab becomes, at best, difficult to interpret, and, at worst, meaningless.

Now consider a polycrystalline anisotropic disc with one easy axis of magnetisation. Suppose the field is again applied at an angle α to the easy direction as shown in Figure 9.1(c). Here, the easy direction is assumed to arise because of a metallurgical texture in which most grains have $\{100\}$ directions near to the easy direction. The dotted lines represent the volume weighted orientation of $\{100\}$ directions in the material. In this case, application of the field causes the same process as for the single crystal to build up the flux density in grains oriented along the easy direction. The total flux density along the field direction ab of the polycrystalline disc is the vector sum of the components of flux density of crystals oriented along ee' together with those of all the mis-oriented grains. The magnitude, B_r, of the resultant flux density and its direction, ϕ_r, relative to the field direction becomes more difficult to define. The component of the resultant flux density along the field direction can be plotted as a B–H curve to give information about the build-up of the flux density along the field direction, but B_r and the ϕ_r of the maximum flux density remains unknown.

If the component of flux density along the field direction is of interest, then this is no problem. The unknown resultant can be found in practice from orthogonally placed search coils whose induced electromotive forces (e.m.f.s) would be $B_r \sin \phi_r$ and $B_r \cos \phi_r$. This approach is common in the analysis of the localised flux distribution in magnetic cores and in the measurement and analysis of rotational magnetisation. It should be noted that the measured value in this case is the average of the orthogonal components in the volume enclosed by the search coils.

In electrical steel sheet, the directional dependence of the magnetisation process is more complex for several reasons, but principally because some grains will be oriented out of the plane resulting in complex surface closure domains which affect the magnetisation process. However, the same basic principle applies, i.e. a B–H curve measured along any direction is formed by the components of B and H in that direction.

The angle between the direction of an applied field and the resulting magnetisation can be found from the basic energy minimisation theory for simple cases. Consider a cubic single crystal magnetised to saturation, M_s, by a large field inclined at an angle, θ_0, to the preferred direction of magnetisation as shown in

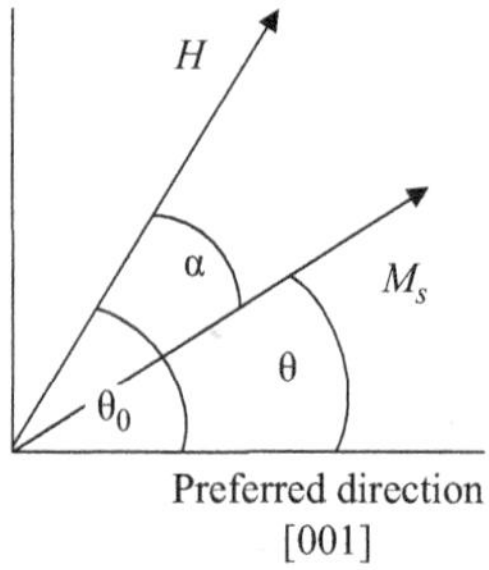

Figure 9.2 Deviation between directions of M_s and H in anisotropic material

Figure 9.2. The resulting magnetisation[2] is directed at an angle, θ, to the preferred direction, which, for convenience, is assumed to be a [001] direction in the (100) plane. Since the magnetisation and field are in the same plane and only the magnetising and magnetocrystalline energies are present, the total energy can be written as

$$E_t = E_k - H\,M_s \cos\,\alpha \,\left(\mathrm{Jm^{-3}}\right) \tag{9.1}$$

where α is the angle between H and M_s. For the energy to be minimum, $dE_t/d\alpha = 0$ and for this condition

$$H = \frac{dE_k/d\alpha}{M_s \sin\,\alpha} \,\left(\mathrm{Am^{-1}}\right) \tag{9.2}$$

The component of magnetisation, M, along the field direction is simply

$$M = M_s \cos\,\alpha \,\left(\mathrm{Am^{-1}}\right) \tag{9.3}$$

Values of appropriate directional cosines of θ can be inserted into the expression for E_k given in (3.1) in order to determine M–H characteristics [2].

The process which occurs in the crystal as the field is increased from zero is shown schematically in Figure 9.3. This assumes that initially the material is completely demagnetised with equal numbers of domains oriented along the [001] and [00$\bar{1}$] directions. This initial condition can be represented simply as the two antiparallel domains equal in volume. When a small field H_1 is applied at an angle to the [001] direction one domain grows at the expense of the other as shown in (a). The flux density B_1 is set up due to the domain wall motion. When the field is increased, as in (b), a domain might be created along the [100] direction so the resultant flux density is now B_2 at an angle θ_2 to the [001] direction. Note that the field has increased in magnitude but is still allied at angle θ_o to the [001] direction. This process continues to reach the condition in (c) where the field is sufficiently high to set up a single domain oriented at θ_3 to the [001] direction. This is equivalent to the situation in Figure 9.2. If a very large field were applied, the

[2]The energy minimisation equation here could be developed in the same way in terms of flux density.

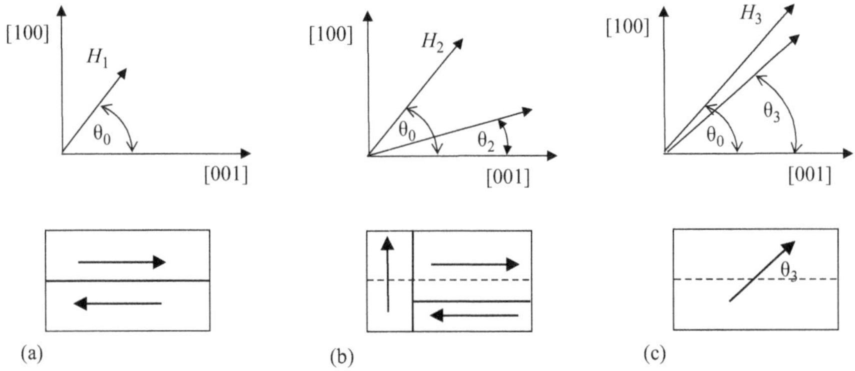

Figure 9.3 Variation of the angle between B and H in a simple domain structure with the field applied at angle θ_0 to [001] direction: (a) low field, (b) intermediate field and (c) large field

magnetisation would become parallel to the field direction at θ_0 to the [001] direction.

The process just described is very simplified and the domain patterns are somewhat artificial but the trend from demagnetisation to saturation at an angle to the easy direction can be clearly seen. However, observations of domains on the surface of GO steel magnetised at angles to the RD infer that internal domain structures do remain oriented along preferred crystal axes [3]. The local magnetisation remains along preferred directions of magnetisation in (b). Only at a very high field in (c) does the magnetisation rotate out of an easy direction of magnetisation.

An approximate estimate of the effect of the anisotropy in relation to the size of an external field in iron can be obtained by considering the difference between additional stored energy when magnetisation is at an angle to the [001] direction in iron and the equivalent stored energy due to misalignment of the magnetisation and the field, simply by putting some values into (3.1) and (3.40). For iron, K_1 is around 4.5×10^4 $\mathrm{Jm^{-3}}$, so (3.1) reduces to

$$E_k = K_1 (\sin 2\theta/2)^2 \ (\mathrm{Jm^{-3}}) \tag{9.4}$$

and the energy stored due to misalignment between the field and magnetisation in Figure 9.2 is $-HM_s \cos \alpha$.

As an example, suppose θ_0, the angle between the applied field and the [001] direction, is 45°. Figure 9.4 indicates components of free energy for different values of field when the magnetisation is at different angles to the [001] direction. At the field of 10 $\mathrm{kAm^{-1}}$, the field energy dominates and the minimum total energy occurs when the magnetisation is along the field direction, i.e. rotated 45° from the [001] direction. At the lower field of 500 $\mathrm{Am^{-1}}$ the minimum total energy occurs when the magnetisation is between 5° and 10° to the [001] direction. This is a very simplified analysis but it does show that the magnetisation is expected to remain close to the [001] direction in iron at typical operating flux densities (a field of

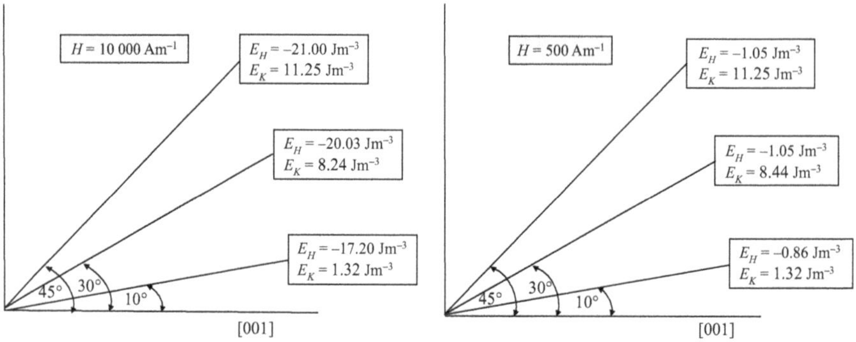

Figure 9.4 Energy stored for magnetisation at different angles to a [001] direction in an iron crystal when a field is applied at an angle θ_0 to the [001] axis ($K_1 = 45$ kJm^{-3} and the K_2 term is neglected, $M_s = 2.1$ T)

500 Am^{-1} produces a flux density of around 1.7 T in GO SiFe and a field of 10 kAm^{-1} is sufficient to bring it close to saturation).

A more accurate solution can be obtained by applying (9.2). There is no point here in obtaining a more accurate solution since there are many approximations in the geometry and the analysis is only included to demonstrate the rough balance between the effect of these two energy terms.

Let us return briefly to the process occurring in the approach to saturation in Figure 9.3(c) towards saturation along the [001] direction. The process is referred to as rotational magnetisation but it should not be confused with magnetisation brought about by the presence of a rotating field as described in Chapter 8 which also is called rotational magnetisation. In the one case, the magnetic dipoles rotate from a preferred crystal axis direction and, in the other case, the dipoles remain in [100] directions of the grains unless a high field is applied as demonstrated in Figure 9.4.

An approximate expression for the angle between B and H can be deduced in terms of the permeabilities along the easy and hard directions in the plane of an anisotropic sheet [4]. Suppose the material has orthogonal easy and hard directions of magnetisation as shown in Figure 9.5, along which the respective permeabilities are μ_1 and μ_2, we can write

$$B_1 = \mu_1 H_1 (\mathrm{T}) \tag{9.5}$$

$$B_2 = \mu_2 H_2 (\mathrm{T}) \tag{9.6}$$

and the actual flux density at angle θ can be resolved into components along the easy and hard directions of magnetisation as

$$B_1 = B \cos\theta (\mathrm{T}) \tag{9.7}$$

$$B_2 = B \sin\theta (\mathrm{T}) \tag{9.8}$$

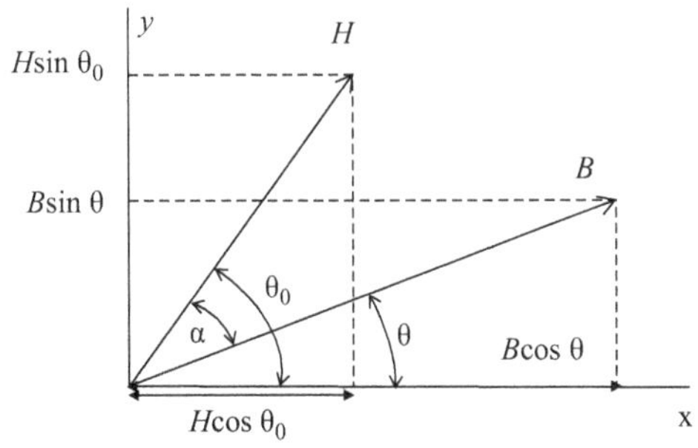

Figure 9.5 Spatial angle α between B and H in the plane of an anisotropic sheet when the flux density is at an angle θ to the easy direction

Hence

$$H_1 = (B_1/\mu_1)\cos\,\theta\,\left(\mathrm{Am}^{-1}\right) \tag{9.9}$$

$$H_2 = (B_2/\mu_2)\sin\,\theta\,\left(\mathrm{Am}^{-1}\right) \tag{9.10}$$

The resultant of these two components of field is at an angle $(\theta + \alpha)$ to the easy direction as shown in Figure 9.5, hence from basic trigonometry

$$\tan(\alpha + \theta) = \left({}^{\mu_1}/_{\mu_2}\tan\theta\right) \tag{9.11}$$

from which

$$\tan\alpha = \frac{\left({}^{\mu_1}/_{\mu_2} - 1\right)\tan\theta}{1 + \left({}^{\mu_1}/_{\mu_2}\right)\tan^2\theta} \tag{9.12}$$

Table 9.1 shows some values of α for different ratios of permeability when B is inclined at an angle to the easy direction. The important trends are, first, that as the permeability ratio increases, i.e. the easy direction gets even more favourable and, second, the angle between B and H has to increase in order to 'drag' the magnetisation away from the easy direction.

From (9.11) it can be deduced that the maximum angle of deviation α_{max} occurs when $\tan\theta = \sqrt{(\mu_2/\mu_1)}$. The values given in Table 9.1 are for an ideal material in which the permeability varies in magnitude sinusoidally from the easy direction to the hard direction of magnetisation which themselves are 90° apart. Although this analysis is quantitatively inaccurate for electrical steels, it does demonstrate the process that occurs.

Misalignment between magnetisation and field can be changed by the presence of mechanical stress which can effectively set up an alternative easy direction in a material which has, or does not have, magnetocrystalline anisotropy. It has been suggested that transverse magnetisation in anisotropic materials can be made use of to reduce losses [5] but it is not clear if this does have any potential benefit in modern materials.

Table 9.1 Values of angle α between B and H when B is at an angle θ to the easy direction of magnetisation in a material with uniaxial anisotropy for permeability ratios μ_1/μ_2 of 2, 5, 10 and 20

	Angle α between *B* and *H* (degrees)			
θ (degrees)	Permeability ratio = 2	Permeability ratio = 5	Permeability ratio = 10	Permeability ratio = 20
0	0	0	0	0
30	19.1	40.1	50.2	58.7
45	18.4	33.6	39.3	42.1
60	13.9	23.4	26.3	28.3
90	0	0	0	0

9.2 Magnetisation at angles to an easy direction under a.c. magnetisation

In Section 9.1, the angle between *B* and *H* was calculated for constant magnetisation. If the magnetising field is controlled to be at a fixed angle to an easy direction of magnetisation, the angle between the field and the arising flux density will vary through the magnetising cycle. It is informative to illustrate the effect in NO and GO electrical steel since they represent the extremes of anisotropy likely to be found in most soft magnetic materials.

Figure 9.6 schematically shows how the angle between *B* and the RD changes as the instantaneous field increases during an a.c. cycle of magnetisation at fixed angles of 30°, 45° and 60° to the RD in sheets of NO and GO steels, respectively. The *B* vector in each case lies within the shaded regions bonded by the dotted lines. The angle between *H* and the average instantaneous *B* is higher in the GO material. The vectors are only approximately to scale.

The characteristics shown in Figure 9.6 were obtained when the magnetising frequency was 5 Hz but similar trends are expected at power frequencies. Significant points are:

- the flux density becomes aligned closer to the field direction as the magnitude of *H* increases;
- the flux density is almost entirely aligned with the RD when *H* is at a small angle (around 30°) to the RD;
- when *H* is oriented at a large angle (say 60°) to the RD, *B* is oriented closer to the TD than *H* and
- the *B* loci are generally elliptical but not aligned with the *H* direction.

Some quantitative results on the deviation of the flux density from the field direction in GO steel, when the field is fixed at 30° and 60° to the RD, are given in Figure 9.7 [7]. When *H* is fixed at 30° the direction of *B* rotates from 0° to 30° to the RD as the field is increased to a maximum value of 60,000 Am^{-1} at close to saturation. When *H* is at 60° to the RD, the angle between *B* and the RD increases

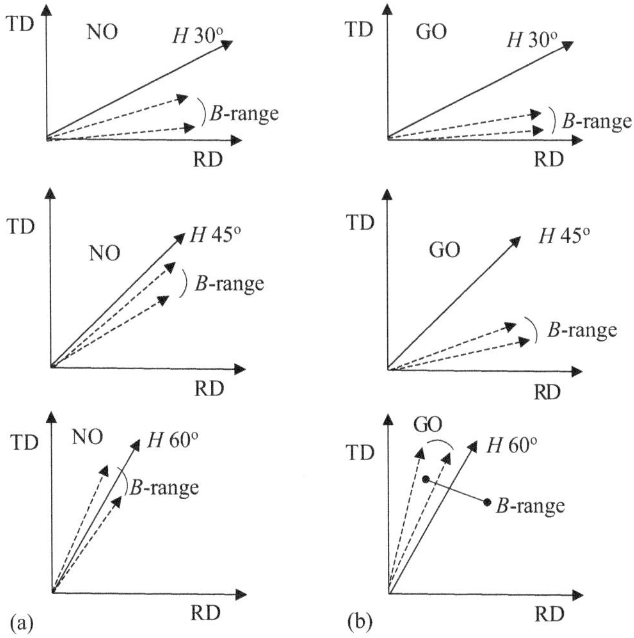

Figure 9.6 Ranges of the variation of the angle between B and H during a magnetising cycle under a.c. magnetisation when the field is fixed at 30°, 45° and 60° to the RD in (a) NO steel and (b) GO steel (Data extracted from [6])

to around 70° before the flux density comes into alignment with H, close to saturation. The angle between B and H is frequency dependent because it depends on the permeability ratio discussed earlier. The measurement results shown in Figure 9.7 agree reasonably well with those in Figure 9.6 in spite of them being made on different grade materials and probably at different frequencies. However, for the reasons given earlier, they do not agree well with values predicted from (9.9).

The main incentive for quantifying the angle between B and H is to find suitable formulations to improve the accuracy of computational electromagnetic analysis involving non-linear anisotropic materials. In the simplest models, the material is assumed to be isotropic so the B–H characteristic measured along the RD is used, but the solution will be inaccurate in most magnetic core geometries, particularly if assembled from GO steel. Often models are made more accurate by incorporating B–H characteristics measured along the RD and TD and then assuming a linear relationship between them to estimate B for a calculated H at some angle to the RD in a given element. This is more accurate, but it does not account for the spatial angle between B and H when H is applied at angles to the RD. Vector methods such as in [8] have been proposed to take this angle into account but the improved accuracy which they should provide does not appear to be quantified and they are not commonly used.

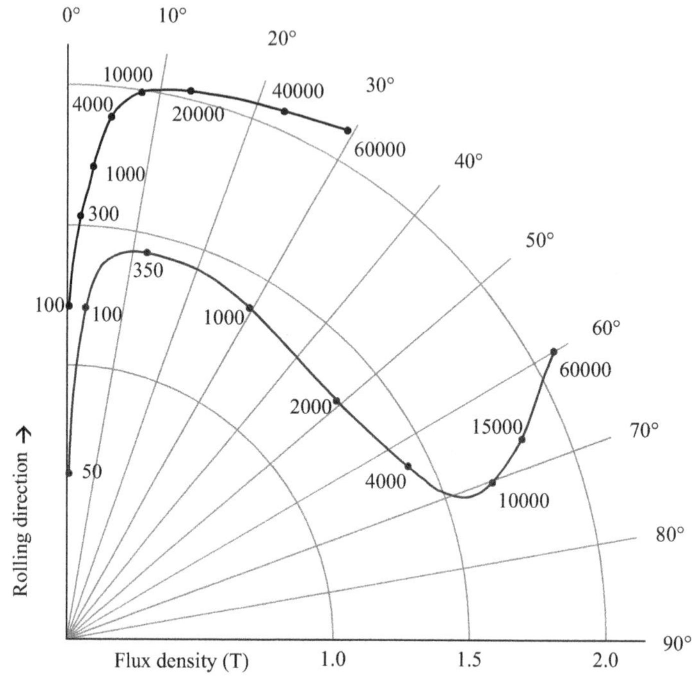

Figure 9.7 Circle diagram showing flux density in a sheet of GO steel when the field is fixed at 30° and 60° to the RD (un-numbered figure in [7], reproduced under free licence CC-BY-4.0)

9.3 Effect of strip width on magnetisation direction in anisotropic material

Figure 9.8(a) shows a strip of anisotropic material cut at an angle θ to the easy direction of magnetisation. Suppose the flux density B_{long} along the strip is controlled, making use of the e.m.f. induced in the search coil cc_1 wound around the strip perpendicular to its long axis. The strip is magnetised by an external field H_a and H_e is the surface field taking into account the demagnetising field H_{dl}. In a practical measurement the e.m.f. induced in search coil CC_1 is proportional to dB_{long}/dt and is not affected by any component of magnetisation which might be present at angles to the longitudinal direction in an anisotropic strip. Hence the magnetisation curve, B_{long} versus H_e, which is often used as a reference measurement in single strip testing, is limited since it does not contain full information about the magnetisation.

Consider the search coils aa_1 and bb_1 shown wound at 45° to the longitudinal direction. The magnitude and direction of the true flux density B_r can be found from the e.m.f.s induced in these coils. Suppose that surface field sensors are also located at 45° to the longitudinal direction as depicted in Figure 9.8(a) to detect components of field H_{45} and H_{-45}.

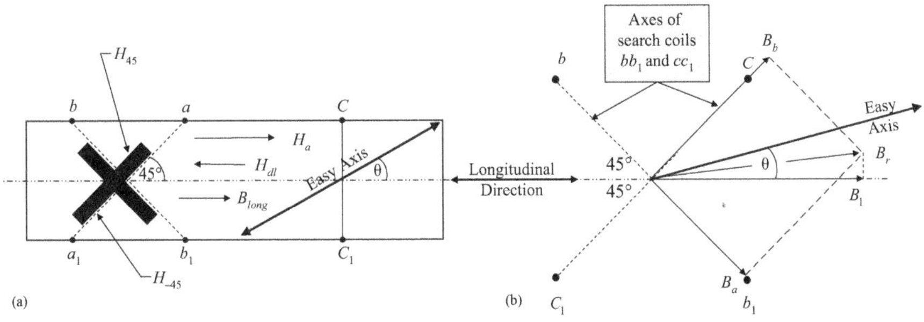

Figure 9.8 (a) Sensor locations and magnetisation direction in a strip of anisotropic magnetic material cut at an angle θ to the easy axis and (b) components of B_r relative to search coils axes

If the strip is cut parallel to the easy direction, $B_r = B_{long}$ and $H_{45} = H_{-45}$ hence the B_{long} versus H_e[3] is the true B–H characteristic of the steel. However, when a strip of anisotropic steel cut at an angle to its RD is magnetised an interesting, and often neglected, phenomenon occurs. The sum of the magnetocrystalline and magnetostatic energy is minimised by rotation of both B_r and H_e away from the longitudinal direction of the strip in a similar manner to that discussed in Section 9.1. However, if the strip has a high aspect ratio (length: width),[4] such rotation does not occur, due to the energy associated with the large demagnetising field in the short direction which is a result of the component of magnetisation in this direction.

The direction of the surface field, which need not necessarily be along the longitudinal direction of an anisotropic strip, has to be determined in order to investigate the angle θ_{bh} in a strip cut at an angle to its easy direction. The orthogonally placed surface field sensors shown in Figure 9.8(a) can be used to obtain the deviation of the surface field H_e from the longitudinal direction as shown in Figure 9.9(a). The outputs from the corresponding flux density sensors can be used as shown in Figure 9.9(b) to obtain the deviation of the resultant flux density B_r from the longitudinal direction and the important angle θ_{bh} between it and the resultant surface field H_r.

The phenomenon is analysed in Figure 9.9(b) by considering the possible effect of the local demagnetising factors caused by the anisotropy. In this case, the resultant B_r is obtained experimentally from the easily accessible components B_a and B_b at 45° to the longitudinal direction. The more useful components of B_r are B_{rrd} and B_{rtd} since we can associate local *opposing* demagnetising fields H_{drd} and H_{dtd} with them, assuming that the magnetic domains forming B_r will be aligned along the $\{100\}$ directions close to the RD or TD, i.e. the applied field H_{appl} is not

[3]In practice H_e can be measured with surface field sensors, in an open magnetic circuit or from the magnetising current in a closed circuit where normally $H_a = H_e$. These techniques are discussed in Section 11.7 of Vol. 2 of this book.

[4]Typically 10:1 for a Epstein strip and many transformer laminations.

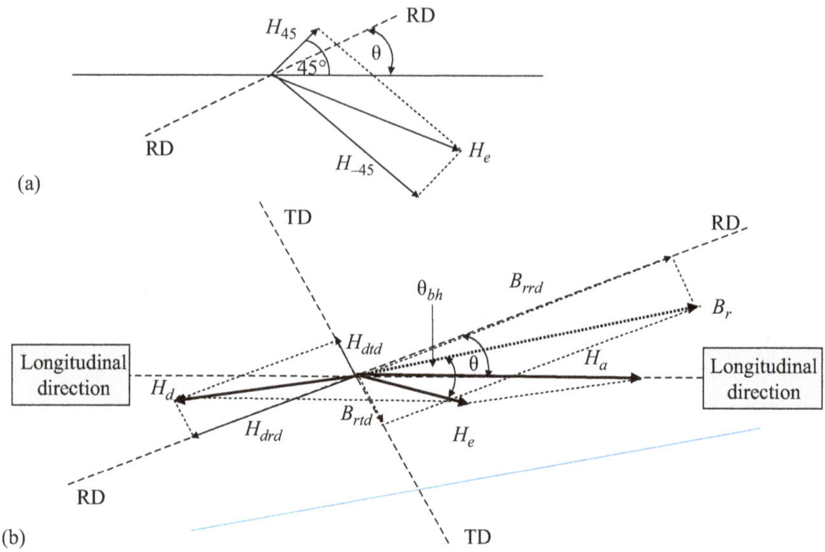

Figure 9.9 (a) Magnetic field components in the anisotropic strip and (b) effect of demagnetising field components on the resultant field on the strip surface

large enough to cause domains to rotate away from {100} directions. The resultant field H_e is then the resultant of H_a and H_d.[5]

The power loss and other magnetic properties of the strip depend on the magnitude of θ_{bh}. This varies with level of magnetisation, the value of θ and the local demagnetising factors which determine H_{drd} and H_{dtd}. The phenomenon, which is often neglected in magnetic measurements, is generally more prominent in anisotropic strips or sheets with low aspect ratios.[6]

The deviation of B and H from the longitudinal direction of strips of GO steel cut at angles to the RD is shown in Figure 9.10(a). The negative angles for H_e indicate that it is directed in the opposite direction, relative to the longitudinal axis, to that of B_r. The strips have a low aspect ratio of approximately 3:1, so significant deviations of B and H are expected and found in strips cut at some angles to the RD [9]. The measurements were made at 50 Hz and the values are indicative of what can be expected from highly anisotropic material.

Figure 9.10(b) shows the angle θ_{bh} between B and H in the strips. This angle has more physical significance in terms of true permeability and loss. These can be compared to those shown in Table 9.1 for an ideal anisotropic material with orthogonal easy axes and linear variation of permeability along directions between these axes. The dotted line on Figure 9.10(b) is the theoretical curve obtained from (9.12)

[5]This demonstrates that the effective field in a magnetised strip is not necessarily parallel to the longitudinal direction or the easy direction of magnetisation.

[6]In this context the aspect ratio is the ratio of strip length to strip width.

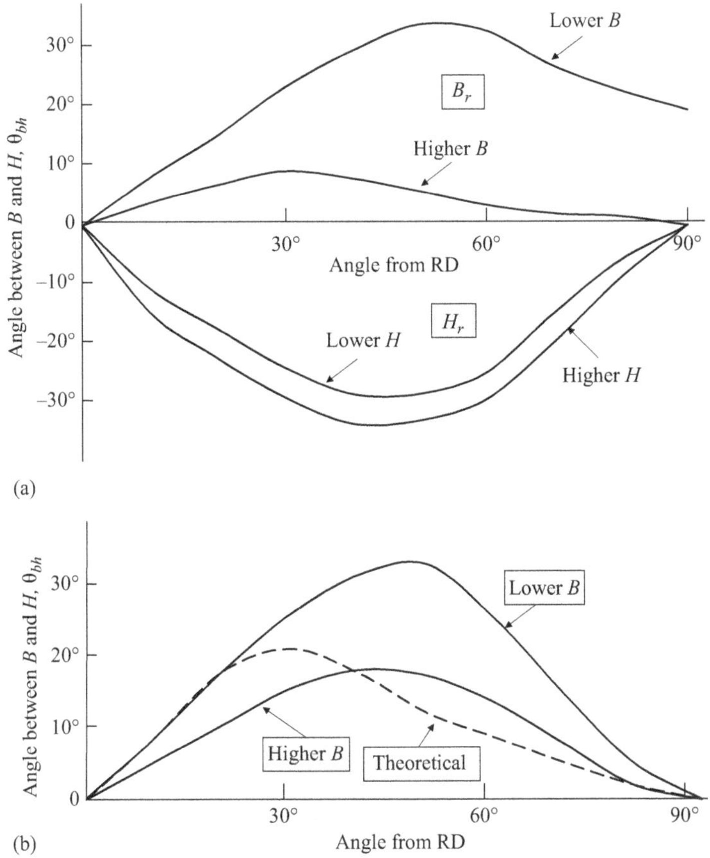

Figure 9.10 Effect of anisotropy on B and H on their directions in wide strips of GO electrical steel cut at angles to the RD: (a) angle between B and H and the longitudinal direction and (b) angle between B and H (Data extracted from results given in [9])

for an anisotropy ratio of 5:1.[7] Although there is no reason why the curves should correspond, it is interesting to note that up to around 30° to the RD, the measured and theoretical curves are comparable, whereas at a higher angle close to the ideal theoretical value, it is lower than measured values, presumably partly because influence of the [111] direction in the GO steel is not included in the theoretical analysis. For a given material, the deviation of B and H from the longitudinal direction is dependent on the overall flux density, the magnetising frequency and the aspect ratio of the strip. The authors are not aware of any method of predicting the angle.

[7]This is a typical value for small transformer core laminations.

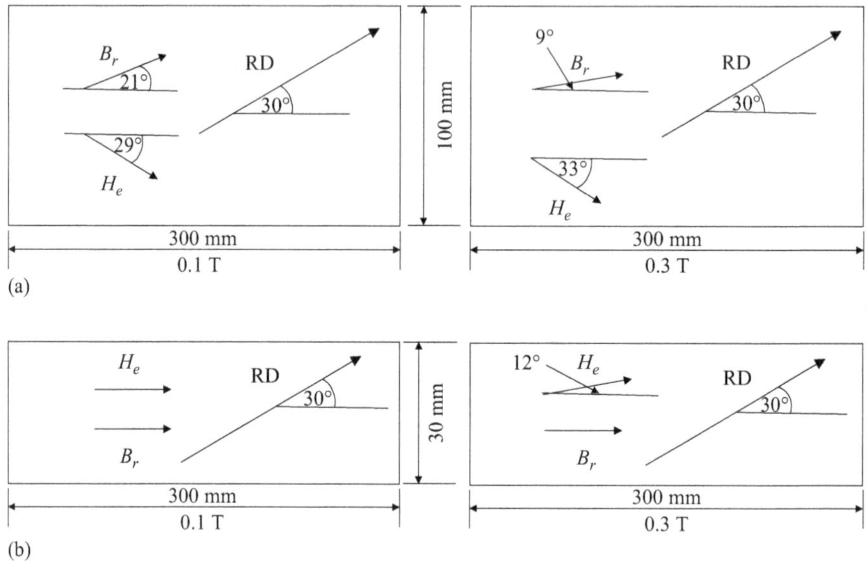

Figure 9.11 Principal directions of B and H in 300 mm long strips of GO silicon steel cut at 30° to the RD: (a) wide strip (aspect ratio 3:1) and (b) narrow strip (aspect ratio 10:1) (Adapted from results in [9])

The phenomenon in GO steel can be approximately quantified using a single grain model. Consider a grain with a 6.0 mm diameter surface area in a 0.30 mm thick sheet of GO steel. The grain can be approximated to an oblate ellipsoid which has an axial ratio of 20. The demagnetising factor is then 0.0369 [10]. If the components of flux density and field are measured, then the relationship in Figure 9.9(b) can be used to obtain the theoretical value of H_e. The approach is not very accurate because of the gross assumptions – first, that the calculation can be based on each grain being perfectly oriented and being magnetically independent of surrounding grains, which in practice is not the case, and, second, that all grains are identical oblate ellipsoids.

Rotation of B_r and H_e measured in 300 mm × 100 mm strips of GO steel cut at 30° to the RD at low flux density is significant as shown in Figure 9.11(a). Figure 9.11(b) shows that H_e and B_r remain oriented along the longitudinal direction of the strip with a low aspect ratio.

The mechanism for this rotational process and its importance as a potential cause of misinterpretation of the magnetic characteristics of strips cut at angles to the RD is not well understood. It will be noticed that the measurements are made at quite low flux densities and the effect becomes lower as the field is increased. This is simply because stored energy associated with the field becomes dominant making the favourable direction of magnetisation lie along the strip direction. Therefore, it can be argued that this effect is of little practical importance in magnetic cores because they normally operate at much higher flux densities.

Even if this is accepted, the results are of academic importance and help in understanding the complex magnetisation phenomena which can occur in anisotropic soft magnetic materials. It is also a reminder that when characterising *B–H* properties of material magnetised at angles to the RD, B_{long} is not necessarily the highest component of flux density in the strip, so the loss can be higher than that obtained using measured values of B_{long} and the component of H_e along the longitudinal direction.

Low magnetisation performance is important in some devices, such as instrument transformer cores, where accurate characterisation of the low field portion of the *B–H* curve is important. Such cores are usually wound from strip always cut parallel to the RD so, again, the effect does not appear relevant, but it is something for investigators of innovative approaches to core configurations to be aware of.

An expression for the theoretical angle δ_b between B_r and the RD has been quoted as [11]:

$$\tan \delta_b = \frac{\mu_{xy}(1 - N_{TD})}{\mu_{xx}(1 - N_{TD}) + N_{TD}\left(\mu_{xx}\mu_{yy} - \mu_{xy}^2\right)} \tag{9.13}$$

where μ_{RD} and μ_{TD} are the permeabilities when magnetised along the RD and TD, N_{TD} is the demagnetising factor of the specimen along its transverse direction. Hence

$$\mu_{xx} = \mu_{RD}\cos^2\theta + \mu_{TD}\sin^2\theta,\ \mu_{yy} = \mu_{RD}\sin^2\theta + \mu_{TD}\cos^2\theta \text{ and } \mu_{xy}$$

$$= \mu_{RD}\mu_{TD}\sin\theta\cos\theta \tag{9.14}$$

where θ is the angle between B_r and the RD. This expression has not been reported to have been tested for Epstein strips, but when applied to square specimens of GO material, θ is found to be similar to that found in Epstein strips reported in [9].

The angular displacement between *B* and *H* in GO steel has been calculated using a numerical technique based on an equivalent magnetic domain model for incorporation into finite element modelling of the magnetic properties of electrical machine cores [12]. However, its accuracy is questionable unless lamination demagnetising factors are taken into account.

9.4 Effect of stacking method on apparent loss of anisotropic strips cut at angles to an easy axis

The Epstein square method for measuring magnetic properties of electrical steels is well documented. Aspects of its standard use are discussed in Chapter 12 of Volume 2 of this book. Investigators are developing computer algorithms to model the variation of permeability, or losses, of anisotropic materials, particularly GO electrical steel, when magnetised at angles to the easy direction. This can be done by testing strips cut at angles to the easy axis of magnetisation in a single strip or single sheet tester. The deviation of *B* or *H* from the longitudinal direction of a strip, as shown in Figure 9.11, causes doubt over the accuracy of such measurements.

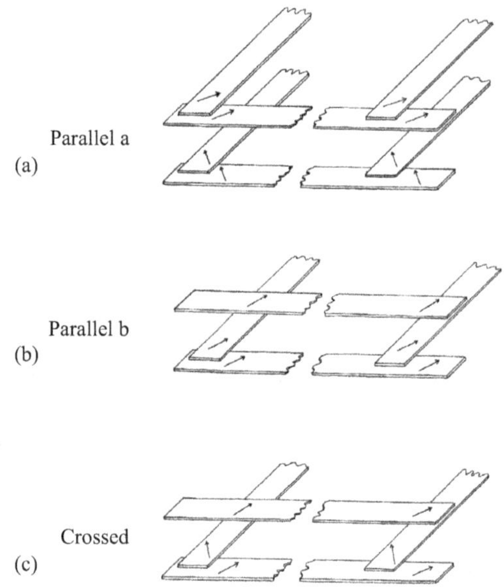

Figure 9.12 Possible configurations of strips cut at an angle to the RD assembled in an Epstein frame: (a) ***II a*** *stacking, (b)* ***II b*** *stacking and (c)* ***X*** *stacking*

Another measurement option preferred by many investigators is to test strips cut at angles to the RD in an Epstein square. Sometimes, characteristics obtained in this way are published in performance data sheets of NO and GO steel. However, there is no agreed specification for testing strips cut at angles to the RD in this way which makes it difficult to compare results from different workers since losses and permeability measured in this way are dependent on the stacking method [13–15].

Figure 9.12 shows some ways in which Epstein strips can be stacked in the frame. The RDs of all strips in each arm can be parallel to each other. This is sometimes referred to as 'parallel' or ***II*** stacking. The alternative is referred to as the 'crossed' or ***X*** configuration. It is expected that similar mechanisms to those proposed in Figure 9.11 may occur in some stacking arrangements, in which case, closure domains will be set up along edges of strips and corner flux transfer mechanisms will vary according to the degree of anisotropy and the angle of cut to the RD [13]. These phenomena have an effect on the losses. Another phenomenon which also comes into play is the interaction between individual domains in adjacent sheets [16]. This has a small influence on the magnetic properties of a stack of laminations cut parallel to the RD so it is expected that this phenomenon would also affect the properties of assemblies of strips cut at different angles to the RD by different measurable amounts [17,18]. It has been verified by experiment that the stacking method has a significant effect on the amount of interlàminar flux [19] but the overall effect on losses has not been quantified.

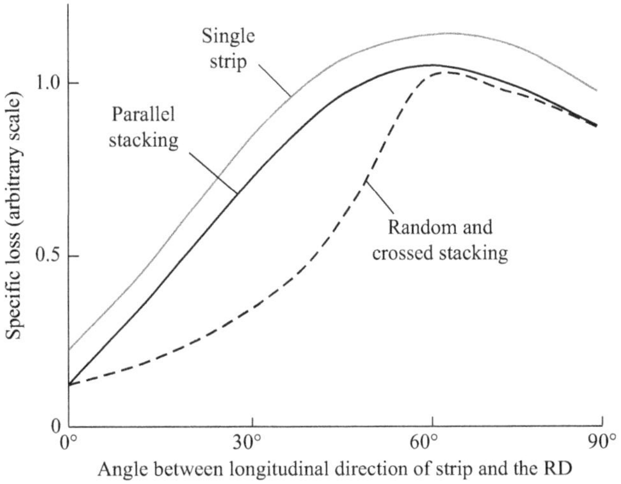

Figure 9.13 Variation of measured power loss in GO steel strips cut at angles to the RD and assembled in a standard Epstein frame, B = 1.2 T, 50 Hz (Figure 3 in [14] reproduced under free licence CC-BY-4.0 and modified)

Figure 9.13 shows typical loss characteristics of GO steel strips cut at different angles to the RD and stacked in these ways. It is not surprising that the losses measured in the ***II*** configurations are similar to the single strip values. It is not so clear why the random stacking and ***X*** configuration are so similar.

The secondary voltage and input current of the Epstein square are used to calculate the losses. These quantities are proportional to the rate of change of *B* along the limb axes and the corresponding component of surface field along the limbs. Clearly, the values of loss or permeability measured using the Epstein Square in this way are very dependent on the flux and field deviations from the longitudinal direction in a similar way to those shown in Figure 9.11.

Corresponding variation of coercivity has been found to occur in Epstein strips of GO steel magnetised under d.c. magnetisation in ***X*** and ***II*** configurations. This has been attributed to magnetostatic interactions between grains in adjacent sheets [19].

It is well known that, when a stack of laminations is tested in the Epstein square, the average flux density throughout the stack might be found to vary sinusoidally with time, but each lamination has a different distorted flux density. The distortion has been attributed to dissimilarity between the *B–H* loops of the individual strips and is quoted as being responsible for increasing loss by around 1% [20], although it is not clear whether this amount varies with stack size and magnetisation level.

Figure 9.14 shows the e.m.f. waveforms induced in search coils wound around individual strips cut at 30° to the RD and stacked in ***X*** and ***II*** configurations. The waveform distortions in the strips in the two different ***II*** configurations are similar to each other, as expected, but they contain a very different harmonic distribution to

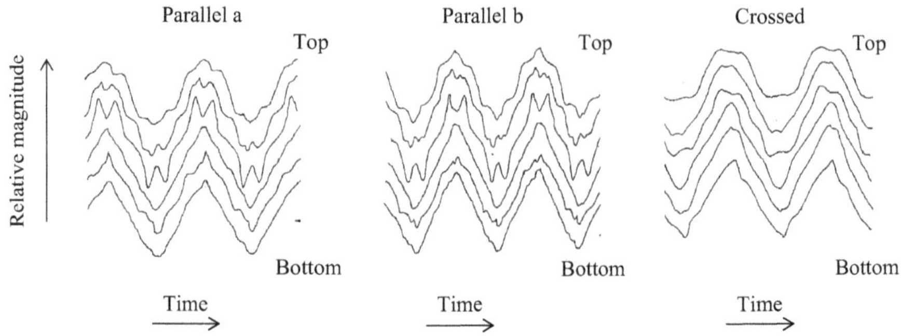

*Figure 9.14 Induced e.m.f. waveforms in strips of GO steel cut at 30° to the RD assembled in an Epstein frame using **X** and **II** type stacking configurations when the overall core flux density (1.2, 50 Hz) is sinusoidal (Figure 6 in [14], reproduced under free licence CC-BY-4.0 and modified)*

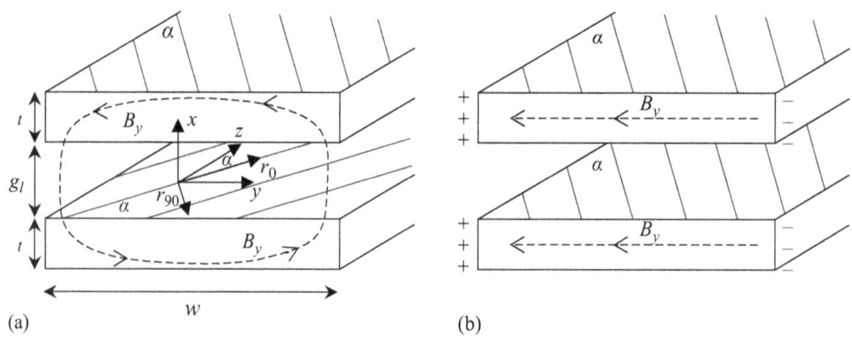

Figure 9.15 Theoretical paths of transverse flux in pairs of laminations of GO electrical steel cut at an angle α to the RD: (a) crossed stacking and (b) parallel stacking (Figure 3 in [21], reproduced under free licence CC-BY-4.0)

that in the strips in the ***X*** configuration. It appears that the distortion is not simply related to material inhomogeneity, but also to the flux deviation from the longitudinal direction. However, no attempt appears to have been made to quantify the effect on the measured magnetic properties in the three configurations.

An explanation for the differences in the magnetic characteristics of cores assembled from strips assembled in these different configurations can be given in terms of a transverse demagnetising field [19]. Figure 9.15 shows the transverse flux in the two configurations in a strip of width w and thickness t with an interlaminar space of g_l. The RDs are indicated by the hatching. The transverse demagnetising factor in the ***X*** configuration is far lower than in the ***II*** configuration

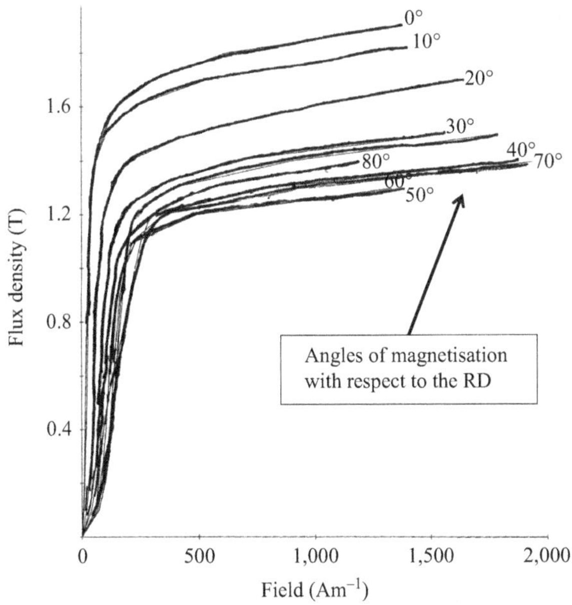

Figure 9.16 *Typical set of B–H characteristics of GO steel magnetised between 0° and 90° to the RD for use in computational electromagnetic analysis*

which leads to lower transverse losses and, hence, lower losses in line with experimental findings.

Figure 9.16 shows the type of *B–H* characteristics often used for modelling magnetisation conditions in computational electromagnetic solvers discussed in Chapter 11 of Volume 2 of this book. The strips of GO steel are cut at angles to the RD and magnetised longitudinally. The figure clearly shows the expected trend in a reduction of permeability as the magnetisation direction used is moved from the RD and the poorest curves are in strips cut at 50° and 60° to the RD. It will be remembered that the most difficult magnetising direction is along the [111] direction which, in a perfectly oriented Goss grain, is at 54.7° to the RD so it is not surprising that the lowest permeability occurs in the strips cut at angles to the RD close to this. Of course, these sorts of characteristics would probably be obtained from an Epstein test, in which case, the measurements would be subject to many of the phenomena which could lead to inaccuracies when extrapolating to wider or different shaped laminations used in transformer and rotating machine cores.

References

[1] Soinski M., and Moses A.J. Anisotropy in iron-based soft magnetic materials. In *Handbook of Magnetic Materials*, vol. 8 (Ed. Buschow K.H.J.), 1995, chapter 5, pp. 325–414

[2] Bozorth R.M. *Ferromagnetism*. IEEE Press, Piscataway, NJ, 1993, p. 578

[3] Yamamoto T., Shiuchi A., Saito A., Okazaki Y., Hasenzagi A., and Pfützner H. 'Magnetostriction and magnetic domain structure changes of grain oriented Si-Fe sheets'. *J. Phys. IV France* 1998, vol. 8, pp. 511–514

[4] Langman R.A. 'Prediction and measurement of rotational magnetisation in an anisotropic polycrystalline material'. *IEEE Trans. Magn.* 1981, vol. 17(1), pp. 1159–1168

[5] Chistyakov V.K., and Mints B.B. 'Influence of transverse magnetization on the magnetic properties of anisotropic electrical steel'. *Bull. Acad. Sci. USSR, Phys. Ser.* 1975, vol. 39(7), pp. 17–19

[6] Fard S.M.B. 'Modelling Anisotropy in Electrotechnical Steels'. *PhD Thesis*, University of Wales, Cardiff, 1992

[7] Weggler P.T. 'Computation of magnetic fields in nonlinear anisotropic media with field dependent degree of anisotropy'. *Proceedings of Compumag*, Oxford, UK, March 1977

[8] Waeckerlé T., and Cornut B. 'Influence of texture in grain oriented steel on magnetization calculated curve in the range of rotational processes'. *IEEE Trans. Magn.* 1992, vol. 28(5), pp. 2793–2795

[9] Layland N.J. 'A Study of Flux and Field Distribution in Electrical Steels'. *MSc Thesis*, University of Wales, College of Cardiff, Sept 1994

[10] Chikazumi S. *Physics of Magnetism*. R E Krieger, New York, 1978, p. 22

[11] Benda O., Bydžovsky J., and Ušák E. 'Combined influence of shape and magnetocrystalline anisotropy on measured magnetisation curves of SiFe sheets'. *J. Phys. IV France* 1998, vol. 8, pp. 627–630

[12] Fujisaki K., and Satou S. 'Angle between ***B*** vector and ***H*** vector in anisotropic electrical steel'. *IEEE Trans. Magn.* 2008, vol. 44(11), pp. 3161–3164

[13] Ferro A., Montalenti G., and Soardo G.P. 'Loss dependence on lamination orientation and stacking'. *Proceedings of the 3rd International Conference on Soft Magnetic Materials*, Bratislava, Czechoslovakia, 1977, paper 9.3, pp. 275–280

[14] Moses A.J., and Phillips P.S. 'Effects of stacking methods on Epstein-square power-loss measurements'. *Proc. IEE* 1977, vol. 124(4), pp. 413–416

[15] Findlay R., Belmans R., and Mayo D. 'Influence of the stacking method on the iron losses in power transformer cores'. *IEEE Trans. Magn.* 1990, vol. 26 (5), pp. 1990–1992

[16] Sláma J. 'Investigation of magnetic interactions between oriented Si-Fe sheets'. *J. Magn. Magn. Mater.* 1984, vol. 41, pp. 275–278

[17] Pfützner H. 'Domain interactions between stacked *Hi-B* SiFe sheets'. *IEEE Trans. Magn.* 1982, vol. 18(4), pp. 961–963

[18] Ebrahimi A., and Moses A.J. 'Correlation between normal flux transfer from lamination to lamination during the magnetization process with static domain structures in grain oriented 3% silicon iron'. *J. Appl. Phys.* 1991, vol. 70(10), pp. 6265–6267

[19] Ebrahimi A., and Moses A.J. 'Influence of the stacking method on the normal flux transfer from lamination to lamination in grain-oriented 3%

silicon-iron samples cut at different angles to the rolling direction'. *J. Magn. Magn. Mater*. 1992, vol. 112, pp. 129–131

[20] Sláma J. 'Coercivity effects in oriented laminations'. *J. Magn. Magn. Mater.* 1982, vol. 26, pp. 40–42

[21] Bishop J.E.L., and Clay P.M. 'The dependence of permeability and losses on direction of magnetization in anisotropic ferromagnetic laminations'. *J. Phys. D: Appl. Phys.* 1971, vol. 4, pp. 1797–1811

Chapter 10

Magnetic circuits

This chapter introduces magnetic circuit terminology and forms of analysis commonly used in electrical steels applications. More detailed approaches which include magnetic circuits related to both hard and soft magnetic materials in general and more complex topologies can be found elsewhere, e.g. [1–3].

An appreciation of magnetic circuits, based mainly on Ampère's circuital law and Faraday's law of electromagnetic induction, is needed to understand design and operation of magnetic measurement systems and the function of magnetic materials in magnetic cores.

The use of computational electromagnetics to its full potential requires an understanding of the practical interaction of magnetic materials with external fields. When only superficial analysis of a magnetic circuit is needed, it can be an unnecessary distraction to turn to a computational solution when basic magnetic circuit theory can be more quickly applied with sufficient accuracy for specific applications. Users of electromagnetic solvers need to know very little about basic magnetic circuit theory when just imputing geometry, material *B–H* or *M–H* characteristics and magnetising conditions before running the programme. However, it is often very useful for users to have a good appreciation of magnetic circuit analysis in order to recognise and analyse reasons for the occasional unexpected solution or when the problem is at the limits of the capability of the solver.

10.1 The basic magnetic circuit

Before considering the basic magnetic circuit, the magnetic field produced by a current-carrying conductor should be more closely defined. Figure 10.1(a) shows the magnetic field, represented by field lines,[1] set up by an electric current I flowing through a single conductor. The field is constant around each circular path shown and, by Ampère's circuital law, the field H_r falls with increasing distance r from the axis of the conductor according to

$$H_r = \frac{I}{2}\pi r \ (\mathrm{Am^{-1}}) \tag{10.1}$$

[1]The lines show the direction of the field and the distance between them is a measure of the field magnitude.

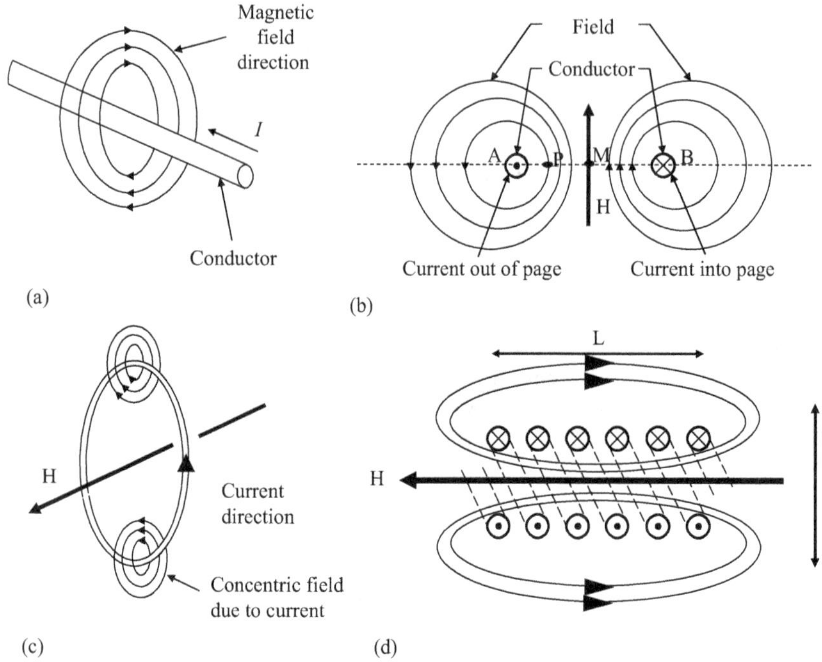

Figure 10.1 Development of a simple magnetic circuit: (a) field around a single current carrying conductor, (b) field around parallel conductors carrying equal currents in opposite directions, (c) field in the vicinity of a single turn of radius r and (d) cross section of the field in a multi-turn circular coil (solenoid)

The field at any point in Figure 10.1(a) is always at right angles to the radius and does not have any components either parallel or radially perpendicular to the conductor. Figure 10.1(b) shows a cross section of the field in the vicinity of two parallel conductors carrying currents flowing in opposite directions.[2] It is easy to see that the field at the point *M* midway between the conductors is double that of the fields produced by each conductor at that point. If the currents were flowing in the same direction, the components of field would be opposing each other and completely cancel out at point *M*.

Suppose a conductor is formed into the shape of a single turn coil of radius *r* as shown in Figure 10.1(c). It is not difficult to visualise that the field flows in the direction shown and that the field on the axis of the coil is directed perpendicular to its plane in exactly the same way as for the two conductors in Figure 10.1(b).

[2]The cross and circle convention is commonly used to indicate the current, flux or field direction in electric and magnetic circuits. Here it indicates one current flowing into the plane of the sheet (×) and the other flowing out (•).

Figure 10.1(d) shows what happens when the single turn is replaced by a circular cross section, multi-turn, air cored, N turn coil.[3] This is, of course, the basis of magnetising materials in many types of magnetic measurement systems and also for energising electrical machine cores.

The fields produced by the currents in these simple geometries can be quantified without any difficulty. Let us return to the parallel conductors in Figure 10.1(b). The magnitude of the field at point P, such that $AP = x$, can be written as the sum of the components from the two conductors as

$$H_x = \frac{I}{2\pi x} + \frac{1}{2\pi(d-x)} \ \left(\mathrm{Am}^{-1}\right) \tag{10.2}$$

where d is the distance between the two conductors. This reduces to

$$H_x = \frac{Id}{2\pi x(d-x)} \ \left(\mathrm{Am}^{-1}\right) \tag{10.3}$$

The variation of H_x between the conductors is shown in Figure 10.2(a). The linear portions represent the linear build-up of field from the centre of each non-magnetic conductor up to its surface.

The field inside square and circular solenoids of the form shown in Figure 10.1(d) are of particular interest since they are commonly used in magnetic test systems. It can be derived from first principles in several different ways. It can be shown that the field at any point Q on the axis on a short, uniformly wound multi-turn coil[3] is given by [4]:

$$H_x = \frac{NI}{2l}(\cos\beta - \cos\alpha) \ \left(\mathrm{Am}^{-1}\right) \tag{10.4}$$

where α and β are the angles indicated on Figure 10.2(b). If the coil is long $\alpha \to 180^\circ$ and $\beta \to 0^\circ$, hence (10.4) reduces to

$$H_x = \frac{NI}{l} \ \left(\mathrm{Am}^{-1}\right) \tag{10.5}$$

The field everywhere in the infinitely long coil is parallel to the axis and it is independent of the coil radius. Equation (10.5) applies anywhere in the solenoid and is does not depend on the shape of the cross section [4].

In a short coil, α and β vary with the ratio of l/r. Figure 10.2(c) shows how the field on the axis of a practical coil might decrease near its ends. This can be minimised in practice by making l/r as large as possible. Extra magnetising turns can be added near the ends of the coil, or turns removed from the central region, to create a more uniform field profile in magnetic test systems.

[3] This is often referred to as a solenoidal coil or simply a solenoid. It is a coil wound as a tightly packed uniform helix. The term normally refers to a helically wound coil of any symmetrical cross sectional shape whose length is far greater than its diameter.

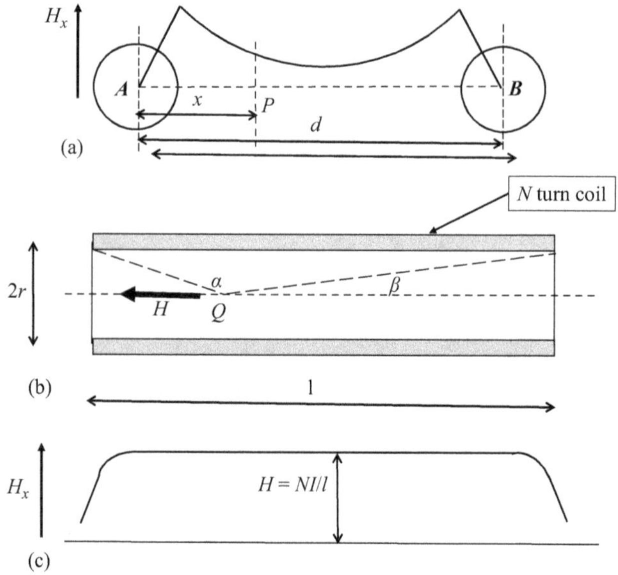

Figure 10.2 (a) Variation of field between two current carrying conductors, (b) cross section through a circular N turn coil of length l and radius r carrying current I and (c) field H_x on the axis of a multi-turn coil

Next, the effect of adding an iron core to a multi-turn coil gives a magnetic circuit identical in form to a single phase transformer if an a.c. voltage is applied to the magnetising coil shown in Figure 10.3. The complexities of the transformer will be dealt with in detail in Chapter 5 of Volume 2 of this book; here some interactions between the components which make up a basic iron-cored magnetic circuit are discussed.

In the magnetic circuit shown in Figure 10.3, a voltage v drives a current i through the N magnetising coils wound around a closed circuit comprising a magnetic material of relative permeability μ_r. The current in the winding creates the magnetic field which, in turn, drives the flux around the magnetic core.

To simplify the analysis, it is assumed that magnetic flux is entirely confined within the core of uniform cross sectional area, A_c, i.e. no leakage flux occurs. These continuous lines of flux all link with the current source producing them. The direction of the flux depends on the direction of the current in the winding or magnetising coil.

The basic relationship between the current and field states that the line integral of H around the closed circuit is equal to the net current enclosed by the path. In this case, the line integral is simply Hl since the field is constant along the entire length of the circuit and the net current enclosed is NI, hence[4]:

$$NI = Hl \text{ (AT)} \tag{10.6}$$

[4]The units are given here as ampère-turns. Strictly speaking, it should be written as just ampère but this is rarely done since in some contexts it could cause confusion with a normal electric current.

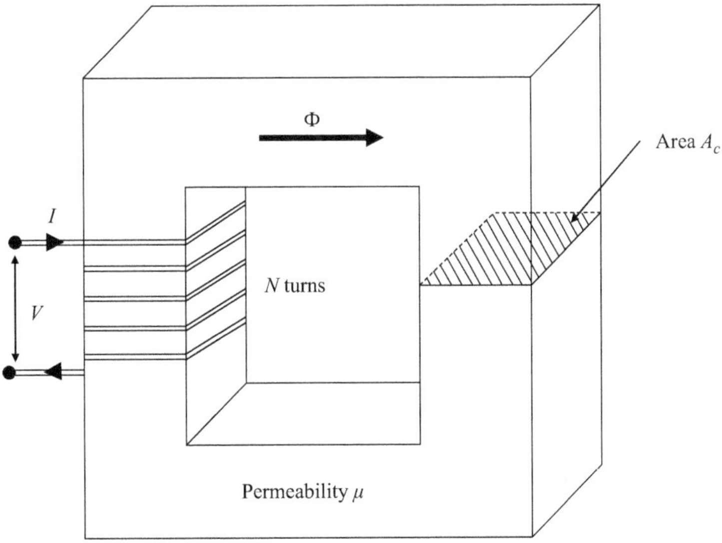

Figure 10.3 A rectangular cross section magnetic core magnetised by an N turn winding

The source of the magnetic field is the product NI which is referred to as the magnetomotive force or MMF. The simplest way of stating Ampère's more generalised law is that the total MMF driving flux around any closed magnetic circuit is equal to the sum, or integral, of all the individual Hl products.[5] Mathematically this is

$$NI = \sum Hl \quad \text{or} \quad NI = \int H dl \ (\text{AT}) \tag{10.7}$$

where dl refers to an incremental length of the path. A simple case is when the magnetic core is a toroid of mean radius r, then the field around the mean circumference is

$$H = \frac{NI}{2\pi r} \ (\text{Am}^{-1}) \tag{10.8}$$

Of course, (10.6) can be applied to the current-carrying conductors shown in Figure 10.1 so along any circular path of radius r in (a) it can be deduced that

$$H_r = \frac{I}{2\pi r} \ (\text{Am}^{-1}) \tag{10.9}$$

[5] $\int Hdl$ is called the line integral of H. Field lines were briefly mentioned in Chapter 2.

10.2 Magnetic reluctance

Consider the magnetic core geometry shown in Figure 10.4 in which the magnetic material path length is l_i and there is an air gap of length l_g in the core. The MMF can be thought of as forcing magnetic flux through the magnetic material, requiring a field component H_i, and through the air gap, requiring a field H_g. Ampère's circuital law can be applied to obtain

$$NI = H_i l_i + H_g l_g \text{ (AT)} \tag{10.10}$$

In the ideal case, the flux Φ and the flux density B each are the same in the magnetic material and in the air gap hence

$$B = \mu_0 \mu_r H_i = \mu_0 H_g \text{ (T)} \tag{10.11}$$

Also, again assuming that there is no flux fringing at the air gap, or leakage elsewhere from the magnetic core,

$$B = \frac{\Phi}{A_c} \text{ (T)} \tag{10.12}$$

Hence substituting for H_i and H_g in (10.10) we obtain

$$NI = \Phi\left(\frac{l_i}{\mu_0 \mu_r A_c} + \frac{l_g}{\mu_0 A_c}\right) \text{ (AT)} \tag{10.13}$$

We can express this equation in words in a way in which we can apply to any magnetic circuit as

MMF = flux × reluctance

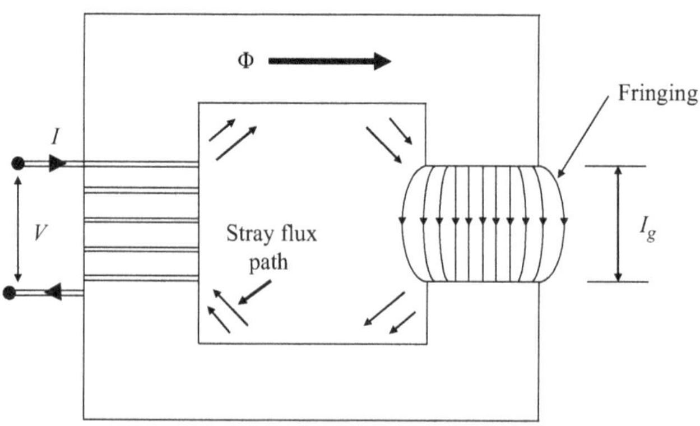

Figure 10.4 Basic closed magnetic circuit comprising an iron yoke of mean length l_i and air gap of length l_g energised by the current I passing through an N turn coil

Also, magnetic reluctance of any circuit component is simply obtained as

$$\text{Reluctance} = \frac{\text{length}}{\text{permeability} \times \text{area}}$$

This can be regarded as the magnetic equivalent of Ohm's law applied to electrical circuits. The MMF is analogous to electromotive force or voltage, the flux to current and the reluctance to resistance. Although complex magnetic circuits contain combinations of series and parallel paths and materials with different permeabilities, they can be analysed in the same way in which Kirchhoff's laws can be used in electric circuits.

Suppose the air gap in the magnetic circuit in Figure 10.4 is very small, such that $l_i > l_g$, then it is useful to find the equivalent permeability μ_e of the core. Equation (10.13) can be rewritten as

$$NI = \frac{\Phi l_i}{A_c \mu_0}\left(\frac{1}{\mu_r} + \frac{l_g}{l_i}\right) \text{ (AT)} \tag{10.14}$$

This can be rearranged as

$$NI = \frac{\Phi l_i}{A_c \mu_0 \mu} \text{ (AT)} \tag{10.15}$$

where

$$\mu = \frac{l_i \mu_r}{(l_i + \mu_r l_g)} \tag{10.16}$$

Consider a practical use of the relationship in (10.16). Suppose a core has a gap length which is 1000th that of the core. In this case, a core made of high permeability material, with, say, $\mu = 10{,}000$, will only give a permeability of around 900. So clearly the use of a high permeability material is not beneficial in such geometry.

Equations (10.10) and (10.11) can be manipulated to obtain the following expressions for the air gap field and the flux density in the circuit as

$$H_g = \frac{NI}{(l_g + l_i/\mu_r)} \text{ (Am}^{-1}\text{)} \tag{10.17}$$

$$B = \mu_o H_g = \frac{\mu_0 \mu_r NI}{(\mu_r l_g + l_i)} \text{ (T)} \tag{10.18}$$

The above MMF equation assumes complete uniformity of B and H in the material and air gap, respectively. In practice, this is not the case for the following reasons:

- leakage flux, where it leaves the magnetic material even before reaching the air gap;
- non-uniformity of permeability due to a shorter magnetic path length at the inside of the circuit compared to that at the outer edges and
- the effective cross sectional area of the air gap is higher towards the centre of the gap because the flux splays outwards during fringing.

These factors can be minimised or eliminated by careful design. For example, the sketches in Figures 10.3 and 10.4 show the magnetising winding located in one small sector of the core. In practice, this could lead to very high leakage flux such that the flux density in the part of the core furthest from the magnetising winding could be far lower than in the region of the core covered by the winding. A magnetising winding should be tightly and closely wound, and uniformly distributed around the full periphery of the core, to avoid this situation.

It is obvious that higher permeability leads to lower reluctance which leads, in turn, to higher flux for a given MMF, or field. Even if the gap length is low compared with the magnetic material length, its reluctance is likely to be much higher and much of the MMF must be used for driving the flux across the air gap.

10.3 Field and flux density distribution in a circular core

Figure 10.5 shows a magnetic annulus of outside diameter radius r_1 and inside radius r_2 magnetised by an N turn magnetising winding carrying a constant current. The MMF of the coil sets up a field at any radius r, according to (10.9) given by

$$H_r = \frac{NI}{2\pi r}\ (\mathrm{Am^{-1}}) \tag{10.19}$$

Hence the field varies inversely with radius, from a minimum value at the outer radius to its maximum value at the inner radius.

If the relative permeability is μ_r, then the flux density B_r at point A is $\mu_0\mu_r H_r$ along the same circumferential direction as H_r, hence

$$B_r = \frac{\mu_0\mu_r NI}{2\pi r}\ (\mathrm{T}) \tag{10.20}$$

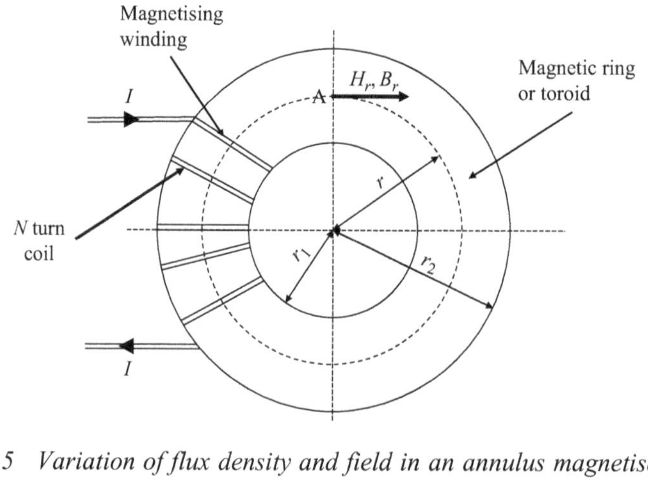

Figure 10.5 Variation of flux density and field in an annulus magnetised by the current I passing through the N turn coil

Therefore, the flux density varies in a similar manner to the field assuming that μ_r is constant. In practice, μ_r varies with flux density so the relationship between B_r and $\frac{1}{r}$ is non-linear. This general principle applies to magnetic cores of any cross-sectional form and shape, e.g. rectangular, toroidal, stacked and solid. However, calculation of the exact distribution can be difficult.

In practice, the uniformity of B and H increases for a narrow annulus ($r_1 \approx r_2$). This is important in magnetic testing where recommended dimensions are widely specified.

10.4 Iron cored solenoid

The field inside an air cored solenoid was briefly described in Section 2.1. Consider now what happens when a magnetic material, such as iron, is placed inside an infinitely long, uniformly wound solenoid with a circular cross section. Suppose the n turn per metre coil is initially empty and a current I is passing through it, then a uniform field H_a is set up in the coil. This creates a low flux density in the air of magnitude $\mu_0 H_a$. Now suppose a solid iron cylinder, whose radius is far less than the internal radius of the solenoid, is placed on the axis as shown in Figure 10.6(a). The current is unchanged so the field which it produces remains the same. If the relative permeability of the material is μ_r, then the iron acquires a magnetic polarisation J and the flux density throughout the iron cross section at AB is given by $\mu_0 \mu_r H_m$, where H_m is the field in the iron.

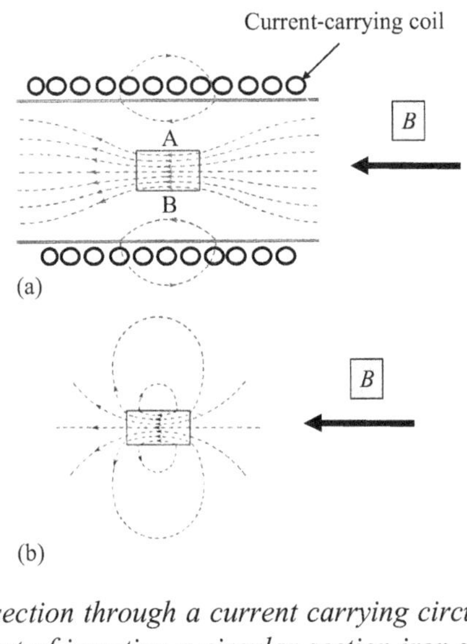

Figure 10.6 Cross section through a current carrying circular solenoid showing the effect of inserting a circular section iron sample on its axis: (a) flux density inside the coil with sample present and (b) flux density due to the magnetisation of the sample

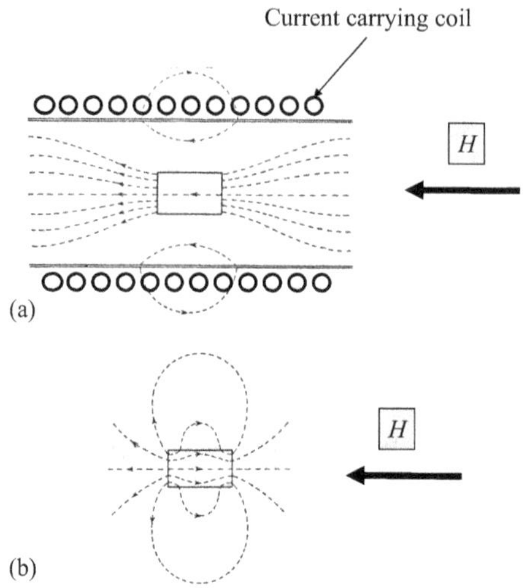

Figure 10.7 Magnetic field associated with the same solenoid as shown in Figure 2.6: (a) field inside the coil with the iron present and (b) field due to the magnetisation of the sample

The resulting flux (or flux density) is distributed as shown in Figure 10.6(a). The flux is uniformly distributed remote from the iron but the flux concentrates closer to the sample hence making the flux density within the sample far higher than in the solenoid, as expected. The B field[6] distribution is, in fact, the vector sum at each point of the B field produced by the MMF of the coil and the B field surrounding the sample if only its internal flux, which must take return paths in the air as shown in Figure 10.6(b), is considered.

The corresponding H-field[7] is shown in Figure 10.7(a). In the air region it is identical to the B-field but within the iron the density of the H-field is now extremely low, but probably sufficient to magnetise the iron to a high flux density. The demagnetising field set up within the sample, as shown in Figure 2.8, is the cause of this low field.

The B and H fields shown schematically in Figures 10.6(a) and 10.7(a) give a good mental picture of the effect of the iron core in the solenoid. In this illustration, if the solenoid and iron have circular cross sectional areas, the area of the iron is only around 10% of the total cross section of the coil. Suppose that the uniform axial field in the solenoid is H_0 when no iron is present and this produces a flux density B_0 in the empty coil. Because of the high relative permeability of the iron, it would be found that the required field within the iron to produce a relatively high flux density, say $10B_0$, would be perhaps $H_0/100$ [5]. The demagnetising field is responsible for reducing the field in the iron to the low value necessary to

[6]The terms B-field and flux density distribution are synonymous.
[7]This is simply another term used for the magnetic field strength.

magnetise to this high flux density. Of course, if the iron was toroidal, the demagnetising field would be zero and a far lower magnetising current would be necessary to magnetise it to the same high flux density. This is one reason why it is far more attractive to carry out magnetic measurements on steels in closed magnetic circuits (such as a toroid, strips arranged into a closed magnetic circuit or single sheets (or strips) provided with flux closure yokes to form a continuous magnetic path).

10.5 Flux density in a magnetic material measured by an enwrapping search coil

This section covers what is essentially a magnetic measurement issue but it is logical to briefly consider it here, since it is related to the field and flux density distribution within an iron cored coil. Related issues in measurements are discussed further in Chapter 12.

Consider a magnetic material of cross sectional area A_m within a coil of area A_{coil} as shown in Figure 10.8. The total flux Φ_m in the magnetic material is given by

$$\Phi_m = \Phi_{\text{coil}} - \Phi_{\text{air}} \text{ (Wb)} \tag{10.21}$$

where Φ_{coil} and Φ_{air} are the total flux in the coil and the flux in the air space between the coil and the material, respectively. Hence

$$A_i B_i = A_{\text{coil}} B_{\text{coil}} - A_{\text{air}} B_{\text{air}} \tag{10.22}$$

where A_{coil}, the total cross sectional area, is the sum of A_m and A_{air}, A_m is the cross sectional area of the material. B_{coil} is the total flux density measured by the search

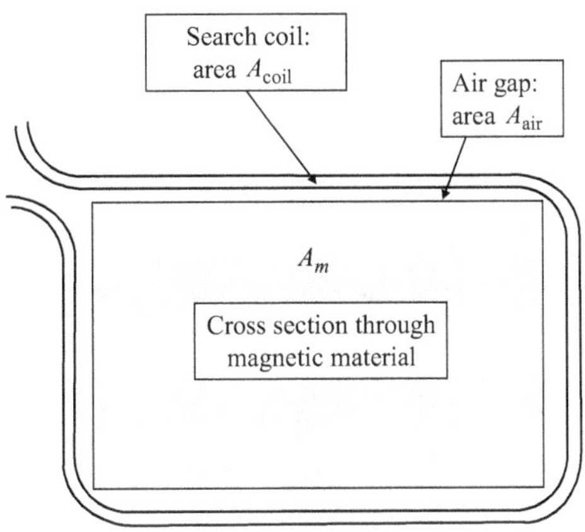

Figure 10.8 Cross section through a square search coil showing area of air interface

Table 10.1 Percentage error in flux density measurement caused by incomplete coil fill

Field	$H_a = 100$ Am^{-1}				$H_a = 50$ kAm^{-1}			
k_c	**1.01**		**1.10**		**1.01**		**1.10**	
Measured flux density B_{coil} (T)	0.100	1.500	0.100	1.500	0.100	1.500	0.100	1.500
Material flux density B_m (T)	0.101	1.515	0.110	1.650	0.1003	1.509	0.104	1.640
$(B_m - B_{\text{coil}})/B_m \times 100\%$	0.99	0.99	9.0	9.0	0.30	0.60	3.8	9.3

coil and B_m and B_{air} are the flux densities in the material and air space, respectively. Equation (2.69) can be rewritten as

$$B_m = B_{\text{coil}}\frac{\text{coil}}{A_m} - B_{\text{air}}\frac{(A_{\text{coil}} - A_m)}{A_m}\ (\text{T}) \tag{10.23}$$

If the magnetic field in the air is H_a, then $B_a = \mu_0 H_a$, so we can finally write

$$B_m = k_c B_{\text{coil}} - \mu_0 H_a (k_c - 1)\ (\text{T}) \tag{10.24}$$

where $k_c = A_{\text{coil}}/A_m$. Table 10.1 shows some examples of the use of this relationship for different values of space fill and field in the air space for low and high flux density and air fields. In practice k_c values as high as 1.1 would be avoided if possible because of the high errors they cause. Also it should be noted that (10.24) is only a guide to trends caused by variation in k_c which itself can only be approximately measured in practice and also factors such as leakage are neglected in the analysis. If k_c is small the percentage error can be simplified to

$$\text{error} \approx -\mu_0 H_a \frac{(A_{\text{coil}} - A_m)}{A_m} \times 100\%\ (\%) \tag{10.25}$$

10.6 Field and flux density at the interface between two media

It is important to know the field and flux density variations at a boundary between two media in many types of measurement and applications of electrical steels. Often, air and an electrical steel are the two media. Consider the simple case of a field causing a flux density B_{ma} in medium a at an angle θ_{ma} to the normal direction to the boundary with medium b directed across the boundary shown in Figure 10.9(a). B_{ma} can be separated into orthogonal normal and tangential components B_{man} and B_{mat}. Likewise, the flux density B_{mb} in medium b is represented by components B_{mbn} and B_{mbt}. It can be shown [6] that the normal components are always equal, i.e. $B_{man} = B_{mbn}$.

If we apply Ampère's law as defined in (10.7) to the infinitely narrow rectangle, whose longer sides are just either side of the boundary between the two

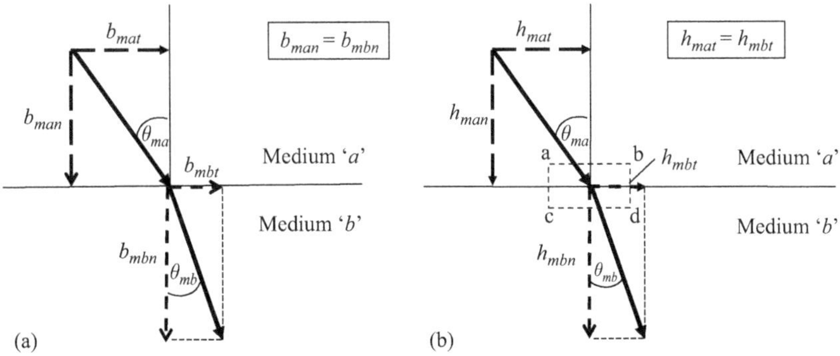

Figure 10.9 Components of flux density and magnetic field at an interface between two media: (a) flux density and (b) magnetic field (not to scale)

media, then since there is no current present, $\int Hdl = 0$ around the path *abcd*. Hence the fields along *ab* and *cd* are equal, simply showing that the tangential components of the field on either side of the boundary are equal, i.e. using the same notation, $H_{mat} = H_{mbt}$.

Let the relative permeabilities of the media be μ_{ma} and μ_{mb}, respectively. Since the normal components of B at either side of the boundary are equal it can easily be shown that

$$\frac{\tan\theta_{ma}}{\tan\theta_{mb}} = \frac{B_{mat}}{B_{mbt}} = \frac{\mu_{ma}H_{mat}}{\mu_{mb}H_{mbt}} = \frac{\mu_{ma}}{\mu_{mb}} \tag{10.26}$$

In the hypothetical case shown in Figure 10.9(a), θ_{ma} is set at 45° and $\mu_{ma} = 5\mu_b$. Hence from (10.26) it can quickly be shown that $\theta_{mb} = 26.5°$.

Suppose medium *a* is air and medium *b* is steel, then $\theta_{mb} \gg \theta_{ma}$ so θ_{ma} will be almost zero whatever the value of θ_{mb}, hence lines of flux will emerge from the steel practically at right angles to the surface. This is depicted in Figure 10.9(b) in which θ_{mb} is fixed at around 20° and $\mu_{mb} = 500\mu_{ma}$, hence making θ_{ma} less than 0.1°. Again, using (10.26), it can be shown that if the relative permeability of the steel is numerically greater than around 10, then $\theta_{ma} < 5°$ when $\theta_{mb} = 45°$.

The flux entering the air from steel will be close to the direction normal to the plane and the direction of flux in the steel itself is affected by the flux direction in the air only if the relative permeability is very low. This applies for some weakly magnetic materials such as austenitic steel whose relative permeability is less than 1.01. It is an interesting exercise to think what happens in electrical steel under a.c. excitation when the relative permeability might vary from several thousand to close to unity during one magnetising cycle, but it is only of academic interest.

It is shown in Section 7.9 that the only practical magnetic parameter of importance at the surface of electrical steel is the tangential component of field, H_t. It is proportional to the permeability of the material and its losses so it is fortunate that it is indeed equal to the *tangential* component of field in the air close to the steel surface because this can be measured with sensors as described in Chapter 12.

The normal component of H at the steel surface and the surface components of B are practically inaccessible quantities which we need not be concerned about since they are normally of no importance in the magnetic assessment of magnetic materials.

10.7 Forces between magnetised laminations

Consider two magnetised laminations carrying the same fixed flux density separated by an air gap of length g as shown in Figure 10.10. Noting the direction of the flux we can imagine the surfaces X and Y as north and south poles of a magnet so there is an attractive force between them.

The flux density B is constant in the air gap if there is no flux fringing. Hence, the energy stored in the air gap is given by

$$E_g = \frac{B^2}{2\mu_0}\ (\mathrm{Jm^{-3}}) \tag{10.27}$$

Suppose the gap is extended by an amount δg and the flux density is unaltered. The extra energy stored in the increased gap would have to be provided by the work done when the force F_g acts through a distance δg hence

$$F_g \delta g = \left(\frac{B^2}{2\mu_0}\right) A\delta g\ (\mathrm{Nm}) \tag{10.28}$$

from which

$$F_g = \frac{B^2 A}{2\mu_0}\ (\mathrm{N}) \tag{10.29}$$

This expression is useful when assessing the forces in transformer joints which are a source of transformer noise.

Now consider two identical steel laminations placed together as in Figure 10.11(a) and suppose each carries a constant flux density B. No eddy currents are present and

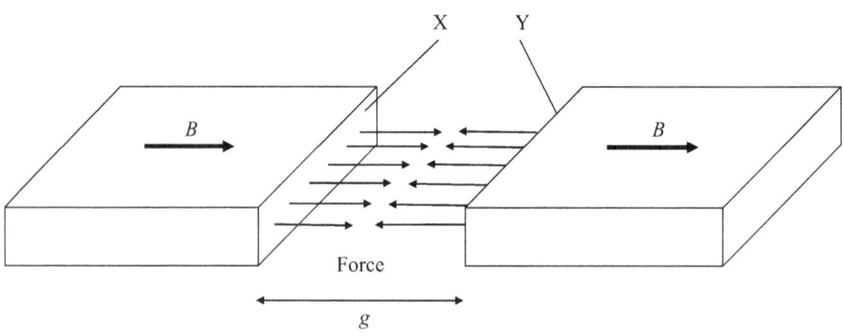

Figure 10.10 Force between the ends of two laminations separated by an air gap in a magnetic circuit

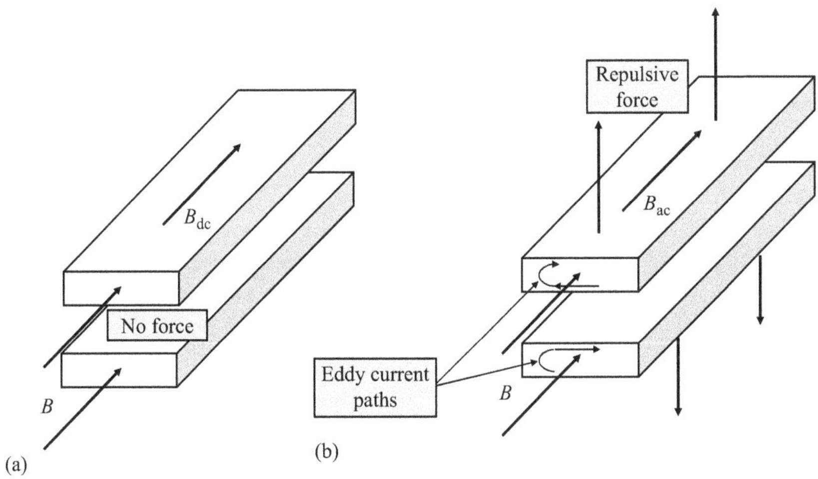

Figure 10.11 (a) Parallel magnetic strips carrying d.c. flux (no force between them) and (b) the same strips magnetised by an a.c. field (repulsive force between them)

there is no magnetic interaction between the laminations, so no mechanical forces are present. Now suppose the flux density is alternating then classical eddy currents will be induced in each lamination as shown in Figure 10.11(b). The eddy currents near the surface of the laminations at the interface flow in opposite directions as shown so a repulsive force[8] is set up between the laminations. The force is proportional to the product of the currents.

References

[1] Fitzgerald A.E., Kingsley C., and Umans S.D. *Electrical Machinery*, 4th edn. McGraw-Hill, New York, 1983

[2] Tumanski S., *Handbook of Magnetic Measurements*. CRC Press, Boca Raton, FL, 2011

[3] MIT Department of Electrical Engineering. *Magnetic Circuits and Transformers*. MIT Press, MA, Nov. 1977

[4] Carter G.W. *The Electromagnetic Field in Its Engineering Aspects*. Longmans, London, 1962, p. 114

[5] Carter G.W. *The Electromagnetic Field in Its Engineering Aspects*. Longmans, London, 1962, p. 123

[6] Carter G.W. *The Electromagnetic Field in Its Engineering Aspects*. Longmans, London, 1962, p. 125

[8]If the currents flowed in the same direction at the interface an attractive force would be set up. The forces can be imagined as being set up due to the concentration of field lines between eddy currents flowing in opposite directions causing the repulsive force and vice versa for currents flowing in the same direction. (Refer to Figure 10.1(b) to visualise the field lines.)

Chapter 11

Effect of mechanical stress on loss, permeability and magnetostriction

It has been explained in Chapter 3 how a mechanical stress causes magnetoelastic energy to be stored in a magnetic material. When this happens the magnetic domain structure changes to minimise the total energy. This, in turn, affects all the structure sensitive magnetic properties such as losses, permeability and magnetostriction.

The magnetoelastic energy in a single domain is proportional to the product of the stress and magnetostriction as given by (3.22). This energy contribution will dominate if other forms of stored energy in a material are low. In this case, even if the magnitudes of the magnetostriction and stress are low, its magnetic properties can be very sensitive to stress.

As already shown in Chapter 3, magnetocrystalline energy in electrical steel is very large if domains are not oriented along an easy crystal axis. In this case, mechanical stress might cause domains to be reoriented to other easy axes to minimise the energy. This, in turn might improve or worsen magnetic properties when a field is applied. Alternatively, the stress might make the easy axes along which domains are initially oriented even more favourable in which case properties will be only slightly stress sensitive.

What is commonly referred to as the *stress sensitivity characteristic* of an electrical steel is determined by measuring magnetic properties while applying a uniaxial tension or compression to sheets or strips of materials using test systems such as those described in Section 12.6. Such characteristics give an indication of how far their magnetic performance might be degraded when they are assembled in a magnetic core where they are subjected to random stresses which are set up during the assembly process. The stress sensitivity defined here is only a guide because the results from the test systems are obtained under controlled, uniform, magnetisation and stress which do not prevail in a magnetic core. However, testing under uniform conditions seems generally acceptable for most purposes. It is the most convenient and reproducible method of assessing stress sensitivity which is available at this time.

This chapter opens with an explanation of the effect of stress on simple domain structures in iron or SiFe single crystals. This knowledge can be used to understand the practical stress sensitivity characteristics of real materials discussed in Chapters 5, 6, and 8 of Volume 2 of this book. Effects of more complex forms of stress which occur in magnetic assemblies are shown in Sections 11.9 and 11.10. Much of the content of this chapter is focussed on electrical steels but effects of stress on other materials are occasionally referred to. Stress effects in nanocrystalline material and cobalt–iron alloys are discussed in Sections 15.2 and 16.5, respectively.

When discussing effects of stress, we need to distinguish between the elastic stresses normally encountered in magnetic cores, which we quantify in terms of the stress sensitivity curves, and the combination of plastic and elastic stress which is produced close to cut edges of steel as a result of shearing, slitting or punching.[1]

11.1 Effect of stress on simple magnetic domain structures

Consider a simple domain structure in a single crystal of iron as shown in Figure 11.1(a). It comprises just anti-parallel domains with a 180° wall between them.[2] The domain wall energy can be neglected since it only has a minor effect on the effect of stress on this basic structure. The domains are oriented along the [001] easy direction so no magnetocrystalline energy is stored.

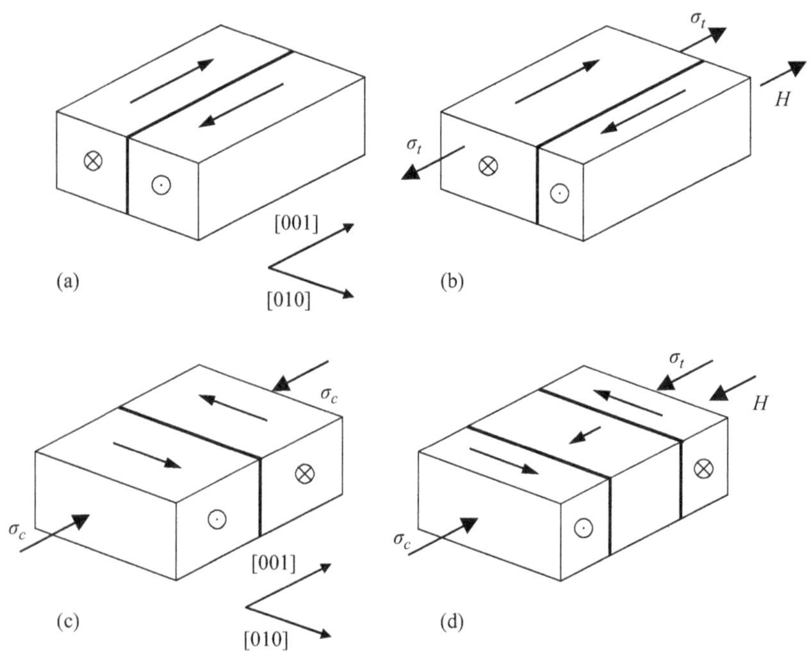

Figure 11.1 Hypothetical effect of stress on a simple domain structure in an iron crystal: (a) domains oriented along [001] in a stress-free, demagnetised state, (b) field applied along the tensile stress direction causing domain wall displacement, (c) effect of compressive stress to rotate the domains into a lower energy demagnetised state and (d) magnetisation in the compressive stress direction by domain rotation

[1]Severe bending also creates plastic stress.

[2]The presence of closure domains needed to eliminate magnetostatic energy as shown in Figure 3.14(d) is ignored here.

Now suppose a tensile stress is applied to the crystal along the [001] direction. The introduced magnetoelastic energy is given by (3.22) which is repeated here as

$$E_{\rm me} = -\lambda_s\sigma\ \left(\mathrm{Jm}^{-3}\right) \tag{11.1}$$

In the case of iron and SiFe alloys, λ_s is positive so using the convention that σ is positive for tensile stress, the magnetoelastic energy introduced into the two domains is negative inferring a decrease in energy. This means the material is still in a minimum energy state so the structure remains unchanged and we can draw the simple conclusion:

Application of a tensile stress along an easy direction of magnetisation along which all domains are already oriented does not have any effect on the domain structure.

In fact, because the tension reduces the free energy, in this case, it makes the [001] direction even more preferred than the other [100] or [010] directions. Suppose we apply an external field H along the [001] direction. Since application of tensile stress does not affect the basic domain structure, magnetisation occurs by simple domain wall motion as shown in Figure 11.1(b) so the permeability is high, losses are low and the magnetostriction is zero.

Now suppose a compressive stress is applied along the [001] direction while the crystal is in the initial demagnetised state. It is easy to deduce that the free energy will increase unless the domains rotate to another easy direction, such as the [010] direction as shown in Figure 11.1(c). It is worth looking at this in a little more detail by making use of the fuller version of the expression for the magnetoelastic energy given in (3.21) which is repeated here as:

$$\begin{aligned} E_{\rm me} &= -(3/2)\lambda_{100}\sigma\left(l^2{l_1}^2 + m^2{m_1}^2 + n^2{n_1}^2 - 1/3\right) \\ &\quad - 3\lambda_{111}\sigma(l m l_1 m_1 + m n m_1 n_1 + n l n_1 l_1)\ \left(\mathrm{Jm}^{-3}\right) \end{aligned} \tag{11.2}$$

Remembering that the direction cosines refer to the stress and magnetisation directions, we can write $l = 0,\ m = 1,\ n = 0$ and $l_1 = 0,\ m_1 = 0,\ n_1 = 1$. On substituting these values into (11.2), the magnetoelastic energy of the structure shown in Figure 11.1(c) is given by[3]

$$E_{me} = -\lambda_{100}{}^{\sigma}\!/\!_{2}\left(\mathrm{Jm}^{-3}\right) \tag{11.3}$$

Hence the free energy has been reduced. Therefore we can draw a second conclusion:

Application of a compressive stress along a preferred direction of magnetisation along which all the domains are initially oriented causes domains to rotate to one or more of the other easy directions.

Now suppose a field is applied along the [001] direction while the compressive stress is present. A higher field is necessary to rotate domains from the [010] direction

[3]Remember that σ is negative since the stress is compressive.

to the [001] direction. The permeability falls and the losses increase due to more complex domain wall motion. Figure 11.1(d) illustrates the hypothetical structure at an arbitrary magnetisation level, in this case, when the magnetisation has increased to around 0.5 M_S. Of course, the rotation of the domains implies the change of magnetostriction shown in Figure 3.11. From this it can be deduced that, if the field is large enough to rotate fully the domains to saturate the crystal in the [001] direction, the magnetostrictive strain is $+3\lambda_{100}/2$. This is the maximum value it can possibly be.

Although the structures in Figure 11.1 are very simplified, they can be relied upon to illustrate the effect of stress without being distracted by complexities of real structures.

These conclusions apply specifically to iron and SiFe alloys.[4] The general effect in other materials can be arrived at in a similar manner taking into account the relative magnitudes of the anisotropy and magnetostriction constants. This analysis gives the ideal static domain structure in the presence of mechanical stress.

11.2 Stress sensitivity derived from domain structures

The effect of stress on the simplest possible domain structure given in Section 11.1 explains the fundamental way in which magnetoelastic energy is introduced into the system, thereby causing the domain structure to change. In this section, we continue in a similar manner by explaining the basic effect of stress on the magnetic properties of a single crystal and polycrystalline iron alloys.

Consider a more realistic domain structure in the central[5] region of a (110) [001] oriented thin disc of iron. Its domain structure in a stress free de-magnetised state is shown in Figure 11.2(a). Just four domains are shown

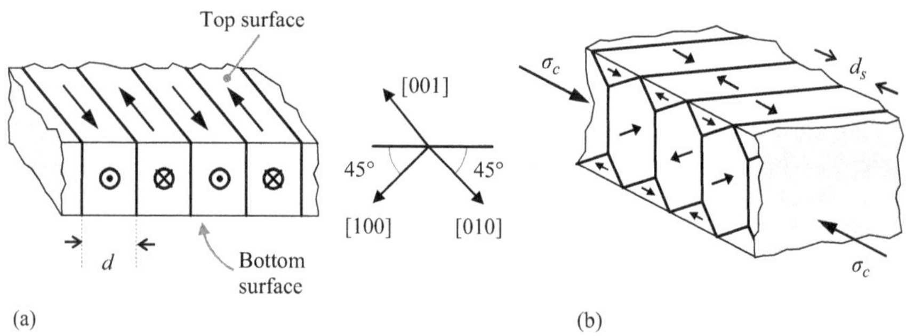

Figure 11.2 Schematic representation of domain structures in a (110)[001] (Goss) oriented single crystal (a) stress-free and (b) under compression applied along the [001] direction

[4]These same two conclusions were obtained in a slightly different way in Section 3.9 but they are so important it is worth repeating them here.
[5]The central region is specified simply to avoid introducing any complication of edge effects.

oriented along the [001] and [00$\bar{1}$] directions. This structure is in a minimum energy state.[6]

Suppose a compressive stress is applied along the [001] direction. This causes the [001] oriented domains to rotate into the [100] or [010] easy directions which are at 45° to the surface plane. Figure 11.2(b) shows the result of this process. The triangular shaped domains on each surface are closure domains. These are formed to avoid magnetostatic energy being stored in the form of free surface magnetic poles which would occur if the internal main domains extended all the way to the surfaces.

The simple structure in Figure 11.2 is representative of what is observed on the surface of a perfectly oriented grain in a (110)[001] oriented steel.

The stress domain structure shown in Figure 11.2(b) was proposed by Dr W. D. Corner and Dr J. J. Mason at Durham University in the UK [1]. It is the foundation on which our understanding of the stress sensitivity of the magnetic properties of soft magnetic materials is based. This particularly applies to well oriented grains in GO electrical steel whose domain structures are similar to the theoretical ones shown in Figure 11.1.

In fact, the first evidence for the existence of domain *stress patterns* was reported 10 years earlier by Dijkstra and Martius [2]. These investigators applied compressive stress to a GO silicon steel specimen and observed surface domain patterns similar to those shown in Figure 11.1. They showed how the stress pattern appeared to propagate from grain to grain across the surface of the material as the compressive stress is increased.

The development of the stress patterns on the surface of GO steel under increasing compressive stress σ_c applied along the RD is illustrated schematically in Figure 11.3. This depicts a simplistic domain structure in a sample comprising

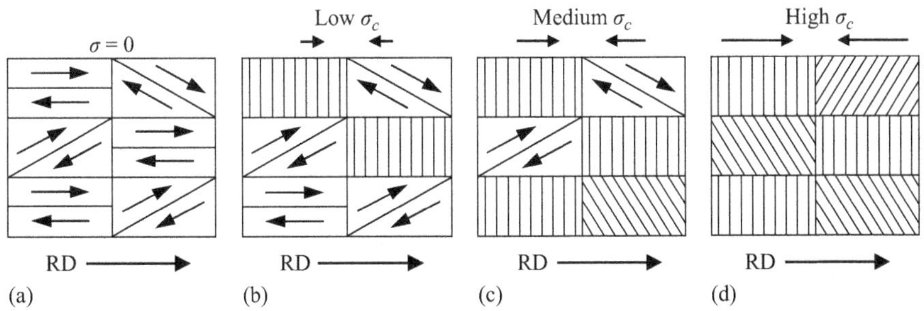

Figure 11.3 Schematic representation of the spread of domain patterns on the surface of grains of GO material with increasing stress: (a) zero stress, (b) low compressive stress, (c) medium compressive stress and (d) high compressive stress

[6]The domains are oriented in this way parallel to the (110) surface to avoid any magnetostatic energy being stored and parallel to the [001] direction to avoid any stored magnetocrystalline energy. Domain wall energy is present and it partly helps to determine the domain width, *d*, which is constant in this simple model. Factors controlling the magnitude of *d* in real crystals are dealt with in Section 1.4 of Volume 2 of this book.

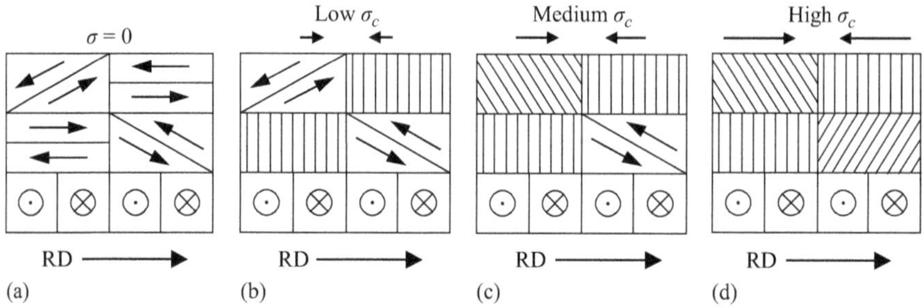

Figure 11.4 Schematic representation of the spread of domain patterns on the surface of grains of NO material with increasing stress: (a) zero stress, (b) low compressive stress, (c) medium compressive stress and (d) high compressive stress

three perfectly oriented grains and three grains which are mis-oriented in the (110) plane. Application of a small compressive stress along the RD causes the stress pattern to be set up in two of the perfectly oriented grains. The pattern is set up in the other perfectly oriented grains at a slightly higher stress. As the compressive stress is increased,[7] stress patterns are set up in the other grains until, finally, the stress pattern covers the whole surface.

Consider next the schematic diagram of domains in a piece of randomly oriented, isotropic non-oriented (NO) material shown in Figure 11.4. In this case, the structure includes extremely mis-oriented grains. The stress pattern builds up in a similar way to that in the GO material but the volume of rotated domains is less than that in the GO material. Therefore, the change in magnetostriction {from state (a) to (d)} is far lower and the increase in power loss under compressive stress should be less than in the GO steel.

The structures and processes shown in Figures 11.3 and 11.4 are over simplified views of what does happen in real materials but they help us to understand stress sensitivity curves and give a basis for quantifying them. For example, let us use the concept to develop the theoretical stress sensitivity characteristic of GO steel. Figure 11.5(a) shows the theoretical variation of magnetostriction with tensile or compressive stress applied along the [001] direction of a perfectly oriented, unconstrained, Goss grain. As we have already seen in the ideal case in Figure 11.1, tension does not affect the structure but compression immediately rotates the domains to produce the large positive magnetostriction.

Now consider a real material made up of mis-oriented grains of various sizes. We will not attempt to work out the stress sensitivity of such a complex system, but we can build up the basic stress sensitivity characteristic from knowledge of the processes developed in Figures 11.3 and 11.4. Suppose a tensile stress is applied. This has a small effect on any closure domains present and on the walls of the main

[7]This is called the transition stage in [11.2].

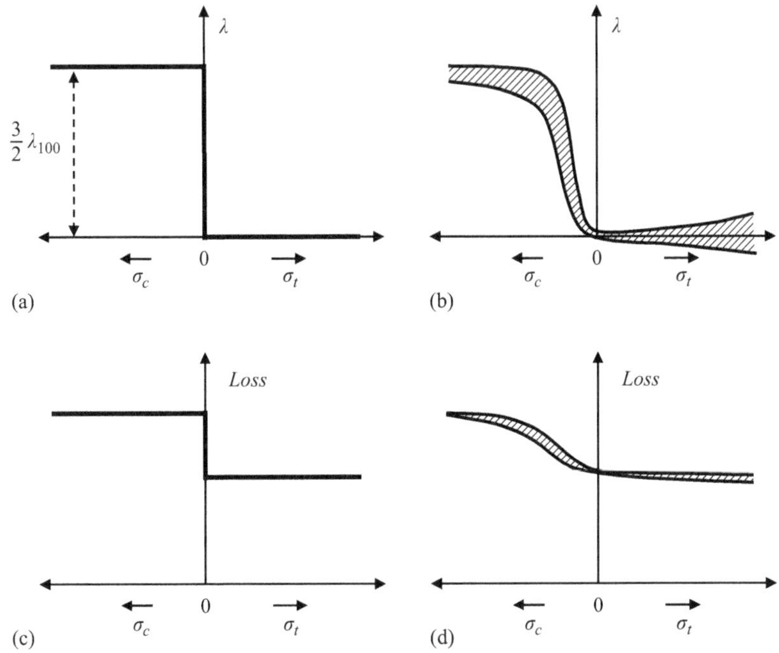

Figure 11.5 *Basic shapes of stress sensitivity characteristics of GO silicon–iron (a) idealised magnetostriction, (b) practical magnetostriction range, (c) idealised loss and (d) practical loss range*

[001] domains which causes low sensitivity of magnetostriction to tensile stress as depicted in Figure 11.5(b).

Application of a very small compressive stress has no effect because its associated magnetoelastic energy cannot initially be balanced by the small increase in domain wall energy or magnetoelastic energy, which would arise if the domains in a perfectly oriented grain switched immediately. This initial state is soon overcome and domains begin to switch as in Figures 11.3 and 11.4 so a gradual increase in magnetostriction occurs as shown. The processes in NO steel could be deduced in a similar manner.

Suppose an a.c. field is applied along the RD of a polycrystalline GO steel while a compressive stress is applied along the same direction. Magnetisation occurs in a far more complex process than when no stress is present. If the same simple argument as for magnetostriction is used, then the stress sensitivity of power loss shown in Figure 11.5(c) and (d) is obtained.

The domain structure shown in Figure 11.2(b) is referred to as *stress pattern I*. If the compressive stress is increased, it can be deduced from an observed decrease of the width of the surface closure domains, d_s, that the spacing of the internal [100] or [010] domains decreases. If a high compressive stress is applied, more complex patterns start to appear on the surfaces of some grains. These are referred to as

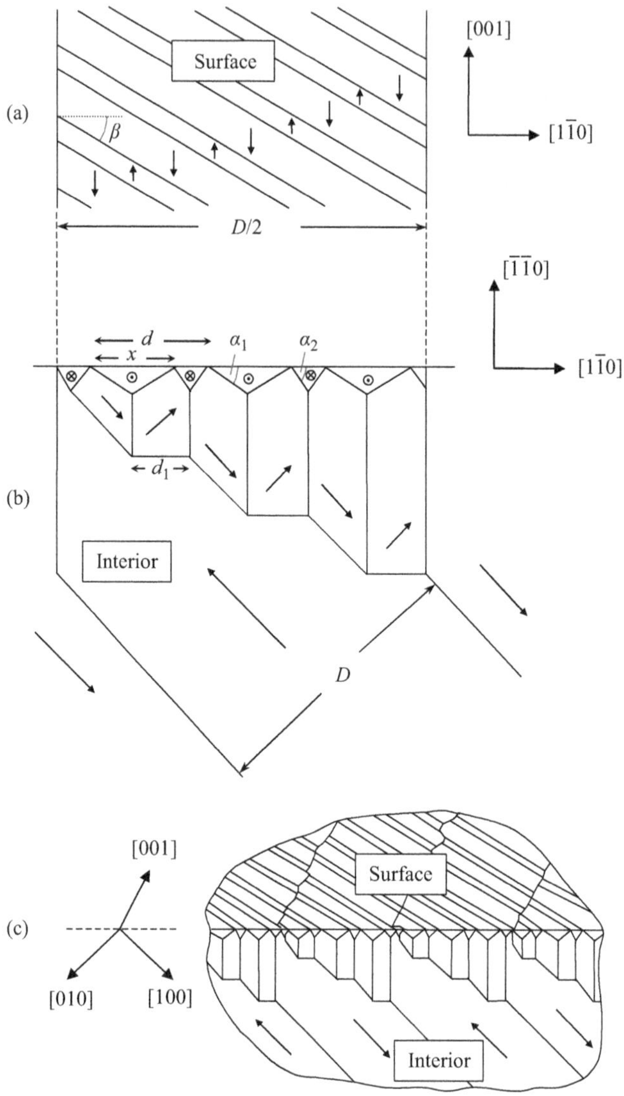

Figure 11.6 Suggested structure of stress pattern II: (a) structure on the (110) [001] surface, (b) section in the (001) plane and (c) diagram of the postulated structure (Figure 8 from [1] reproduced under free licence CC-BY-4.0)

stress pattern II. The domains appear as chain-like structures formed along the walls of stress pattern I. Figure 11.6 shows a suggested structure of stress pattern II [1]. The main sub-surface domains are oriented along the [100] and [$\bar{1}$00] directions, as in the case of stress pattern I, but in this case, the normals to the walls lies in a [010] direction. The existence of these sub-surface structures has still not been experimentally verified.

Most basic analyses of the stress sensitivity of electrical steels are based on stress pattern I partly because it occurs at low stress. It is most often encountered and it is far easier to use in models or predictions. Stress pattern II comes into play at around 18 MPa and immediately covers around 60% of the surface of a typical GO steel [1]. This increases to almost 90% at 35 MPa so it dominates at stress levels greater than normally encountered. However, the internal structures of the domain structures are similar, so an analysis based solely on the presences of stress pattern I should not introduce gross inaccuracies.

On application of a low compressive stress the shape of the surface closure domains, such that those oriented along the magnetising direction, expand and the others contract without changing the volume of the [100] main domains; hence no magnetostriction occurs [3]. At a slightly higher stress the stress domain structures begin to develop so magnetostriction occurs. On the application of a field, the magnetisation at which this transition begins to occur is sometimes referred to as the critical flux density[8] B_{cr} given by [3]

$$B_{cr} = d_s B_s / 2t \ (\mathrm{T}) \tag{11.4}$$

where d_s is the width of the surface closure domains, t is the sheet thickness and B_s is the saturation flux density.

The magnetic field H_{cr}, which has to be applied to switch the stress domain structure to the field direction, can be estimated for an ideal (110)[001] oriented crystal. If the presence of closure domains is again neglected, on the application of a field H along the [001] direction the free energy can be expressed as

$$E = -HM \cos\theta - \lambda_s \sigma \ \left(\mathrm{Jm^{-3}}\right) \tag{11.5}$$

where M is the magnetisation, σ is the stress and θ is the angle between H and M. The stress pattern is unaffected if the field is low so $\theta = 90°$, hence $\lambda_s = -\lambda_{100}/2$. Therefore, the energy becomes

$$E_1 = -\lambda_{100}\, \sigma/2 \ \left(\mathrm{Jm^{-3}}\right) \tag{11.6}$$

Suppose the field is increased to the stage at which the main [010] or [100] stress domains rotate to the [001] direction, then the energy is given by

$$E_2 = -HM + \lambda_{100}\sigma \ \left(\mathrm{Jm^{-3}}\right) \tag{11.7}$$

The switch of domain structure only occurs when E_2 is slightly less than E_1. This coincides with

$$H_{cr} = 3\lambda_{100}\, \sigma/2\, M_s \ \left(\mathrm{Am^{-1}}\right) \tag{11.8}$$

[8]This should not be confused with the critical flux density defined in connection with overlap corner joints discussed in Chapter 5 of Volume 2 of this book.

It will be noted that the critical field,[9] as derived here, is independent of peak flux density and magnetising frequency. Also, it is directly proportional to the stress.

The structure of the 90° closure domains caused by compressive stress can be quantified using basic domain theory already outlined in this chapter. The critical stress for the occurrence of the domains (i.e. the inception of stress pattern I) increases as the sheet thickness is reduced [4]. This has implications on the effectiveness of the domain refinement of GO steel when subjected to compressive stress [4].

An interesting finding, which does not appear to have been directly exploited, is that external tensile stress of the order of 20 MPa to 30 MPa can reduce the volume of closure domains around inclusions which effectively increases the permeability [5].

11.3 Effect of biaxial stress

The building stress within stacked cores is randomly distributed, so perhaps an alternative method of characterisation of stress sensitivity is to apply independent orthogonal forces to a sheet of steel to set up what is commonly referred to as biaxial stress. It could be argued that biaxial stress is a more accurate way of characterising stress sensitivity if building stresses are randomly distributed in magnitude and direction.

Much of the interest in biaxial stress originates in its application in non-destructive analysis of residual stresses in steels. Early experimental findings were promising but there was no theoretical basis capable of quantifying the effect on magnetic properties. However, a simple domain model can be used to show why equal biaxial components of stress do not cause equal and opposite changes in permeability compared to the application of the same magnitudes of uniaxial tension or compression to isotropic mild steel [6]. This difference can be substantial, even in isotropic mild steel, as shown in Figure 11.7.

Figures 11.8 and 11.9 show the effect of transverse stress on the loss and magnetostriction of GO steel when subjected to simultaneous stress along its RD and TD and magnetised at 1.5 T, 50 Hz along the RD [7]. The characteristic shown as a dotted line on each figure represents equal RD and TD stresses over the full stress range. This could be more representative of actual conditions in many parts of randomly stressed laminated cores. In this case the stress sensitivity is far less than assumed from uniaxial stress characterisation. The effectiveness of insulating coatings on the surfaces of GO steels depends on the biaxial stress characteristics of the steel [8]. This is discussed in detail in Section 2.10 of Volume 2 of this book.

The characteristics of NO steel presented in Figure 11.7 show that the permeability is lower under uniaxial compression than under equivalent biaxial stress, whereas under tension, the property is better under uniaxial stress. Figures 11.8 and 11.9

[9]Note that the so-called critical field and critical flux density have no corresponding points on the *B–H* curve.

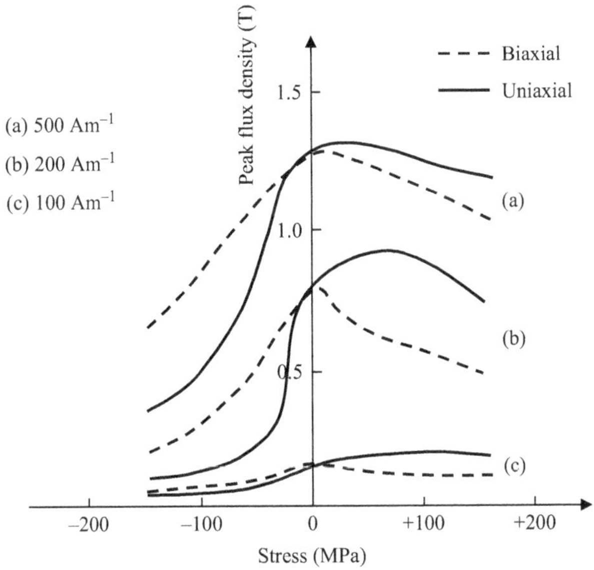

Figure 11.7 Variation of peak flux density at three values of d.c. field under uniaxial stress and biaxial stress of the same values of longitudinal and transverse stress. (Modified version of data extracted from Figure 2 and Figure 6 in [6])

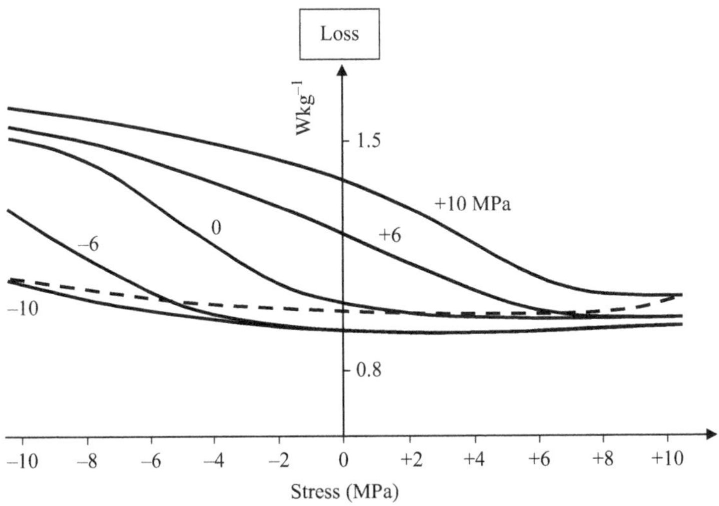

Figure 11.8 Variation of loss with different combinations of biaxial stress applied to conventional GO steel magnetised at 1.5 T, 50 Hz (Figure 2 from [7] reproduced under free licence CC-BY-4.0)

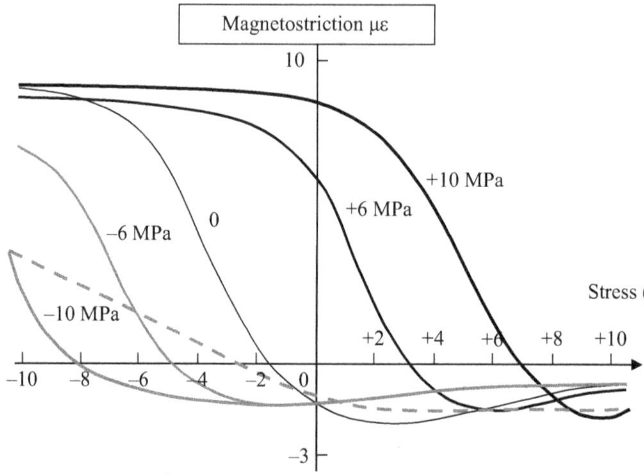

Figure 11.9 Variation of magnetostriction with different combinations of biaxial stress in conventional GO steel magnetised at 1.5 T, 50 Hz (Figure 3 from [7] reproduced under free licence CC-BY-4.0)

show the same trends in that loss and magnetostriction of GO steel are both considerably lower under biaxial compressive stress than when the uniaxial stress is applied along the RD. However, in the GO material, there is little difference between the properties under tensile uniaxial or biaxial stress. Analysis of very simplified models of the domain structures incorporating the onset of the domain stress pattern shown in Figure 11.3 can explain the trends in the effect of biaxial stress in the two types of materials.

As just inferred, it is difficult to produce accurate models to quantify the relationships between the effects of biaxial stress and an equivalent uniaxial stress with the magnetic properties, even of isotropic soft magnetic materials. Various energy minimisation approaches have been attempted to predict *B–H* curves under complex stress at domain, grain and lamination levels using multiscale modelling particularly for predicting the behaviour of rotating machine cores under very high complex stress. One method based on minimisation of the so-called Helmholtz free energy density is claimed to be able to predict the hysteresis loss of NO electrical steel under multiaxial stress conditions from measurements of uniaxial stress sensitivity [9]. The approach is being developed as a tool in numerical analysis of electrical machine performance.

Measurements and modelling of the effect of biaxial stress in 49Co49Fe2V over the range ± 60 MPa show that the largest effect on the magnetisation curves occurs when one stress is tensile and the orthogonal one is compressive [10]. Significant discrepancies between modelled and measured results were attributed to conventional modelling approaches not working well enough because of the non-linear effect of stress. It was concluded that a non-monotonous and multiscale model should be further developed to cope with this problem [10]

11.4 Stress sensitivity of GO steel

Fundamental aspects of the stress sensitivity of the magnetic properties of all soft magnetic materials including electrical steel, were covered in Section 11.3. More practical aspects of commercial GO steel are included in this section and NO steel is covered in Section 11.5.

The stress sensitivity of magnetic properties of an electrical steel depends on many factors including the material texture, surface topology, coating and internal imperfections. As implied in Figure 11.5(a) and (c), tensile stress applied along the [001] direction of a perfect single crystal has no effect on magnetostriction or losses.

In practise, it has long been known that tensile stress tends to reduce the spacing of main domains and to change the volume and form of closure domains [11,12]. The widths of bar domains in a grain of GO steel are never all the same[10] due to factors such as the irregular shape and variable internal stress distribution in the grain. Tensile stress tends to equalise the size of main domains, which is a potential contributor to lower losses. The effect of tensile stress is related in a complicated way to the grain misalignment and grain size. The width of the main domains is dependent on the relationship between tension and grain size in electrical steel which is a useful relationship to be aware of in material development, particularly in the optimisation of the effect of the coating stresses [13]

The refinement of the main domain walls under tensile stress continues under a.c. magnetisation at power frequency. This was shown in one of the first studies of real time domain wall motion in GO steel in which the Kerr magneto-optic effect was used in conjunction with high speed photography to capture images during a few magnetising cycles when magnetised at 60 Hz [14].

Tensile stress applied along the RD does not always improve the magnetic properties of GO steel. Development of demagnetising fields at grain boundaries in grains whose [001] directions tilt out of the sheet plane can reduce the a.c. permeability, produce negative magnetostriction and decrease any tendency for the core loss to drop with increasing tension [15].

Insight into the effect of compression on magnetic properties in general can be gained by considering the effect of stress on the a.c. *B–H* loops of GO steel. The application of a uniaxial compressive stress causes a constriction of the loops [16]. This is illustrated in Figure 11.10. It is caused by the permeability being drastically reduced by the onset of the domain stress pattern shown in Figure 11.2.

The critical field, which overcomes the effect of transverse stress domains in an ideal (110)[001] crystal of GO steel, is given in (11.8). The calculated critical fields are 180 Am^{-1} and 720 Am^{-1} at stresses of −10 MPa and −40 MPa, respectively. The corresponding measured values are 175 Am^{-1} and 630 Am^{-1} [17]. This shows an approximate proportionality between the field and the compressive stress as

[10]The length to width ratio of a grain has an influence and also the length of individual domains and their interaction with surrounding grains. Aspects of such controlling factors are discussed in Chapter 1 of Volume 2 of this book.

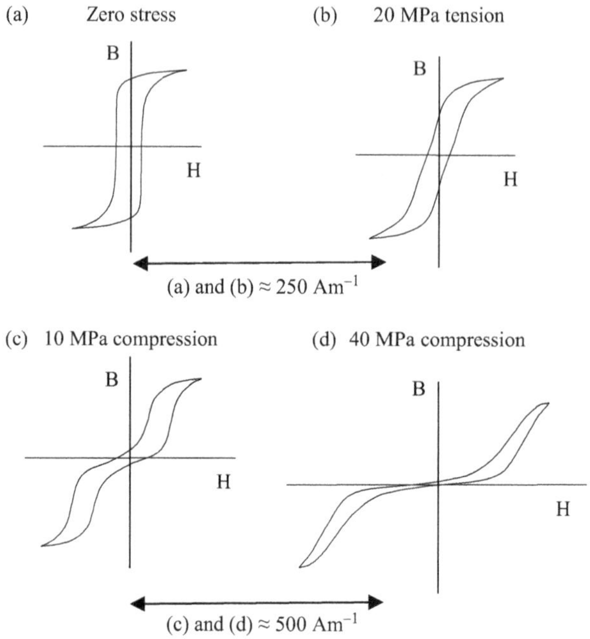

Figure 11.10 Effect of compressive stress on the B–H loop of typical GO steel magnetised at 1.5 T, 50 Hz: (a) unstressed, (b) 5 MPa, onset of stress domain pattern, (c) 15 MPa, stress pattern I dominates and (d) 40 MPa, stress patterns I and II present (Adapted from Figure 4 in [17])

implied in (11.8) which is valid for an ideal (011)[001] grain assuming ideal stress free and stress pattern I domain structures. This assumption is not valid for polycrystalline material.

Typical stress sensitivity characteristics of power loss and magnetostriction of a range of grades of GO steel magnetised at 1.5 T, 50 Hz are shown in Figures 11.11 and 11.12 respectively. They follow the trends shown in Figure 11.5 deduced from domain observation. The loss increases by 150% to 250% at the highest compressive stress. At around −5 MPa, which is often assumed to be typical of the value of random building stress expected in transformer cores, the increase is up to around 50%, although it is far lower in better grades of steel. The spread of magnetostriction characteristics is wider. The factor which mostly determines the stress sensitivity of GO steel is its coating which, as shown in more detail in Section 2.10 of Volume 2 of this book, tends to shift the whole characteristic of a GO steel to the left.

The bands shown in Figures 11.11 and 11.12 are a guide to what is expected in today's materials. Apart from the coating, the stress sensitivity of GO steel depends on factors such as peak flux density and sheet thickness. These are quantified in detail in Chapter 2 of Volume 2 of the book.

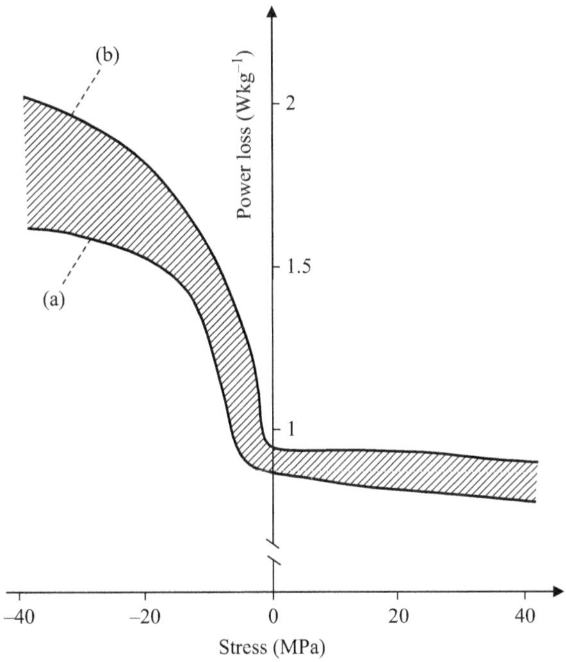

Figure 11.11 *Spread of stress sensitivity of power loss of typical grades of GO electrical steel magnetised at 1.5 T, 50 Hz: (a) and (b) show low and high stress sensitivity, respectively*

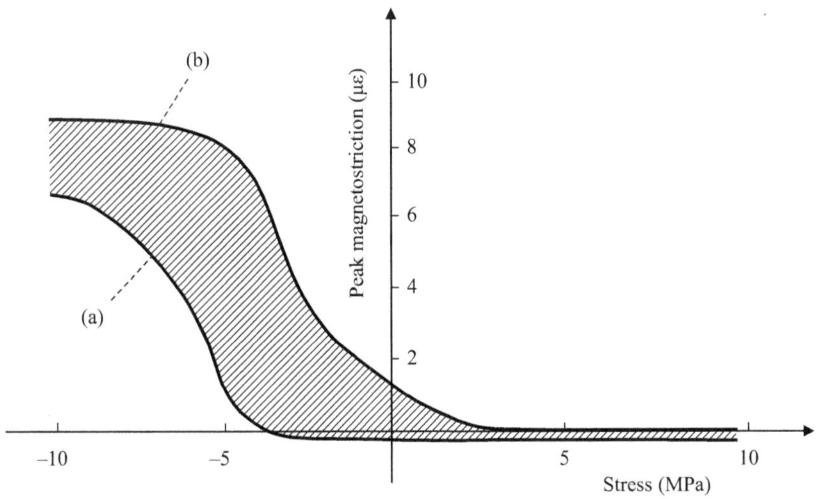

Figure 11.12 *Spread of stress sensitivity of peak magnetostriction of typical grades of GO electrical steel magnetised at 1.5 T, 50 Hz: (a) and (b) show low and high stress sensitivity, respectively*

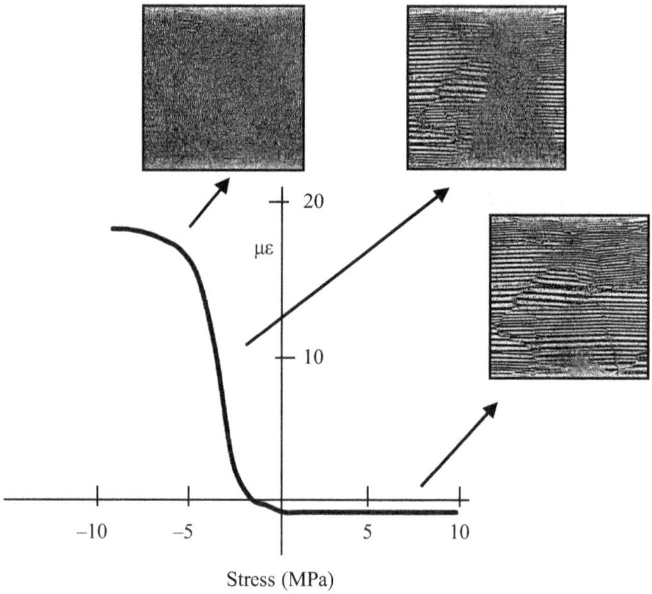

Figure 11.13 Occurrence of stress domain patterns at points on the stress sensitivity characteristic of a high permeability GO steel [18]

Figure 11.13 shows where stress patterns appear on a magnetostriction stress sensitivity characteristic[11] of a high permeability GO steel [18]. There is little change in magnetostriction until the compressive stress reaches −1 MPa. By −5 MPa, stress pattern I can be seen on the surface of a few grains as a dark region, unfortunately the spatial resolution is not high enough to see clearly the very narrow surface closure domains. At a compression of around −10 MPa, stress pattern II covers around 75% of the surface and the magnetostriction is reaching its limit of around 20 με.

It will be noted that [1] showed that a stress of around −20 MPa which was necessary to produce the 75% surface coverage by the stress pattern whereas in [18] it is −10 MPa. This is likely to be due to the improved alignment of the grains in steels studied more recently in which [001] directions are on average closer to that of the stress axis when applied along the RD.

General effects of temperature on properties of soft magnetic materials are covered in Chapter 18 but it is interesting to look at one aspect of the effect of temperature on magnetostriction of GO steel here. Figure 11.14 shows the stress sensitivity of the magnetostriction of a GO steel over a temperature range from 20 °C to 200 °C [19]. A significant factor here is that stress sensitivity curves are significantly shifted to the right as the temperature is increased because the

[11]The peak to peak magnetostriction is shown in this figure. The peak value is shown in earlier figures. The relationship between these terms is explained in Table 12.1.

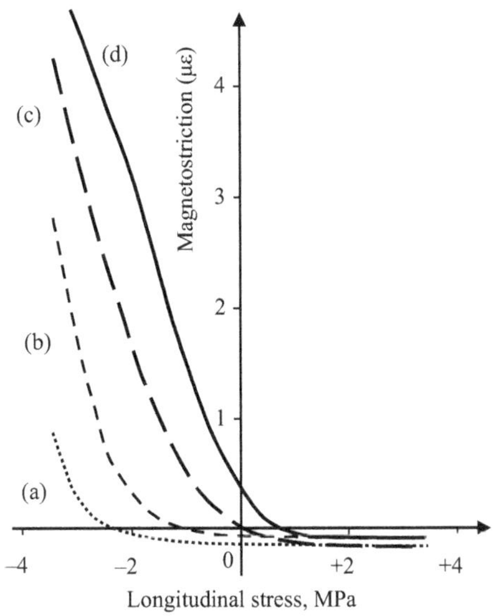

Figure 11.14 Effect of temperature on the stress sensitivity of magnetostriction of a conventional GO steel magnetised at 1.5 T, 50 Hz: (a) 20 °C, (b) 100 °C, (c) 160 °C and (d) 200 °C (Figure 11 in [19], reproduced under free licence CC-BY-4.0 and modified)

effectiveness of the beneficial strain in the steel, caused by the different natural contractions of the steel and the coating set up while the coating forms during the production process, reduces.

The strong influence of anisotropy on the spatial angle between B and H when magnetised at angles to the RD is discussed in Chapter 9. Stress also affects the angular displacement in GO steel. Figure 11.15(a) depicts a square sheet of GO steel in which stress is applied along the RD while the flux density is controlled to be directed at an angle α to the RD by placing a magnetising coil at an angle β to the RD. Figure 11.15(b) shows how β can be larger, or smaller, than α depending on the magnetisation angle and the stress. There are of course infinite combinations of stress and magnetisation direction but these sorts of stress characteristics can be used to represent stress more realistically than in other methods of electromagnetic modelling of electrical machine cores.

Magnetic properties of GO steel under rotational magnetisation are also stress sensitive. Under uniaxial tensile stress of around 30 MPa along the RD rotational loss increases by 30% under rotational magnetisation whereas compression along the same direction has little effect on the loss due to the effect of the stress on the spatial angle between B and H [21]. The same trend has been found with magnetostriction [22].

The effect of stress on the angle between B and H under rotational magnetisation can be deduced from basic energy minimisation theory to confirm observed changes in loss and magnetostriction but they have not been quantified.

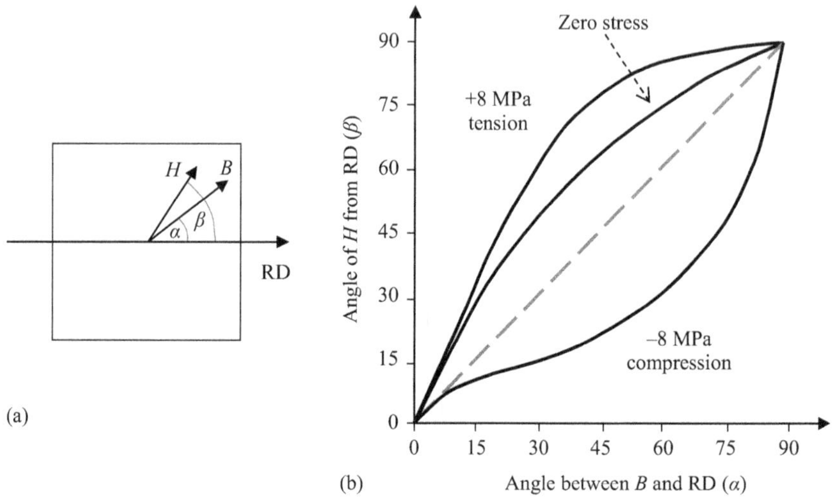

Figure 11.15 (a) Stress and magnetisation directions in a sheet of GO steel and (b) effect of ± 8 MPa stress applied parallel to the RD on the angle between B and H when B is set at between 0° and 90° to the RD (Figure 4 in [20], reproduced under free licence CC-BY-4.0 and modified)

11.5 Stress sensitivity of NO steel

The very simple model shown in Figure 11.4 was used to explain the general stress sensitivity characteristic of NO steel. This ignores several factors in the steel but the illustration is kept as simple as possible just to illustrate the general trend in stress sensitivity which might be expected[12] [23].

Tension applied along the RD has a negligible effect unless it is so high that it overcomes the magnetocrystalline energy and rotates domains in the mis-oriented grains out of their <100> directions to align closer to the RD. Of course, if grains are only slightly oriented this process will occur at a lower stress. This process could tend to reduce losses when magnetised along the RD.

If a compressive stress is applied along the RD, stress domain patterns appear in some grains making magnetisation along that direction more difficult as well as increasing the losses and magnetostriction. The losses do not increase as much as in an equivalent GO steel because the unstressed material has a relatively lossy initial domain structure and a smaller relative volume of grains will exhibit stress patterns. Likewise, the increase in magnetostriction under compression is less than in GO steel because its peak value is less than $3/2\lambda_{100}$.

[12]This model could be easily extended to take account of other factors such as a grain size distribution, closure domains and texture in an actual steel, using approaches such as in [19]. This is not included here since the simple model is sufficient to show the general trend.

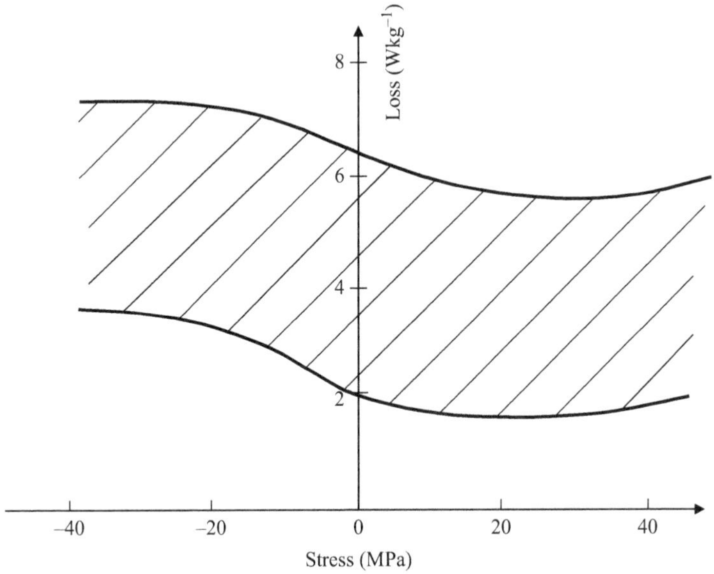

Figure 11.16 Range of stress sensitivity of power loss of typical NO steel magnetised and stressed along the RD magnetised at 1.5 T, 50 Hz

Figure 11.16 shows the spread of stress sensitivity of loss of typical grades of NO silicon–iron. In absolute terms, the losses are higher than in GO steel, although the stress sensitivity is lower [24]. Generally speaking, the increase in loss under high compressive stress is lower in large grain material [23,25] whereas, at lower compressive stress, the loss is higher in large grain material [24]. This is attributed to changes in excess loss and hysteresis loss under stress [26].

General effects of stress on the magnetostriction of typical grades of NO steel are shown in Figure 11.17. As with losses, the stress sensitivity of magnetostriction is lower than that of GO steels, but the absolute value in the stress-free state is higher.

The stress sensitivity of electrical steels varies with silicon content. The drop in the magnitude of λ_{100} with increasing silicon content implies that stress sensitivity drops also. This is indeed the case, as can be seen from Figure 11.18 [27]. The characteristic shown is measured at 400 Hz in 0.2 mm thick steel but it is expected to be representative of NO steels in general. The same materials were tested in the form of stator laminations. The shrink fitting process[13] increased the stator loss by around 25% when 3% Si laminations were used, but the deterioration using the highest silicon grades was minimal, the loss even falling in a 6.7% Si Fe alloy [27].

The small, randomly oriented grains in NO steel make the effect of stress on its domain structure difficult to observe or analyse. For example, the type I and type II domain stress patterns appear on the surface of grains with close to (110)[001] or (100)[001] orientations from which localised internal structure can be inferred, but

[13]See Section 11.10.5.

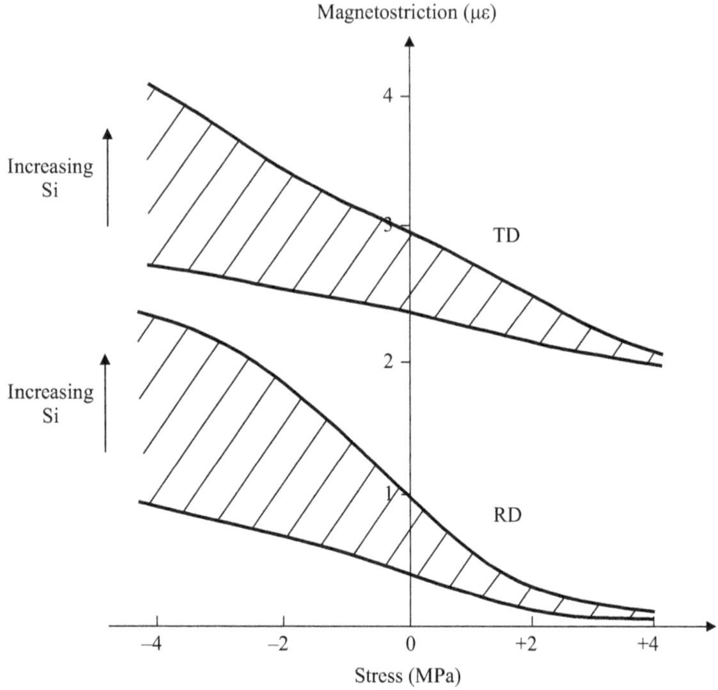

Figure 11.17 Range of stress sensitivity of the peak value of the fundamental component of peak magnetostriction of strips of typical NO steels with silicon content between around 1.0% and 2.5%. Strips cut along the RD and TD and stressed along the longitudinal direction when magnetised at 1.5 T, 50 Hz

there will be many more grains oriented in far different directions whose complex surface patterns make it difficult to predict sub-surface structures. The problem is compounded by the fact that grains in NO steels are of the order of 20 μm to 200 μm diameter so structures are difficult to observe.

Typical domains on the un-stressed surface of a 0.1% Si NO with an average grain size of around 20 μm are shown in Figure 11.19(a) [25]. Bar domains are apparent on the surface of a few domains but complex closure patterns are present on the surface of many domains and no patterns are distinguishable on some. Figure 11.19(b) shows the effect of surface tensile and compressive stress set up by bending the steel.[14] Tensile and compressive effects are shown on the same

[14]Note that the stress is induced by bending the sample which is practically easier to do than by applying an in-plane force to set up a uniform longitudinal stress. The magnitudes given here might appear to be very large but they are surface values, the internal stresses drops linearly to zero at the centre of the sample. Since the grains in this NO steel are small compared with the sample thickness, the surface stress patterns can be assumed to be representative of those which would be obtained under uniaxial longitudinal tensile or compressive stress. Section 11.6 covers the effect of bending stress in more detail.

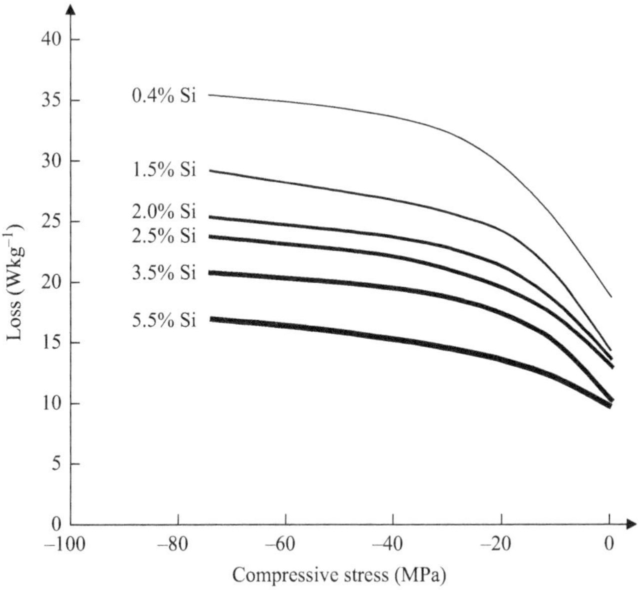

Figure 11.18 Variation of the stress sensitivity of loss (1.0 T, 400 Hz) of 0.2 mm thick NO steels with silicon content (Figure 2 in [27] reproduced under free licence CC-BY-4.0 and modified)

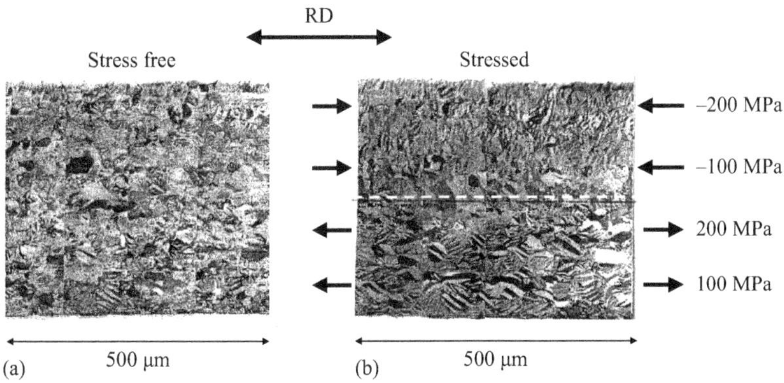

Figure 11.19 Effect of stress on domain structure of a demagnetised sample of 0.1%Si NO steel: (a) stress-free and (b) effects of surface tension and compression due to bending of the same (Reproduced under free licence CC-BY-4.0 from Figure 4 [25])

figure so that the differences can be more easily seen. When the stress is tensile, more bar domain structures are visible in far more grains implying an improvement in magnetic properties when magnetised along the RD. It is difficult to quantify the effect of these complex structures. Under compression, most of the bar domains are replaced by stress patterns, so magnetic properties will deteriorate. Under less severe bending of GO steel, the 180° bar domains are present throughout the thickness of the steel, but many lancet domains are present on the compressive side and all the supplementary domains are suppressed on the tensile surface [28].

The rotational loss and magnetostriction of NO steels vary with stress in a similar way to GO materials described in Section 11.4, but to a lesser extent.

11.6 Effect of bending stress

One of the most common causes of mechanical stress in laminated cores is the bending of the laminations. It has a significant effect on the magnetic properties. A common cause of bending stress is the inherent waviness of electrical steel strip. Although largely very low in good quality modern materials, it can cause additional losses or increased magnetostriction when flattened in an assembled core. Even a waviness of amplitude 3 mm in a 1 m length of GO steel can set up surface bending stress of around ± 2 MPa [29]

The trends of some early experimental measurements [30] of the effect of bending stress on the magnetic properties of GO material are still valid today. Table 11.1 shows the averaged variation of magnetic properties of 0.35 mm thick high, medium and low permeability GO steel. The deterioration of all properties with reducing curvature is significant but there is a spread across the material grades. The only definite trend is that the highest permeability material had the highest stress sensitivity of permeability, but there are no identifiable trends with loss or with coercivity.

Table 11.1 Average variation of the change of magnetic properties with bending stress of three grades of GO steel with maximum permeabilities of 1,04,000, 70,000 and 54,000 (Deduced from data given in [30] measured at 1.5 T, 60 Hz magnetisation)

Radius of curvature (m)	Curvature (m^{-1})[a]	Surface bending stress (MPa)[b]	μ_{max} as a percentage of that of a flat lamination	H_c as a percentage of that of a flat lamination	Power loss as a percentage of that of a flat lamination
∞	0	0	100	100	100
3.30	0.30	6.7	80–92	80–92	102–115
1.73	0.58	12.7	59–70	59–70	107–129
1.12	0.89	19.6	47–62	47–62	117–138
0.38	2.63	57.1	25–51	25–51	–
0.20	5.00	108.5	17–29	145–192	–

[a]The curvature is simply the reciprocal of the radius but it is included here simply because several investigators use it as the reference measure of bending.

[b]This is calculated from the results in [30] assuming a value of Young's modulus of 124 MPa.

Other measurements [31,32] agree reasonably well with [30] considering that they were carried out using different types of apparatus and on different grades and thicknesses of steel.

It is not straightforward to find relationships between the effect of a bending stress and that of an equivalent uniaxial stress on magnetic properties. If the magnetic properties of a lamination are improved and degraded equally under uniaxial tension and compression along the axis along which bending occurs, the bending would have no overall magnetic effects. Clearly, this is not the case with most soft magnetic materials whose properties are stress sensitive.

It is useful to consider a simplified analysis of the bending processes which occur in thin laminations. Consider a strip of length $2l$, initially in a stress-free state, before bending into an arc of a circle of mean radius r within its elastic limit. Its convex surface is in a state of tension, because it is elongated, and the concave surface is in a state of compression.

The stress varies linearly through the thickness of the lamination shown in Figure 11.20. The position where the stress is zero is the neutral axis. In the case of a homogeneous strip, the neutral axis is identical to the geometric axis.

The bending stress is given as [33]

$$\sigma_b = tEr/(2r + d) \text{ (Pa)} \tag{11.9}$$

where E is Young's modulus and t is the thickness of the strip. In practice $r \gg d$ so the expression can be rewritten as

$$\sigma_b = aE/r \text{ (Pa)} \tag{11.10}$$

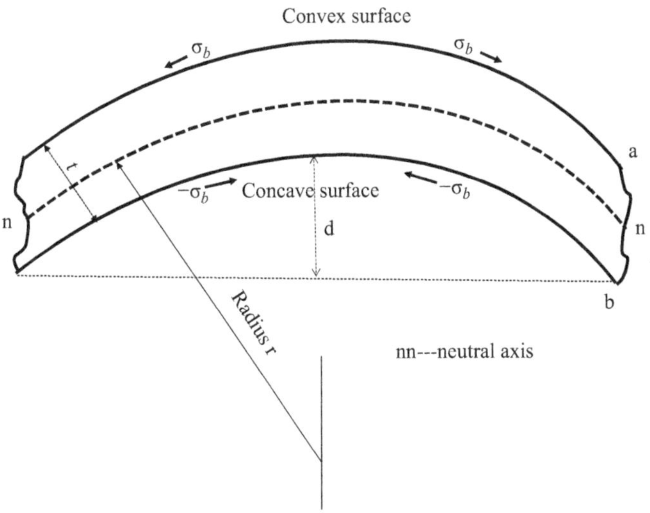

Figure 11.20 Cross section through a strip bent into an arc of a circle and corresponding bending stress distribution

The presence of magnetostriction in a magnetic strip changes the internal stress distribution. If no magnetostriction is present the total stored mechanical energy is zero (as can be inferred from the symmetrical, linear stress curve). When magnetostriction is present magnetoelastic energy is introduced so the internal stress distribution changes. What, in effect, happens is that the position of the neutral axis changes. This can be deduced from the fact that the bending modulus[15] becomes non-linear due first to the interaction between magnetostriction and the elastic bending effects, referred to as the ΔE effect [34], and, second, the effect of anticlastic bending[16] in a strip [34,35].

The effect of bending on domain structure will be considered later in this section but it is instructive to consider what might happen to domains in a strip of GO steel cut parallel to its RD when it is bent longitudinally. Suppose the strip is artificially thick such that the individual grains are far smaller than the strip thickness. In this case, it is easy to visualise stress patterns being developed in grains close to the concave surface and antiparallel bar domains in grains close to the convex surface being unaffected. In GO steel most grains are far larger than the strip thickness, so the effect of bending becomes a little more difficult to predict. If stress domains are set up on the compressed side of the strip and bar domains are still present on the tensile side, then a complex structure would be present between them.

This complex combination of magnetostrictive strain and elastic strain combined with complicated changes in domain structure in GO steel do have practical consequences. A moderate bending stress can lead to changes in bending modulus of over $\pm 15\%$, from which it can be predicted to change by over $\pm 30\%$ in large sheets used in transformer cores [35]. Under a.c. magnetisation, this can possibly cause distortion resulting in acoustic noise generally attributed to core resonances. It is interesting to note that although bending is probably the largest source of stress in large transformer cores, as stated earlier, the consequences of its effect on mechanical vibration and noise still do not appear to have been widely reported.

The complex bending stress has a large effect on the domain structure in GO steel as inferred earlier in this section. In the un-magnetised state, characteristic bar domains are observed on the surface of a bent sample in a state of tension which is very similar to what is found in the material when subjected to uniaxial tension along the RD [31,33].

Figure 11.21 shows surface domain patterns on either side of a bent lamination of GO steel compared with those of the stress-free state and with linear stress applied to the same region of the steel [33]. Figure 11.21(a) shows the domain structure in the stress-free, demagnetised state comprising bars domains as normally found on the surface of a well oriented GO grain. Figure 11.21(b) shows the anticipated large domain refinement in the same region when a uniaxial tensile

[15]This is equivalent to Young's modulus in a bending mode [34].

[16]When a lamination is bent along its longitudinal direction a curvature is also created in the transverse direction. This anticlastic bending becomes more significant in laminations with high width to length ratios.

stress of 35 MPa is applied along the RD.[17] Figure 11.21(c) shows transverse closure domains under uniaxial compression of 5 MPa.

Figure 11.21(d) shows the image on the tensile side of the bent sample when a surface stress of 35 MPa is created by bending to a radius of curvature of 3.88 m. Although this image is not clear, very wide transverse domains are present. However, it is clear that the domain structures on either surface are not only different, but also quite different to those which are observed under uniaxial stress. This confirms that the effect of bending stress on the magnetic properties is not just a simple combination of ordinary tensile and compressive stress on either side of the sample.

A further finding in [33] is that the magnetostriction (measured with resistance strain gauges) is different on either side of a bent sample. At around 1.8 T the magnetostriction on the concave and convex surfaces were 16 $\mu\varepsilon$ and 11 $\mu\varepsilon$, respectively. A difference is to be expected, since the surface domain structures are different, but the magnitudes do not correspond with simple stress domain patterns. It is possible that this inherent differential expansion causes further flexing of bent laminations which could occur in stacked transformer cores. However, it not known whether this phenomenon is large enough to have any practical significance.

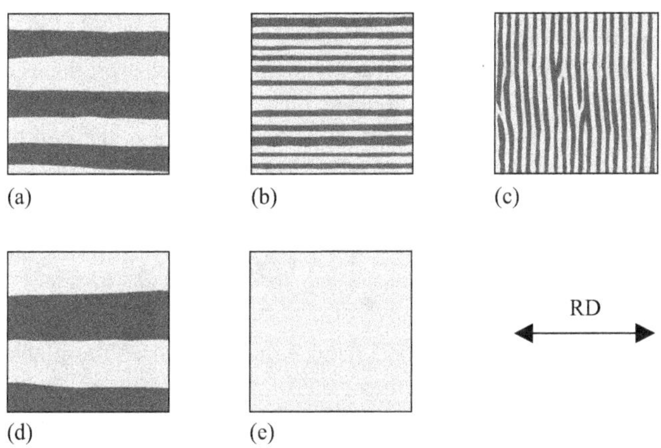

Figure 11.21 Stress patterns on either side of a bent strip of GO steel: (a) unstressed, (b) uniaxial tensile stress of 35 MPa along the RD, (c) uniaxial compressive stress of 5 MPa along the RD, (d) convex (tensile) side of the bent strip (surface stress of +35 MPa) and (e) concave side of the strip (surface stress of −35 MPa)[18] (RD is directed across the page) [33]

[17]No scales are included on the original versions of these images but they are all believed to be the same scale so direct comparisons can be made.
[18]Faint, blurred transverse domains can be seen in the original image of this in [33].

It was shown in Section 11.5 that the properties of NO electrical steel are stress sensitive, but not as much as those of GO steel. Hence its properties should also be susceptible to bending stress. Because of the much smaller grain size than in GO materials, it is anticipated that stress patterns very similar to those found under uniaxial compressive stress will be present on the concave side of a bent strip. Likewise, normal domain refinement is expected in grains close to the convex surface. There would be some interaction between the grains in tensile and compressive regions, but not to the extent found in GO steel where, of course, the tensile and compressive stress pattern regions occur mainly in single grains. We have not found any published evidence for this, or quantification of the effect of bending stress in NO steel or other materials, apart from GO steel.

A more complex practical form of deformation is twisting of laminations forming a part, or all, of a limb or yoke of a stacked magnetic core. The magnetic effects of twisting thin ribbon or wire have been widely investigated in relation to magnetic sensor applications, but the effects of twisting, or torsion on laminations have not been widely reported, probably partly because of the difficulty of making magnetic measurements on twisted laminations.

One of the few reports on the phenomenon does propose that it is a simple test to carry out and could be used more easily than the current ways of applying uniaxial compressive stress described in Section 12.6. It is claimed that this gives a more meaningful assessment of the practical stress sensitivity of a steel [36]. However, it was also pointed out that a given angle of twist produces a far higher compressive stress in wide sheets than in narrow ones and this could affect the magnetic characteristics considerably [36]. If this is the case, the building factor of large transformer cores could be more affected by stress than is realised at present, but the phenomenon has not been quantified.

11.7 Effect of normal stress

By normal stress we mean stress occurring perpendicular to the plane of laminations in a stacked core. It is sometimes called interlaminar stress. It can be caused by many factors, the main one being core clamping. Normal stress is randomly distributed throughout a core.

Laboratory testing is usually carried out by applying a uniform force to a stack of strips or laminations. Occasionally, normal stress is applied to a single lamination in order to obtain basic material information independent of any other building effects which occur when laminations are stacked together.

Some causes of normal stress are shown in Figure 11.22. In (a), the stress is uniform and easy to calculate from the applied force. An exaggerated side view of a stack of laminations is shown in (b). This illustrates inherent curvature due to coil set[19] and the effect of strip thickness variation. If it is assumed that the laminations are unconstrained, they are actually in a state of low stress and the magnetic

[19]See Section 13.2.

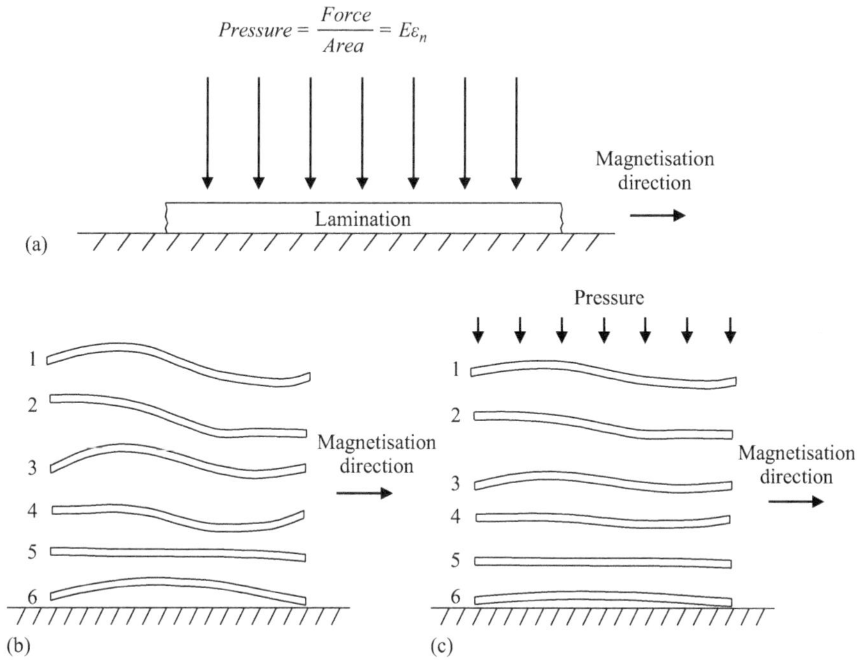

Figure 11.22 Schematic diagrams showing the occurrence and effect of normal stress: (a) cross section showing it applied to a single, flat, strip, (b) cross section through a stack of random non-flat strips located on a flat bed and (c) effect of normal stress on the stack of laminations in (b) (Not to scale and the waviness is exaggerated)

properties of the stack could be measured to give what might be called the stress-free characteristics.

The normal force applied in (c) flattens the laminations effectively causing random bending. The structure sensitive properties deteriorate partly due to the bending component and partly due to the more fundamental effect of normal stress on the domain structure in individual grains. If the normal stress is sufficiently large, the strips become completely flattened and a further increase of normal stress will produce the same effect as it would on a single flat strip.

One of the first quantifications of the effect of normal stress was carried out by Eingorn in the mid-1960s. Figure 11.23(a) shows some of Eingorn's findings [37]. It can be seen that the normal stress is most harmful in strips cut parallel to the RD and when the mis-orientation is greater than around 55° the loss increase is less than around 2% at any stress level and is independent of angle. The angular dependence of loss appears to show that the Goss texture has an influence on the characteristics.

It is suggested [38] that the normal stress causes different relative changes in in-plane dimensions such that the elongation in the length is ten times greater than

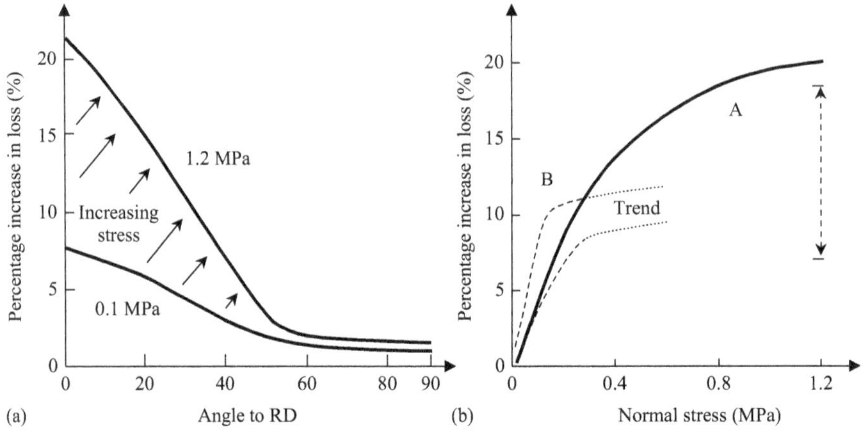

Figure 11.23 Percentage increase of loss at 1.5 T, 50 Hz, due to flattening stress: (a) in 660 mm × 30 mm × 0.35 mm thick GO strips cut at angles to the RD. (Derived from results given in [37].) (b) Comparison of loss increase given in [37] (a) magnetised along the RD with measurements on larger (990 mm × 150 mm × 0.33 mm) sheets of GO material (b) (Figure 8 of [39], reproduced under free licence CC-BY-4.0 and modified)

in the strip width.[20] This sounds unlikely but if it were the case it would in turn decrease the influence of the normal stress at high angles of magnetisation to the RD. To the authors' knowledge this has not been verified but it implies that clamping stress has an insignificant effect in regions of GO cores in which the flux is directed at a high angle to the RD. However, a theoretical investigation on the influence of the degree of texture on the loss increase in GO steels showed that the effect of normal pressure on domain structures, and hence power loss and magnetostriction, was quite different from that of uniaxial stress and also that a stress over 1.0 MPa was necessary to favourably affect the domain structure [39]. This implies that the loss increase shown in Figure 11.23 will be suppressed at high normal stress. However, this would not be expected to have any noticeable effect in transformer cores assembled in a conventional manner.

Domain observations show that the main domain widths in grains of GO steel can be refined (narrowed) on the application of normal stress of up to around 3 MPa to a single sheet or a stack [40]. The width of domains more than 1.0 mm wide was reduced by at least 50% whereas the normal stress had little effect on narrow domains, around 0.5 mm wide.

The effect of normal stress in transformer cores is most relevant in regions where the in-plane flux is parallel to the RD (the majority of the core surface).

[20]This is completely based on elastic deformation, any further influence of magnetostriction is not included.

It can be seen in Figure 11.23(b) that in strips cut parallel to the RD the loss initially increases rapidly with increasing pressure and then increases far less rapidly above a pressure of around 0.8 MPa. The figure also shows some results of applying pressure to different grades of larger GO laminations assembled as single-phase transformer cores. The trends are the same despite the stress being applied to different materials using different stressing processes over different stress ranges.

Further general agreement was reported in [41] but it also reported a significant influence of the coating and surface topology on the effect of normal stress; the loss increase at 1.5 T, 50 Hz being from 8% to over 20% according to surface treatment. Random bending stress, which could be caused by variation in insulating thickness, was simulated and shown to increase localised losses by up to 30%.

The effect of normal stress up to 28 MPa on the loss and permeability of stacks of ring specimens of NO steel varies significantly with the grade of steel [42]. The underlying reasons why they should be so dependent on the grade are not clear.

It will be apparent in this section that several apparently conflicting findings on the trends in changes in magnetic properties due to normal stress have been reported. There might be doubt over the accuracy or validity of some findings but to summarise it is safe to conclude that normal stress applied to a perfectly flat single sheet of electrical steel has very little effect on its magnetic properties as already stated. However, the effect of normal stress applied to stacks of laminations or full size laminated cores can be influenced by many assembly factors which can change hence and contribute in different ways to observed changes in magnetic performance.

11.8 Effect of stress on components of loss

As discussed in Section 7.8, the whole concept of splitting losses into hysteresis and classical eddy current components has little physical justification so analysis of the stress sensitivity of the individual components needs to be considered with this in mind. Both hysteresis and excess components, derived as described in Section 7.8, are stress dependent [43]. It is clear, however, that the changes to the domain structure due to applied stress described in Section 11.2 would have a significant impact on both the static and dynamic magnetisation mechanisms.

The influence of stress on a.c. and rotational hysteresis of NO electrical steel can be estimated from an energy minimisation model although it is difficult to extrapolate predictions to high stress levels [44]. A significant fall in static hysteresis loss occurs in NO steel with increasing tensile stress, whereas only a slight fall occurs in GO steel [45]. At the same time, the Steinmetz coefficient increased by up to 25% in the GO steel thus inferring a measureable increase in hysteresis loss unless other factors change. The reason for the unexpected stress dependence of static hysteresis on tension in the NO steel is attributed to impurities [45], but how the stress affects the impurities is unclear.

It can be deduced from (7.34) that the classical eddy current loss is not dependent on elastic stress. However, traditional two component analysis of losses in a ring-shaped specimen of GO steel showed the eddy current component increasing with bending stress but no explanation has been given [43]. It is possibly due to the fact that internal magnetisation in bent GO material is extremely non-uniform so many of the assumptions and approximations in loss analysis become invalid [46].

It has also been reported that the eddy current anomaly factor η falls with increasing tension in GO steel, but not in NO steel [45]. It falls more so in uncoated GO steel showing the close association between external tension and built-in tension caused by the stress coating [45].

It has been proposed that knowledge of one stress dependent parameter, obtained by measurement along with the stress dependent hysteresis loss, is sufficient to estimate accurately the stress dependence of the excess loss [47]. This approach only accounts for tensile stress whose effect is normally far less than compressive stress. But it is possibly a useful model for use in rotating electrical machine core problems.

It was hinted at the beginning of this section that loss separation under applied stress is challenging. Published results are somewhat contradictory, perhaps because many are difficult to compare since experimental details are not always sufficiently specified and tests are carried out under wide ranging conditions and on many types of materials.

11.9 Effects of building stresses in electrical machine cores

The previous sections in Chapter 11 in this chapter have focused on the magnetic effects of easily quantified forms of mechanical stress mainly related to measurements under closely controlled stress and magnetisation conditions. These provide a useful basis for understanding, or modelling, overall effects of stress. However, the presence of a complex combination of stresses is possibly the largest contributor to the Building Factor.[21] In this section, common practical situations in which such stresses act together in stacked and wound electrical steel cores are discussed. Other aspects of the effects of stress on motor stator cores assembled from CoFeV laminations and distribution transformer cores assembled from amorphous ribbon are discussed in Sections 16.7.2 and 9.1 of Volume 2 of the book, respectively.

11.9.1 Clamping stress

An important aspect of the successful design of laminated cores is determining the appropriate pressure which does not cause undue deterioration of magnetic properties, but is also capable of producing a tight core which remains perfectly stable in service. In Section 11.7, it is shown that normal stress introduced by pressing or clamping a core generally increases no-load losses. Excessive clamping also implies extra higher manufacturing cost.

[21]See Chapter 8 of Volume 2 of this book.

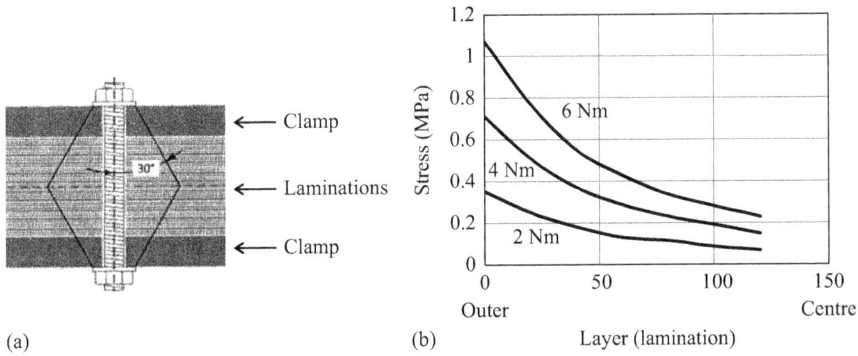

Figure 11.24 Localised normal stress set up by a tightened bolt clamping a stack of laminations: (a) bolt geometry and area experiencing a normal force and (b) normal stress variation from outer lamination to the centre of the core [51]

Ineffective clamping of a power transformer core can lead to excessive vibration and acoustic noise. In a rotating machine core, excessive stress can even lead to fatigue and damage to stator teeth, which in turn can break off laminations and penetrate the winding to cause electrical failure. It has been suggested that an average pressure on the stator core of a large rotating machine should be of the order of 0.3 MPa [48]. A lower stress is probably sufficient in transformer cores. Values used in practice are mainly confidential to the manufacturer but probably range from 0.2 MPa to 0.5 MPa.

The clamping stress in a laminated core depends on the method of application. Clamping is carried out in many ways, partly determined by the core size and partly by a manufacturer's in-house experience and developments. Large cores can be clamped making use of tightly wound steel or glass fibre bands. If the pressure is applied by limb clamping bars bolted together externally, the pressure should be reasonably uniform over the clamped area. Sometimes, coils and windings are clamped separately to avoid any effect of winding stress on the core. However, at other times the winding stress is transferred to tie bars which are designed to induce favourable tension in the core.

Sometimes, clamping bars are fixed in position using bolts through holes in the limb or yoke laminations as shown in Figure 11.24(a).[22] These are tightened to a given torque to set up the required core clamping pressure. The clamping pressure is then non-uniform, not just in the plane of the laminations in the stack, but also into the depth of the stack as shown in the following.

[22]This form of clamping is not used today in power transformers because, when bolts are tightened, electrical shorts can occur at the edges of the holes due to burrs. However, it is still of academic interest particularly when assessing the findings of earlier research on core stresses in general.

The relationship between the torque T, to which a bolt is tightened, and the resulting stress in the nth layer of the clamped stack can be obtained from the following expression [49].

$$\sigma_n = F_n/A_{c,n} \text{ (Pa)} \tag{11.11}$$

$$F_n = T/J_T d_b \text{ (N)} \tag{11.12}$$

where F_n is the force applied to a lamination, J_T is a torque coefficient, $A_{c,n}$ is the cross sectional area covered by the force on lamination n and d_b is the diameter. The value of J_d is usually assumed to be 0.45 for a steel bolt [50]. Figure 11.24(a) shows the force distributed over the 30° region to create a different stress in each layer of laminations [51].

Equations (11.11) and (11.12) can be applied to the stack to obtain the internal normal stress distribution. Figure 11.24(b) shows the magnitude of the normal stress in laminations from the surface to the centre of a 240 layer GO steel core. The clamping torques of 2 Nm, 4 Nm and 6 Nm cause average pressures of 0.16 MPa, 0.33 MPa and 0.49 MPa, respectively [51].

The normal stress set up in this way varies significantly both over the cross sectional area of layers of laminations and into the depth of the core. Such through bolts are rarely used in small cores but a related effect can occur in externally clamped cores if the bolts are close to the core edges. The phenomenon is not only of academic interest since it applies to various core testing situations where experimentalists assume uniform normal stress is present in a stack of laminations.

It has been reported that better grades of GO steels are least affected by clamping pressure [39]. This is possibly due to the better grades being a little flatter, but the findings came from measurements made several years ago and it is not known whether the same finding applies for modern grades of steel. Another study has shown that domain refined (DR) steel is less susceptible to clamping stress than conventional GO steel [52]. The reason for this seems most likely to be the lower stress sensitivity of the DR steel. It seems unlikely that there would be any effect of any difference between the flatness of the two materials in laboratory cores on which these findings were obtained.

The same authors investigated the effect of clamping over different regions of a core [52]. Table 11.2 summarises the results of the study on two single phase model cores.

The quantity ΔW in Table 11.2 is the percentage increase in building factor per 1% surface area per MPa. The increase in the core building factor is by far the lowest when the pressure is only applied to the corners. When these are 'normalised' to a standard unit area, the increases in ΔW are equivalent in the three cases. This implies that the total clamping force determines the increase in building factor, not the magnitude of the localised pressure, nor the location of the clamped areas [52].

The general effect of clamping pressure on the building factor of transformer cores is discussed in Chapter 8 of Volume 2 of this book.

Table 11.2 Effect of position of clamps on loss increase in single phase model cores assembled from conventional GO steel (CGO) and domain refined GO steel (DR) [52]

Stress application	Material	% of surface area directly stressed	Maximum pressure (MPa)	% Increase in building factor		Normalised increase in core loss, ΔW	
				1.5 T	1.7 T	1.5 T	1.7 T
Uniform		70	0.8	8.0	6.7	0.14	0.12
	DR	70	0.8	6.7	4.8	0.12	0.09
Localised over	CGO	5	4.8	3.1	2.6	0.13	0.11
corner joints	DR	5	4.8	2.4	1.6	0.10	0.07
Localised over	CGO	5	10.3	8.6	8.1	0.17	0.16
whole yoke	DR	5	10.3	7.6	5.6	0.15	0.11

11.9.2 Wound cores

Bending stress degrades the magnetic properties of strip wound cores. This so-called winding stress is normally removed by a stress relief anneal. If a wound core has very tight corners, for example the 90° bends in rectangular cores, plastic deformation may be present causing irreversible deterioration of localised magnetic properties. Plastic deformation at the edges of narrow strip wound cores can cause a degradation of properties. This is discussed in Section 11.10.

Many cores are cut to produce C-cores which are impregnated with epoxy resin to provide structural integrity. Many such resins shrink on curing and set up random compressive stress which can increase losses by 20% or more [53]. However, this is tolerated in many applications where magnetic performance is far less critical than cost.

11.9.3 Stacked cores

Uniaxial, bending, normal and even torsional stress can simultaneously occur in a laminated core. The influence of each can be estimated when applied separately, as discussed in Sections 11.6 and 11.7 but when they occur together in a core their localised interaction and random distribution are so complex that they do not appear to have ever been quantified in systematic experiments or by computational stress analysis.

From experience of stacked transformers cores, it is usually assumed that random core stresses lie in a region ±20 MPa. Experimentalists tend to characterise steels under uniaxial stress up to around ±40 MPa to cover this range. Even higher uniaxial stresses which are applied in studies of the performance of high speed rotating machines are of interest.

Figure 11.25 shows the static stress distribution in laminations in an un-magnetised 100 kVA, single phase transformer core assembled from GO steel and tested in a horizontal position [54]. These were measured with uniaxial foil

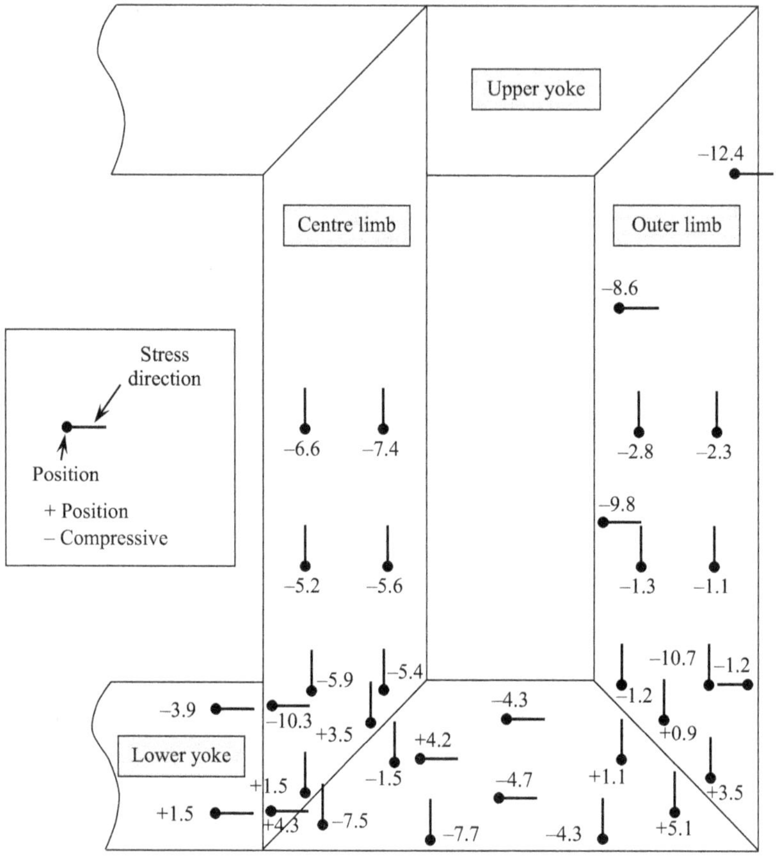

Figure 11.25 Localised building stress measured at locations on laminations in a layer of a 100 kVA, three-phase core [54]

resistance gauges distributed on laminations in the centre region of the core stacks. The distribution appears to be completely random but it is worth noting that the highest compressive stress is only just over -12 MPa. Of course, only one component of stress is shown at each position so a full picture is not given. The gauges themselves introduce some stress because of their finite thickness. The average stress increased by 20% when the measurements were carried out with the core in a vertical position, as it would be commonly used in practice, rather than a horizontal position. This could be due to changes in stress distribution from predominantly flattening stress in the horizontal position to increased lower in-plane joint stresses in the vertical position. Of course, this could also be a particular characteristic of the one core tested.

Another possible cause of localised stress in a stacked core is temperature gradients causing differential expansion throughout individual core laminations. Although the temperature distribution throughout a core can be calculated reasonably accurately accounting for the cooling methods, there seems to be little published data on temperature distribution from which the associated mechanical stress can be estimated. The stress set up by a temperature gradient might be harmful or beneficial depending on its orientation with respect to the localised magnetisation direction. It can easily be calculated that a stress of around 0.5 MPa can be set up in the limb laminations of a large transformer core due to a temperature difference of 10 °C from the surface of the limb to its centre. Far higher localised stresses are possible where additional eddy current heating takes place due to burred laminations.

To conclude this section it can be said that the change of magnetic properties of a stacked core due to building stress depends on core assembly, core size, lamination shapes and the operating flux density and magnetising frequency. However, more specific quantification of the stress distribution in cores is needed to determine true relationships between the well quantified stress sensitivity characteristics obtained from laboratory tests and what is found in stacked cores in general.

11.10 Slitting and punching stress in electrical steel

11.10.1 Background

Electrical steel is normally produced in coils from 700 mm to 1200 mm wide. For most power transformer applications these are mechanically slit at high speed into narrower coils normally in the range over 300 mm wide to 50 mm wide prior to further cutting to produce laminations or wound cores. Much narrower coils are used in some wound cores. Slitting can create narrow regions of damaged material close to the cut edges leading to a localised fall in permeability and increase of loss referred to as the *cut edge effect*. Figure 11.26 shows a commercial slitting tool in action.

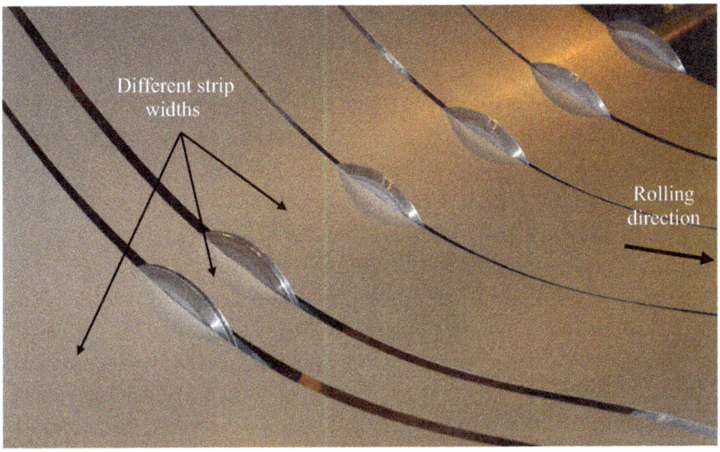

Figure 11.26 Slitting of a coil of NO electrical steel into various strip widths (Photo courtesy of Cogent Power Surahammar)

A wound core is directly produced by cutting off the required length of the slit coil. To produce transformer laminations, the coil must be slit to the required width prior to cutting each lamination to size. Rotating machine core laminations are stamped from slit to width coil. The three processes are shown diagrammatically in Figure 11.27.

Possible stressed regions due to slitting and pressing or cropping can be recovered to some extent by a stress relief anneal. The benefit of annealing a wound core is that the harmful effect of winding and slitting stress can be at least partly recovered. Stacked power transformer cores are not normally annealed because the cost of the extra step does not justify the benefit of improved magnetic properties over what is usually a very small proportion of the core steel. The effect of the relatively large stressed regions around the periphery of the laminations of small stator cores would be reduced considerably by a stress relief anneal but this is rarely done because of the increased cost.

After slitting, laminations for stacked cores can be formed in several ways. The most common method is to stamp, or punch, single laminations directly from the coil. Basically, a hardened steel punch is forced at high pressure through the material to form the required shape. This method is universally used for large volume manufacture of laminations of all sizes. The process can be carried out at very high speed. Dimensional accuracy can be very high, and the cost per lamination is low. The cutting tools can be very complicated, particularly for complex shaped rotating machine laminations which are punched in multiple steps. Because of the high surface hardness of silicon steel alloys, the cutting process causes greater tool wear compared to many other steels, so it needs to be carefully monitored since, as will be seen later in this section, cutting laminations with worn tools or incorrect lubricant leads to a deterioration of magnetic properties, or possibly complete core failure.

Punching tools can be extremely expensive so it is not cost effective to carry out the process on small volumes of laminations or for producing laboratory prototypes. In these cases several other cutting techniques are used, but before considering these, some fundamental effects of cutting which affect the magnetic performance of laminations need to be considered.

Any form of punching or mechanical slitting causes a complicated mechanical stress distribution near the cut edges of a lamination. This is schematically illustrated in Figure 11.28(a). Even a sharp cutting blade will produce a burr which can be tens of microns high. The cutting process produces two forms of stressed regions near the cut. The elastically stressed zone extends over the whole cut edge region and the plastically stressed zone is present closer to the cut edge.

The sharp, bare point at the tip of the burr can make contact with a bare region of an adjacent region as shown in Figure 11.28(b). This is an electrical short circuit between laminations (or between turns in a wound core) which can increase eddy current losses over their full widths which in the most severe cases could lead to catastrophic core failure on test or even in service as discussed in Section 5.13.2 in Volume 2 of this book.

It is sometimes sufficient just to have an approximate estimate of the variation of the deterioration of magnetic properties due to slitting strip or laminations.

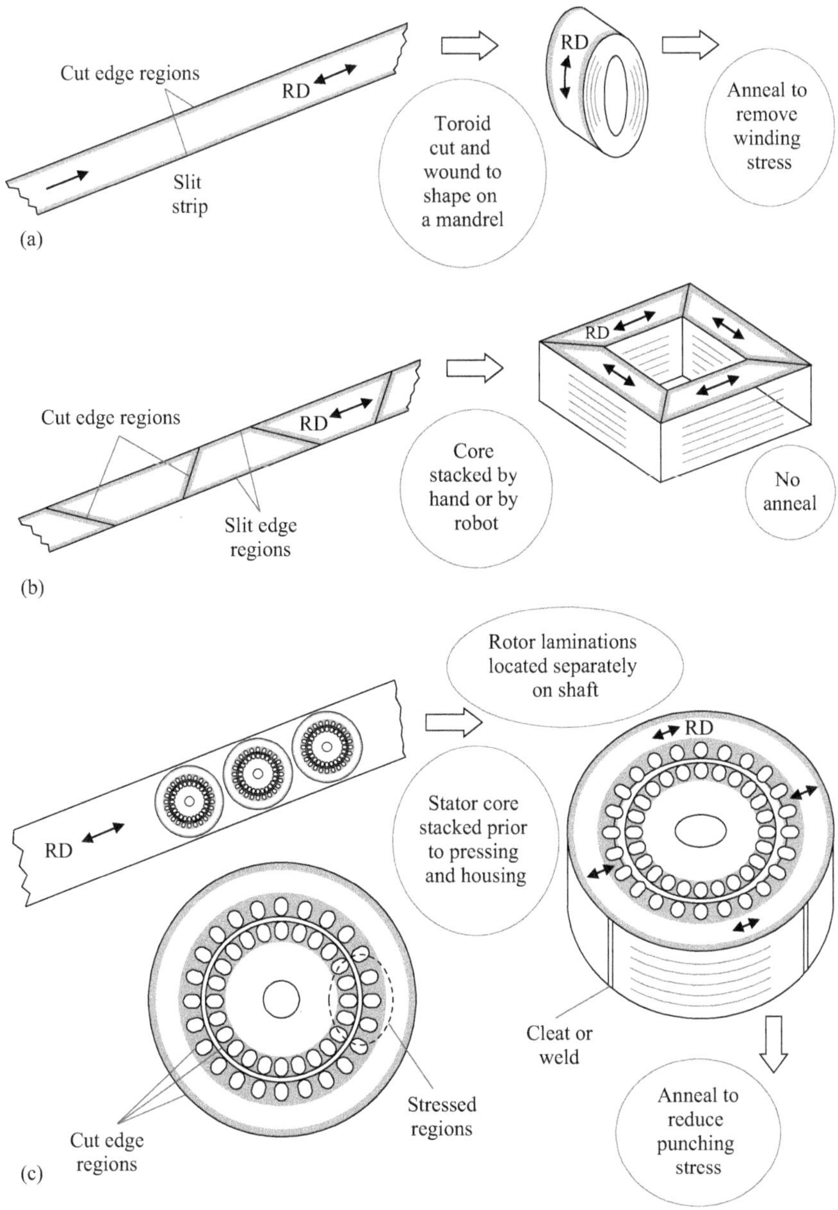

Figure 11.27 Stressed regions caused by cutting and assembling basic forms of magnetic cores: (a) wound core, (b) transformer core and (c) rotating machine core

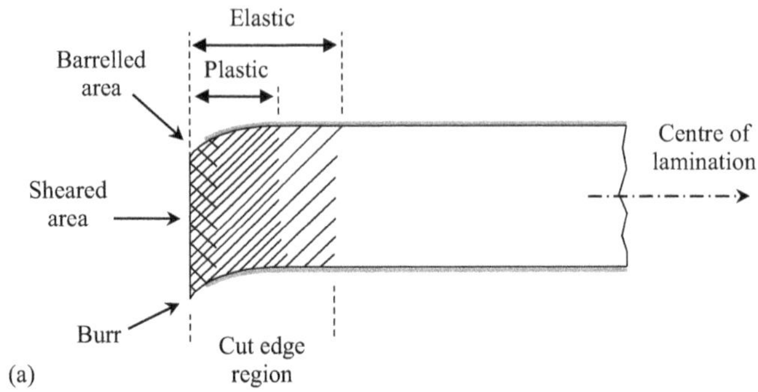

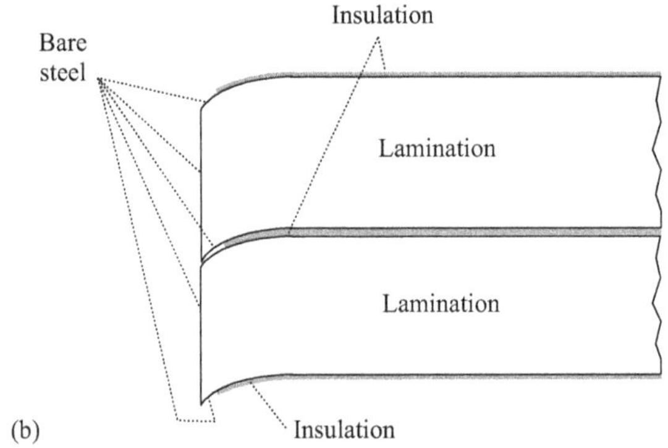

Figure 11.28 Effects of lamination cutting (a) cross section through the thickness of a lamination showing possible stressed and burred areas near the cut edge and (b) laminations shorted together at the bare cut edges

A simple relationship found from experience is that the increase in loss ΔP of an ideal strip of width w_c due to the cut edge effect is approximately given by [55].

$$\Delta P = 0.2/w_c(\%) \tag{11.13}$$

This is based on measurements showing that, when an effective cutting tool is used, the damaged cut edge region is 1.0 mm wide, and effectively has a relative permeability of unity. Despite the gross assumptions, this simpler relationship has been found to be reasonably accurate for GO steel strip width from around 100 mm to 350 mm.

Figure 11.29 shows some general effects of cutting of 30 mm wide strips of grain oriented steel on the losses over this width range and the further degradation caused by tool wear [55]. The reduction of loss after a stress relief anneal is also shown. The most significant points are that the increase in loss is less than 1% in strips wider than 220 mm and that the stress relief anneal does not fully remove the effect of the damaged region in GO steel. Similar increases in loss found in several NO steels produced by three other manufacturers fall in the shaded range in Figure 11.27(a) [56].

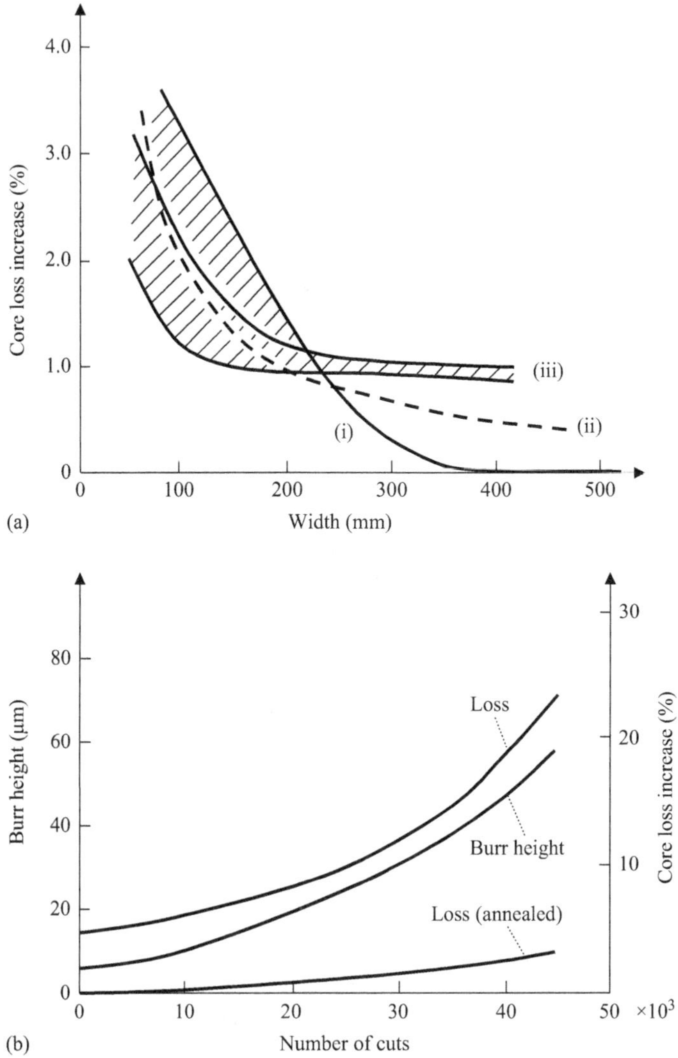

Figure 11.29 *(a) Percentage loss increase due to slitting stress in GO steels (1.5 T, 50 Hz): (i) 0.3 mm strips [54], (ii) predicted from (11.13) and (iii) range from strips made by other manufacturers [55]; (b) increase of loss and burr height of narrow GO strip due to tool wear (Adapted from data in [55])*

The results in Figure 11.29 demonstrate three important trends in the cut edge effect. First, the size of the phenomenon is similar in NO and GO steels; second, the effect begins to become significant in lamination or strip widths under 100 mm to 200 mm wide and third, the harmful effect can be significantly reduced by a

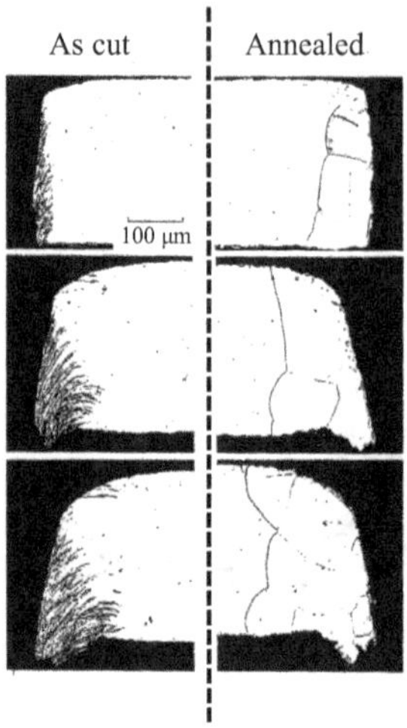

Figure 11.30 Sections of cut edges of 0.30 mm thick GO steel showing variation of burr height with age of the tool and the beneficial effect of stress relief annealing (Adapted version of Figure 7.1 in [57] by permission of IET)

suitable stress relief anneal. The deterioration due to slitting is practically independent of the grade GO steel, but it increases with increasing flux density.

Photographs of burrs taken after cutting GO steel strip with a new and over used punch are shown in Figure 11.30 [57]. The left hand images show the development of more severe edge burrs as the cutting tool becomes worn and the right hand images show the same burred regions after a 830 °C stress relief anneal. It will be noticed that after using the tool, which has already made 2,000 cuts, no burr is visible and the damage to the loss of the steel is fully recovered by the stress relief anneal as indicated in Figure 29(b). The burr height increases in a very similar fashion to that shown in Figure 11.29(b). It can be clearly seen that new grain boundaries are produced after the stress relief anneal which increase in volume as the tool wears. These are recrystalised regions which although stress free will no longer be {110}<001> oriented. Therefore, there will be some degradation of the magnetic properties due to the poorer orientation and smaller grain size in this region.

The IEC specification standard for grain oriented electrical steels [58] specifies that the burr height should be measured, relative to the sheet thickness, 10 mm from the sheet edge. It allows for a burr height of 25 μm, although users often specify far lower than this. A good quality cutting blade should cause the burr height to be less than 10 μm. The total coating thickness on GO steel is normally in the range 2 μm to 5 μm, which is thinner than the anticipated lowest burr height, so there is always a possibility of the coating being penetrated by a spurious burr even in the best of circumstances.

11.10.2 Practical aspects of the cut edge region

It has long been appreciated that plastic deformation takes place near a cut edge up to around 1.0 mm from the surface. The elastic deformation can extend up to around 10 mm into a lamination from its edge [59–62]. The exact value depends on the material composition and cutting process.

The elastic stress can be largely eliminated by a stress relief anneal which is sometimes used for NO steel applications. The plastic zone might recrystallise during high temperature heat treatment to improve the properties of NO steels but the Goss texture cannot be recovered in GO steel laminations [55]. Annealing solely to reduce the cut edge effect is not always cost effective. The magnetic properties of wide laminations are only marginally improved as can be deduced from Figure 11.29(a) but annealing of laminations in strip less than around 100 mm wide is very beneficial magnetically, so there can be an economic breakeven point for some applications.

It is impossible to eliminate the cut edge effect completely, but regularly identifying and replacing worn cutting tools at least minimises the problem. Full de-burring can be carried out, but this is at the risk of producing regions of bare steel which might even be more susceptible to making contact with adjacent strip or sheet. The extent to which the cut edge effect deteriorates the magnetic properties of a core, or the extent to which a given burr size and distribution might be safe and reliable in a core, is difficult to quantify. The deterioration when stamping laminations of NO steel can be minimised by careful optimisation of the die tool clearance relative to the grain size of the steel [63].

Figure 11.31 [60] shows some computed influences of material thickness and hardness on the plastic strain distribution along the centre of a lamination. It can be seen immediately that the range and magnitude of the strain increases, both with material thickness and with surface hardness. The theoretical analysis also shows that the elastic region is longer in the thicker materials. In practice, the extent of the plastically stressed region depends greatly on the sharpness of the cutting tool which is difficult to quantify [64]. However, Figure 11.31 gives a reasonable guide to the general extent of the damage.

Surface domain observations help confirm the extent of the cut edge region. Domain structures near cut edges of motor lamination teeth are complex and difficult to interpret before and after stress relief annealing but some trends have been identified [65]. The shear stress appears to force domains into easy directions normal to the lamination plane. This drastically reduces permeability when magnetised in the sheet plane. This normal-to-plane structure occurs in a region up to around one strip thickness from the cut edge which is comparable with the theoretical results in Figure 11.31. In a study of the effect of slitting GO steel from full width of 660 mm down to 40 mm, it was deduced that the length of the zone adjacent to the cut where plastic deformation occurs is around double the strip thickness [66]. This is higher than that deduced from the domain studies in NO steel but they both appear within the theoretical plastic deformation range in Figure 11.31.

Domains near the cut edge are more mobile after stress relief annealing, indicating an increased localised permeability which implies reduced hysteresis. This

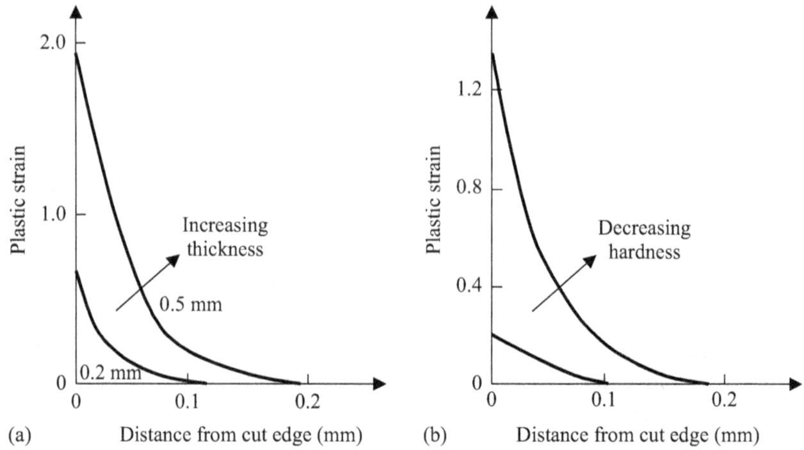

Figure 11.31 Influence of (a) lamination thickness and (b) surface hardness on the magnitude and extent of the plastic strain region deduced from 2D finite element analysis of the cut edge region on non-oriented steel (Figure 10 in [60], reproduced under free licence CC-BY-4.0 and modified)

has been indirectly confirmed from measurements showing that the increase in loss is insignificant at 400 Hz magnetisation when wide (100 mm) NO sheet is cut into narrow strips (12 mm) [67]. This is because the high frequency eddy current loss is not dependent on the cutting process and masks the hysteresis loss increase.

Figure 11.32 [68] shows the increase in loss of material due to cutting from 100 mm to 12 mm wide. Cutting strip in stages from 100 mm down to 25 mm produces a small, gradual degradation but this noticeably increases in the 25 mm strip when the damaged region extends to about 10 mm on either edge of the strip. The 12.5 mm strip is seriously damaged by the cut edge effect.

In a similar study on 1.0% Si NO steels it was found that at a length of 100 m kg^{-1} the loss increased by around 30% when a sharp cutting tool is used, but by 40% when it was worn[23] [59]. The influence of the cutting process on the magnetic properties becomes most important for large values of cutting length per unit mass and high silicon steels operating in the range 0.4 T to 1.5 T [69]. There are many concerns about the effect of damage due to lamination punching on the performance of rotating machines assembled from NO steel [70].

The extent of the cut edge region can be deduced from measurement of the localised flux density in the region using the needle probe technique.[24] Table 11.3 shows some results obtained from measurements on a 0.35 mm thick NO steel with a burr height of just 7.3 μm [71]. Measurements of this type are inherently

[23]The quantity metre.kilogram^{-1} is effectively the cut length per unit mass of lamination.
[24]See Section 12.2.

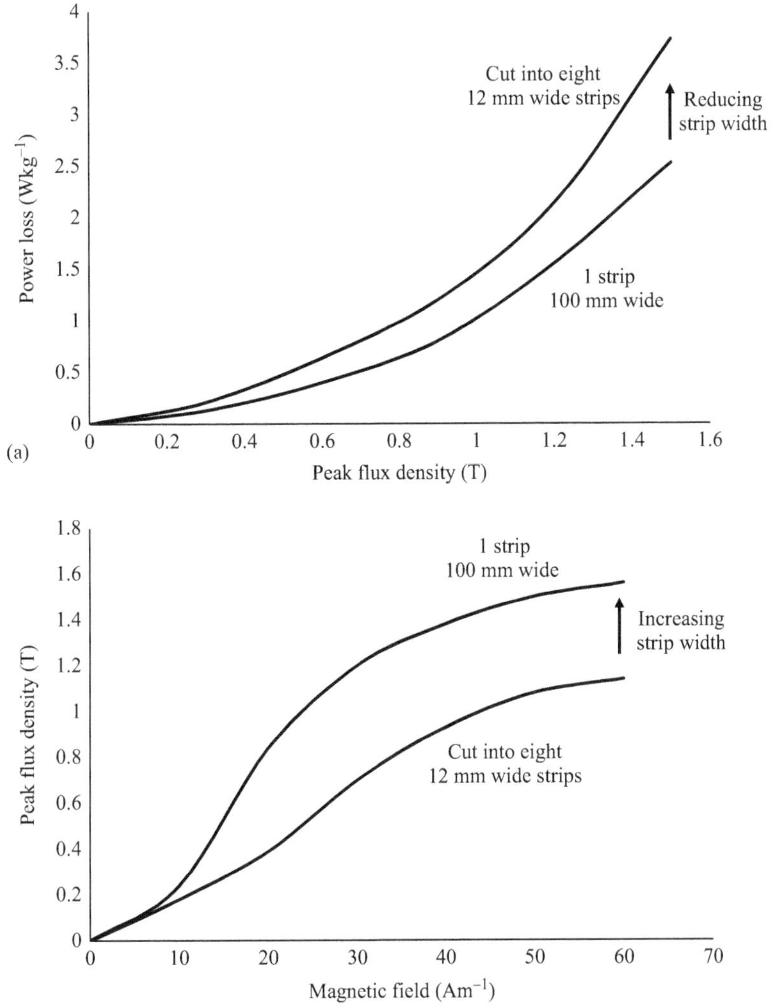

Figure 11.32 Effect of successive cutting of un-annealed NO steel into strips from 100 mm to 12 mm wide on (a) loss, (b) B–H curves (50 Hz magnetisation) (modified versions of Figures 5 and 7 from [68])

Table 11.3 Characteristics of the cut edge region of an un-annealed NO electrical steel at different nominal peak flux densities [71]

Nominal flux density B_p (T)	Distance from cut edge at which B_p has dropped by 10% (mm)	B_p at cut edge (T)	Approximate % of strip value (%)
1.49	0.5	1.35	87
1.17	3.0	0.8	68
0.91	5.0	0.4	50

inaccurate so the values given in the table should be treated with caution. The distance from the edge, where a detectable drop of 10% in flux density occurs, is effectively the length of the cut edge zone. It can be seen that it increases substantially with reducing flux density. At a high field the domains oriented along easy directions close to the normal to the plane of a sheet can switch relatively easily to easy directions close to the field direction so the flux density builds up here and the length of the region affected by the cut edge stress is reduced.[25] The lamination was not annealed so elastic and plastic zones are present. At high flux density the flux has only fallen to 90% of the nominal strip value. This is expected since the permeability of the bulk of the strip is low. Conversely, at low flux density, the permeability of the bulk of the steel is higher, so far more flux is moved out of the stressed area.

The stress distribution, hence degradation in magnetic properties, is a function of the deformation near the cut edge. This, in turn, may be affected by the surface hardness and the grain size of the steel. A comprehensive study of cutting low, medium and high Si content NO steels with grain size of 15 μm to 150 μm concluded that the significant variation of the degradation from material to material depends far more on the grain size than the silicon content [72].

What appear to be conflicting findings on the effect of a cut edge have been widely reported in the past [69]. The difficulty in comparing results from different authors is similar to that encountered in assessing conclusions from studies of parameters such as magnetostriction and rotational magnetisation. The problem is that results are reported from many different materials using different techniques with unquantified validations often with insufficient information to make direct comparisons. Although experimental and analytical techniques have greatly advanced in recent times, quantified data should still be treated accordingly.

11.10.3 Other cutting methods

Mechanical cutting is most effective for large scale production, but the cost of producing a punching tool is prohibitive for short runs, or prototyping of complex shapes. In these circumstances other methods such as YAG or CO_2 laser cutting, water jet cutting, wire cutting, electrical discharge machining and even chemical etching can be used. Wire cutting has least impact but it is only convenient for very small scale applications. Many comparisons of the effectiveness of these techniques have been reported. All cause some degradation of the magnetic properties [73,74].

[25]This implies that the cut edge region quantified by the fall in flux density is less than the actual stressed region, particularly at high magnetising field. If it were possible to measure the loss variation from the cut edge towards the undamaged part of the lamination, we would find the non-uniformity extending further from the edge than the flux density variation. This is not often recognised in publications quantifying the cut edge zone, often incorrectly. The inconsistency is of no consequence because of the large number of factors which influence the effect, approximations made in its measurement and interpretation in general.

Laser cutting is a popular alternative to mechanical punching for small scale production. It is slow and can cause more harm to magnetic properties due to the localised heating which has a different effect on the properties of the cut edge region than mechanical cutting [75]. However, an appropriate choice of the laser source and processing conditions can produce more consistent properties, avoiding the potential variability due to tool wear in mechanical cutting [77]. The optimum choice to be compatible with a specific material can lead to no more damage than the best mechanical treatment [75–78].

The cutting method is very important when preparing laboratory specimens, such as Epstein strips. The stress produced by laser cutting can create stress patterns over the bulk of the surface area of the strip which, in turn, has a strong detrimental effect on the *B–H* characteristics and the stress sensitivity of magnetostriction [63]. The stress distribution even depends on the direction of the cut relative to the RD of GO steel. Fortunately, the stress patterns disappear after heat treatment and the only harmful effect which remains is the plastic damage to the Goss structure close to the edge of a strip.

11.10.4 Modelling the effect of the cut edge effect

Modelling the effect of the cut edge on magnetic properties is important in computational studies of the flux distribution in a magnetic core, particularly if the degraded region is large in relation to the critical core dimensions. If it is not taken into account near stator lamination teeth, the computed prediction of the flux density, hence the overall performance of the machine, could be seriously in error.

The essential quantity to input to an electromagnetic field solver is a realistic term to represent the variation of the *B–H* relationship in the damaged area. No method has yet been developed to theoretically predict its localised variation. A simple, approximate way of dealing with the problem is to assign a relative permeability of unity to the whole cut edge region based on a typical plastically deformed length of 1 mm and an un-annealed elastic deformation of 10 mm. This is only loosely based on the true damage but it is simple to apply provided the damaged area is small compared to the critical dimensions of the core being investigated.

A slightly more accurate method is to assume a linear decrease of permeability across the cut edge zone. The actual variation, such as shown in Table 11.3, could be used to make this more relevant for a given material. We are not aware how well these simple approaches work in practical electrical machine models.

Much effort has been put into producing more complex mathematical expressions for the flux density variation in the cut edge region. These are mainly based on measured or computed permeability distributions to produce the *B–H* data needed in the computational analysis [79–84] or on models for the *B–H* variation based on magneto-mechanical coupled modelled [85–87]. It is beyond the scope of this book to discuss these approaches. It is sufficient to say that the cut edge problem is being shown to have a larger influence on machine performance than previously believed so there is a growing drive to produce more effective models and ways of quantifying their accuracy in practice [88].

11.10.5 Shrink fitted stator cores

The magnetic properties of non-oriented steels can be substantially degraded during lamination punching and assembly into rotating machine stator and rotor cores. The stator laminations of many types of small industrial motors are partly secured by the common *shrink fitting* process. An aluminium frame is simply heated to a necessary temperature and the cooler stator lamination pack, at ambient temperature, is placed into the hot frame to produce a tight fit when the complete assembly cools to ambient temperature.

Manufacturers use various methods to carry out this operation. Ideally the process should set up an interference fit between the frame and the stator laminations partly to secure the laminations against high starting torques or locked rotor conditions.[26] Harmful compressive stress of up to more than 100 MPa is claimed to be possible in localised regions of shrink fitted stator cores [89]. In other cases the permeability of NO steel has been shown to fall by 50% or more and the iron loss increase by 30% [90].

The effect of shrink fitting can be simulated in a stator pack by applying uniform hydraulic pressure to its outer surface and magnetising the yoke as it would be magnetised as in a three-phase induction motor. Some results of such an exercise using a stator core assembled from 165 mm outer diameter stator laminations cut from a low quality grade of NO steel are shown in Figure 11.33 [91]. This shows the effect of radial compressive stress on the local loss measured on individual laminations behind slots and teeth in positions shown using the initial rate of temperature technique.[27] It seems significant that the localised loss increases sharply at a stress of around 2.0 MPa to 2.5 MPa. This is probably due to the onset of the stress domain patterns in this steel. Obviously, if a lower stress is sufficient to fulfil the purpose of shrink fitting, it should be used although it is likely that this would not be adequate in high power density machines.

It will be noticed in Figure 11.33 that the loss behind the teeth is lower than behind the slots. This is due to larger components of rotational flux being present behind the teeth and mainly transverse flux behind the slots as is shown in Section 6.10 of Volume 2 of this book. The effect is summarised in Table 11.4. The ranges given indicate the spread of loss measurements behind different teeth and slots. There is only a small difference in ellipticity ratio between the stressed and unstressed state so only values for the unstressed state are shown.

Points to note from the results in Table 11.4 are:

- a radial stress of 2.7 MPa causes a substantial increase in stator loss using this particular material;
- there appears to be some correlation between the localised loss and the degree of rotational flux and
- the increase in loss due to stress is highest in the central region of the lamination where the stress and rotational flux are high.

[26]The locked rotor test is briefly described in Section 6.17 of Volume 2.

[27]See Section 11.5 of Volume 2 for a description of this technique.

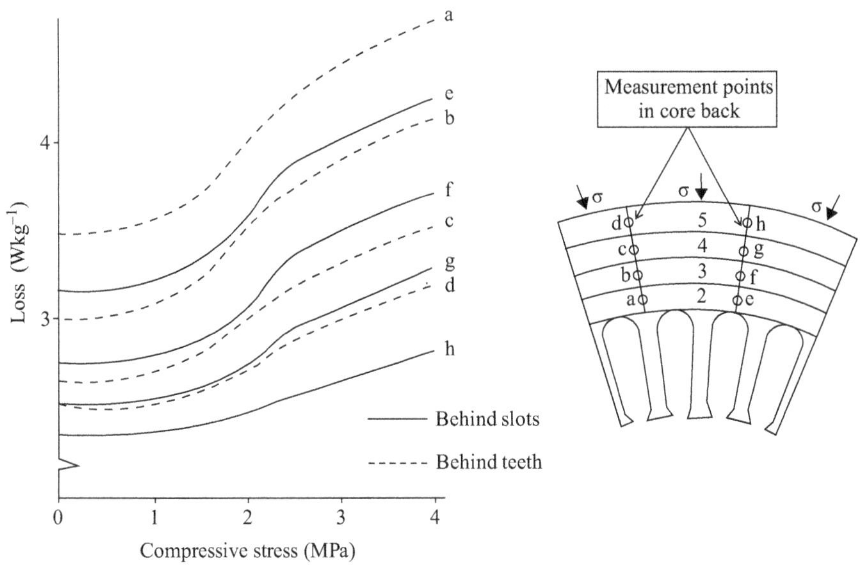

Figure 11.33 Variation of localised loss behind teeth and slots of a 165 mm pack of NO steel laminations magnetised as in a three-phase induction motor and subjected to uniformly distributed radial compressive stress at a back-iron magnetisation of 1.0 T, 50 Hz (Modified version of Figure 3 in [91])

Table 11.4 Variation of localised loss and in-plane flux density behind teeth and slots of the NO laminations in stress-free and stressed state at back-iron magnetisation of 1.0 T, 50 Hz (data extracted from [91])

		Loss (Wkg^{-1})				Ellipticity ratio[a] of flux density			
Position		**2**	**3**	**4**	**5**	**2**	**3**	**4**	**5**
Unstressed	Behind teeth	3.5–3.8	3.0–3.4	2.7–3.0	2.5–2.7	0.37	0.33	0.16	0
	Behind slots	3.2–3.8	2.8–3.4	2.5–3.0	2.4–2.7	0.23	0.20	0.13	0
2.74 MPa compression	Behind teeth	4.5–4.7	4.0–4.2	3.5–3.9	3.2–3.5	Same as unstressed			
	Behind slots	4.2	3.7–4.0	3.3–3.6	2.8–3.1				
Percentage increase due to stress		23	26	28	23				

[a]The ellipticity ratio is the ratio of the radial to circumferential components of the first harmonic of the flux density in this context. The higher the ratio, the more circular is the spatial locus of the flux density.

The results can be extrapolated to other materials, flux density and core dimensions provided the relative rotational flux and internal stress distribution are taken into account. However, the authors are not aware of any quantified published results.

Stress due to shrink fitting experimentally simulated in a three-phase induction motor geometry shows the compressive stress can produce only a small change in the flux distribution resulting in a small increase in rotational loss [92]. This broadly verifies the more detailed analysis in [91].

Computer simulations of the increase in iron loss due to change in depth of the shrink ring for a laboratory scale machine stator core with 120 mm outer diameter NO electrical steel stator laminations are compared with an experimental value obtained without a ring in Table 11.5 [93]. 2D computational stress analysis was carried out and changes of *B–H* loop shape due to localised stress throughout the stator core were used in electromagnetic FEA to obtain the losses. The wider shrink fit ring produces higher computed loss although, as in [92], no significant change in flux distribution due to shrink fitting was found. The computed loss due to shrink fitting is 25% to 35% for different width rings and the experimental finding was around 15% to 20%[28] [93]. It is interesting to note that in this case computed percentage increase in loss due to punching is far higher than that due to shrink fitting.

The results in Table 11.5 can be compared with simulations of total loss before and after shrink fitting shown in Table 11.6. This shows results of a 2D static stress analysis combined with a 2D field analysis of the effect of shrink fitting of an induction motor stator into an aluminium housing. The measured loss of the motor with the housing in place is far higher than the computed value which is not surprising given the difficulty in estimating the true and computed shrink fit interferences as well as the normal difficulties in accounting for all the loss sources in a computer model. In spite of this, it is interesting to note that shrink fitting changes the no-load loss distribution, particularly caused by an increase in the stator copper loss [93]. This additional harmful effect of shrink fitting seems to have been mainly overlooked in many studies; its significance needs to be verified.

Table 11.5 Computed iron loss of a simulated stator core showing effect of punching and shrink fitting [93]

	Loss (W kg^{-1})				
	No shrink fit Simulation	**10 mm shrink fitting ring**	**20 mm shrink fitting ring**	**No punching stress**	**Punching stress**
Measured		–	–	2.35	3.51
Computed	2.19	2.75	2.96	2.21	3.19

[28]Material properties, stress levels and core – back flux density are not fully quantified in [95], so these cannot be directly compared but they show a level of agreement expected.

Table 11.6 Calculated components of loss in a 750 W four pole induction motor with and without shrink fitting stress (losses are normalised with respect to total loss without stator housing) [94]

	Stator Cu loss (%)	**Stator Fe loss (%)**	**Rotor Fe loss (%)**	**Stator housing loss (%)**	**Total computed loss (%)**	**Measured on motor (%)**
Without housing stress	53.4	34.6	9.5	2.5	100	–
Including housing stress	60.3	41.3	9.5	9.8	119.9	147

It is interesting to note that the overall loss increases due to shrink fitting shown in Tables 11.5 and 11.6 are comparable despite the fact that entirely different computational techniques were used and the shrink fit parameters are probably quite different.

References

[1] Corner W.D., and Mason J.J. 'The effect of stress on the domain structure of Goss textured silicon-iron'. *Br. J. Appl. Phys.* 1964, vol. 15, pp. 709–718

[2] Dijkstra L.J., and Martius U.M. 'Domain patterns in silicon-iron under stress'. *Rev. Mod. Phys.* 1953, vol. 25(1), pp. 146–149

[3] Swift W.M., and Wolfe G.F. 'Influence of compressive stress on magnetic properties of a (110)[001] single crystal of 3% Si-Fe'. *IEEE Trans. Magn.* 1976, vol. 12(3), pp. 244–248

[4] Iwata K., Ishiyama K., Suzuki M., *et al.* 'Quantitative analysis of 90° closure domains occurring by compressive stress in Fe3%Si(110) steels'. *IEEE Trans. Magn.* 2014, vol. 50(11), Art. No. 2007004

[5] Makarov V.P., Molotilov B.V., Moskvin A.S., Rytvin V.M., and Gorlach I.A. 'Effect of external stresses on the domain structure near inclusions in Fe-3% Si crystals'. *Bull. Acad. Sci. USSR Phys. Ser.* 1975, vol. 39(7), pp. 46–49

[6] Langman R. 'Magnetic properties of mild steel under conditions of biaxial stress'. *IEEE Trans. Magn.* 1990, vol. 26(4), pp. 1248–1251

[7] Moses A.J., and Phillips P.S. 'Some effects of stress in Goss-oriented silicon-iron'. *IEEE Trans. Magn.* 1978, vol. 14(5), pp. 353–355

[8] Banks P.J. 'Dynamic magnetostriction and mechanical strain in oriented 3% silicon-iron sheet subject to combined longitudinal and transverse stresses'. *Proc. IEE* 1967, vol. 114(10), pp. 1537–1546

[9] Aydin U., Rasilo P., Martin F., *et al.* 'Modeling the effect of multiaxial stress on magnetic hysteresis of electrical steel sheets: a comparison'. *IEEE Trans. Magn.* 2017, vol. 53(6), Art. No. 2000904

[10] Hubert O. 'Influence of biaxial stresses on the magnetic behaviour of an iron-cobalt sheet – experiment and modelling'. *Prezeglad Elektrotechniczny* 2007, vol. 83(4), pp. 70–77

[11] Zaikova V.A., and Shur I.S. 'Change in the magnetic structure of silicon iron crystals under the influence of elastic stress'. *Bull. Sci. USSR* 1958, vol. 22, pp. 1177–1182

[12] Kirenskii L.V., Savchenko M.K., and Rodichev A.M. 'Domain structure dynamics in transformer iron crystals under the influence of stresses'. *Izv. Akad. Nauk SSSR. Ser. Fiz.* 1958, vol. 22, pp. 1171–1176

[13] Bichard V.M., Coombs A., and Daniels M.R. 'The effect of applied tensile stress on the domain structure of grain oriented Si-Fe'. *IEEE Trans. Magn.* 1978, vol. 14(5), pp. 782–784

[14] Houze G.L. 'Effect of longitudinal tensile stress on the static and 60 Hz domain structures of grain-oriented silicon steel'. *J. Appl. Phys.* 1969, vol. 40(3), pp. 1090–1091

[15] Shilling J.W. 'Domain structure during magnetization of grain-oriented 3% Si-Fe as a function of applied tensile stress'. *J. Appl. Phys.* 1971, vol. 42(4), pp. 1787–1789

[16] Swift W.M., and Wolfe G.F. 'Influence of compressive stress on magnetic properties of a (110)[001] single crystal of 3% SiFe'. *IEEE Trans. Magn.* 1974, vol. 12(3), pp. 244–248

[17] Moses A.J., and Davies D. 'Influence of compressive stress on magnetic properties of commercial (110) [001] oriented silicon-iron'. *IEEE Trans. Magn.* 1980, vol. 16(2), pp. 454–460

[18] Anderson P.A. *Internal Report*. Wolfson Centre for Magnetics, Cardiff University.

[19] Moses A.J. 'Effects of applied stress on magnetic properties of high permeability silicon-iron'. *IEEE Trans. Magn.* 1979, vol. 15(6), pp. 1575–1579

[20] Moses A.J. 'Effects of stress on the magnetic properties of grain-oriented silicon-iron magnetised in various directions'. *IEEE Trans. Magn.* 1981, vol. 17(6), pp. 2872–2874.

[21] Basak A., and Moses A.J. 'Influence of stress on rotational loss in silicon iron'. *Proc. IEE* 1978, vol. 125(2) pp. 165–168

[22] Krell C., Baumgartinger N., Krismamic G., Leiss E., and Pfützner H. 'Stress effects on the multidirectional magnetic behaviour of grain-oriented silicon iron sheets'. *J. Magn. Magn. Mater.* 2000, vol. 215–216, pp. 63–65.

[23] Bengtsson C., and Hörnfeldt S. 'Field and stress dependence of the magnetization in a ferromagnetic polycrystalline materials of cubic structure'. *J. Appl. Phys.* 1982, vol. 53(11), pp. 8311–8313

[24] Shirkoohi G.H., and Moses A.J. 'Effect of stress on iron loss and apparent power in low silicon non-oriented electrical steel'. *Anales de Fisica* 1990, Ser. B 86, pp. 265–267

[25] Senda K., Fujita A., Honda A., Kuroki N., and Yagi M. 'Magnetic properties and domain structures of nonoriented electrical steel under stress'. *Electr. Eng. Jpn.* 2013, vol. 182(4), pp. 10–18

[26] LoBue M., Sasso C., Basso V., Fiorillo F., and Bertotti G. 'Power losses and magnetization process in Fe-Si non-oriented steels under tensile and compressive stress'. *J. Magn. Magn. Mater.* 2000, vol. 215–216, pp. 124–126

[27] Oda Y., Toda H., Shiga N., Kasai S., and Hiratani T. 'Effect of Si content on iron loss of electrical steel sheet under compressive stress'. *IEEE Trans. Magn.* 2014, vol. 50(4), Art. No. 2002304

[28] Hubert A., and Schäfer R. *Magnetic Domains: The Analysis of Magnetic Microstructures*. 3rd edn. Springer, Berlin, Germany, 2009, p. 427

[29] Moses A.J. 'Effects of stresses on magnetic properties of silicon-iron laminations'. *J. Mater. Sci.* 1974, vol. 9, pp. 217–222

[30] Cole R.W. 'Effect of elastic bending on magnetic properties of oriented silicon iron'. *J. Appl. Phys.* 1958, vol. 29(3), pp. 370–37

[31] Pfützner H., Futschik K., and Luo Y. 'Effects of bending on G.O. silicon iron sheets'. *IEEE Trans. Magn.* 1982, vol. 18(6), pp. 1499–1501

[32] Perryman R. 'The effects of bending stress on 3% grain-oriented silicon-iron'. *J. Phys D. Appl. Phys.* 1974, vol. 7, pp. 1047–1052

[33] Saito A., Yamamoto T., and Iwasaki H. 'Magnetization properties and domain structures of grain-oriented silicon steel sheets due to bending stress'. *IEEE Trans. Magn.* 2000, vol. 36(5), pp. 3078–3080

[34] Banks P.J. 'Effect of magnetostriction on the elastic properties of Goss-oriented Si-Fe sheets subject to bending'. *IEEE Trans. Magn.* 1972, vol. 8 (2), pp. 205–211

[35] Banks P.J. 'Anticlastic curvature in anisotropic beams'. *Aeron. J.* 1974, vol. 78(767), pp. 525–528

[36] Pfützner H., Bengtsson C., and Zehetbauer M. 'Effects of torsion on domains of coated Si-Fe sheets'. *IEEE Trans. Magn.* 1984, vol. 20(1), pp. 1554–1556

[37] Eingorn Y. Ya. 'The effect that anisotropy of the magnetic properties and the pressure normal to the plane of the sheet has on the electromagnetic characteristics of transformer iron'. *Elektrotechica* 1970, vol. 3, pp. 53–54

[38] Eingorn Y. 'Influence of pressure and deformation on magnetic properties of transformer steel'. *Electricity (Russia)* 1965, vol. 2 (in Russian)

[39] Joslin P.C., Moses A.J., and Thompson J.E. 'Some aspects of the effects of longitudinal and normal stress on the power loss and flux distribution within the core and corners of a single phase transformer core'. *Proc. IEE* 1972, vol. 119(6), pp. 709–716

[40] Bengtsson C., Pfützner H., and Hoffman K. 'Domain modification due to compression perpendicular to the rolling direction of coated Hi-B sheets'. *IEEE Trans. Magn.* 1984, vol. 10(6), pp. 2116–2119

[41] Lykov E.L., Starodubtsev Yu.N., and Druzhinin V.V. 'Influence of surface state of sheets of anisotropic electrical steels on the sensitivity of magnetic losses to normal pressure'. *Elektrotekhnika* 1979, vol. 50(7), pp. 38–40

[42] Yabumoto K., and Yanase S. 'Magnetic properties of non-oriented electrical steels under compressive stress normal to their surface'. *Przeglad Elektrotechniczny* 2011, vol. 87(9b), pp. 97–100

[43] Miyagi D., Maeda N., Ozeki Y., Miki K., and Takahashi N. 'Estimation of iron loss in motor core with shrink fitting using FEM analysis'. *IEEE Trans. Magn.* 2009, vol. 45(3), pp. 1704–1707

[44] Rasilo P., Steentjes S., Belahcen A., Kouhia R., and Hameyer K. 'Model for stress-dependent hysteresis in electrical steel sheets including orthotropic anisotropy'. *IEEE Trans. Magn.* 2017, vol. 53(6), Art. No. 2001004

[45] Foster K. 'Stress and induction dependence of hysteresis losses in electrical steels'. *J. Appl. Phys.* 1984, vol. 55(6), pp. 2127–2129

[46] Hagihara H., Takahashi N., Fujiwara K., Ishihara Y., and Masuda T. 'Magnetic properties evaluation of grain-oriented electrical steel sheets under bending stress'. *IEEE Trans. Magn.* 2014, vol. 50(4), Art. No. 2002104

[47] Singh D., Rasilo P., Martin F., Belahcen A., and Arkkio A. 'Effect of mechanical stress on excess loss of electrical steel sheets'. *IEEE Trans. Magn.* 2015, vol. 51(11), Art. No. 1001204

[48] Walker J.H., Rogers G.J., and Jackson R.L. 'Pressing and clamping laminated cores'. *Proc. IEE* 1964, vol. 111(3), pp. 565–577

[49] Budlynas R.G., and Nisbett K.J. *Shigley's Mechanical Engineering Design*, 9th edn. McGraw-Hill, New York, 2011

[50] Ray D., Gnanamoorthy R. 'Friction and wear behaviour of vinylester resign matrix composites filled with fly ash particles'. *J. Reinf. Plast Compos.* 2007, vol. 26, pp. 5–13

[51] Phophongviwat T. 'Investigation of the Influence of Magnetostriction and Magnetic Forces on Transformer Core Noise and Vibration'. *PhD Thesis*. Cardiff University, UK, 2013

[52] Girgis R.S., teNijenhuis E.G., Gramm K., and Wrethag J.-E. 'Experimental investigation on effect of core production attributes on transformer loss performance'. *IEEE Trans. Magn.* 1998, vol. 13(2), pp. 526–531.

[53] McQuade F., Moses A.J., and Ling P.C.Y. 'Correlation of internal stress and power loss with domain motion in strip wound magnetic cores'. *J. Magn. Magn. Mater.* 1994, vol. 133, pp. 136–139

[54] Abdul Qader A.A.A. 'Flux and Power Loss in a 3 Phase Distribution Transformer'. *PhD Thesis*, University of Wales, 1982

[55] Carlberg P.M., and Lewin B. 'The influence of edge effects and stress on losses'. *ERA Technology Report* N089-0278, *Conference on "Modern Magnetic Materials"*, Session 4, Paper 2, London, 1989, pp. 4.2.1–4.2.10

[56] teNyenhuis E., and Girgis R. 'Effect of slitting electrical core steel on measured iron loss'. *J. Magn. Magn. Mater.* 2000, vol. 215–216, pp. 110–111

[57] Beckley P. *Electrical Steels for Rotating Machines*. The Institution of Electrical Engineers, London, 2002

[58] IEC 60404-8-7 (1997) Magnetic materials – Part 8-7: Specifications for individual materials – Cold-rolled grain-oriented electrical steel strip and sheet delivered in the fully processed state

[59] Schmidt K.H. 'Influence of punching on the magnetic properties of electrical steel'. *J Magn. Magn. Mater.* 1976, vol. 2, pp. 136–150

[60] Omura T., Zaizen Y., Fukumura M., Senda K., and Toda H. 'Effect of hardness and thickness of nonoriented electrical steel sheets on iron loss deterioration by shearing process'. *IEEE Trans. Magn.* 2015, vol. 51(11), Art. No. 2005604

[61] Emura M., Landgraff F.J.G., Ross W., and Barreta J.R. 'The influence of cutting techniques on the magnetic properties of electrical steels'. *J. Magn. Magn. Mater.* 2003, vol. 254–255, pp. 358–360

[62] Klimczyk P., Anderson P., Moses A.J., and Davies M. 'Influence of cutting techniques on magnetostriction under stress of grain oriented electrical steel'. *IEEE Trans. Magn.* 2012, vol. 48(4), pp. 1417–1420

[63] Wang Z., Shuhui L., Ronggao C., Wang X., and Wang B. 'Influence of grain size and blanking clearance on magnetic properties deterioration of non-oriented electrical steel'. *IEEE Trans. Magn.* 2018, vol. 54(5), Art. No. 2000607

[64] Harstick H.M.S., Ritter M., and Riehemann W. 'Influence of punching and tool wear on the magnetic properties of nonoriented electrical steel'. *IEEE Trans. Magn.* 2014, vol. 50(4), Art. No. 6200304

[65] Takezawa M., Kitajima K., Morimoto Y., Yamasaki J., and Kaido C. 'Effect of strain by mechanical punching on nonoriented Si-Fe electrical sheets for a nine-slot motor core'. *IEEE Trans. Magn.* 2006, vol. 42(10), pp. 2790–2792

[66] Godec Z. 'Influence of slitting on core losses and magnetization curve of grain-oriented electrical steels'. *IEEE Trans. Magn.* 1977, vol. 13(4), pp. 1053–1057

[67] Moses A.J., Derebasi N., Loisos G., and Schoppa A. 'Aspects of the cut edge effect stress on the power loss and flux density distribution in electrical steel sheets'. *J. Magn. Magn. Mater.* 2000, vol. 215–216, pp. 690–692

[68] Tamaki T., Fujisaki K., and Yamada T. 'Hole effect on magnetic field uniformity in single sheet tester'. *Prezeglad Elekrotechniczny (Electric. Rev.)* 2009, vol. 85(1), pp. 71–73

[69] Schoppa A., Scheinder J., and Roth J.-O. 'Influence of the cutting process on the magnetic properties of non-oriented electrical steels'. *J. Magn. Magn. Mater.* 2000, vol. 215–216, pp. 100–102

[70] Chiang C.-C., Knight A.M. Hsieh M.F., *et al.* 'Effects of annealing on magnetic properties of electrical steel and performances of SRM after punching'. *IEEE Trans. Magn.* 2014, vol. 50(11), Art. No. 8203904

[71] Nakata T., Nakano M., and Kawahara K. 'Effects of stress due to cutting on magnetic characteristics of silicon steel'. *IEEE Trans. Magn. Jpn.* 1992, vol. 7(6), pp. 453–457

[72] Rygal R., Moses A.J., Derebasi N., Schneider J., and Schoppa A. 'Influence of cutting on magnetic field and flux density distribution in non-oriented electrical steels'. *J. Magn. Magn. Mater.* 2000, vol. 215–216, pp. 687–689,

[73] Hoffmann M., Naumoski H., Herr U., and Herzog H.-G. 'Magnetic properties of electrical steel sheets in respect of cutting: micromagnetic analysis and macromagnetic modeling'. *IEEE Trans. Magn.* 2016, vol. 52(2), Art. No. 2000114

[74] Kurosaki Y., Mogi H., Fujii H., Kubota T., and Shiozaki M. 'Importance of punching and workability in non-oriented electrical steel sheets'. *J. Magn. Magn. Mater.* 2008, vol. 320, pp. 2474–2480

[75] Baudouin P., Belhadj A., Breaban F., Deffontaine A., and Houbaert Y. 'Effects of laser and mechanical cutting modes on the magnetic properties of low and medium Si content nonoriented electrical steels'. *IEEE Trans. Magn.* 2002, vol. 38(5), pp. 3213–3215

[76] Loisos G., and Moses A.J. 'Effect of mechanical and Nd: YAG laser cutting on magnetic flux distribution near the cut edge of non-oriented steels'. *J. Mater. Process. Tech.* 2003, vol. 161, pp. 151–155

[77] Dickmann K. 'Influence of the laser cutting process on the magnetic properties of electrical sheets'. *Anales de Fisica* 1990, vol. 86, pp. 82–84

[78] Siebert R., Schneider J., and Beyer E. 'Laser cutting and mechanical cutting of electrical steels and its effect on magnetic properties'. *IEEE Trans Magn.* 2014, vol. 50(4), Art. No. 2001904.

[79] Schoppa A., Scheinder J., and Wuppermann C.-D. 'Manufacturing steps of electromagnetic components and their influence on the magnetic properties'. UK Magnetics Society, Seminar: *Stress, Magnetostriction and Losses in Magnetic Materials*, Newport, Gwent, UK, 23rd February 2000

[80] Loisos G., and Moses A.J. 'Variation of magnetic flux density with distance from the cut of non-oriented electrical steel'. *Stahleisen* 2004, pp. 317–322

[81] Bali M., De Gersem H., and Muetze A. 'Finite-element modeling of magnetic material degradation due to punching'. *IEEE Trans. Magn.* 2014, vol. 50(2), Art. No. 7018404

[82] Lazari P., Atallah K., and Wang J. 'Effect of laser cut on the performance of permanent magnet assisted synchronous reluctance machines'. *IEEE Trans. Magn.* 2015, vol. 51(11), Art. No. 8114305

[83] Crevecoeur G., Sergeant P., Dupré L., Vandenbossche L., and Van de Walle R. 'Analysis of the local materials degradation near cutting edges of electrical steel sheets'. *IEEE Trans. Magn.* 2008, vol. 44(11), pp. 3173–3176

[84] Toups T., Lorenz S., Kimmich R., and Dopprlbauer M. 'FEM investigation of magnetic properties of electrical steels influenced by cutting process'. *Proceedings of the 3rd International Conference* on Magnetism and Metallurgy, Ghent, Belgium, 4th–6th June 2008, pp. 273–296

[85] Ossart F., Hug E., Hubert O., Buvat C., and Billardon R. 'Effect of punching on electrical steels: experimental and numerical coupled analysis'. *IEEE Trans. Magn.* 2000, vol. 36(5), pp. 3137–3140

[86] Hubert O., and Hug E. 'Influence of plastic strain on magnetic behaviour of non-oriented Fe-3Si and application to manufacturing test by punching'. *Mater. Sci. Technol.* 1995, vol.11, pp. 482–487

[87] Chiang C.-C., Hsieh M.-F., Li Yu-H., and Tsai M.-C. 'Impact of electrical steel punching process on the performance of switched reluctance motors'. *IEEE Trans. Magn.* 2015, vol. 51(11), Art. No. 8113304

[88] Linder M., Kluge T., and Lippmann M. 'Determination of the cutting influence on soft magnetic materials and its effect in electrical machines: Part 2: Consideration of the cutting influence in magnetic calculations'. *MagNews* 2017, issue 4, pp. 20–22

[89] Miyagi D., Miki K., Nakano M., and Takahashi N. 'Influence of compressive stress on magnetic properties of laminated electrical steel'. *IEEE Trans. Magn.* 2010, vol. 46(2), pp. 318–321

[90] Takahashi N., Morimoto H., Yunoki Y., and Miyagi D. 'Effect of shrink fitting and cutting on iron loss of permanent magnet motor'. *J. Magn. Magn. Mater.* 2008, vol. 320, pp. e925–e928

[91] Moses A.J., and Rahmatizadeh H. 'Effects of stress on iron loss and flux distribution of an induction motor stator core'. *IEEE Trans. Magn.* 1989, vol. 25(5), pp. 4003–4005

[92] Enokizono M., and Fujiyama S. 'The influence of compressive stress on the magnetic flux density in the three-phase induction motor core model'. *J. Phys. IV France* 1998, vol. 8, pp. 801–804

[93] Fujisaki K., Hirayama R., Kawachi T., *et al.* 'Motor core iron loss analysis evaluating shrink fitting and stamping by finite-element method'. *IEEE Trans. Magn.* 2007, vol. 43(5), pp. 1950–1954

[94] Yamazaki K., and Fukushima W. 'Loss analysis of induction motors by considering shrink fitting of stator housings'. *IEEE Trans. Magn.* 2015, vol. 51(3), Art. No. 8102004

Chapter 12
Magnetic measurements on electrical steels

12.1 Introduction

This chapter will give an overview of the most currently used measurement methods applicable to electrical steels with a particular focus on methods which have been standardised or are in the process of standardisation. A more complete discussion of magnetic measurement techniques is available in Fiorillo's excellent *Measurement and Characterisation of Magnetic Materials* [1], Tumanski's *Handbook of Magnetic Measurements* [2] and the many sources referenced throughout this chapter.

Although termed *magnetic measurements* almost all of the methods discussed are electrical measurements from which the magnetic properties can be calculated. The most fundamental of these is commonly referred to as the *B–H* curve which characterises the response of the magnetic material to an applied field. This non-linear characteristic, seen many times in this book, describes the material in a precisely defined state. The definition and control of that state can be as important as the measurement itself since, as for example shown in Chapter 11, the properties of ferromagnetic materials including electrical steels are sensitive to a wide range of parameters. In operation, soft magnetic materials are unlikely to be subjected to the ideal conditions defined in measurement standards. Therefore, as well as considering standard measurements, this chapter also discusses methods which take other factors into account such as stress, temperature and magnetisation waveform.

Fundamentally, all of the measurements can be thought of in four distinct steps:

1. Pre-processing – calculation of the required values for measured quantities required to set the test conditions;
2. Initialise test – set up the test conditions to the prescribed state;
3. Measure – sense and record all of the required quantities and
4. Post processing – convert the measured quantities to required parameters.

It is clear that steps 3 and 4 are the most critical in terms of the accuracy and repeatability of the measurement and for soft magnetic materials the uncertainty of measurement can be very high. Establishing the required state is particularly difficult, the size and shape of the test piece can have a significant influence on its magnetisation and the method of sample preparation is similarly critical. These materials are rarely homogenous which means that the measurement area must be carefully considered to be representative of the material in use and the non-uniformity of the magnetisation conditions (influenced by both the material and the magnetisation method) should be considered.

The non-linearity of the magnetic properties adds further complications to establishing a well-defined magnetisation state (and defining what the control parameters should be) whilst the historical dependence makes the initial conditions difficult to assess.

When defining new measurement methods, all of these factors must be considered and users of magnetic measurement equipment should have an awareness of the influence of both the controlled and uncontrolled parameters on their final results.

12.2 Effect of sample geometry (toroids, single strips, rings and single sheet)

Of particular importance is the magnetic circuit and how the sample is included within that. Although electrical steels are only available in strip form, they can be cut and measured in a variety of different geometries which include:

- Epstein frame
- Single sheet
- Single Epstein
- Ring
- Toroid

This is not an exclusive list – any closed circuit with a region of uniform cross section could be used – but represents those most commonly used in industry and academia. Measurements made on any of these geometries will have intrinsic systematic errors and, as a result, correlating magnetic measurements from different geometries is extremely problematic. Care should always be taken when making such correlations that the range of test conditions and sample properties over which the correlation has been made are clearly identified.

12.2.1 Epstein frame

The most well-known and widely used of these test geometries is the Epstein frame, being the standard test upon which all of the material specifications of electrical steels are currently based. The test is described in detail in the international standard IEC 60404-2 [3] but the geometry, and later the test, will be summarised here.

The Epstein frame is a simulation of a simple transformer constructed from test strips of 30 mm width and 280–305 mm length with 4 equal limbs and double overlapped corners as shown in Figure 12.1. Typically, 24 strips are used but any multiple of 4 is valid (although less than 8 is not recommended). Due to the narrow width, the effect of the cut edge on the performance of the material is large and thus samples are nearly always subjected to a stress relief anneal. This has the further advantages of producing flatter samples and returning them to a virgin, unmagnetised, condition. However, by performing this anneal the sample is likely to be in a better state than that seen by the final customer having had any residual stresses removed. The anneal also makes it impossible to test grain oriented (GO) steels

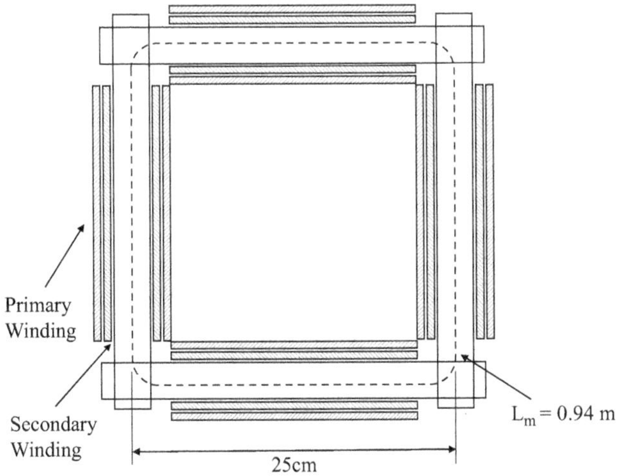

Figure 12.1 Epstein frame geometry

which have been laser domain refined since the influence of the domain refinement will have also been removed.

Each limb of the Epstein frame has a primary and secondary winding wound on to non-magnetic and non-conducting formers. The secondary winding is evenly wound onto the former and the primary over this. The four sets of windings are then connected in series. Typically 700 turns are used for both windings at power frequencies and 100 or 200 turns for medium frequencies. Frames with tapped windings are also used to provide flexibility.

Due to the fact that the formers are necessarily of a larger cross sectional area than the cross section of the test specimen a mutual inductor is used for compensation of the air flux in the winding formers. This is normally placed at the centre of the Epstein frame although linkage with the single circulatory turn of the main windings can lead to errors so it is best to have the mutual inductor remote from the main winding.

For GO materials all strips are cut in the rolling direction. For non-oriented (NO) materials half of the strips are cut parallel to the rolling direction and half in the transverse direction. When stacking NO materials two opposite limbs of the frame are stacked with material cut in the rolling direction and the other two limbs are used for material cut in the transverse direction. The Epstein frame test method has proven to be very repeatable but repeatability can be further improved by numbering the samples and stacking them identically each time they are tested and by placing 1 N weights at each corner to minimize air gaps.

The geometrical formulae for field strength, cross sectional area (required for B measurement) and effective mass (required for calculations of specific total loss and VA) are given in Formulae (12.1)–(12.3).

$$\text{Magnetic field strength, } H = \frac{N_1 I_1}{0.94} \quad \left(\text{Am}^{-1}\right) \tag{12.1}$$

$$\text{Cross sectional area, } A = \frac{m}{4\rho l} \quad (\text{m}^2) \tag{12.2}$$

$$\text{Effective mass, } m_a = \frac{0.94m}{4l} \quad (\text{kg}) \tag{12.3}$$

where N_1 = number of primary turns, I_1 = primary current (A), m = mass of pack of Epstein samples (kg), ρ = conventional density of sample (kg m^{-3}),[1] l = sample length (m) and 0.94 m is the effective magnetic path length of the Epstein frame.

The suitability of the Epstein frame has been considered by Antonelli *et al.* [4] who used finite element modelling to investigate the non-uniformity and estimate the errors in the measurement of specific total loss. They used the model to extrapolate 'true' values of magnetic parameters from the central limb area, thus neglecting the influence of the corners and showed errors in loss of between 8% and 12% which varied with frequency (between 10 Hz and 400 Hz). In addition to this, large errors at low frequency (0.1 Hz) in the *B–H* characteristics were also noted. Marketos *et al.* [5] proposed a measurement approach to the same problem utilising two Epstein frames of different limb lengths but identical corner areas which was developed from an approach by Sievert [6]. The results were presented as a variation in the mean magnetic path length but showed similar findings to Antonelli. Additionally, material permeability was shown to have a significant bearing on the path length with high permeability materials giving a longer than standard path length and low permeability materials giving a shorter than standard path length. This technique was later used by Qingyi [7] who utilised a weighted path length to take into account both the uniform limb regions and the corner loss in a more widely applicable manner.

The work described above highlights the major flaw with the Epstein frame: that its dependence on a mean path length makes it difficult to compare materials of differing permeability and anisotropy (or even to compare the same materials under different conditions where the permeability will change, e.g. flux density and frequency). There is mounting pressure to replace the Epstein frame as the standard method for classifying GO steels due to this and the unsuitability of the test to domain refined electrical steel. For that reason, it has been largely replaced in industry by the single sheet test with the results correlated to the Epstein frame. The methods and applicability of these correlations are subject to considerable discussion as the errors in each geometry independently vary with many factors. It is clear that the Epstein frame is a highly reproducible measurement with international round robin tests using transfer samples giving excellent agreement between laboratories [8]. However, the sample preparation procedure including cutting, stress relief annealing and handling can introduce much larger variations. This will be discussed further in section 12.2.2.

[1]The symbol ρ is used here since it is commonly used in SI terminology. The symbol D is used in other parts of this book D since it is widely used in other sections of the magnetics community.

12.2.2 Single sheet tester

The single strip tester (SST) offers a more rapid test than the Epstein frame eliminating both the stacking procedure and the need for a stress relief anneal (due to the sample size and the negligible influence of the cut edges). The sample and frame are described in IEC 60404-3 [9] and take the form of a 500 mm by 500 mm sample, enwrapping windings mounted on a former and a pair of flux closure yokes as shown in Figure 12.2. Single yoke systems are also seen but these have inherent systematic errors attributed to eddy current loss in the sample area around the pole face.

Primary and secondary windings are wound on a non-magnetic, former which enwraps the sample between the poles of the yokes and a mutual inductor is used as with the Epstein frame, to compensate for air flux.

The yokes are generally manufactured from a high permeability material (such as GO steel) and have an area much greater than that of the sample so that the operating flux density in the yokes is low and they offer a near zero reluctance to the flux, thereby avoiding any drop in the magneto-motive force (MMF). There are two construction methods for producing the yokes. Either they can be strip wound and then cut into two C-cores or stacked with a mitred joint to best utilise the anisotropic properties of the GO steel.

The yokes are the critical elements in minimising the systematic errors in this test and the utmost care should be taken in their preparation. In order to minimise in-plane eddy currents in the sample when flux transfers from sample to yoke, the flux transfer should be equal between the two yokes and uniform across their width (if perfectly symmetrical flux transfer is achieved the in-plane eddy currents will cancel to zero). This requires the pole faces to be flat and coplanar which is usually achieved by machining the pole faces to a fine tolerance. These faces must then be acid etched to remove short circuits between the laminations introduced by the machining in order to minimise eddy currents in the yokes. The importance of the inter-laminar resistance of the yoke is highlighted in a Chinese study [8] which shows that the error in the measured loss increases with conductivity measured at the yoke faces.

With large yokes, such as in this frame, they are usually counterbalanced to provide only a 'kissing' force when in contact with the sample although Drake [10] showed the effect of this pressure was negligible. Practically, however, these yokes are heavy and require a mechanism for safely raising and lowering the top yoke with many modern systems utilising an interlocked pneumatic system.

The conventional path length is defined to be the distance between the pole faces (450 mm) but, as with the Epstein frame, this will vary with the relative reluctance of the sample and yokes.

Even with well-prepared yokes, the measured magnetic properties of the material will be influenced by both their reluctance and their loss. This will give an overestimate of losses (up to 4% of the measured loss could be contributed by the yokes [11]) and an underestimate of permeability.

The increased acceptance of the SST by industry due to improved testing efficiency and wider applicability required a correlation with the Epstein frame

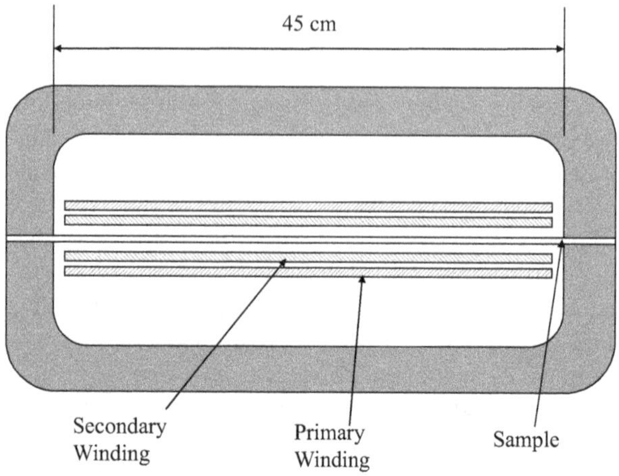

Figure 12.2 SST geometry

upon which electrical steel specifications are based. As noted previously, the two methods have fundamental differences in the homogeneity of the magnetic circuit and it is not possible to define a simple general correlation factor for all materials and magnetising conditions whilst a complex model, which compensates for the systematic errors in both arrangements, would reduce transparency. Detailed intercomparisons between them using matched sample pairs highlighted the range of correlation factors resulting in a wide variety of factors being adopted by manufacturers and users of electrical steels. At the time of publishing, detailed discussions are taking place within IEC TC68 to establish a single correlation factor giving complete transparency for the user.

The basic formulae governing the use of the SST are given in (12.4)–(12.6):

$$\text{Magnetic field strength, } H = \frac{N_1 I_1}{l_m} \quad \left(\text{Am}^{-1}\right) \tag{12.4}$$

$$\text{Cross sectional area, } A = \frac{m}{\rho l} \quad \left(\text{m}^2\right) \tag{12.5}$$

$$\text{Effective mass, } m_a = \frac{l_m m}{l} \quad (\text{kg}) \tag{12.6}$$

where N_1 = number of primary turns, I_1 = primary current (A), m = mass of the sample (kg), ρ = conventional density of sample (kg m^{-3}), l = sample length (m) and l_m = mean magnetic path length of the SST (m).

Many other sizes of SST are in use within magnetics laboratories which are not defined by international standards. Most common is the single Epstein tester (SET) whilst a range of square sheet testers between 100 mm and 400 mm are seen in published work mostly for their applicability to available samples.

Also common is variations in the H sensing methodology with incorporation of H-coils (see Section 12.3.2) perhaps the most common. These remove the influence

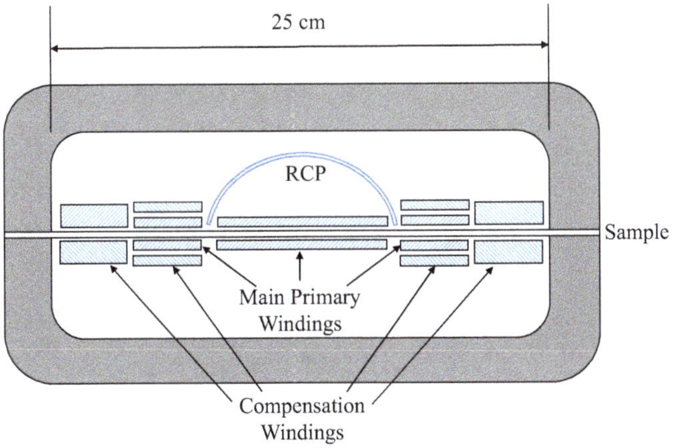

Figure 12.3 CSST schematic adapted from [13]

of the yokes in the measured properties but, historically, have not been used in industry due to their fragile nature. Modern printed circuit board (PCB) manufacturing techniques are used for the fabrication of large H-coils for SSTs, which are highly reproducible and relatively rugged, which should increase their application. A single H-coil offers significant improvements over the magnetising current approach whilst two H-coils would see further improvements [12]. The use of two (or more) H-coils allows extrapolation of the measured magnitude to that on the surface of the sample if (in the case of two coils) it is assumed that the field gradient is linear over these short distances.

An alternative to using an H-coil in this situation is to retain the use of the magnetising current for calculation of applied magnetic field but use a set of supplementary windings to compensate for the effect of the yokes, air gaps and variation of the sample properties. A compensated single strip tester (CSST) was described by Nafalski *et al.* [13] and is shown in schematic in Figure 12.3. The CSST makes use of a magnetic potentiometer (see Section 12.3.2) on the central section of the test sample whose output drives the compensation windings via a high gain amplifier. All other aspects of this tester remain identical to the conventional SST. Since the magnetic potentiometer is seeking a null value it does not require calibration, however it is important that it is uniformly wound and that its ends are in close proximity to the sample surface. The relative complexity of this arrangement means that it is not widely used currently, although its advantages could make it well suited to measurements of very soft materials such as iron-based amorphous strip.

The measurement of iron-based amorphous strip is more challenging than that of electrical steels due to its small cross sectional area, high permeability and low coercivity. Measurement methods generally take the same form as the SST for electrical steels with ASTM A932/A932M [14] describing a 300 mm SST test with the magnetising current being used for the determination of the magnetic field

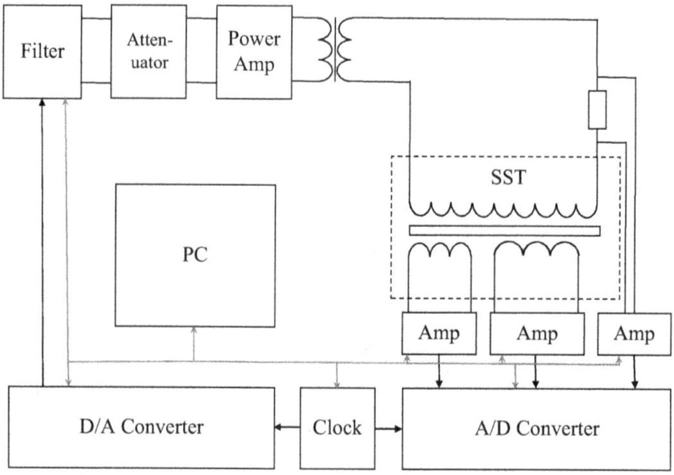

Figure 12.4 ASST schematic adapted from [15]

strength. The principal differences are that this standard utilises a single sided yoke and permits multiple strips in order to reduce the error in the cross sectional area (and therefore in flux density setting) and increases the signal amplitude in the secondary winding.

The use of multiple strips in any SST type test adds complexity to the magnetic circuit resulting in a non-uniformity of flux density through the height of the stack (which will change with peak flux density) and an increase in the non-uniformity along the length of the stack (and thus an increase in the error of the mean magnetic path length).

Hagihara *et al.* [15] describe a modified SST test method for single amorphous strips. A frame with interchangeable yokes (soft ferrite or GO electrical steel) and the capability for measuring applied field by magnetising current or H-coil was constructed and an in-depth analysis of its performance conducted. A schematic of the measuring circuit used is shown in Figure 12.4. Signal amplifiers for the B-coil, H-coil and magnetising current are clearly shown which are necessary to increase the signal-to-noise ratio on the very small signals. By using three identical amplifiers any phase shift introduced should be common across all of the input channels.

It is very clear from this work that the yokes have a large effect on the measured properties of the strip with a significant extra MMF evident in the measured *B–H* loops over that from the H-coil measurement as shown in Figure 12.5. However, it should be noted that due to the very high permeability of the strip under test, and thus the large field gradient at the surface, the H-coil could be presenting a significant under-estimate. The paper's findings recommend a method utilising an H-coil and a single sided ferrite yoke for development towards a standard.

A round robin test was later carried out using this test method [16] by circulation of a test frame and samples which confirmed that the reproducibility was within 3%.

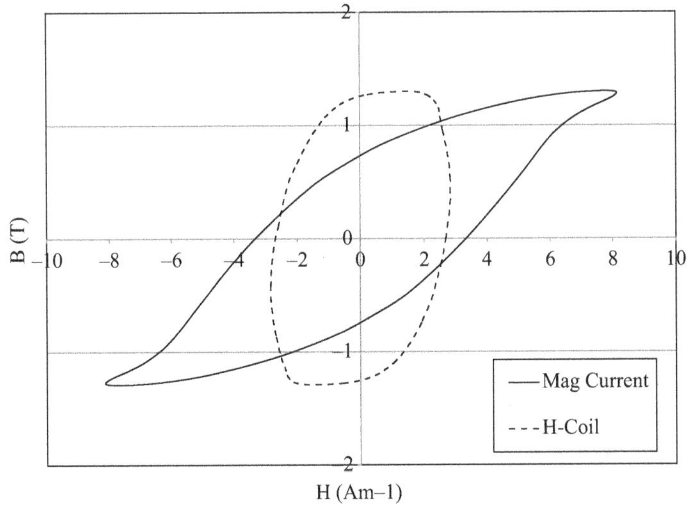

Figure 12.5 ASST B–H loop measurement for magnetising current and H-coil measurement adapted from [15]

12.2.3 Rings and toroids

Rings or toroids are perhaps the simplest geometries for a magnetic test. Rings are usually stamped from sheet material and stacked. They can only be used to test nominally isotropic sheet material since flux will travel in all directions within the plane. They will provide properties which are averaged for all directions of the sheet.

Toroids are strip wound (like a clock spring) from slit steel sheet. The magnetisation occurs mostly along the length of the strip although some interlaminar flux propagation occurs.

Rings or toroids are often considered the perfect magnetic circuit but they have several drawbacks. If the material is not ordinarily to be made into a ring or toroid, the preparation of such a sample for test purposes is laborious. In most cases, each test sample has to have magnetising and sensor windings applied to it manually. Magnetising and/or sensor windings can take the form of bundled harnesses which can save time in the winding operation, although large magnetising fields can still be difficult to generate. As with preparing Epstein samples by guillotine, stamping laminations, and particularly winding toroids, creates a lot of stress in the sample which needs to be annealed out. Other cutting methods such as electro-discharge machining (EDM) have been shown to be less damaging to the magnetic properties. As well as enwrapping the core, the primary and secondary windings each describe a single turn which would generate a field and detect a flux respectively normal to the core's magnetisation direction. This will create an error which may be significant when the permeability of the test piece is low.

The circumferential field at the inner diameter of the core is not the same as that at the outside diameter and, in fact, the field decreases as you move from the

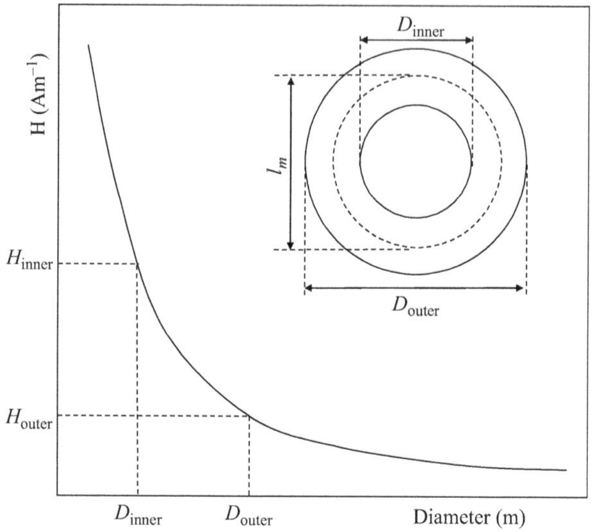

Figure 12.6 Influence of ring diameter on magnetic field strength (adapted from [17])

inside out (see Figure 12.6). It is generally accepted that, in order to maintain reasonable uniformity and hence make a measurement of the material properties, the test core should not be one which is geometry dependant. The ratio of outer diameter to inner diameter should be less than 1.25 as defined in the IEC standard for these measurements [17].

The secondary winding is the first to be wound on to the sample. This is evenly spaced around the ring and wound tightly to minimise any air flux (although not so tightly so as to stress the sample). The primary winding is then wound over the secondary, again taking care that it is evenly spread around the core. The number of primary windings is dictated by the required field and the voltage and current capabilities of the generator:

$$\text{Magnetic field strength, } H = \frac{N_1 I_1}{2\pi r} \quad (\text{Am}^{-1}) \tag{12.7}$$

$$\text{Mean magnetic path length, } l_m = \pi\left(\frac{d_1 + d_2}{2}\right) \quad (\text{m}) \tag{12.8}$$

$$\text{Cross sectional area, } A = \frac{m}{\rho l_m} \quad (\text{m}^2) \tag{12.9}$$

where N_1 = number of primary turns, I_1 = primary current (A), ρ = conventional density of sample (kg m^{-3}), r = radius of the ring (m), d_2 = outer diameter (m) and d_1 = inner diameter (m).

12.3 Sensing methods

12.3.1 Flux density sensing

All of the techniques for measuring flux density (or polarisation) described here will utilise the voltage induced in a coil enwrapping some part of the sample under test. As such, they rely on an alternating or transient flux as defined by Faraday's law (12.10). By the rearrangement of this, it can be shown that the flux density is proportional to the integral of the voltage and that the constant of proportionality is the inverse of the product of the number of turns on the B-coil and its area:

$$\text{Induced voltage, } V = -N\frac{d\emptyset}{dt} \quad (\text{V}) \tag{12.10}$$

$$\text{Therefore, } V = -NA\frac{dB}{dt} \quad (\text{V}) \tag{12.11}$$

$$\text{Hence, } B = -\frac{1}{NA}\int V\,dt \quad (\text{T}) \tag{12.12}$$

where N = number of turns on the B-coil and A = cross sectional area of coil (m^2).

The cross sectional area of the B-coils is usually greater than the sample which they enwrap. Therefore, flux from the air surrounding the sample will be included in the measurement thereby introducing an error:

$$B_{\text{measured}} = B_{\text{actual}} + B_{\text{air}} \quad (\text{T}) \tag{12.13}$$

$$= B_{\text{actual}} + \mu_0 H\frac{(A_{\text{air}} - A_m)}{A_m} \quad (\text{T}) \tag{12.14}$$

$$= J + \mu_0 H \quad (\text{T}) \tag{12.15}$$

where B_{measured} = flux density calculated from integrated voltage of a B-coil using (12.12) (T), B_{actual}= real flux density in the sample (T), B_{air} = flux density in air enclosed within the B-coil (T), A_{air} = cross sectional area of the air enclosed by the B-coil (m^2) and A_m= cross sectional area of the sample (m^2).

This error can be compensated for by connecting a mutual inductor with its primary winding in series with the magnetising winding and its secondary winding in series opposition with the B-coil. The value of the mutual inductance can be calculated through (12.16) (for B measurement) or (12.17) (for J measurement). Practically it is usually necessary to trim the mutual inductor coil by winding a few turns on or off whilst the mutual inductor is connected to the sensing coil and the magnetising winding energised without the presence of the test sample. An oscilloscope or voltmeter can be used to determine the null condition when compensation is perfect in the case of J compensation.

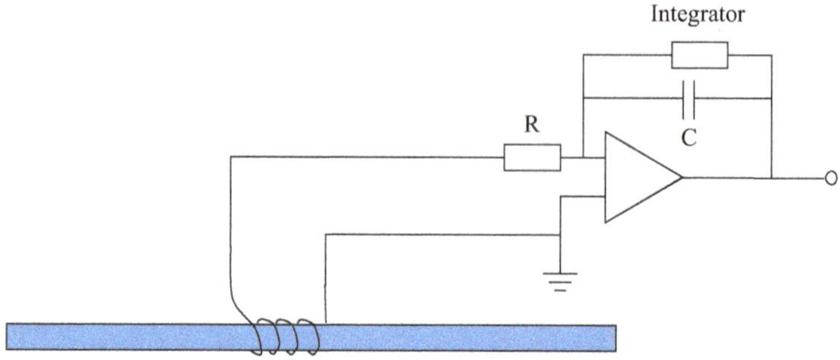

Figure 12.7 B-coil and integrator

If B_{actual} is required, the mutual inductance is given by

$$M = \frac{\mu_0 N_1 N_2 (A_{coil} - A_m)}{l_m} \quad \text{(H)} \tag{12.16}$$

If J is required, the mutual inductance is given by

$$M = \frac{\mu_0 N_1 N_2 (A_{coil})}{l_m} \quad \text{(H)} \tag{12.17}$$

where N_1 and N_2 = number of primary and secondary turns respectively in the mutual inductor.

Alternatively, digital sampling techniques can perform this compensation computationally using these equations but this technique is based on the assumption that the measured H is both accurate and uniform over the area in which B is measured.

Integration of the B-coil voltage was traditionally performed through hardware integration circuits, such as the simple example shown in Figure 12.7, where the output voltage is determined by the values of the discrete components (12.18) such that B can be determined from (12.19):

$$\text{Output voltage, } V_{\text{out}} = \frac{1}{RC}\int V_{in}\, dt \quad \text{(V)} \tag{12.18}$$

$$\text{Therefore, using (12.13), Flux density, } B = \frac{RC}{NA} V_{\text{out}} \quad \text{(T)} \tag{12.19}$$

Manufacturers of flux integrators, instruments which integrate and measure the voltage from a B-coil, and scale it accordingly, have highly refined designs which minimise the errors due to drift inherent in their use (particularly at very low frequency). However, the majority of measurement systems developed in recent times utilise computational methods for integrating a digitally sampled voltage waveform using the trapezoidal rule as shown in (12.20):

$$B = -\frac{1}{2NAf_s}\sum_{i=1}^{n} (V(i) + V(i-1)) + k \quad \text{(T)} \tag{12.20}$$

where f_s = sampling frequency of the data acquisition hardware (Hz), n = number of samples per cycle and k = a constant which is required to make the average of the waveform over one cycle zero (V).

A measured waveform of B over a whole period (or several periods) allows the measurement of a wide range of parameters such as the peak value (often a measurement set-point) and analysis of harmonic components or plotting of a complete B–H hysteresis loop.

However, in many measurements only the peak magnetic flux density is required and this can be afforded through a simple voltage measurement as shown below. Suppose the flux density is given by

$$B(t) = \sum_{r=1} a_r \sin(r\omega t + \varphi_r) \quad \text{(T)} \tag{12.21}$$

where r = harmonic number, a_r = constant giving the amplitude of each harmonic, ω = angular frequency (rads^{-1}), and φ_r = phase of each harmonic relative to the fundamental (rad).

Suppose also that, at some time (S), the flux density reaches a maximum given by

$$B(S) = \widehat{\mathrm{B}} = \sum_{r=1} a_r \sin(r\omega S + \varphi_r) \quad \text{(T)} \tag{12.22}$$

At a time $^{\pi}/_{\omega}$ (half the period of the fundamental frequency), B is given by

$$B\left(S + \frac{\pi}{\omega}\right) = \sum_{r=1} a_r \sin(\mathrm{r}\pi + r\omega S + \varphi_r) \quad \text{(T)} \tag{12.23}$$

If it is assumed that r is always odd, i.e. if the flux waveform contains only odd harmonics then

$$B\left(S + \frac{\pi}{\omega}\right) = -B(S) \quad \text{(T)} \tag{12.24}$$

Therefore the positive and negative peaks are separated by half a period:

$$\frac{dB}{dt} = \sum_{r=1} a_r \cos(r\omega t + \varphi_r) \quad \left(\mathrm{Ts}^{-1}\right) \tag{12.25}$$

$\frac{dB}{dt}$ will have zeroes when $B = \widehat{\mathrm{B}}$, i.e. $t = S + \frac{n\pi}{\omega}$

The mean value of $^{dB}/_{dt}$ taken over half a period is therefore

$$\left(\frac{dB}{dt}\right)_{\text{ave}} = \frac{\omega}{\pi}\int_S^{S+\frac{\pi}{\omega}} \sum_{r=1} a_r \cos(r\omega t + \varphi_r)\mathrm{dt} \quad \left(\mathrm{Ts}^{-1}\right) \tag{12.26}$$

$$= \frac{\omega}{\pi}\sum_{r=1} a_r[\sin(r\omega t + \varphi_r) - \sin(r\pi + r\omega t + \varphi_r)] \quad \left(\mathrm{Ts}^{-1}\right) \tag{12.27}$$

Since r is only odd this reduces to:

$$\left(\frac{dB}{dt}\right)_{\text{ave}} = \frac{2\omega}{\pi}\sum_{r=1} a_r \sin(r\omega t + \varphi_r) = \frac{2\omega\hat{B}}{\pi} \quad \left(\text{Ts}^{-1}\right) \tag{12.28}$$

$$V_{\text{ave}} = NA\left(\frac{dB}{dt}\right)_{\text{ave}} = \text{NA}\frac{2\omega\widehat{\text{B}}}{\pi} = 4\widehat{\text{B}}ANF \quad (\text{V}) \tag{12.29}$$

Therefore, through a measurement of the average voltage (by this we mean the rectified mean and not the numerical average over the whole period) the peak flux density can be determined. It is common for voltmeters to feature r.m.s. measurement and not average and these may still be used but only in the specific case of pure sinusoidal voltage waveforms:

$$\widehat{V} = \sqrt{2}V_{\text{rms}} = \frac{\pi}{2}V_{\text{ave}} \quad (\text{V}) \tag{12.30}$$

$$V_{\text{rms}} = \frac{\pi}{2\sqrt{2}}V_{\text{ave}} = \frac{2\pi}{\sqrt{2}}\widehat{\text{B}}ANF \approx 4.44\widehat{\text{B}}ANF \quad (\text{V}) \tag{12.31}$$

The design and placement of B-coils is often critical to an accurate measurement. Clearly, the number of turns and area (which are somewhat dependent on sample size and shape) are directly proportional to the voltage output so it is often the case that, when the measurement frequency is low, many turns are required to achieve a measurable voltage, whilst high frequencies can lead to un-measurably large voltages.

12.3.2 Magnetic field measurement

Magnetic field strength may be measured by several different methods. The most common is to rely on the proportionality of the field and the magnetising current which is derived from Ampere's law. Thus, with a known mean magnetic path length (defined with knowledge of the magnetic circuit), current and number of magnetising windings, the field acting on a sample within the coil can be calculated. This simple measurement is usually realised by measuring the voltage across a known resistance in the current return path of the coil. It does, however, suffer from several sources of error. Most notable of these is the dependence on the mean magnetic path length which, as referred to previously, is not a constant for a given experimental set-up.

In a closed magnetic circuit, the field can usually be defined with reasonable accuracy. However, the uniformity of the field relative to the size of the sample should be considered. Often, only sections of test piece are subject to a uniform field at the calculated value.

When the magnetic circuit is open the demagnetisation field in the test piece can be significant, depending on the specific geometry. Where possible, open circuit measurements should be avoided.

Figure 12.8 Large format H-coil for an SST manufactured on a PCB

To determine the field independent of the magnetic path length we can use an H-coil which measures the tangential field component close to the surface of the sample. This relies on the fact that the tangential field at the surface is the same as that inside the material but neglects the error due to the coil area, which is significant due to the rapid drop off in the magnitude of the field away from the surface. The coils are usually made from enamelled copper wire, as fine as can be practically wound (typically 50 μm), on to an insulating, non-magnetic former which should be as thin as possible, whilst being rigid enough to maintain its dimensional stability.

These H-coils are special examples of B-coils and, as such, are subject to the same set of equations describing their behaviour such that the flux density is proportional to the integral of the voltage and the field can be found by dividing by the permeability of free space. This also means that the sensitivity of the coil is determined by the number of turns and cross sectional area of the coil. Since the cross sectional area is small, by design, large numbers of turns must be used (typically several thousand). Recent examples of H-coils for use in large SSTs have utilised etched PCBs, accuracy such as that shown in Figure 12.8 which is approximately 30 cm × 20 cm, to produce a flat coil with higher dimensional accuracy.

The *Chattock coil* or *magnetic potentiometer* uses the principle that a coil of small cross sectional area taking any path between two points on the surface of a sample (typically a magnetic potentiometer is arch shaped) measures the magnetic potential difference due to the tangential flux between those points [1]. The coils are tightly wound on non-magnetic (and usually non-metallic) formers which must have a uniform cross section. The voltage induced in the coil can be calculated from (12.32):

$$V = \mu_0 \frac{N}{l} A \frac{d}{dt}(Hx) \quad (\mathrm{V}) \tag{12.32}$$

$$\text{Therefore, } H = \frac{l}{\mu_0 N A x} \int V \, dt \quad \left(\mathrm{Am}^{-1}\right) \tag{12.33}$$

where l = length of the coil (m) and x = distance on the surface of the material between the ends of the coil (m).

These coils have many of the same drawbacks and advantages as H-coils but often find use where the field needs to be measured over a small region or where the coil needs to be shaped around other windings.

Hall effect sensors are widely used due to their compact size and wide frequency range. These solid state devices utilise the Hall effect to create a voltage proportional to the magnetic field when supplied with a constant current. They are cheap and conveniently packaged into integrated circuits which often include compensation for temperature, which is critical due to their high temperature sensitivity, current control and offset compensation. These sensors are most often used when a localised field measurement is required, for instance, when determining the distribution of field in a solenoid, or when making a measurement of local magnetic properties in a material (see Section 12.4.3). It can be difficult to position the Hall element close to the surface due to the packaging.

12.4 A.C. magnetic measurements of losses and permeability

12.4.1 The wattmeter method

The most common method of performing a.c. magnetic measurements is by using the wattmeter method with a two winding configuration [3]. This can be realised for a range of geometries as described in Section 12.2 with a two winding arrangement such that the primary winding carries the magnetising current and the secondary winding (usually that closest wound to the sample under test) is unloaded and therefore carries a negligible current.

An equivalent circuit for any two coil closed circuit a.c. measurement circuit is shown in Figure 12.9. This ideal circuit has no stray capacitance or leakage inductance and the resistance of the secondary winding is negligible compared to the input resistance (R_2) of the measuring instrument.

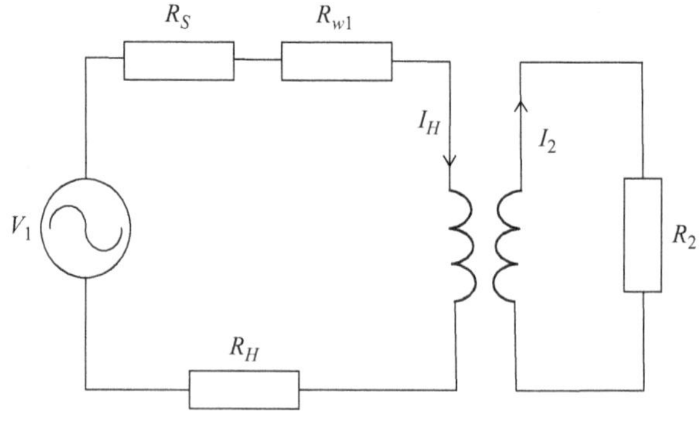

Figure 12.9 Equivalent circuit of a.c. magnetic measurement

where R_S = the source resistance (Ω), R_{W1} = the resistance of the primary winding (Ω) and R_H = resistor utilised for measurement of the primary current (Ω).

The average power delivered into the magnetising winding can be calculated from the integral of the multiplication of the instantaneous primary current, i_H, and the voltage, U_L and can be written as

$$P = \frac{1}{m_a}\frac{1}{T}\int_0^T U_L(t)i_H(t)dt \quad \left(\mathrm{Wkg}^{-1}\right) \tag{12.34}$$

where m_a = active mass of the sample[2] (kg).

Since it is not possible to measure U_L in a real circuit, the properties of the ideal transformer allow us to substitute the secondary voltage with knowledge of the primary and secondary turns to give

$$U_2(t) = \frac{N_2}{N_1}U_L(t) \quad (\mathrm{V}) \tag{12.35}$$

Therefore

$$P = \frac{1}{m_a}\frac{N_1}{N_2}\frac{1}{T}\int_0^T U_2(t)i_H(t)dt \quad \left(\mathrm{Wkg}^{-1}\right) \tag{12.36}$$

A commercial wattmeter or power analyser can be used to perform the current and voltage measurement and integrate their product over a cycle (or averaged over a fixed number of cycles) with the constants for the number of turns and sample mass being used by the operator to calculate the specific total loss in watts per kg.

Equation (12.36) is exactly equivalent to (12.37):

$$P = \frac{1}{T\rho}\int_0^T H\frac{dB}{dt}dt \quad \left(\mathrm{Wkg}^{-1}\right) \tag{12.37}$$

$$\text{Since, } i_H = \frac{H(t)l_m}{N_1} \quad (\mathrm{A}) \tag{12.38}$$

$$\text{and } U_2 = -N_2A\frac{dB}{dt} \quad (\mathrm{V}) \tag{12.39}$$

The measurement circuit diagram for the wattmeter method is given in Figure 12.10.

The primary circuit comprises a signal generator which is fed into a power amplifier via a pre-amplifier circuit. The power amplifiers utilised for this purpose are often laboratory class amplifiers which are designed to handle highly inductive non-linear loads with low power factors and current waveforms with high peak amplitudes. This is complemented with robust protection circuitry to protect the amplifier from collapsing fields generating large reverse voltages. As a lower cost option, many laboratories utilise power amplifiers designed for audio or public

[2]The active mass is defined as the mass of the part of the test sample which is magnetically active and is usually calculated from the mean magnetic path length. For example, for an Epstein frame it is $^{ml_m}/_{4l}$ where l is the length of each sample.

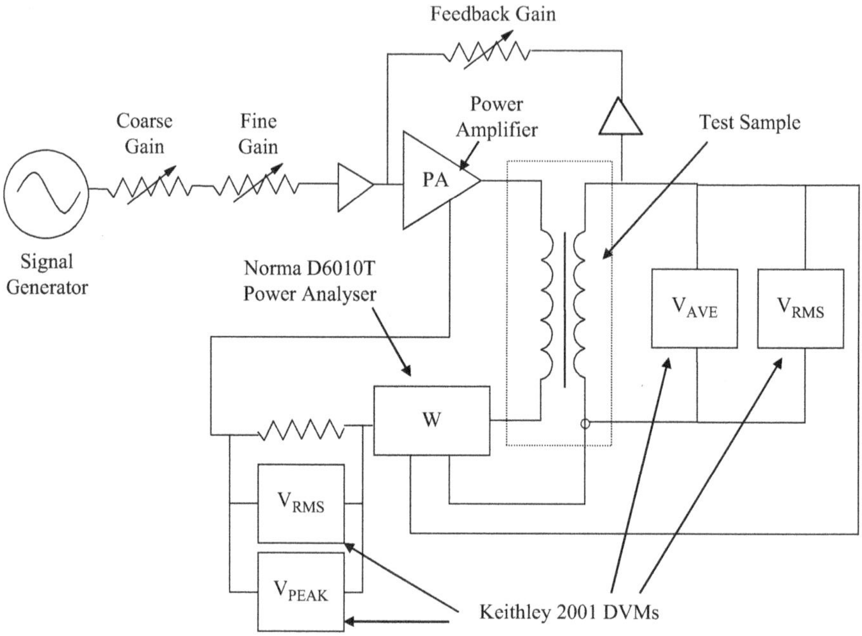

Figure 12.10 Measurement circuit of an implementation of the wattmeter method using discrete instruments

address systems. Whilst these are often perfectly adequate, it is good practice to include isolation through a matching transformer and add simple protection circuits to the outputs whilst monitoring of current and voltage waveforms with an oscilloscope can usually provide sufficient warning of instability.

The pre-amplifier's basic function is to give fine control of the input to the power amplifier and, as such, can sometimes be eliminated if the signal generator has a sufficiently sensitive control. However, in most cases, this circuit also includes the additional capability for proportional control of the secondary voltage in order to maintain the waveshape of the secondary voltage.

Waveshape control is essential due to the highly non-linear nature of the test circuit. Ensuring that it is controlled within certain limits (often determined by ideal values of functions such as the form factor or total harmonic distortion) allows fair comparison between tests. The non-linearity can be due to a large number of factors including:

- the non-linear *B–H* characteristic for the material under test,
- leakage reactances and parasitic capacitances,
- the non-linear response of the power amplifier and
- distortion due to mismatch between the power amplifier and the magnetic circuit (which changes with *B* due to non-linear permeability and therefore impedance).

It is relatively simple to realise this control using a low cost operational amplifier circuit such as that shown in Figure 12.11. Here an op-amp is configured

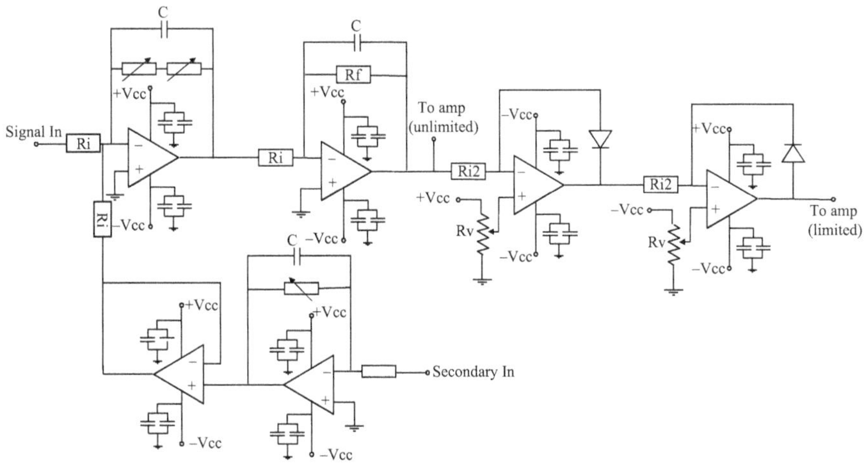

Figure 12.11 Typical analogue feedback circuit for control of secondary voltage

as a summing amplifier with coarse and fine variable resistors giving a variable gain of the sum of the signal generator input and a proportion of the secondary voltage from the inverting amplifier with a third variable resistor. The remainder of the circuit comprises two buffer stages and a voltage limiting stage designed to prevent too large a voltage being applied to the power amplifier.

The primary circuit is completed with the current channel of the wattmeter and a non-inductive resistor of a low value, typically 0.1 Ω to 1.0 Ω, across which a voltmeter is used to measure the r.m.s. and peak current (note: modern power analysers will allow all these functions to be performed in a single unit). The secondary circuit is connected to a pair of parallel voltmeters, giving measurement of r.m.s. and average voltage (this should be a true rectified mean which is not available on many commercial voltmeters), and the voltage channel of the wattmeter. Again, some power analysers will accomplish all of these functions.

The measurement is usually taken at a set value of peak magnetic flux density determined using (12.30) from the average measured voltage. Care should be taken not to overshoot the set point but if this does occur then a demagnetisation procedure should be followed whereby the peak flux density is increased beyond the highest value reached and then slowly reducing it to zero. Measurements at set peak magnetic field strengths are also reasonably common and the same procedure is followed, substituting the field determined from equations in Section 12.2 using the measured value of the peak primary current.

12.4.2 Digital interpretation of the wattmeter method

The realisation of an a.c. magnetic measurement through the use of discrete measurement instruments has several advantages with the most important being that all of the individual instruments can be easily calibrated and a history of the

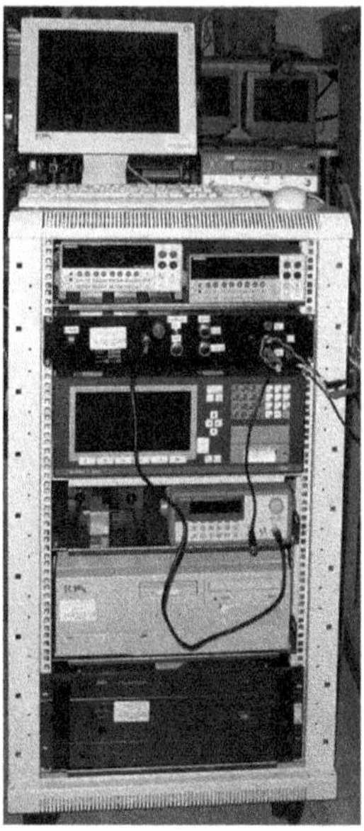

Figure 12.12 Photograph of a test system for the wattmeter method using discrete instruments

instruments' stability recorded, thus allowing for a high level of confidence in the measured values and therefore the declared uncertainties. However, a system such as this (pictured in Figure 12.12) can be very costly and the measurement procedure is extremely time consuming with each test point being manually set and controlled before recording the instrument values and entering them into a spreadsheet for post processing.

Digital measurement techniques can provide several levels of automation to this process. Most are simply a PC fitted with suitable hardware, normally a GPIB interface although, increasingly, instruments are offering a USB interface and can be used to capture the data from the discrete instruments. This can greatly speed up the procedure, prevent recording errors and allow immediate post processing to create all of the calculated parameters and graphs required.

Most current implementations, however, are realised by a completely digital system eschewing discrete instruments in favour of a PC containing, or linked to, high speed data acquisition and generation hardware (ADDA). This hardware can be in the form of an expansion card or externally in its own chassis linked to the PC by a high speed bus. The cost of such hardware has reduced considerably whilst acquisition speeds and channel numbers have increased. The result is a system of similar capability to the discrete instruments with greater flexibility at a fraction of the cost.

Figure 12.13 shows a schematic of such a system. The magnetic and electrical circuits remain identical to Figure 12.10 with the ADDA and associated software algorithms replacing all of the instruments. Clearly, the complication in building a

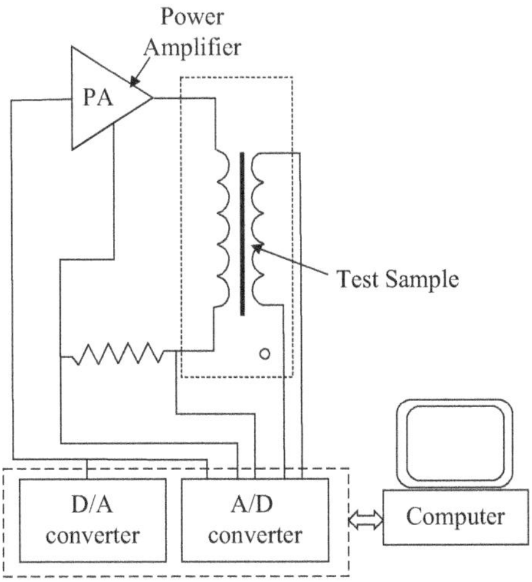

Figure 12.13 Schematic of a digital interpretation of the wattmeter method, compare with Figure 12.10

system of this type is in coding the required functions which would achieve the following basic tasks:

- control the output voltage such that the secondary voltage is at the required value,
- measure secondary voltage and primary current waveforms simultaneously and
- calculate H_{pk}, loss, VA, etc.

It is a logical next step to add additional functionality into the software such that the following tasks are undertaken:

- automatically set a range of flux densities at predefined frequencies and waveshapes,
- control the waveshapes within tightly defined parameters,
- output live *B* (through software integration of secondary voltage) and *H* waveforms along with *B*–*H* loops and
- perform harmonic analysis of *B* and *H* waveforms.

The data acquisition hardware will need to incorporate at least two analogue input channels (more if additional control parameters or measured quantities are required) and the inter-channel delay should be as small as possible in order to minimise phase errors in the calculation of power loss. If possible, simultaneous sampling is preferable.

The synthesised magnetisation waveform is usually generated from the same hardware and, as such, the sampling is governed by the same clock. If this is the case, then the sampling should be configured such that the sampling frequency is an integer multiple of the magnetising frequency and that this multiple satisfies the Nyquist criterion, i.e. it is at least twice the frequency of the highest frequency harmonic. The only way this can be guaranteed is if an anti-aliasing filter[3] is used to limit the bandwidth of the incoming signals (some hardware includes these filters).

Most data acquisition hardware operates with an amplitude resolution of at least 16 bits which is perfectly sufficient for most measurements depending on the available ranges but users should be aware of the size of the quantization error (range/$2^{\text{resolution}}$) relative to the amplitude of minor loops [18].

In this method the instantaneous polarisation, $J(i)$ at time $t = i/nf = i/f_s$, can be found from the derivative of the instantaneous secondary voltage using

$$J(i) = -\frac{1}{N_2 A n_s f}\sum_{k=1}^{i}\frac{[U_2(k) + U_2(k+1)]}{2} - J_0 \quad (\mathrm{T}) \tag{12.40}$$

where n_s is the number of samples per cycle.

The instantaneous magnetic field strength, H(i), is calculated from

$$H(i) = \frac{N_1}{l_m R_H} U_H(i) \quad (\mathrm{Am}^{-1}) \tag{12.41}$$

where U_H is the measured voltage across the resistor in the primary circuit (V).

The instantaneous values of B and H can then be used for simple determination of peak and r.m.s. values of each. The specific total loss is calculated from (12.38) and given as

$$P = -\frac{N_1}{N_2 A n_s l_m R_H \rho}\sum_{i=1}^{n} U_H(i) U_2(i) \quad (\mathrm{Wkg}^{-1}) \tag{12.42}$$

The digitised H and dB/dt waveforms can also be utilised to implement digital waveform control to a prescribed ideal. This can be significantly more powerful than analogue methods allowing operation at higher frequency and more complex waveforms without the auto-oscillations which are so often incurred under analogue control. Many control methods can be implemented, from simple proportional feedback to an optimised full PID controller. Methods which use a parametric representation of the *J–H* loop to predict the input voltage can offer faster convergence.

Proportional control is the easiest method to implement but is only stable for linear systems. Non-linearity in the *B–H* characteristic, amplifier and magnetic circuit will increase considerably with increasing frequency and non-sinusoidal waveforms. Zurek *et al.* [19] presented a refinement of a proportional controller with the addition of an adaption module which compensated for the non-linear frequency response by calculating and adjusting the phase lag at each frequency component. A block diagram of this algorithm is shown in Figure 12.14.

Later developments of this algorithm [20] replaced the adaption module with feedback of the reference magnetising voltage and adding a differential term to

[3]Aliasing is an artefact of digital sampling whereby higher frequency harmonics appear at lower frequencies mirrored about the Nyquist frequency.

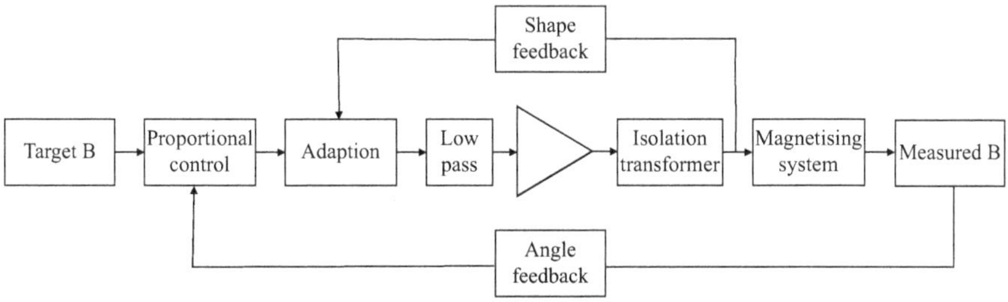

Figure 12.14 Block diagram of adaptive digital feedback algorithm (adapted from [19])

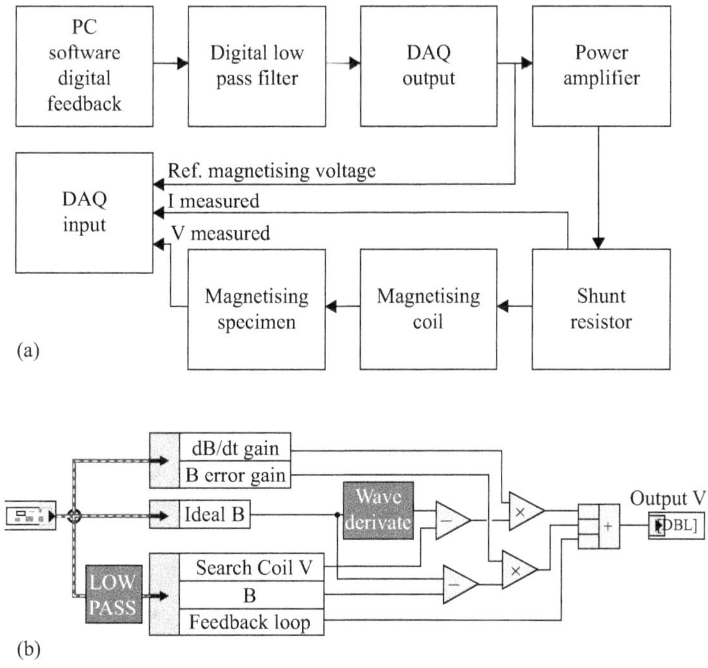

Figure 12.15 PD control with feedback of reference voltage

improve stability. A block diagram and the implementation in National Instruments LabVIEW™ is shown in Figure 12.15. This algorithm has proven to be fast and stable for a wide range of test conditions, materials and geometries.

12.4.3 Localised measurements

Localised magnetic measurements are useful in a range of investigations and many published works make use of them, particularly when analysing the variation of flux density and power loss.

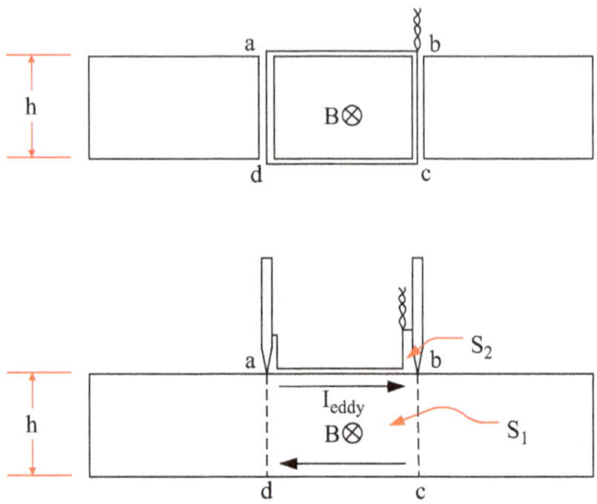

Figure 12.16 Comparison of search coil and needle probe methods

In common with bulk measurement of B, local measurement requires integration of the induced voltage of a coil enwrapping an area of the material. The difference in local measurement is that the coil is formed from only a few turns (often only one) of fine enamelled wire passed through a pair of holes in the material. Preparation of these coils is extremely difficult and requires a great deal of care in the drilling of the holes and winding of the coil. Short circuits between the coil and test specimen are difficult to avoid, although using a varnish to insulate the edges of the holes can help.

Since both the number of turns and the cross sectional area are small, the induced voltage is extremely small and requires extremely sensitive voltage measurement or pre-amplification. Induced voltages from unwanted sources can be minimised through tight twisting of the lead wires. Determination of cross sectional area is difficult and prone to large errors for small coils.

A less destructive alternative to the search coil is the needle probe method, first proposed by Werner [21], whereby a pair of needles make electrical contact with the sample to simulate a single (or half) turn search coil. The principles are illustrated in Figure 12.16.

For the search coil method the induced voltage in the coil is given by

$$V_{\text{coil}} = Nhl_{ab}\frac{dB}{dt} \quad \text{(V)} \tag{12.43}$$

In the case of the needle probes, using Faraday's law along the path *abcda* where S_1 is the area enclosed by *abcd*, leads to

$$V_{ab} + V_{bc} + V_{cd} + V_{da} = \int_{S_1} \frac{\partial B}{\partial t} dS \quad \text{(V)} \tag{12.44}$$

If we now assume that $V_{\text{needle}} = V_{ab} = V_{cd}$ and $V_{bc} = V_{da}$ (for a uniform flux distribution) we obtain

$$V_{\text{needle}} = \frac{1}{2}\int_{S_1} \frac{\partial B}{\partial t} dS - V_{bc} \quad (\text{V}) \tag{12.45}$$

V_{bc} is assumed to be equal to zero if the measured area is not at a sample edge (distance to edge should be greater than h/2 [22]) so

$$V_{\text{needle}} = \frac{1}{2} h l_{ab} \frac{dB}{dt} \quad (\text{V}) \tag{12.46}$$

As with search coils, the voltages generated from needle probes are very low and other sources of pick-up should be mitigated where possible. As shown in Figure 12.16 the connecting wires to the probes should be routed such that the loop constructed between the strip surface and the wire is minimised [23] whilst the leads should be tightly twisted. The signal size will also be influenced by the probe spacing with Loisos and Moses [24] comparing 25 mm and 4 mm spacing with errors of 2% and 12.5%, respectively.

An alternative geometry with needles placed on the top and bottom of the sample forming a loop to the sample edge was proposed by De Wulf [25] which was further developed into a hybrid of the two methods (with two needles on each side) by Loisos *and* Moses [26] which gave significantly lower errors (compared to search coils) for small needle spacings. This modified needle probe approach, however, comes at the expense of requiring access to both sides of the sample which is often problematic.

Complete local characterisation can be implemented by making a simultaneous measurement of magnetic field strength by the inclusion of a Hall probe [27] or H-coil [28] in the probe gap such that the field sensor is in close proximity to the surface when the probes (which are usually sprung) are depressed (Figure 12.17).

Figure 12.17 Photograph of needle probe with Hall sensor

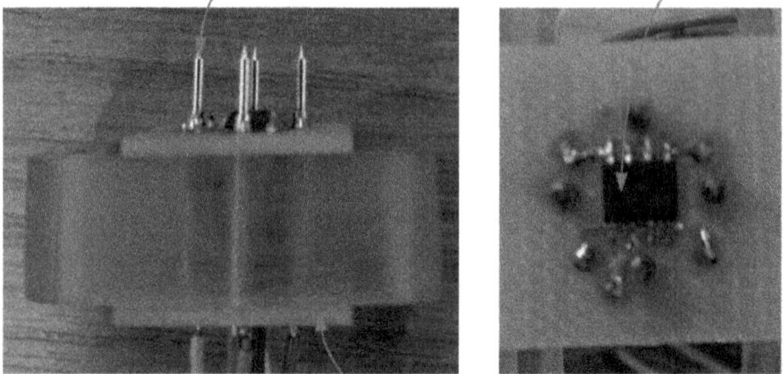

Figure 12.18 Photograph of 2D needle probe and H-coil

Enokizono *et al.* [28] described a two-dimensional (2D) arrangement of needle probes and two layer H-coil (similar to Figure 12.18) such that the components of flux density and surface field could be measured in two orthogonal directions enabling the vector properties to be ascertained. This measurement can be extremely useful even in the case of uniaxially magnetised materials as the angles of the flux and field vectors are likely to vary considerably on a local scale influenced by domain structures and other variations in permeability.

Many local magnetic measurements on laminated materials take place on material inside laminated stacks and whilst wire-wound coils can be constructed from ultra-thin (say 20 μm diameter) wire, these will still introduce a significant extra air gap resulting in changes to the flux distribution in the stack. For this reason thin film sensors have been developed.

Basak *et al.* [29] used physical vapour deposition (PVD) for the production of search coils with a thickness of less than 0.05 μm. As for wire-wound coils, holes were drilled in the lamination and an insulating layer of zinc sulphide was deposited followed by a pair of aluminium tracks either side of the strip which created conductive loops via the holes. Since masks could be produced of many sensors, the time consuming deposition process was partially offset. Ilo *et al.* [30] demonstrated normal flux density sensors which were chemically etched from a 1 μm thick sputtered copper film on top of a silica layer such that flux distribution around transformer joints could be studied. A development of this approach was inkjet printed coils of conductive inks onto a polyamide film with a total thickness of 50 μm [31].

Utilizing the advantages of both the needle probe technique and thin films Pfutzner [32] proposed sensors with conducting tracks running to spots of bare steel negating the need for through holes. Again, a silica layer was used to supplement the steel's insulation with a total sensor resulting in a total sensor thickness of 10 μm. This technique was built on by Mazurek [33] who used PVD to deposit copper needle probe sensors (<0.5 μm thick) onto a silica base layer as shown in Figure 12.19.

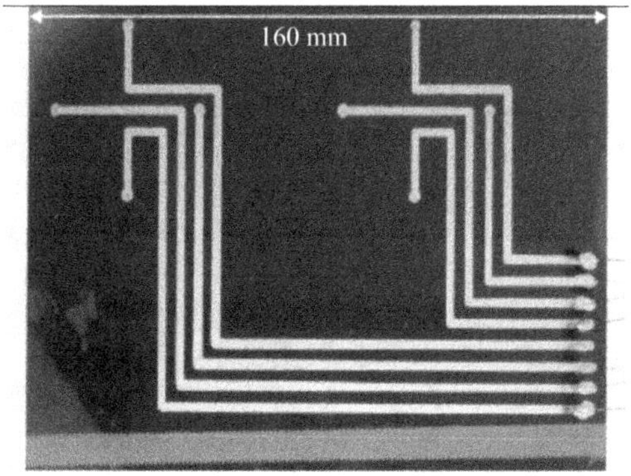

Figure 12.19 Thin film copper probes using the needle probe technique

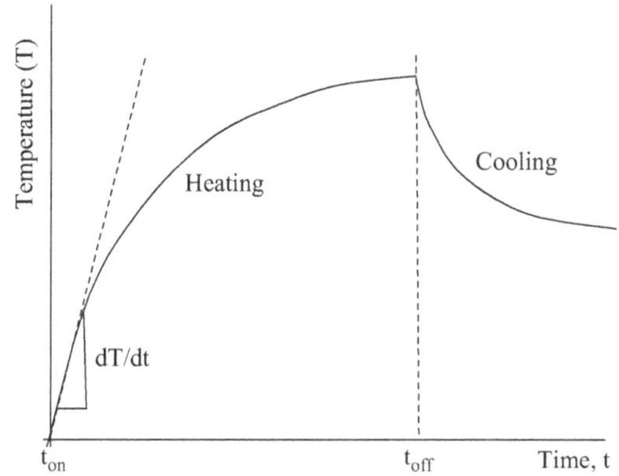

Figure 12.20 Typical heating and cooling curves showing the initial rate of rise of temperature

The fluxmetric methods of determining power loss discussed up to this point rely on difficult measurements of H and dB/dt. A potentially simpler alternative is thermometric in the form of the initial rate of change of temperature. This is a measurement of the energy absorbed or released by the magnetic material when the source of magnetisation is turned on or off. Figure 12.20 shows a typical heating and cooling curve together with a linear plot showing the heating of the body in perfect adiabatic conditions (where no heat dissipation to the environment occurs).

The power loss can be determined directly from the initial rate of rise of temperature using

$$V_{\text{needle}} = c_p \frac{T - T_0}{t - t_0} + K(T - T_0) \quad \text{(V)} \tag{12.47}$$

where c_p is the specific heat capacity of the magnetic material (461 $Jkg^{-1}K^{-1}$ for 3.2% SiFe), K is the heat transmission coefficient (determined by regression analysis of the cooling curve) and T, T_0 are the temperatures at times t and t_0, respectively.

The measurement period should take into account the time constant of the measurement system. Thermocouples are commonly used for these measurements as originally proposed by Ball and Lorch [34]. Basak [29] and later Mazurek [33] developed thin film, interlaminar thermocouples by depositing tracks of chromel and alumel to form a type K thermocouple at the junction.

Non-contact optical pyrometers or thermal cameras may also be used for surface loss measurements if line of sight access is available but it can be difficult to achieve the required resolution unless an adiabatic chamber can be used.

12.4.4 Measurements under simulated operational conditions

The influence of stress on the magnetic properties of soft magnetic materials is of significant interest due to the high sensitivity of these materials to manufacturing and operational stress. It is relatively common to find equipment to measure the stress sensitivity in magnetic measurement laboratories and there are even commercially produced tests now available. Application of uniaxial tension and compression is most common but the application of biaxial stress has also been investigated.

Simulating in-service or manufacturing stress in a laboratory test is very difficult, some rotating machines operate with sections of the magnetic cores approaching the elastic limit and local variations in the stress condition can be large. For a laboratory test, it is important to establish a uniform and repeatable stress such that materials can be compared. Most tests follow the form of an SST with a sample which extends beyond the flux closure yokes in order to attach the stressing mechanism.

Most simply, stress can be applied by attaching weights to the sample end by a cable [35] whilst the other end of the sample is clamped in place. Often pulley systems are used to route the connecting cable such that tension or compression can be applied to the strip [36]. Extremely high stresses (up to 700 MPa) can be applied by placing an SST frame within large mechanical test rigs [37] or by measuring large stacks of material within a hydraulic stress frame. It is critical that the sample and stress application retain their alignment throughout the test and most systems include alignment rods and sample restraint [38]. When applying compression to single laminations it is particularly difficult to avoid buckling the sample and often a mechanism for applying a small amount of uniform pressure normal to the surface is incorporated to minimise this (when considering magnetostriction measurements this should be avoided as the increased friction will tend to reduce the measured magnetostriction).

Kai *et al.* [39] demonstrated a system for applying biaxial stress to a cross-shaped sample similar to that shown in Figure 12.21 to create a controlled 2D stress tensor in the central part of the sample. In this case the tester is combined with a 2D magnetic test as described in Section 12.5.

A typical uniaxial stressing system is shown in Figure 12.22. One end of the sample is clamped to a large non-magnetic block which, along with the yoke and windings, is

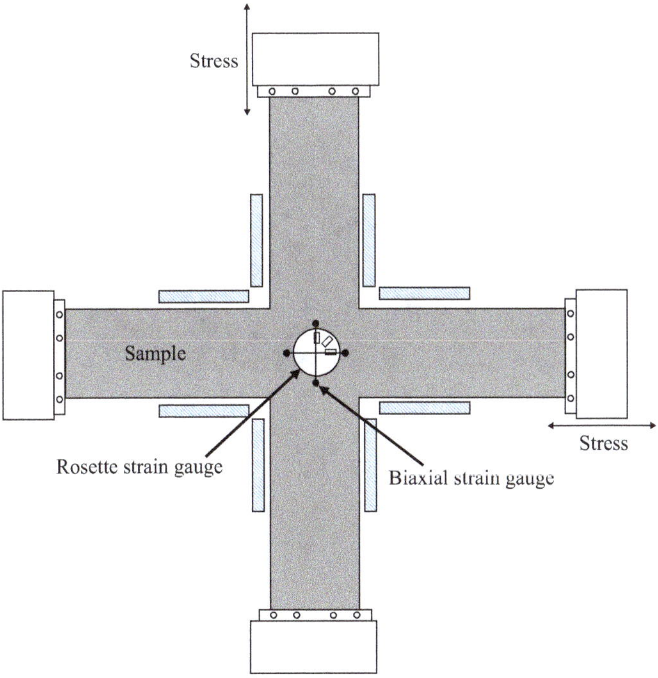

Figure 12.21 Biaxial, 2D tester with cross-shaped sample

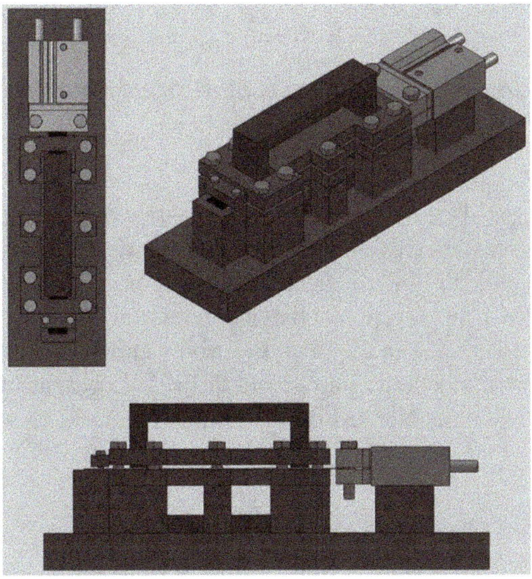

Figure 12.22 Drawing of a typical uniaxial stressing rig incorporated into an SST

Figure 12.23 Photograph of a typical uniaxial stressing rig incorporated into an SST

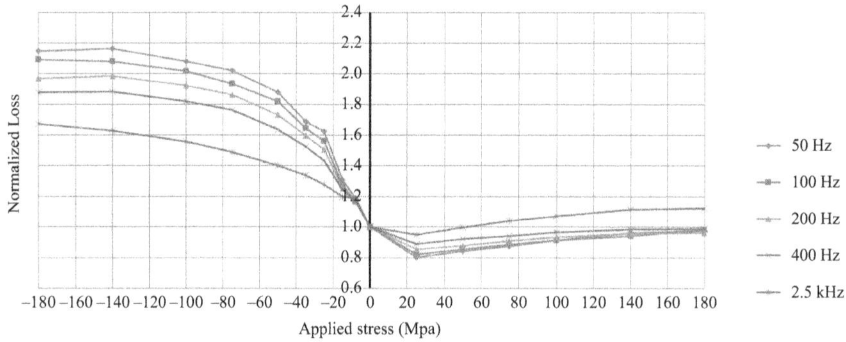

Figure 12.24 Typical plot loss against applied stress for a NO electrical steel at 1.0 T and a range of frequencies

firmly fixed to the base of the system. A clamp at the free end of the strip is connected to a pneumatic linear actuator. This actuator serves two purposes. First, guides and linear bearings within the actuator limit the freedom of movement of the free end clamp such that it may only move in the direction of the strip extension. Second, the cylinder allows the application of linear stress to the strip. Electro-pneumatic regulators modulate the pressure in the actuator under the control of a LabVIEW virtual instrument utilizing an in-line load cell. A photograph of this rig is shown in Figure 12.23 whilst Figure 12.24 shows a typical set of measured characteristics.

12.4.5 D.C. biased a.c. measurements

Simulating d.c. biased a.c. magnetisation conditions is becoming more important with the proliferation of power electronic converters and the threat of geo-magnetically induced currents due to solar events.

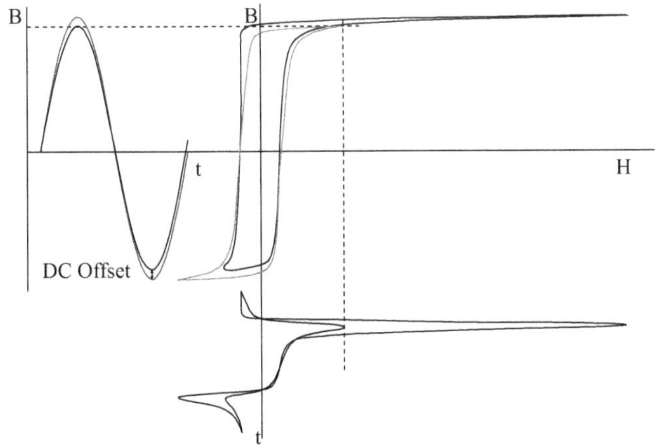

Figure 12.25 Effect of d.c. offset B on B–H loop and field (current) waveform

The d.c. bias will result in a non-symmetrical *B–H* loop and can lead to half-cycle saturation as well as high peak currents as shown in Figure 12.25.

There are generally two methods employed to achieve these tests in laboratories: injection of d.c. flux via a second excitation coil or modification of the magnetisation waveform to include a d.c. component. The first of these methods is the most common, being used for material tests in an Epstein frame or SST and for simulated transformer tests. The principle is a simple one whereby an additional winding is made at some position in the magnetic circuit which is excited by a d.c. supply current. Since this winding will couple with the main primary winding a choke should be included in the d.c. winding to prevent the induction of a.c. voltages in this winding [40]. A variation on this approach was developed by Miyagi *et al.* [41] who proposed an open SST (without flux closure yokes) with an a.c. winding which was contained within a large Helmoltz coil which imposed the d.c. field and does not couple with the a.c. coil. The hardware for such an arrangement is relatively complex.

By taking advantage of the digital measurement and control methods as described in Section 12.4.2, d.c. flux can be super-imposed on the a.c. flux simply by controlled modification of the magnetising waveform. The approach taken can vary [42–44], but all rely on the establishment of a d.c. flux density which the a.c. flux density is then superimposed upon.

The system described in [45] utilises an SST 500 which has been modified to include a large format H-coil (see Figure 12.8), which is essential due to the high flux densities possible in one half-cycle. The magnetising and control system is identical to that described in Section 12.4.3 but magnetises the material with the controlled flux density waveform proposed by [44] and shown in Figure 12.26(c) which is constructed by the addition of waveforms (a) and (b). The peak of waveform (b) is the d.c. offset level. Plotting of the *B–H* loop (Figure 12.27) and calculation of power loss and so forth is performed with the section of waveform in Figure 12.26(c).

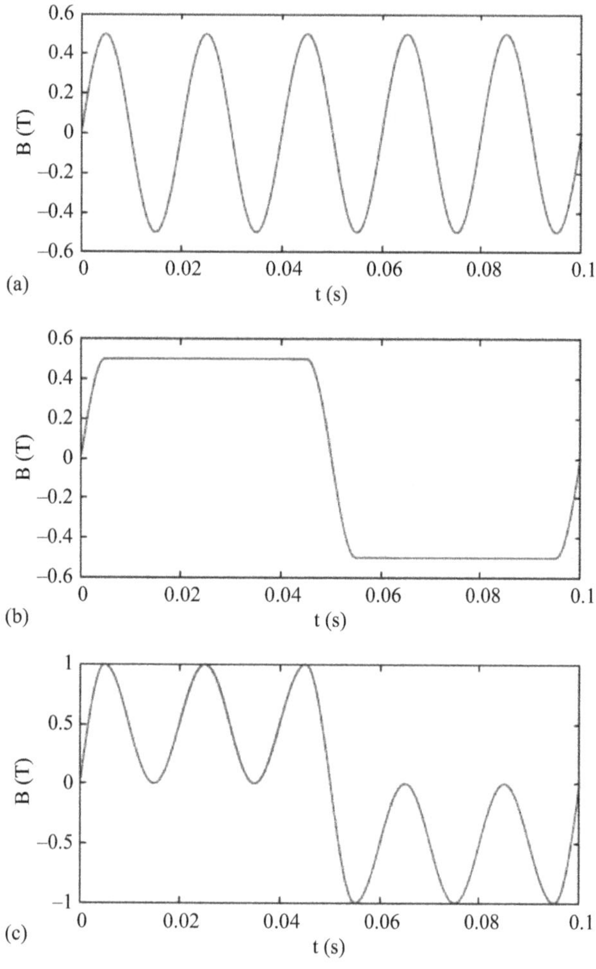

Figure 12.26 Construction of waveform for d.c. offset magnetisation (a) is superimposed on (b) to create (c)

12.5 2D and rotational magnetic measurements

12.5.1 Measurement principles

As described in Chapter 8, 2D and rotational magnetisation are extremely important due to their occurrence in the cores of many common electrical machines. It is worth revisiting the terminology before continuing to discuss their measurement. The a.c. measurements discussed previously in this chapter are uniaxial or one dimensional (1D). The MMF and measurement of B and H are all applied to a single axis and it is assumed that there are no orthogonal components of either quantity. By adding orthogonal sensors a measurement can be made of both

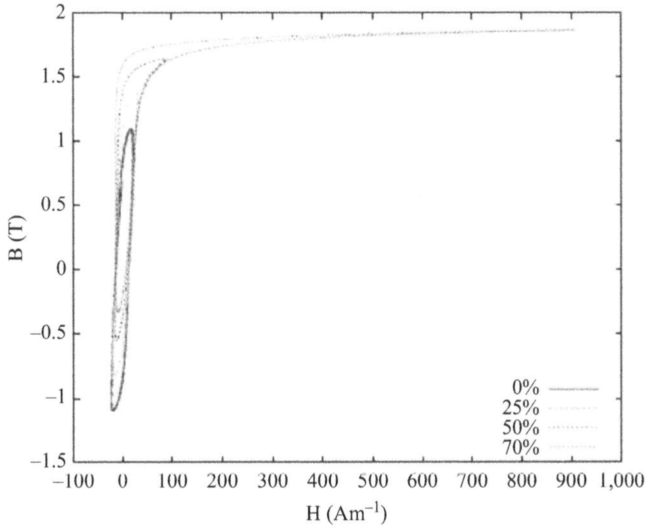

Figure 12.27 Plotted B–H loops for increasing d.c. offset at $B_{AC} = 1T$, 50 Hz

components and a resultant vector can be calculated for the a.c. (i.e. the magnitude is time varying) field and flux density. If the angle of these vectors rotates with time then we have rotational magnetisation. If the magnitude of one of these vectors is invariant with time then it is defined as pure rotational B or H (their loci will describe a circle). For cyclic rotational magnetisation the vectors would rotate through 360° in one period. The measurement methods summarised below are generally suitable for measuring both conditions. More thorough descriptions can be found in work dedicated to this subject such as [46].

12.5.2 Magnetisation systems

The magnetising is required to establish the vector magnetisation conditions described in section 12.3.2. This could be achieved by using a uniaxial source of MMF such as a high permeability yoke wound with a coil connected to an a.c. source and rotating either the yoke or sample in the horizontal axis such that the required angle of magnetisation is achieved. A system with a rotating yoke is proposed by Tumanski [47] which, whilst simple in principle, is mechanically complex.

Systems used for 2D magnetisation today exclusively use a multi-phase winding with a.c. voltages applied to each. The applied MMFs are aligned such that the angular spacing is identical to the phase shift between the synchronised a.c. signals on the windings. The first published example of such a system was from Boon and Thompson [48] who utilised a pair of orthogonal C cores with pure sinusoidal magnetising voltages with a relative phase shift of 90° applied to coils enwrapping the yokes. Double yoke frames on the same principle such as [49], featuring an 80 mm square sample, or [50], with a large 300 mm square sample and flux closure path closer to an SST, appeared much later. A similar yoke arrangement has also been

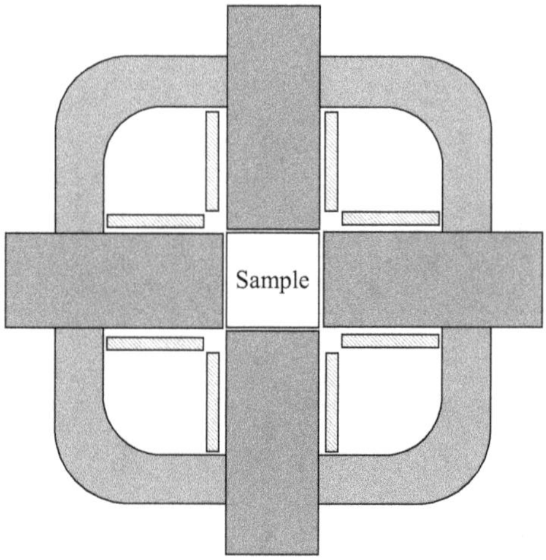

Figure 12.28 In-plane yoke construction for a square rotational tester (adapted from [54]). Grey sections indicate flux closure yokes

utilised with a cross-shaped sample [51] giving benefits in field uniformity but at the expense of more complex sample preparation. These types of frames all have significant differences in the length of the flux closure paths for the two magnetisation directions since the yokes must overlap and this asymmetry can cause issues in controlling the uniformity of the applied field [52]. A recent variation on this geometry utilised a cubic stacked iron core which was rotationally magnetised in the plane of the laminations whilst stress was applied in the normal direction [53].

A more symmetrical geometry was proposed in [54] whereby an in-plane closure yoke was constructed which is shared by MMF generated from the two sets of orthogonal windings as shown in Figure 12.28. The 80 mm square sample was placed at the centre of the frame. Small air gaps and chamfered yokes both improved the field uniformity in the centre of the sample. The investigators estimated that the magnetisation conditions were sufficiently uniform over the central 20 mm of the sample and this ratio of sample size to measurement area is typical for most of the geometries considered here. Magnetisation arrangements of this type are now the most commonly found in magnetics laboratories.

The approach of utilising an integrated flux closure path was further developed in [55] where a laminated stator stack with a three-phase winding was used to magnetise a round sample, see Figure 12.29(a). It was found that significantly higher flux densities were achievable in this arrangement than those used previously. A similar approach was also made in [56] and [57] using a two-phase winding which offered slightly reduced complexity but retained the capability for rotational measurement at very high B (up to 2.0 T). The comparative ease of controlling the flux loci in this arrangement goes some way to compensating for the additional time and cost of preparing round samples which is usually achieved by EDM or laser cutting.

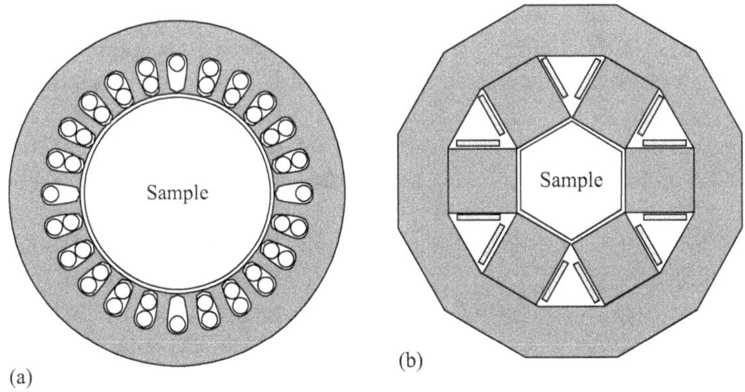

Figure 12.29 Rotational magnetisation systems using (a) a stator stack and (b) a three-phase hexagonal arrangement

A compromise between the round multi-phase magnetising arrangement and the simpler square sample is the use of a three-phase system with three pole pairs as in [58]. In this case the sample is hexagonal (Figure 12.29(b)) which, though more complex to cut than a square sample, can be cut with a traditional guillotine. This geometry was also applied to a tester for GO electrical steel on a large (160 mm) sample with the benefit of improved control in the highly anisotropic material [59].

Modern examples of these systems all utilise digital waveform control techniques similar to that described in Section 12.4.2 where each phase is controlled simultaneously. It places a large demand on the software and hardware and leads to costly waveform generation and acquisition hardware with at least double the number of inputs and outputs than uniaxial systems (this is a further drawback of the three-phase systems). Most common is B (or dB/dt) control where arbitrary B loci can be defined in terms of the relative dimensions of an ellipse or, as with uniaxial measurements, more complex magnetisation where each orthogonal component can be constructed from its harmonic content. Some systems also employ H control, e.g. [58], which adopts exactly the same approach but with the orthogonal H components as the reference waveforms.

12.5.3 Loss measurement

Measurement of loss in rotational systems largely mirrors that in uniaxial systems. The wattmeter, fieldmetric (with H sensors) and thermometric (with a temperature sensor) methods are all applicable but present greater challenges than in the uniaxial case, the most significant of which is the non-uniform magnetisation of the sample.

All types of 2D system require a measurement of the B vector. Due to the non-uniform magnetisation the measured area is confined to the central area of the sample which leads to measurement areas of typically 20 $\times$ 20 mm but with examples significantly larger [56], particularly when using round or hexagonal type magnetising fixtures. Regardless of the magnetising method, the variation towards the sample edges is extremely large and hence the errors in using a fully enwrapping B-coil would be large and difficult to compensate. Thus, localised methods of B measurement are applied to the uniform area – specifically search coils and needle probes.

Local search coils, as discussed in Section 12.4.3, are fabricated from fine wire (typically 0.1 mm) wound through a pair of drilled holes made with a fine drill bit (typically 0.3 mm diameter). For the purposes of 2D measurements a pair of orthogonal search coils are wound such that they cross in the centre of the sample. The alignment of these coils is critical as will be discussed later in this section. In order to keep close contact with the sample (and minimise left off H sensors) as well as limiting the damage of larger holes, most investigators utilise single turn B-coils. A consequence of this and the small cross sectional area is that the induced voltages can be extremely small[4] and, often, small signal amplifiers are employed. A potential drawback of this approach is the additional non-linear effects of the amplifiers which can be difficult to calibrate on multi-channel systems. Signal averaging is also used but precise triggering is required to avoid additional errors [60] which can be achieved post acquisition.

Needle probes offer an alternative to B-coils which is non-destructive and significantly reduces the sample preparation time. Pre-fabricated measurement heads with needle probes arranged to take orthogonal measurements are shown in Section 12.4.3 and are directly applicable to 2D measurements.

The H vector, if required, is usually measured using a pair of orthogonal H-coils. Since the accuracy of these coils is dependent on their distance from the sample surface, both coils are normally wound onto the same former (see Figure 12.30) and require coil winding equipment with high positional accuracy and fine control of wire tension. It is also important to take great care in handling as the fine overlaid wire is prone to damage and short circuits.

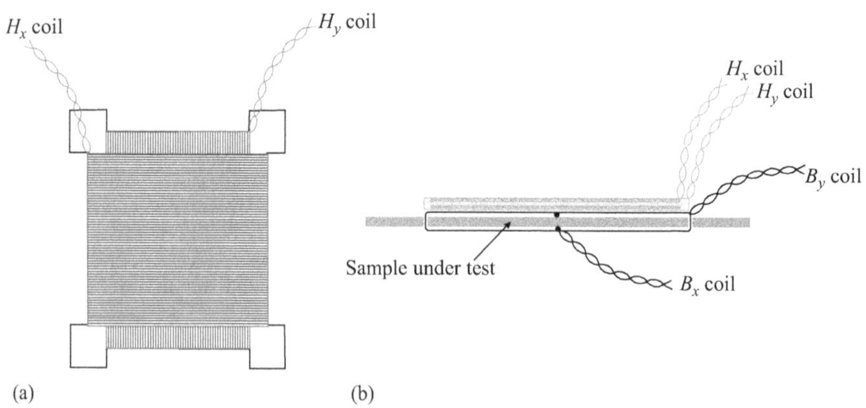

Figure 12.30 Orthogonal H-coils

[4]Zurek has estimated this to be as low as 12 μV at 1 Hz, 0.5 T [46] but a more manageable 0.6 mV at 50 Hz.

The H-coils are designed to cover the same measurement area as the orthogonal B-coils and are usually placed directly on top of the B-coils. More highly developed systems incorporate the sensors into a plastic sample cover such that their positioning is repeatable. As with uniaxial systems, multiple H-coils at fixed distances from the sample surface can be utilised to extrapolate to the H value at the surface which is particularly useful when the H-coil is encapsulated but has a significant overhead in additional processing required. At least one commercial system is known to employ this technique.

The H vector can also be determined from solid state field sensors such as a Hall sensor following careful calibration and compensation for amplitude and phase errors although these do not usually allow the *measurement* areas of H and B to be coincident. It is also possible to determine field from the magnetising current as in an SST; however, the use of yoke wound coils and the sample geometry make correlation of the field and current problematic.

From measurement of the orthogonal components of dB/dt (from the B-coil or needle probe voltages) and H (from the integrated H-coil voltages or other direct methods), the specific total loss can be calculated as described in Section 8.7.2 and repeated below:

$$P_r = \frac{1}{T\rho}\int_0^T \left(\frac{dB_x}{dt}H_x + \frac{dB_y}{dt}H_y\right)dt \ \left(\mathrm{W\,kg^{-1}}\right) \qquad (12.48)$$

This fieldmetric-based loss measurement method is probably the best compromise between accuracy and complexity but it still gives quoted total uncertainties in the range of 8%–12% [61]. Whilst many of the sources of error have been addressed above, one particular source (and an artefact of it) is the source of several investigations. Researchers found large differences in the rotational loss measured when rotating the B vector clockwise (CW) or counter clockwise (CCW) at high flux densities, with extreme cases showing negative loss in one direction and positive in the other. It was found that the majority of this difference could be explained by alignment errors in the B- and H-coils. For the B-coils, even with accurately drilled holes, the larger diameter of the hole relative to the wire allowed angular errors of the order of $\pm 0.01°$ to $\pm 0.02°$. With the H-coils, the angular errors could be calculated precisely by placing the H-coil on a rotational stage within a solenoid and measuring the angle between the zero crossings for each winding [62]. Errors in this angle were typically found to be an order of magnitude greater than those in the B-coils. Once these angles, and the relative angle between the B-coil and H-coil, were known (for an H-coil the value would be constant) the loss values could be compensated as shown in Figure 12.31.

A more straightforward approach, however, is simply to average the CW and CCW data since this can be shown to have the same effect as compensation [63]. It should be noted, however, that errors in orthogonality of the B-coils will result in an error in the shape of the B loci which is not easily recoverable.

Thermometric methods have also been applied to the measurement of rotational loss using the same method as outlined in Section 12.4.3 [64] with thermocouples

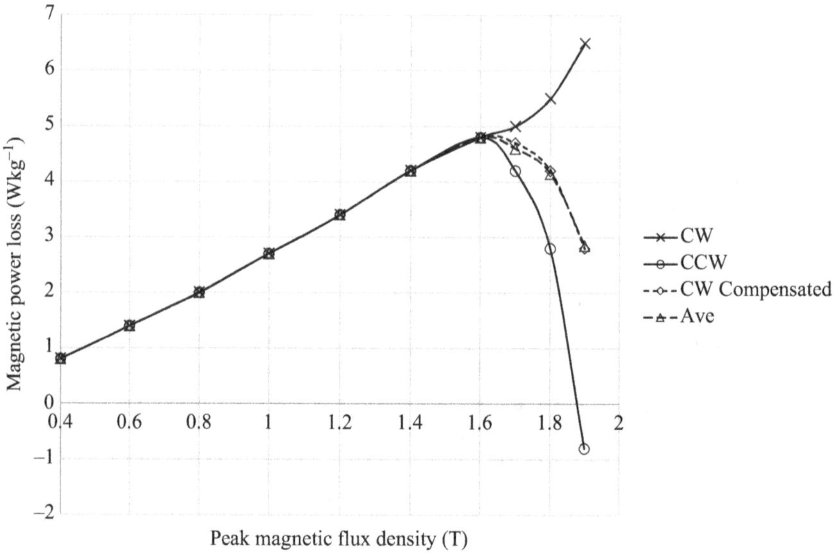

Figure 12.31 Rotational power loss values in CW and CCW directions (adapted from [62])

being applied to round samples which were enclosed in an adiabatic chamber under partial vacuum. This method suffers from fewer sources of uncertainty than the fieldmetric method but it is costly and complex to implement, such that the desired level of accuracy is achieved.

The variety of geometries and sensing methods inevitably leads to poor agreement between laboratories with a round robin between six European laboratories giving a relative standard deviation of 15% [65]. With no convergence towards a single geometry there is unlikely to be a standardised approach in the near future unless there is a significant demand from industry.

12.6 Magnetostriction measurements

Magnetostriction, as described in Section 3.16, is a strain induced by the magnetisation of a material and, as such, its measurement can be made by any strain or displacement transducer which can give suitable resolution. For the ranges of materials covered in this book typical measured magnetostriction may be of the order of 0.01×10^{-6} to 50×10^{-6}. Therefore, depending on the sample size (or more correctly the magnetised length) the measured displacement could be significantly less than 1 μm.

Magnetostriction can be measured under d.c. or a.c. excitation with some transducers being applicable to both. The range of test systems and geometries utilised mirror those described above for a.c. and quasi-d.c. measurements, with a single sheet arrangement being the most popular.

Also of interest is the stress sensitivity of magnetostriction as it is, perhaps, the magnetic parameter which is most sensitive to changes in the magnetic domain structure and, therefore, to applied stress. It is thus common for magnetostriction measurement systems to incorporate a mechanism for applying linear stress.

Currently, no standard exists for magnetostriction measurement although a technical report was published by the IEC [66] reviewing the main techniques employed in magnetics laboratories. Due to its strong link to transformer noise, IEC TC68 is working towards an a.c. magnetostriction measurement standard for electrical steels.

12.6.1 Magnetostriction parameters

For a.c. magnetostriction, there is a wide range of parameters which can be measured or calculated; the main ones are summarised in Table 12.1. Figure 12.32 shows a plot of instantaneous magnetostriction against instantaneous polarization, typically known as a butterfly loop due to its characteristic shape.

12.6.2 Magnetostriction measurement transducers

There are many techniques used for measuring magnetostriction including: capacitive plates [67,68], piezo-electric displacement transducers (record stylus) [69,70], inductive or LVDT (linear variable differential transducer).

One of the most widely used methods of measuring displacement is the resistance strain gauge [71–73]. The operating principle of these devices is well known. Upon extension or contraction of the substrate the length of the gauge wire or foil is altered and, since the resistance is proportional to the length and inversely proportional to the cross sectional area, the resistance will change in proportion with

Table 12.1 Summary of measured magnetostriction parameters

Parameter	Description
Peak or zero-peak	The magnitude of the peak value of magnetostriction which can be positive or negative. Care should be taken with this value as it can be affected by d.c. offsets in the measured characteristic due, for instance, to integrators. An alternative definition of zero-peak is the instantaneous value of the magnetostriction at the peak value of magnetic polarization which can be very different to the definition above
Peak to peak	The magnitude from peak to peak of the magnetostrictive strain waveform
Harmonic amplitude	The amplitude of each of the harmonics of the magnetostrictive strain calculated by Fourier transform
Magnetostriction velocity	This can be measured directly, integrated from acceleration, or differentiated from displacement and can be represented in each of the three forms above
Weighted summations	When seeking correlation with machine noise it is popular amongst investigators to use a weighted summation of the harmonic amplitudes. The most popular of these is the A weighted summation as it takes into account the average response of the human ear

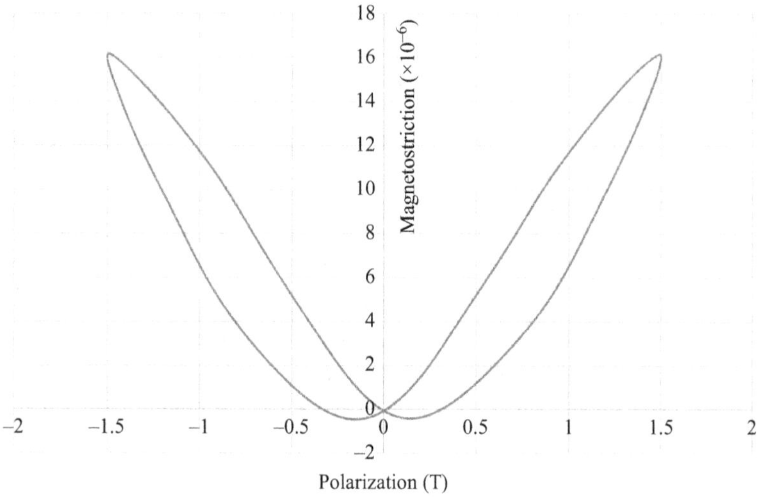

Figure 12.32 Typical measured butterfly loop for uncoated GO electrical steel at 1.5 T, 50 Hz

the strain. Highly sensitive strain gauge bridges are readily available commercially which can be used to attain the required resolution. It can be further improved with the use of 'dummy' gauges, a second identical strain gauge connected into the opposite arm of the bridge to the active gauge and used to compensate for temperature and other variations. This gauge should also be attached to a sample of the same type as that being magnetised but sitting outside the magnetic circuit.

The strain gauges are attached to the sample with an adhesive (cyanoacrylate or similar) and are available in a wide range of sizes and types including rosette gauges which enable the measurement of 2D strain tensors for rotational magnetostriction. They are incredibly useful for low volume d.c. or a.c. testing and, particularly, for localised measurements of magnetostriction (investigations have included grain to grain variation in GO steel [74]). Care should be taken when using them on lamination materials due to their high sensitivity to strip bending (unless a differential measurement on both sides of the strip is possible) and also when using on inhomogeneous materials when the gauge dimensions are of the same order as the feature size.

Several laboratories have constructed bespoke test systems for testing larger volumes of electrical steel samples in an SST arrangement. These have tended to converge on a limited number of sample geometries, magnetisation arrangements and vibration transducers which are reviewed in an International Electrotechnical Commission (IEC) technical report [66].

Samples are typically 500 mm × 100 mm, 600 mm × 100 mm or 305 mm × 30 mm. The larger samples tend to be favoured due to the smaller influence of sample preparation, meaning that they can be used 'as cut' and are therefore more representative of the finished product. This also means domain refined material can be tested.

The magnetic circuits used for magnetisation of the single strip samples take the form of single or double return yokes which are either normal to the plane of the sample (the standard SST geometry) or in the same plane as the sample. The advantage of the in-plane yokes is that they allow a double yoke arrangement without imparting a force on the sample surface. However, there are more complex 3D flux paths around the poles of the yokes than in the standard SST.

The majority of test systems also incorporate the application of linear stress due to the high stress sensitivity of magnetostriction. In almost all cases, this is achieved using a guided pneumatic cylinder such that torsion cannot be applied to the sample which is driven via an electro-pneumatic valve which modulates the input pressure to the cylinder in proportion with a d.c. control voltage. Combined with an in-line load cell, a closed loop control system can be easily implemented using a PC and data acquisition hardware. A more detailed description of typical stressing systems is given in Section 12.4.4. Combining this with magnetisation control, as described in Section 12.4.2, and acquisition of the vibration signals creates an all-in-one system for the automatic assessment of magnetostriction. Systems such as [75] operate in compression only, up to -5 MPa, whilst others [76,77] are designed for a more complete assessment of stress sensitivity over a stress range of $+10$ MPa to -10 MPa.

The transducers used in these systems are primarily based on optical methods or piezo-electric accelerometers with the accelerometers used in those systems used for wide stress range measurements. Optical methods have the significant advantage of being non-contact, save for the bonding of extremely low mass mirrored targets to the samples avoiding any loading of the sample which could dampen the magnetostrictive vibrations. When using contact measurements, careful evaluation of the influence of the additional mass must be made.

The most common optical sensors utilise in-plane laser vibrometers or laser displacement meters. Vibrometers utilise the Doppler effect to measure the velocity of the vibrating target with a laser-beam typically generated by a helium-neon laser but, more recently, with solid state lasers. The measurement is usually the differential between two target mirrors some distance apart in the uniformly magnetised region of the sample and uses two laser heads for simultaneous measurement at the two points [75,78], as shown in Figure 12.33.

The laser displacement sensor uses an optical heterodyne interferometer which measures displacement from the beat signal phase difference due to a reference beam and that reflected from the target. References [79] and [80] demonstrate the use of these sensors for in-plane measurement of the magnetostrictive displacement and find good agreement with a laser vibrometer, although [80] shows some differences in the butterfly loop at lower flux density (0.5 T).

Other optical methods include a laser linear encoder (utilising a fine grating attached to the sample in a vertical SST) [81] and a speckle interferometer [82] both of which position the laser such that the beam is normal to the plane of the sample. Ultra-thin materials such as amorphous ribbon can be measured using a fibre-optic dilatometer where the sample acts as a partial shutter in a cut optical fibre [83]. More information on magnetostriction measurement in amorphous materials is available in detail from Squire [84].

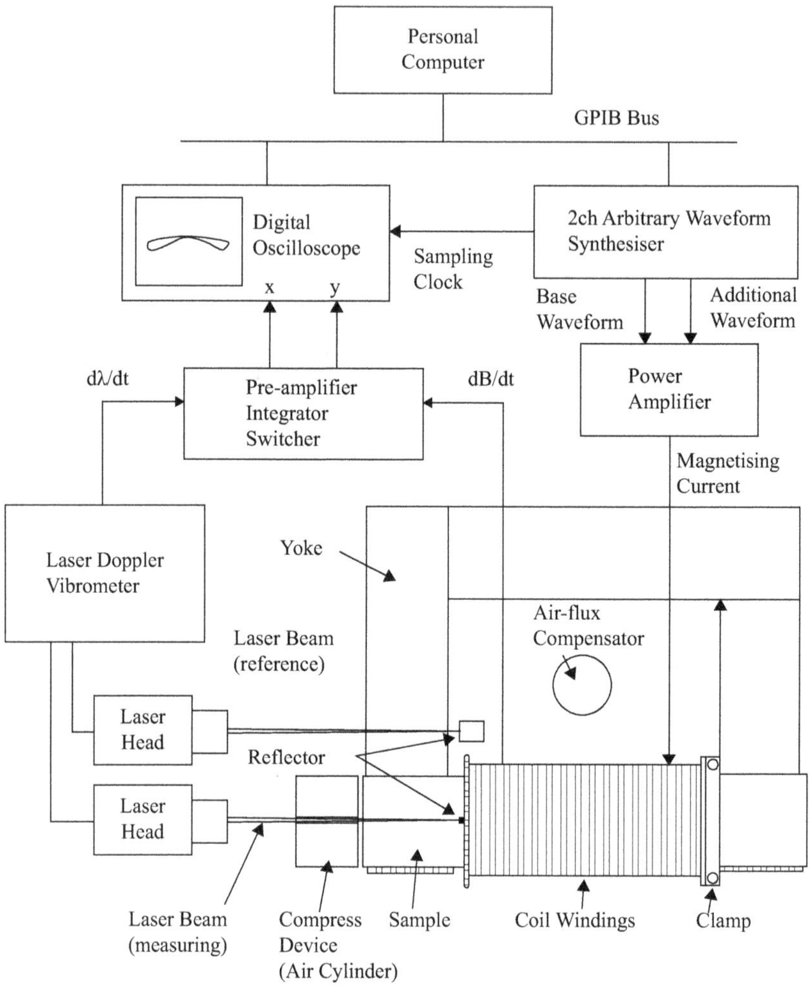

Figure 12.33 Magnetostriction measurement system based on twin head laser vibrometer [75]

Piezo-electric accelerometers are widely used in the field of industrial vibration measurement. They consist of a piece of artificially polarised ferroelectric ceramic which, when mechanically stressed, generates an electric charge across its pole faces proportional to the applied force. Therefore, when the sensor is accelerated a charge is developed which is proportional to the acceleration due to either a seismic mass attached to the ceramic element or through the bending of the element itself.

A magnetostriction measurement system utilising a pair of piezo-electric accelerometers was described in [76] and [77]. The differential output from these sensors is measured and double integrated to provide a displacement signal free from any external vibrations. The output from accelerometers is usually in the form of a

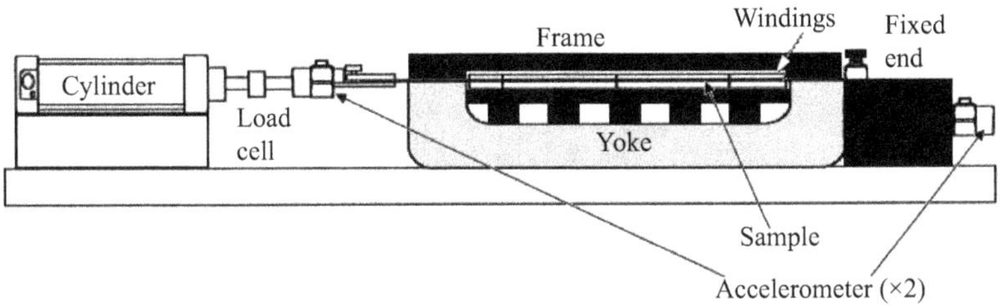

Figure 12.34 Diagram showing placement of accelerometers in magnetostriction measurement system [77]

charge. However, lightweight (a few grams) accelerometers are now available which include line drive electronics and so produce a voltage output, offering a low impedance output and therefore high immunity to noise and insensitivity to cable motion.

Systems utilising accelerometers have included them within the clamping arrangements for the stressing mechanism with one mounted near the fixed end of the strip and the other integral to the clamp which connects the strip to the stress actuator. When using this arrangement the strained length is usually assumed to be the same as the mean magnetic path length for the SST.

A system developed from [77] incorporates a single sided SST for 500 mm × 100 mm samples with fully enwrapping secondary and primary windings (Figure 12.34). The closure yokes are designed such that the sample protrudes at either end. One end of the sample is clamped to a large non-magnetic block which, along with the yoke and windings, is firmly fixed to the base of the system. At the other end of the sample a lightweight clamp houses a low mass, high sensitivity accelerometer which, along with a reference accelerometer at the fixed end of the sample is used to measure the magnetostrictive vibrations.

The clamp at the free end of the strip is connected to a pneumatic linear actuator. This actuator serves two purposes. Guides and linear bearings within the actuator limit the freedom of movement of the free end clamp such that it may only move in the direction of the strip extension (as discussed in Section 12.4.4). Second, the cylinder allows the application of linear stress to the strip. Electro-pneumatic regulators modulate the pressure in the actuator under the control of a LabVIEW virtual instrument utilizing an in-line load cell. The virtual instrument (VI) runs on a PC equipped with a high speed, multi-channel, simultaneous sampling, analogue input and output card. The same VI also performs all of the other system control and data acquisition including:

- magnetization control utilizing fast digital feedback algorithms (identical to those shown in Figure 12.15) to allow the flux density waveform to be controlled to an arbitrary shape;
- measurement of magnetostrictive acceleration and digital integration for measurement of velocity and displacement (magnetostriction) together with harmonics and weighted summations and
- calculation of power loss and permeability.

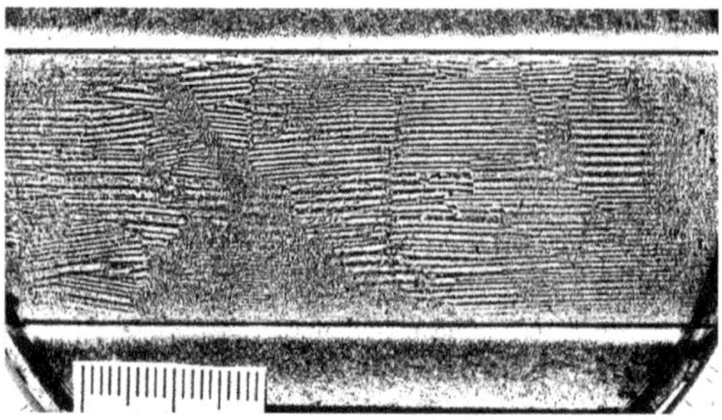

Figure 12.35 Domain image acquired under compressive stress using the Bitter technique

This allows the system to perform fully automatic tests at a range of applied stress from +10 MPa to −10 MPa. At each stress point, the measurements described above are performed to give the simultaneous evaluation of all of the important magnetic parameters.

As discussed above, measurement of magnetostriction (particularly under applied stress) is extremely challenging with several possible sources of error. Verification of the measurement with particular reference to these challenges is essential to provide confidence in the measured characteristic and several experiments have been undertaken to verify the results from the system described above [85].

Stress uniformity was assessed through the application of strain gauges along the length of the measured strip (a modified top plate was created to accommodate the gauges and connecting wires). No variation in stress could be observed along the magnetized length of the sample after following the steps concerning alignment outlined above.

Confirmation of measured magnetostriction under applied stress was achieved through static domain observations with no magnetising windings in place. A magnetic domain viewer utilizing the Bitter technique was used together with a digital camera to acquire domain images at applied stresses from −10 MPa to +10 MPa. These images were analysed by LabVIEW Vision software (as shown in Figure 12.35) to calculate the percentage increase of stress patterns in comparison to the total area of the observed surface. To calculate magnetostriction ($\lambda_{measured}$) under stress the following equation, based on the coverage of stress patterns (C%) was used.

$$\lambda_{measured} = \frac{\lambda_{maximum} C\%}{100\%} \tag{12.49}$$

where $\lambda_{maximum}$ is the maximum magnetostriction of a sample subjected to a compressive stress just sufficient to produce the maximum magnetostriction.

It is assumed that the sample is populated entirely by grains which exhibit a domain configuration, first described by Dijkstra and Martius [2], named Stress Pattern I and that it behaves as a single grain which has an angle of yaw from the rolling direction, ϕ, which represents the average angle of misorientation for the polycrystalline material. With reference to the cross section through the grain

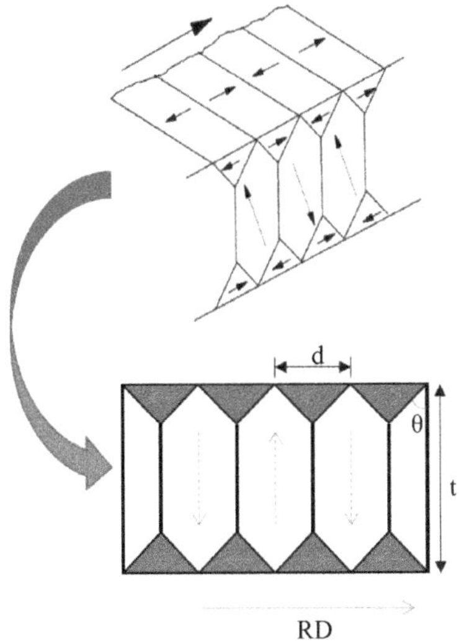

Figure 12.36 Longitudinal section of domain structure through a grain exhibiting Stress Pattern I

shown in Figure 12.36, the magnetostrictive strain, when an external magnetic field is applied, can now be calculated if it is assumed that, at magnetic saturation, B_s, the grain consists only of [001] domains aligned in the magnetising direction.

At a magnetic flux density, B, the volume fraction of [001] domains is given as

$$\text{Volume fraction} = \frac{B}{B_s}\cos\phi \tag{12.50}$$

Therefore the maximum magnetostriction is given by

$$\lambda_{\text{maximum}} = \frac{3}{2}\lambda_{100}\cos 2\phi\left(\frac{B}{B_s}\cos\phi - \frac{d}{2t}\tan\theta\right) \tag{12.51}$$

An example is shown in Figure 12.37 for a 0.30 mm thick conventional GO silicon steel magnetised to 1.70 T. $\lambda_{100} = 2.4 \times 10^{-5}$, $B_s = 2.03$ T. The average value of the closure domain width, d, was measured using LabVIEW Vision software to be an average of 0.07 mm and the angle the closure domain makes with the surface plane, ϕ, was assumed to be 55°.

As can be seen from Figure 12.37, the magnetostriction measured by the mms agrees well with the results from the calculation based on the domain patterns.

12.6.3 Rotational magnetostriction

Rotational magnetostriction is the strain measured when the sample is rotationally magnetized as defined in Chapter 8 and measurements of it are usually made by the systems described in Section 12.5.2.

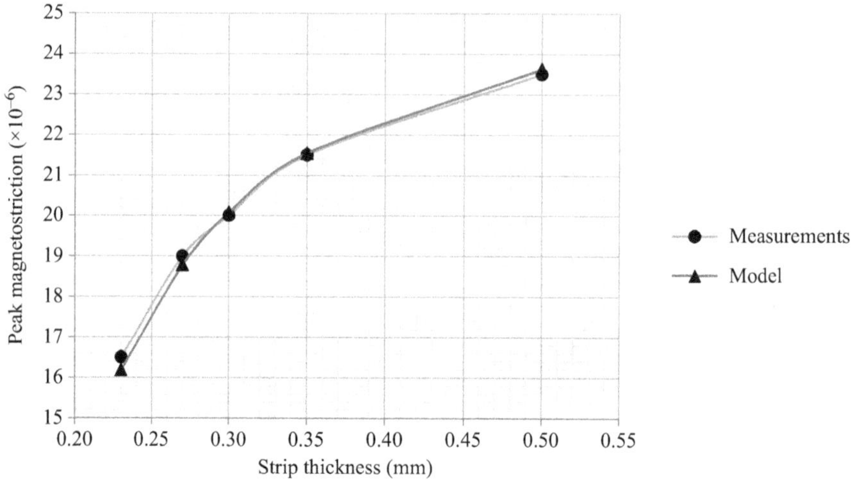

Figure 12.37 Measured versus calculated magnetostriction for a conventional GO silicon steel under applied stress

Most measurements are taken with strain gauges in the form of a rosette gauge which is three overlaid gauges normally positioned at 0°, 45° and 90°. Using the three gauges the strains (ε_a, ε_b and ε_c) can be resolved into two orthogonal components (ε_x and ε_y) and a shear component (γ_{xy}) using (12.52) which completely describes the strain on the surface [86] (Figure 12.38):

$$\begin{bmatrix} \varepsilon_a \\ \varepsilon_b \\ \varepsilon_c \end{bmatrix} = \begin{bmatrix} \cos^2\theta_a & \sin^2\theta_a & \cos\theta_a \sin\theta_a \\ \cos^2\theta_b & \sin^2\theta_b & \cos\theta_b \sin\theta_b \\ \cos^2\theta_c & \sin^2\theta_c & \cos\theta_c \sin\theta_c \end{bmatrix} \begin{bmatrix} \varepsilon_x \\ \varepsilon_y \\ \gamma_{xy} \end{bmatrix} \tag{12.52}$$

where θ_a, θ_b and θ_c are the angles of the three gauges relative to some reference.

These gauges are normally positioned under the B-coils (often on the opposite side of the sample to the H-coil) and connected to a three channel strain gauge bridge.[5] This may be configured as a full, half or quarter bridge with the full or half bridges having the possibility of including dummy gauges for compensation of temperature effects.

Examples of these measurements on square [87], hexagonal [88] and round [89] samples show the effectiveness of the method.

[5]Data acquisition hardware incorporating strain gauge bridges is now commonly available and in many cases will allow internal synchronisation with other hardware for capturing B and H.

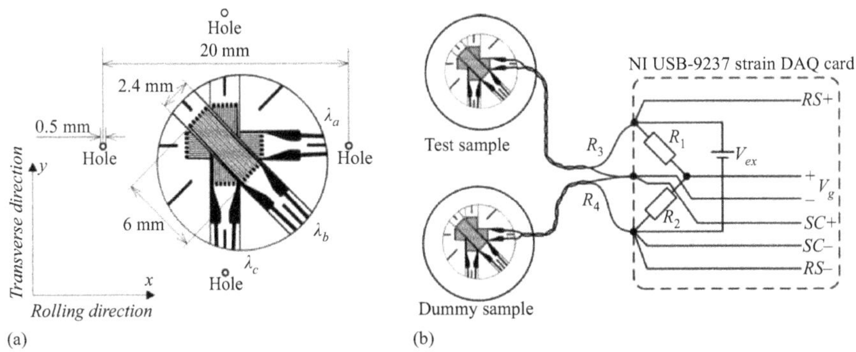

Figure 12.38 Rosette strain gauge for (a) measurement of rotational magnetostriction and (b) half bridge connection [86]

12.7 On-line measurements

12.7.1 Practical challenges

All of the measurements discussed previously are made off-line and are generally made on a small sample which is considered representative of a much larger batch of material. In the case of electrical steels, approximately <1 kg of material is tested from coils weighing up to 9 tonnes. Ideally the whole batch of material would be magnetically tested as part of the production process and the off-line tests discarded saving considerable time and effort. Whilst it is the case that an on-line test is made routinely in almost all soft magnetic material production, the challenges associated with the measurement mean that they cannot replace the off-line test with the same degree of confidence. These challenges include:

- line tension
- line speed
- strip deformation and joining
- strip dimensions
- strip temperature
- earth's field effects
- material anisotropy

The majority of on-line testers for strip material utilise a similar geometry to the SST with enwrapping windings and a pair of flux closure yokes aligned in the rolling direction of the strip. Non-enwrapping solutions have been investigated (and is discussed later in this section) which offer significant practical advantages but magnetisation uniformity and correlation with off-line measurements are significantly affected.

These testers are normally placed towards the finishing end of the final process line[6] where the strip is subject to significant line tension due to the nature of the coil

[6]Testers can also be used earlier in the process for control and prediction purposes.

to coil process. Typically, this will be in the region of 10 MPa, but could be as high as 50 MPa close to the coiler for some materials. As discussed previously, all of the magnetic properties are sensitive to applied stress (although tension usually leads to an improvement in steel properties) and, depending on the material grade, changes of several per cent could be observed in loss measurement. This can be compensated for based on, grade specific, stress sensitivity data from off-line trials but requires monitoring of the line tension which can be estimated from the roller motor drive currents. Additionally, variations in strip flatness or asymmetry in the final coil can create large variations in the tension across the strip or even create through-thickness stress gradients which would have a much greater influence on the measured properties. Alternatively, loop sections can be introduced to create a low tension region for the test [90], negating the need for compensation at the expense of additional cost, complexity and line length.

The strip is also in constant motion with line speeds typically in the range 45–120 metres per minute (0.75–2 ms^{-1}) and some newly installed facilities are reportedly much faster. Due to non-uniformity in the applied field, the strip velocity can introduce significant extra eddy currents in the strip. This has been investigated by Havlicek [91] who estimated that at line speeds below 1 ms^{-1} the effect is negligible but becomes significant above approximately 2 ms^{-1}.

If the tester is at the finishing end then it is likely that the strip is flat and edges have been trimmed. However, due to the line speed small edge waves or other defects carry significant kinetic energy and could potentially damage test equipment. Due to the continuous nature of the process, coils are also joined together by a welded joint which will have to pass through the tester. For this reason the flux closure yokes require a significant air gap between the pole faces (typically 20–40 mm) with the internal clearance within the winding former being of the same order. Both the pole faces and the former also tend to be protected with sacrificial wear layers which are routinely maintained.

The introduction of these air gaps leads to increased flux leakage and therefore reduces the uniformity of magnetisation along the magnetised length of the material. This, in turn, increases the errors in the magnetic path length used for calculation of the magnetic field strength. Mitigation of these effects could be achieved through the use of an RCP and compensation windings as demonstrated in [92], although it was explained in [93] that control of the compensation winding current in a live trial was problematic, sometimes resulting in increased errors. A simpler solution was proposed [93] whereby the magnetising winding was separated into five identical coils distributed along the length of the sample with each having the same controlled voltage applied giving a more even distribution of MMF and reducing the flux density drop off at the coil ends. Although this method showed an improvement in flux density distribution it increased the error in measured H due to the relatively high currents in the outer-most coil sections. As with the SST, the accuracy of H measurement can be improved with an H-coil. However, these will be bound by the same requirements for strip clearance as the pole faces which violates the fundamental principle of H-coil measurements, i.e. close proximity to the surface. Multiple H-coils will require former clearances (and therefore winding area) to be further increased. Despite this, static strip tests showed that H-coil measurements offered significant accuracy improvements over the magnetising current but this advantage was negated by a much larger variability in dynamic tests [93]. This variability was not explained but was likely to be due to dynamic variation in the strip pass line.

Establishment of the correct test induction (typically 1.5 T) is dependent on accurate determination of the sample area. In an off-line test this is achieved by measurement of the sample mass and, with knowledge of the conventional density, the area can be determined to a high level of accuracy. For the on-line test, this is usually determined with a thickness gauge (typically based on ionising radiation) close to the centre line of the strip which averages the thickness over the conventional path length of the tester. These gauges are regularly calibrated and offer reasonable levels of accuracy but variation of thickness across the width cannot be determined.

Compensation for strip temperature can also be important depending on the ambient temperature and the proximity of furnaces, with temperature ranges of more than 100 °C possible. As with tension compensation, this can be achieved with compensation factors determined off-line.

The orientation of material production lines, with respect to the earth's magnetic field, is not normally considered when new plant is commissioned but, in the case of magnetic materials, it can have an impact on the on-line measurements. Unless oriented perpendicular to the static field[7] the strip will be subject to a d.c. magnetic field which can be observed as an asymmetric *B–H* loop. The influence of this field will be dependent on the permeability and geometry of the material so the test equipment must include this as an additional control parameter. This is usually accomplished with a simple proportional control algorithm.

A rolling direction on-line tester is sufficient for GO electrical steel, whose properties are graded along this direction, but for NO steel the properties are an average of the rolling and transverse directions. There can be significant texture variations of NO materials and a measurement based on one direction only does not give the complete picture. On-line testers capable of measurement in the transverse direction have been proposed but are extremely challenging to implement. Clearly, a standard enwrapping winding is not applicable to the transverse direction but Z-shaped coil geometries such as those shown in Figure 12.39 have been proposed [93,94] where magnetisation in the centre section is in the transverse direction.

With this measurement geometry, the yokes are also designed to provide flux closure over the same area but, given that the air-gap requirements are the same as for the longitudinal test, the ratio of the effective winding length to the air gap is small resulting in an inhomogeneous *B* distribution. Multi-section or compensating end windings can again mitigate this to some extent but presents a much greater challenge than in the rolling direction. It should be remembered that, in most cases, the permeability will be higher in the rolling direction and this further compounds the difficulty of uniaxially magnetising the material in the transverse direction.

[7]The orientation of the earth's field could be influenced by local structures which are ferromagnetic.

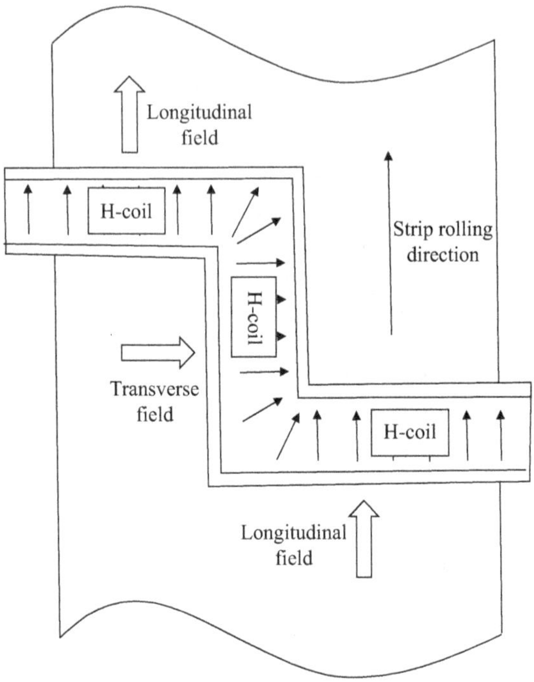

Figure 12.39 Proposed configuration for a transverse on-line tester for electrical steels (adapted from [93])

12.7.2 Non-enwrapping systems

The dynamics of a continuous moving strip mean that passing it through a closed winding former is limiting in many respects. Maintenance, in particular, is difficult as the strip must be threaded through the tester. A tester which does not enwrap the strip, or ideally is on one side of the strip only, therefore offers significant advantages.

The flux due to a single row of straight conductors all carrying current in the same direction will produce a field perpendicular to the current direction. Normally, in an enwrapping winding, this is mirrored with a second row of conductors with current flowing in the opposite direction (of course it is a continuous wire wound as a solenoid but we can consider them as separate conductors) and the strip runs between the two rows of conductors. If, instead of the current returning on the other side of the sheet, the returning conductors are bundled together and returned on the same side of the sheet but over an infinitely long path then a similar field profile (albeit of approximately half the magnitude for the same current) will be created. Figure 12.40 shows this arrangement where PQRS is the current return path which is of the order of several metres (as an approximation of infinity). A further iteration of this approach is to use two sheets of conductors with opposing polarity such that they create regions of opposing flux which is forced normal to the sheet at the midpoint as a so-called 'flux fountain'. This enables measurement of

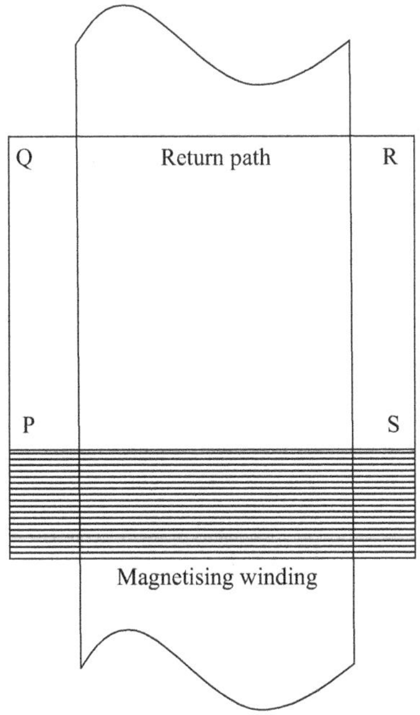

Figure 12.40 Single sided, non-enwrapping magnetising winding with far remote return conductors [94]

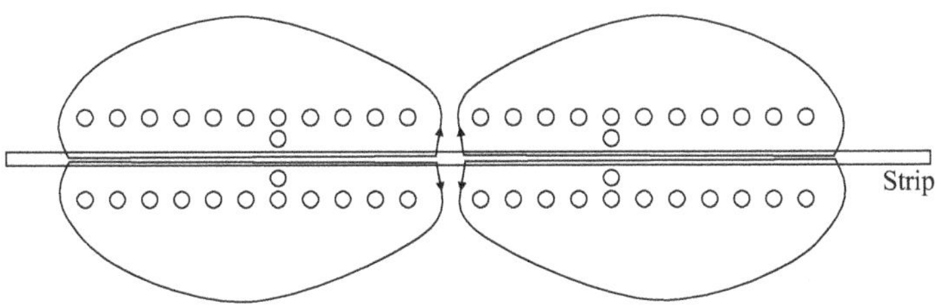

Figure 12.41 A pair of single sided, non-enwrapping magnetising windings with normal flux sensing coils [94]

flux by a normal coil (a pair of normal coils would be preferable as they would account for variations in the strip pass line) about the centre line of the magnetising coils (Figure 12.41).

Further development of this approach leads to the arrangement in Figure 12.42 whereby flux closure yokes have been introduced to better control the return flux path and H-coils for field sensing.

A frame of this type was also investigated by Khanlou *et al.* [95] who estimated a power loss prediction uncertainty as within 5%. Further development of this test by the same investigators saw the addition of an RCP and compensation windings with an improved prediction uncertainty of 2.5% [96].

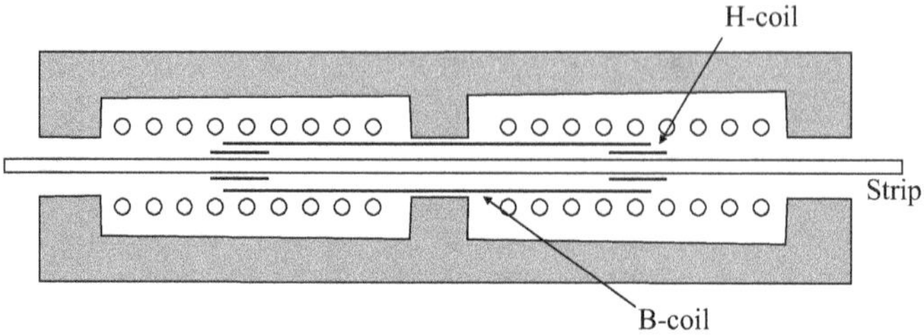

Figure 12.42 A pair of single sided, non-enwrapping magnetising windings with normal flux sensing coils, flux closure yokes and H-coils [95]

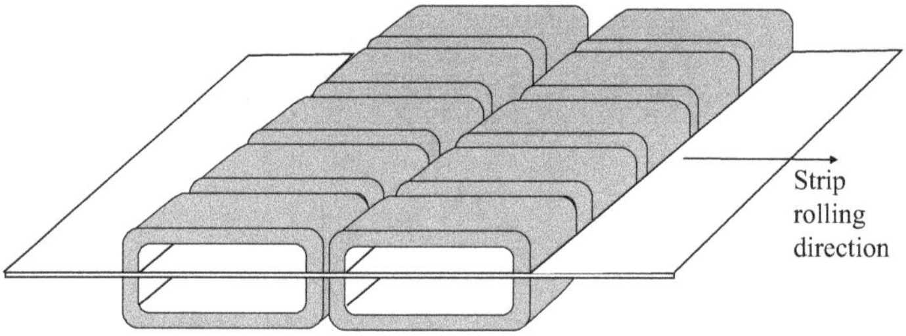

Figure 12.43 Array of flux injection probes for on-line power loss testing [98]

An alternative to a full width on-line test was proposed by Iranmanesh *et al.* [97]. An array of 20 flux closure yokes, arranged as shown in Figure 12.43, were each wound with magnetising windings and connected in parallel such that the flux passed from the outer pole faces to the inner, i.e. the two rows were oppositely magnetised. Flux sensing was achieved by a combination of search coils on the paired centre limbs and top of each yoke. Although these search coils were all interconnected in this case, an obvious further development of this arrangement would be the sensing of each yoke individually allowing a measure of cross strip variation in properties. Being non-enwrapping, the test frame was able to be rotated such that it could measure in the rolling or transverse directions.

12.8 The d.c. magnetic measurements

Measurement of the magnetic properties of ferromagnetic materials under d.c. conditions offers significant additional insight into the behaviour of the material and is essential data for many modelling approaches. All of the methods discussed below are used to obtain the complete four quadrant *B–H* or *M–H* loop and the normal magnetisation curve. Several of the methods can also be used to obtain d.c. minor loops or reversal curves.

As described in Section 12.3.2, there are several methods of measuring the magnetic field strength which are independent of frequency and can be used for d.c. measurements. For the measurement of flux density or magnetisation there is no escaping Faraday's law of electromagnetic induction and a B-coil. For d.c.

measurements however, it is clear that there will be no voltage in the coil if there is no change in B with time.

There are several techniques which are employed to allow measurement of the d.c. induction, which will be discussed in this section together with measurement instruments which utilise them. They are:

- employ a very slow varying field to introduce a very small dB/dt (quasi static);
- use the transient which occurs when a d.c. field is applied or removed (point by point);
- vibrate the sample within a d.c. magnetic field and
- vibrate the B sensing coil.

12.8.1 Quasi-static measurements

The simplest way to measure the d.c. properties is actually to use an alternating signal at a frequency low enough so that classical eddy current loss is negligible and to allow the magnetisation to respond to the field. In reality, this means cycle times of greater than 60 s (sometimes up to 600 s) and, possibly, some form of control over the rate of change of magnetisation.

The result of very slow magnetisation changes is that the voltages to be measured at the B-coil are incredibly small and require highly sensitive measuring instruments. The flux density in the sample is often measured by an integrating fluxmeter which, in modern instruments, is a low drift, high accuracy integrator coupled with a sensitive digital volt meter.

The integrating fluxmeter produces an output directly proportional to the flux density. The integrator constant is usually compensated for internally. Most instruments also allow the input of N and A to give direct B output.

Standard methods for quasi-static d.c. measurements are described in ASTM A773/A773M – 14 [98] and IEC 60404-4 [99]. These standards describe tests on ring samples (as described in Section 12.2.3) or with a permeameter.

The permeameter frame is similar to an SST with a pair of flux closure yokes and a magnetising winding that either enwraps the sample (type A) or the yokes (type B). Flux density sensing is achieved via a 10–50 mm long B-coil wound directly onto the sample using a fine, insulated, wire. If the sample has sharp edges, it is often necessary to protect the B-coil's insulation with insulating tape or paper. If the error in B due to the additional air flux is too great this can be compensated for using

$$B_{\text{Corrected}} = B - \mu_0 H \frac{A_c - A}{A} \quad \text{(T)} \tag{12.53}$$

where A_c is the cross sectional area of the B-coil (m^2) and A is the cross sectional area of the sample (m^2).

The coil is connected to an integrating fluxmeter with tightly twisted lead wires. Alternatively, a J-coil can be used which consists of three concentric windings of which one is the sensor winding and the remaining two are the compensation windings. The ratio of the area of the compensation windings is determined from the area and number of turns on the sensor winding (as described in annex C of [99]). The magnetic field strength is measured using either an H-coil and a second integrating fluxmeter or a longitudinal Hall probe with appropriate gaussmeter. The H sensors are placed on top of the B-coil at the centre of the sample.

The permeameter is designed to accommodate strips or bars through sets of pole pieces machined from a high permeability, low coercivity material to accommodate the sample shape and allow access for the required probes and coils.

Measurement of the ring samples is largely in accordance with a.c. measurements where two sets of windings are evenly wound on to the sample. H is determined from the voltage across a standard resistor in the ground return path of the magnetising winding and B is measured by connecting the secondary winding to an integrating fluxmeter.

Other sample geometries which are not covered in these standards include Epstein and SST for which the test method is identical to rings (note that these frames are normally compensated to measure J and not B). An electromagnet can also be used for the generation of much larger fields and/or smaller samples whilst open circuit measurements can be made in a solenoid where the sample shape is such that the demagnetising field is negligible in the magnetisation direction. The measurement method for these geometries is broadly similar to the permeameter method.

A schematic of a typical quasi-static measurement system [100] is shown in Figure 12.44. This system enables all of the geometries described above to be tested with H measurement by gaussmeter and Hall probe or series resistor and voltmeter. The magnetising windings are driven by a bipolar power supply (with a second connected in parallel for higher power requirements). All the measurement equipment is connected to a PC using a GPIB bus; the power supply output is proportional to a control voltage generated by the PC interface.

The test begins with a demagnetisation cycle which uses a low frequency (typically 1 Hz or lower) sinewave function modulated by a decay function as defined by [101] which decays by 10% every 1.5 cycles until a breakpoint is reached whereafter it decays by 1% of the breakpoint value every 1.5 cycles until it reaches zero. These parameters, along with the peak field, are user definable and are usually the subject of experience or a number of iterations of the test (until a symmetrical B–H loop is achieved). Demagnetisation is covered in more detail in Section 12.8.5.

The software then measures the B–H loop measurement. The simplest mode of operation is where dH/dt is kept constant (i.e. the H waveform follows a saw-tooth pattern). Here, the normal magnetisation curve and all four quadrants of the major magnetisation loop are split into a user defined number of equal H increments. The output current is then ramped to each of the H increments and the readings from the fluxmeter and gaussmeter (or voltmeter) are acquired.

On completion, the remanence and coercivity are calculated by interpolation around $H = 0$ and $B = 0$ axes. The normal curve is fitted using the Cholesky algorithm and differentiated to find the maximum permeability. The loss per cycle is calculated from the area of the loop using the trapezoidal rule.

The influence of the loop speed on the measured B–H loops (and consequently the calculated parameters) can be considerable for very soft materials (coercivity $<100\ \mathrm{Am^{-1}}$) and result in large errors [102]. Figure 12.45 shows a zoomed-in section of the measured B–H loop of a high permeability GO electrical steel for a range of loop speeds (with constant dH/dt) from 240 s down to 60 s. It is clearly

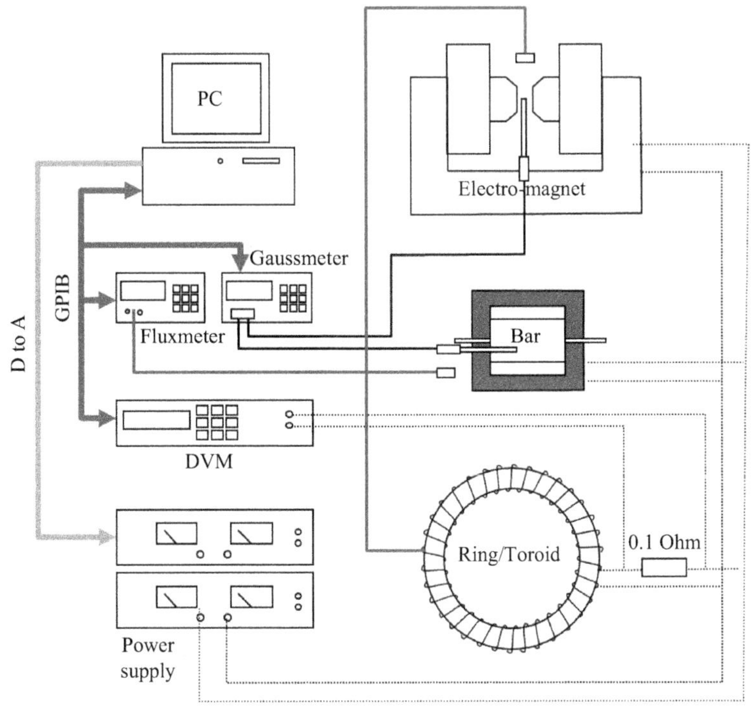

Figure 12.44 Schematic of a typical quasi-static magnetic measurement system [100]

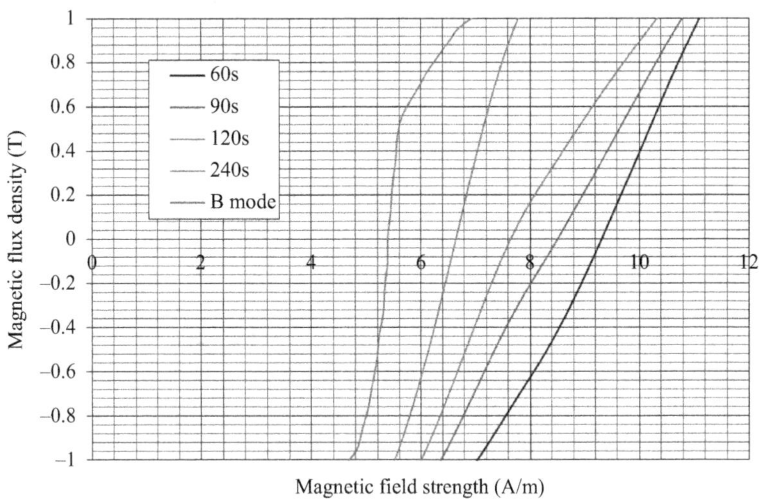

Figure 12.45 Plot of section of B–H loop showing influence of loop speed on the loop shape for low coercivity materials

shown that the loop width reduces significantly with increasing loop time. The plot of 'B mode' describes a simple proportional control algorithm which modulates *dH/dt* such that *dB/dt* closely follows a discretised target flux density versus time waveform.

Other methods for controlling the time derivative of *B* or *H* have been investigated. Squire *et al.* [103] used a truncated tangent function for *H(t)* which acts to spread out the peaks in the *dB(t)/dt* function. Alternative methods using proportional control algorithms are very slow due to the number of iterations required to achieve acceptable control but approaches utilising an inverse function have been shown to converge in just a few iterations [104]. In this method the *H* waveform generated by the previous iteration is used to plot the *B–H* loop which is pre-configured as a sawtooth on the first iteration. The descending half of the *B–H* loop is then isolated and shifted along the *H* axis so that zero magnetic field strength corresponds with zero magnetic flux density (it approximates an anhysteretic curve). This is then fitted to the general polynomial given in (12.54) using a least squares linear fit generated by the Householder algorithm:

$$H(B) = a_1 B + a_2 B^2 + a_3 B^3 + \ldots + a_{30} B^{30} \quad \left(\mathrm{Am}^{-1}\right) \tag{12.54}$$

A phase shift is then introduced between *B* and *H* to allow the formation of a hysteretic curve which is derived from the phase shift in the previous iteration and the resulting function is used to generate the *H* waveform for a sawtooth *B* waveform (constant *dB/dt*) by feeding it back to the beginning of the process. Further iterations are then processed until the measured *B* waveform describes a perfect sawtooth wave within pre-defined limits of the ideal form factor and THD. This algorithm is illustrated in Figure 12.46.

The other challenge of quasi-static measurements is drift of the magnetic flux integrator. Most modern instruments include automatic algorithms for tracking and compensating for integrator drift but these must be used with caution as the slowly increasing *B* close to saturation can be interpreted as drift by the software. Algorithms which track the drift over a period prior to the measurement (when the applied field is zero) and extrapolate the drift function over the measurement period have been proven to work successfully.

Drift can be minimised further with attention to the experimental set-up, specifically:

- connections between the B-coil and the fluxmeter should be minimised and all junctions made copper to copper to minimise thermoelectric potentials;
- temperature gradients and draughts should be avoided and junctions encased in a good thermal insulator (polystyrene) and
- power up the fluxmeter several hours before the measurement to ensure all components are temperature soaked and stable.

If these precautions are insufficient to prevent drift and the *B–H* loop does not close, the data can be corrected by using a method described by Garcia and Rivas [105] where a third-order polynomial representing the drift is solved assuming a symmetrical *B–H* loop.

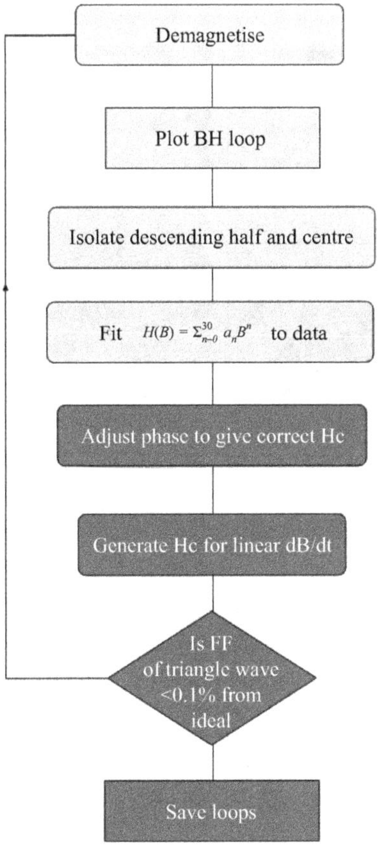

Figure 12.46 Block diagram of algorithm used to achieve controlled dB/dt waveform in quasi-static measurement system

12.8.2 Point-by-point measurement

The point-by-point measurement is an alternative method to the quasi-static method described by [98] and [99] but is rarely used due to the proliferation of computer controlled quasi-static systems.

The normal curve is plotted by incrementing the applied field and, at each stage, reversing the field using a reversing switch (with the integrating fluxmeter shorted). After approximately ten reversals the shorting switch is opened and the values of B and H recorded. This process is continued up to the peak magnetising field.

The full B–H loop is plotted in a similar manner starting at the peak positive field and reducing in multiple steps down to the negative peak field before reversing the process back to the positive peak field. At each stage the fluxmeter is shorted before the field is changed and opened after multiple reversals.

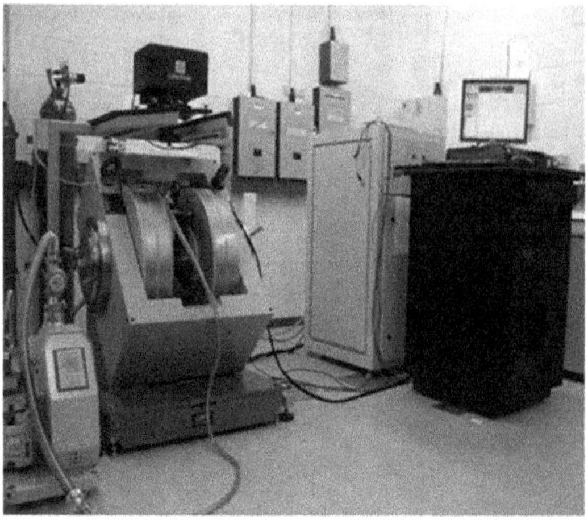

Figure 12.47 Photograph of a VSM showing the electromagnet (left), power supply (back right) and control unit (front right)

12.8.3 Vibrating sample magnetometer

The vibrating sample magnetometer (VSM) is perhaps the most commonly used method of characterization for magnetic materials in laboratories but is not often used for electrical steels or other bulk soft magnetic materials due to the small sample size and open circuit methodology. It does not offer an absolute magnetic measurement and requires calibration by transfer standards (usually nickel spheres [106]). It does, however, offer extremely high sensitivity and, depending on the size of electromagnet, high fields and is used specifically for the determination of the saturation magnetisation of soft magnetic materials.

In a typical system, such as that shown in Figure 12.47, the sample is mounted at the end of a rigid rod attached to a mechanical resonator which oscillates the sample, usually in a vertical direction, at a fixed frequency. Surrounding the sample is a set of pick-up coils. As the sample moves, its magnetic field, which is proportional to its magnetic moment, M, alters the magnetic flux through the coils, dM/dt. This induces a current directly proportional to dM/dt, which can be amplified and detected using lock-in amplifiers.

The external magnetising field is provided by a horizontal electromagnet; however, larger fields can be generated using a superconducting solenoid. Hysteresis loops can be traced by sweeping the field through all four quadrants (this can take many minutes).

VSMs are amongst the most sensitive magnetometers. However, there are several disadvantages:

- vibration of the sample can produce vibration of the pick-up coils resulting in large spurious signals (This problem is reduced if the magnet's field is very homogeneous and thus most VSMs use large electromagnets with large pole piece diameters.);

- the measurement is open circuit. Thus demagnetisation fields must be compensated for and
- the sample must be small to stay within the homogenous region of the electromagnet.

12.8.4 Coercimeters

The final class of d.c. magnetic measurement equipment is the coercimeter. As the name suggests these systems are designed exclusively to make measurements of coercivity and can be applied to both soft and hard magnetic materials where the application requires grading of materials by their coercive field (e.g. relay steels). Since the coercivity is determined at the point where the polarisation of the material is zero the measurement can be made in an open circuit configuration without compensation for demagnetising fields (as they are also zero).

All coercimeters follow the same basic procedure. First, the sample is driven to saturation, usually in a large solenoid energised either by a d.c. power supply or a capacitor bank which can generate a high pulsed field. The applied field is then slowly reduced back to zero. A reverse field is then slowly applied whilst the stray field due to the test sample (in a direction normal to the applied field direction) is measured. Once the detection of a null field, or the detection of the changing polarity of the field, is made the measurement is stopped and the field at the coercive point recorded (often by interpolation about $J = 0$). A schematic of the measurement arrangement is shown in Figure 12.48.

The methods of measurement of the stray field vary considerably with several detailed in the standard IEC 60404-7 [107]. The first is a coil wound onto a former which overlaps the end of the sample and is attached, via a rigid non-magnetic rod,

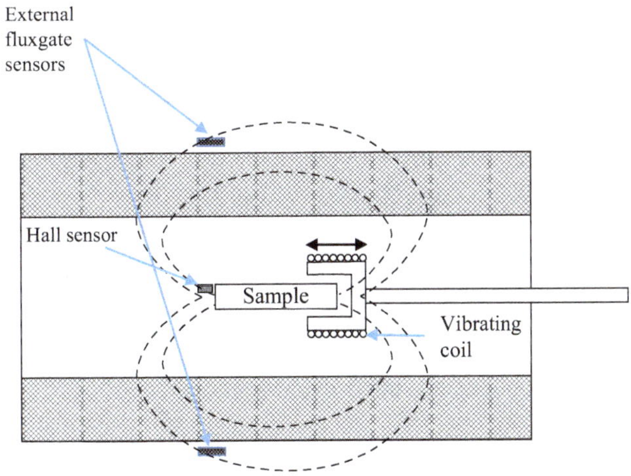

Figure 12.48 Schematic of a coercimeter showing possible arrangements of stray field sensing: vibrating coil, internal Hall sensor or external fluxgate sensors

to a vibrator (this is usually a speaker coil type actuator and therefore should be a considerable distance from the sense coil to avoid interference with the measurement). The sensitivity of this method can be greatly improved using a lock-in amplifier synchronised to the vibration source. A field sensor, e.g. a Hall sensor, can be placed at one end of the sample oriented such that it is sensitive to the field in the direction normal to the magnetising coil axis. Finally, a pair of differentially connected field sensors (fluxgates are often used due to their high sensitivity) can be mounted at either side of the sample and can be mounted inside or, more usually, outside the magnetising coil.

The nature of this measurement makes it very sensitive to stray fields and even to the earth's magnetic field, so several precautions should be taken. The simplest is to align the coercimeter such that the axis of the solenoid is perpendicular to the earth's field [108], i.e. in an east–west direction (local measurement of the earth's field should be made as it can be influenced by nearby ferromagnetic materials). Most coercimeters also rely on some form of shielding, either passive in the form of a high permeability shield (e.g. containing the coercimeter within a box or tube made from NiFe sheet) or active where a multi-axis Helmholtz coil is used to cancel out static and time varying external fields [109]. Fiorillo [1] cautions against the use of a shielded enclosure due to the possibility of distortion of sample stray field but this is often out-weighed by the additional cost and complexity of an active compensation system.

12.8.5 Demagnetisation

Ensuring that the sample is in a reference magnetisation condition prior to testing can be critical to enable repeatable magnetic measurements and, in particular, for d.c. measurements where that reference condition is required to calibrate the measured B. In almost all cases that reference condition is chosen to be the demagnetised state.

The demagnetised state is usually achieved in one of two ways: by raising the temperature of the sample above its Curie point and cooling back to room temperature or by cyclically magnetising the material with a decaying field amplitude. The thermal method is the most reliable but can be inconvenient (particularly if many tests are required on the same sample) and requires access to a high temperature furnace with a controlled atmosphere to protect the sample from oxidation. Therefore, most situations require a field-based demagnetisation procedure.

The simplest way of creating the decaying a.c. field required to demagnetise a sample is to connect a solenoid to a mains connected variac and establish an a.c. field with a peak amplitude greater than that previously experienced by the sample (use the saturation field if the history of the sample is unknown). Place the sample in the solenoid and then slowly withdraw it to a significant distance from the solenoid (say 1 m) before turning off the variac. For best results this should be done in an east–west direction in the absence of other stray magnetic fields.

A more convenient and repeatable method is to perform the same procedure in situ, i.e. with the sample inside the magnetising system to be used for the testing.

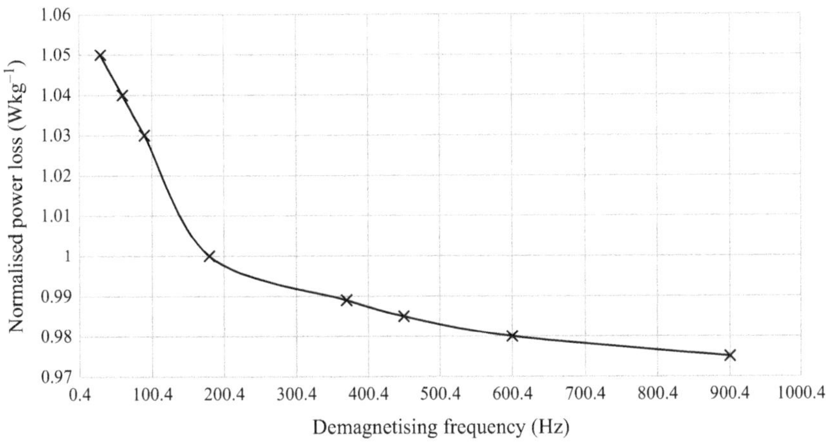

Figure 12.49 Measured power loss at 180 Hz, 0.5 T (normalised to loss at 180 Hz demagnetising frequency) in electrical steels for a range of demagnetising frequencies (adapted from [110])

Here the decaying field is produced by controlling the amplitude of the a.c. magnetising current. For standard a.c. power loss testing, for example, using the Epstein frame [3], it is recommended that the sample is demagnetised before measurement but the frequency of the demagnetising waveform is not defined. It is therefore implicit that the test frequency is used for the demagnetisation cycle which has been shown to have a significant influence on the measured properties. It is clear that the demagnetisation field should penetrate the full thickness of the material without significant attenuation and therefore the skin depth should be considered. Even for frequencies where the skin depth is greater than the material thickness Takada [110] and Bertotti [111] show that the measured loss at lower flux densities (e.g. 0.7 T) decreases with increasing demagnetising frequency. This is thought to be due to an increase in the number of domain walls being nucleated at the higher demagnetisation frequencies resulting in a narrower domain wall spacing and hence lower dynamic losses. At higher flux densities, these additional walls tend to be annihilated, hence the effect is reduced [111] (Figure 12.49).

As discussed in Section 12.8.1, when designing digital measurement systems for quasi-static measurement it is possible to have fine control over the demagnetisation cycle with the majority of systems being based on an algorithm similar to that described in [103]. This utilises a slowly decaying a.c. field at a frequency of 0.1 Hz. The demagnetisation begins with the generation of 1.5 periods at an initial peak field determined by the user. Following this step, the loop symmetry was assessed using the function in (12.55) where the parameters are determined as shown in Figure 12.50:

$$S = \frac{|B_{m+} - B_{r+}| - |B_{m-} - B_{r-}|}{|B_{m+} - B_{r+}| + |B_{m-} - B_{r-}|} \tag{12.55}$$

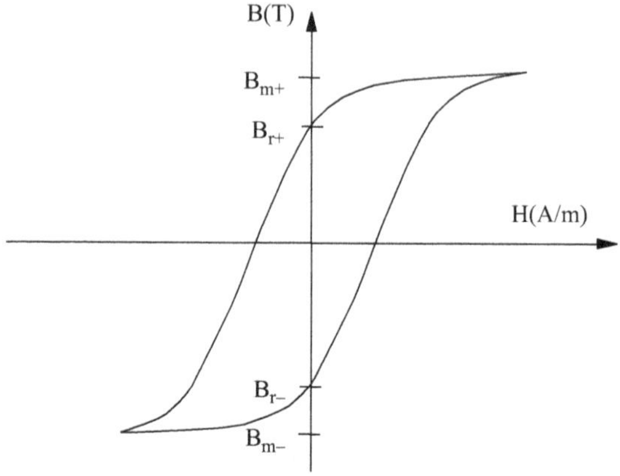

Figure 12.50 Assessment of parameters for determination of loop symmetry (adapted from [101])

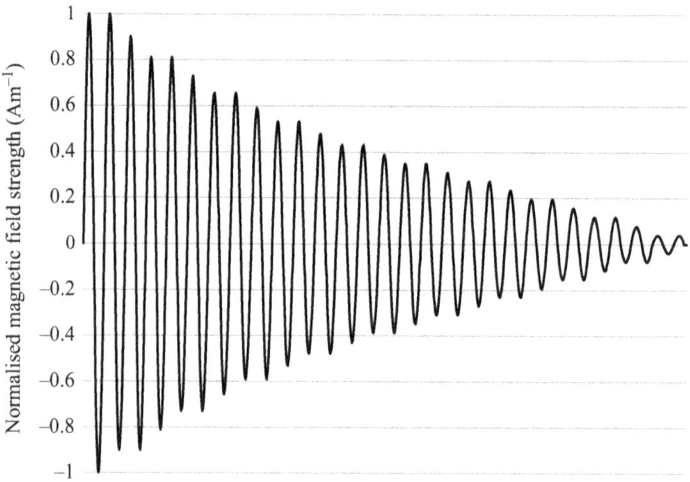

Figure 12.51 Example demagnetisation cycle using the method from [102]

The authors assigned an acceptance value of 2% on the value of S. If this was achieved, the peak field was automatically reduced by 10% and the procedure above repeated. If the value of S was greater than 2% then the value of the peak field was increased by 3% and the procedure above repeated. A threshold field value was set (200 Am^{-1}) and below this value the peak field was reduced in fixed increments of 10% of the threshold. The resulting demagnetisation waveform is shown in Figure 12.51 for a low starting field value in order to observe the individual cycles.

12.9 Surface insulation testing

As described in Section 20.15, electrical steel laminations have a coating system applied to the surface with the primary function of providing inter-laminar resistance. It is critical that this property is specified and it is therefore necessary to define a suitable measurement technique. Several methods have been proposed which include stack measurements (see Figure 12.52) and surface measurements of one sample (Figure 12.53) and two samples (Figure 12.54). All of these methods suffer from the difficulty of one coating surface having a large effect on the overall result. For example, if just one surface of one lamination was coated well in Figure 12.53 a high resistance would be recorded.

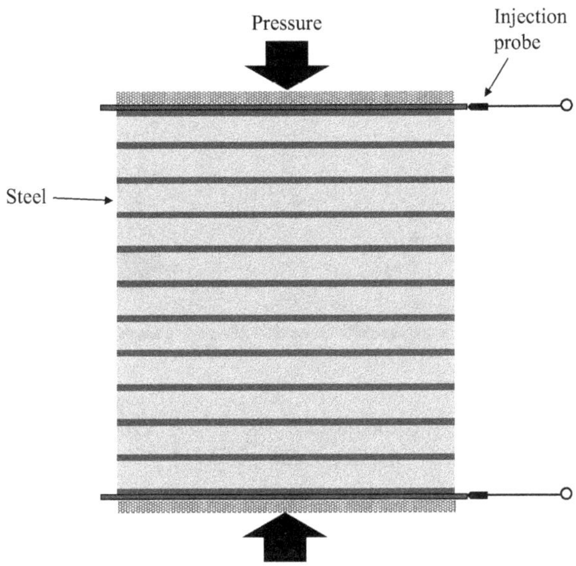

Figure 12.52 Stack measurement of insulation resistance in electrical steels (adapted from [112])

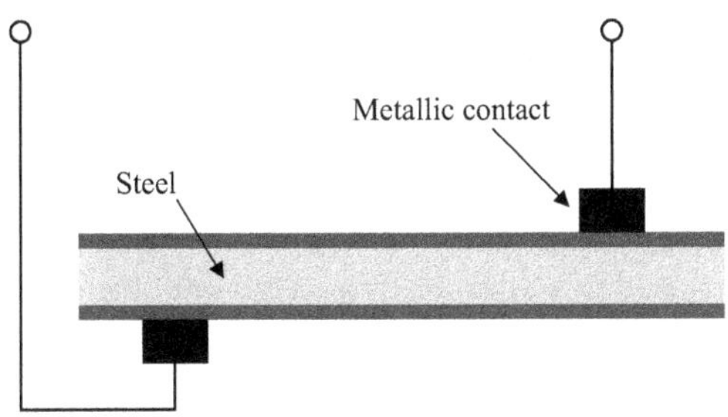

Figure 12.53 Measurement of insulation resistance in a single electrical steel lamination between two coated surfaces (adapted from [113])

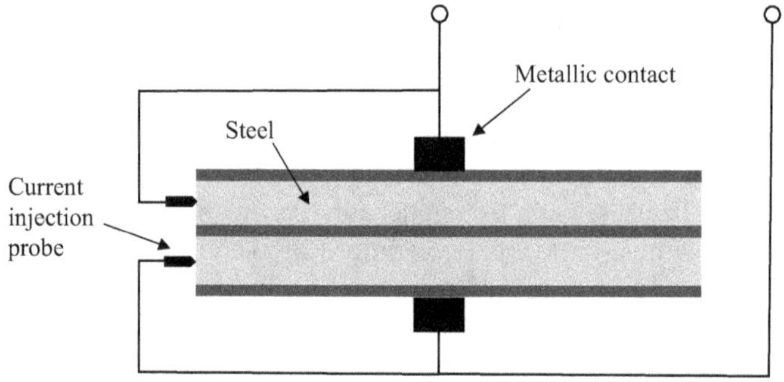

Figure 12.54 Measurement of insulation resistance with two electrical steel laminations (adapted from [113])

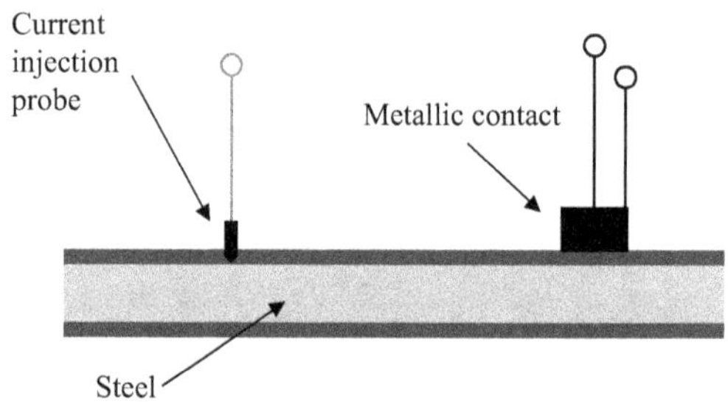

Figure 12.55 Measurement of insulation resistance between the steel substrate and the coated surface (adapted from [113])

For this reason, coatings are usually assessed through the resistance between a point electrode in contact with the substrate and a plate electrode in contact with the coated surface as shown in Figure 12.55.

This method, known as the Franklin method, is the subject of standard IEC 60404-11 [114] where the resistance is measured between the steel substrate and ten contact buttons on the coated surface.

Contact to the steel substrate is through a pair of twist drills which pierce the coating and provide a reference for the probe voltage and a current return path. The contact buttons are usually manufactured from brass or stainless steel and have a contact area of 64.5 mm^2 each. They are typically formed into a single assembly via sprung mountings and this assembly is pressed onto the surface (usually with a pneumatic system) such that a pressure of 2 Nmm^{-2} is applied over the total contact area.

The standard allows two modes of operation. Mode A (Figure 12.56) has a 5 Ω resistor in series with each contact and 0.5 V is applied across all of the contacts in series. The current from the stabilised supply is measured (usually by means of a

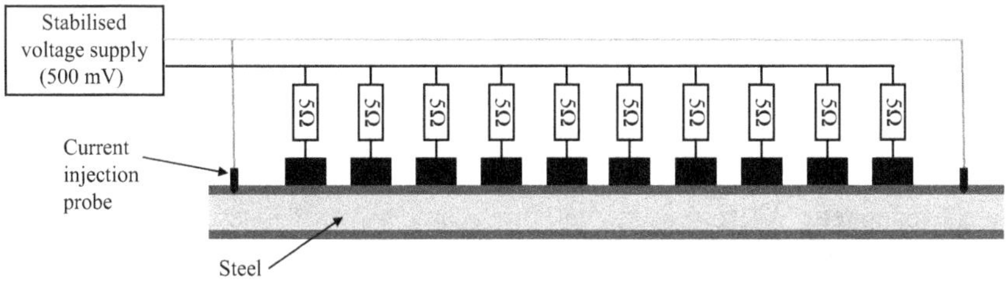

Figure 12.56 Franklin method surface insulation test: mode A

voltage across a known resistance) and a coefficient for the surface resistance is determined from

$$C = A\left[\frac{U}{I_{ave}} - \frac{R}{10}\right] \quad (\Omega\mathrm{m}^2) \tag{12.56}$$

where U is the stabilised voltage (0.5 V), I_{ave} is the average current measured (over 10 measurements typically), R is the series resistance (5.0 Ω) and A is the total contact area (645 mm^2).

Mode B uses a stabilised voltage of 0.25 V across each contact individually. Here, relays switch between each of the contacts and the current is measured so that the measured resistance at each contact is given by

$$R_s = \frac{U}{I_c} \quad (\Omega) \tag{12.57}$$

where U is the stabilised voltage (0.25 V) and I_c is the measured current.

Ten measurements are recorded for each test and, if repeated ten times, 100 resistance measurements in total are taken. The standard requires these values to be arranged in order of magnitude and the 16th and 50th value reported.

Both tests require a calibration check prior to measurement by placing a copper sheet on the test bed and ensuring the measured current and voltage are correct. Further calibrations to confirm the linearity (using calibrated resistors) and the magnitude and uniformity of the test pressure (using pressure sensitive paper for instance) can also be performed. The nature of the test means that it is very sensitive to contamination of the surface through dirt or grease so the samples and contacts should be clean. Mode A is particularly sensitive to small bare patches, or regions of thin coating which may not be representative of the whole sheet so care should be taken in its use. Additionally, the variation in the coating thickness has been shown to be significant and statistical methods based on large sample sizes would best approximate stack performance [115].

It should be noted that this method should only be used to give a comparative indication of the performance of the surface insulation and is not a true measurement of inter-laminar resistance which would not only incorporate two coating layers but also the lower likelihood of two areas of thin coating coming into contact [116].

12.10 Barkhausen noise measurement

The phenomenon of Barkhausen noise (BN) and an example of an e.m.f. waveform caused by BN was introduced in Section 3.17. Because of the stochastic nature of the waveform it is challenging to quantify the effect independent of random sources of electromagnetic interference. There are many ways of measuring BN in electrical steels using surface probes or enwrapping flux sensors [117]. Figure 12.57 shows a basic system for measuring BN from regions of a strip of electrical steel at power frequencies. The specimen is magnetised under controlled flux density at power frequency. The multi-turn sensing coils are wound in opposite directions and located close together. They are connected in series opposition in order to remove the Faraday e.m.f. from the output signal. Figure 12.58 shows a schematic diagram of the component parts.

A high pass filter is necessary to remove unwanted low frequency interference and environmental noise despite being located in an electromagnetically shielded chamber. Usually the BN signal is contained over a frequency range of about 3 kHz to 300 kHz. Outside this range it is difficult to distinguish other types of noise from genuine BN.

Because of the stochastic nature of BN, a variety of statistical methods can be used to analyse the signals. The sum of the amplitudes of BN peaks is convenient but does not count simultaneously occurring peaks or take account of the contribution of peaks of different magnitudes. Measurement of the power spectrum does contain such information. Kurtosis is a parameter that describes the peakiness of a BN distribution relative to a normal distribution. The r.m.s. value of BN is repeatable from cycle to cycle. Generally, similar trends in measurements seem to occur whichever method is used.

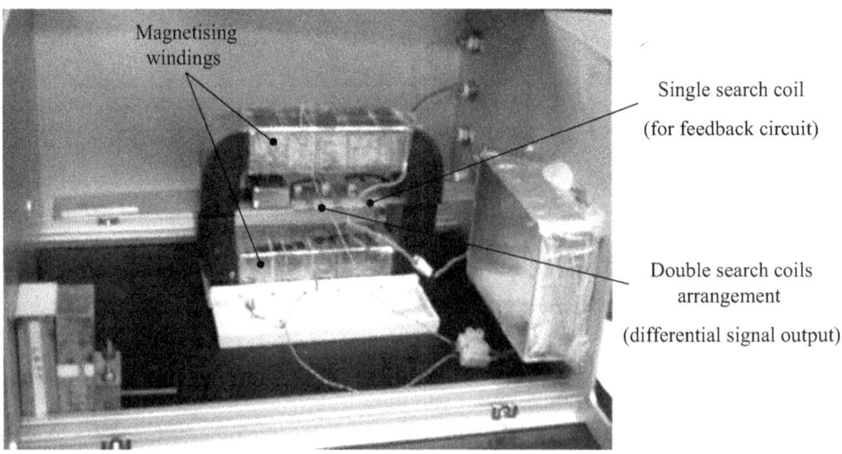

Figure 12.57 Single Strip Tester with double search coil pick-up for the measurement of Barkhausen noise

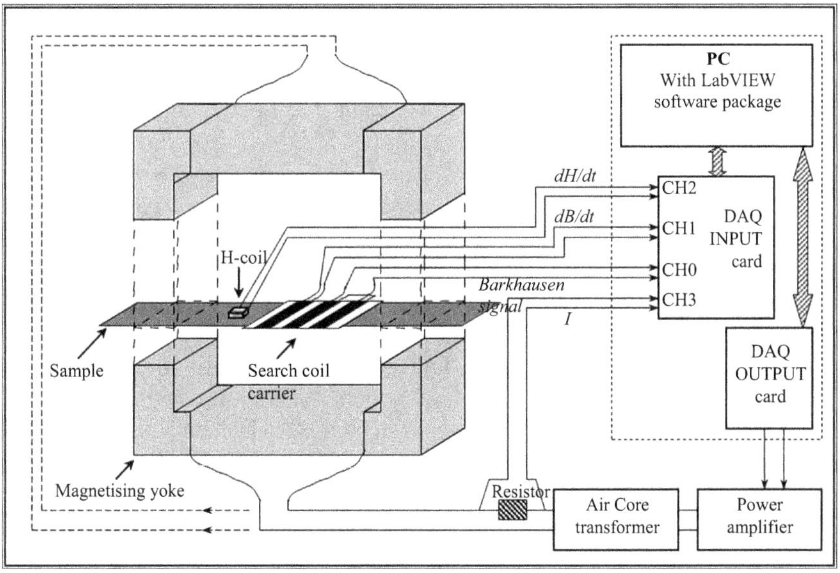

Figure 12.58 Schematic diagram of Barkhausen noise measurement system

The measured noise is very sensitive to the design of the pick-up coils and their physical separation [118]. The BN signal is also dependent on the signal sampling frequency which means some small or simultaneously occurring BN events may not be detected but this is not usually a problem provided that the measurement is used for comparative analyses [119].

Although BN measurements on electrical steels can be accurate and repeatable it is still unclear what proportion of events are picked up at any depth beneath the surface of the steel.

References

[1] Fiorillo F. *Measurement and Characterization of Magnetic Materials*. Elsevier Academic Press, 2004

[2] Tumanski S. *Handbook of Magnetic Measurements*. CRC Press, Boca Raton, FL, 2011

[3] IEC. *IEC60404-2, Magnetic materials – Part 2: Methods of measurement of the magnetic properties of electrical steel sheet and strip by means of an Epstein frame*, 1996

[4] Antonelli E., Cardielli E., and Faba A. 'Epstein frame: how and when it can be really representative about the magnetic behaviour of laminated electrical steels'. *IEEE Trans. Magn*. 2005, vol. 41(5), pp. 1516–1519

[5] Marketos P., Zurek S., and Moses A.J. 'A method for defining the mean path length of the epstein frame'. *IEEE Trans. Magn*. 2007, vol. 43(6), pp. 2755–2757

[6] Sievert J.D. 'Determination of AC magnetic power loss of electrical steel sheet: present status and future trends'. *IEEE Trans. Magn.* 1984, vol. 20(5), pp. 1702–1707
[7] Qingyi K. 'Determination of the weighted mean path length of the epstein frame'. *COMPEL* 2014, vol. 33(1), pp. 224–233
[8] IEC. *IEC TR 62981, Studies and comparisons of magnetic measurements on grain-oriented electrical steelsheet determined by the single sheet test method and Epstein test method*, 2017
[9] IEC. *IEC 60404-3, Magnetic materials – Part 3: Methods of measurement of the magnetic properties of magnetic sheet and strip by means of a single sheet tester*, 1992
[10] Drake A., and Ager C. 'Specific total power loss measurements using an IEC single-sheet tester'. *IEE Proc.* 1987, vol. 134(7), pp. 601–604
[11] Belgrand T. 'Measurement of grain oriented electrical steel properties using IEC 404-3 standard'. *J. Phys. IV France* 1998, vol. PR2, pp. 737–740
[12] Nakata T., Kawase Y., and Nakano M. 'Improvement of measuring accuracy of magnetic field strength in single sheet testers by using two H coils'. *IEEE Trans. Magn.* 1987, vol. 23(5), pp. 2596–2598
[13] Nafalski A., Moses A.J., Meydan T., and Abousetta M. 'Loss measurements on amorphous materials using a field compensated single strip tester'. *IEEE Trans. Magn.* 1989, vol. 25(5), pp. 4287–4291
[14] ASTM A932/A932M - 01(2012). *Standard test method for alternating current magnetic properties of amorphous materials at power frequencies using the Wattmeter-Ammeter-Voltmeter method with sheet specimens*, 2012
[15] Hagihara H., Tanaka M., Takahashi Y., Fujiwara K., and Ishihara Y. 'Standard measurement method for magnetic properties of Fe based amorphous magnetic materials'. *IEEE Trans. Magn.* 2014, vol. 50(4), Art. No. 6100604
[16] Watakabe R., Tanaka M., Takahashi Y., Fujiwara K., Ishihara Y., and Azuma D. 'Study of standard measurement method of magnetic property of Fe-based amorphous strip – round robin test results'. *IEEE Trans. Magn.* 2016, vol. 52(5), Art. No. 6000204
[17] IEC. *IEC 60404-6, Magnetic materials – Part 6: Methods of measurement of the magnetic properties of magnetically soft metallic and powder materials at frequencies in the range 20 Hz to 200 kHz by the use of ring specimens*, 2003
[18] IEC. *IEC TR 62383, Determination of magnetic loss under magnetic polarization waveforms including higher harmonic components – measurement, modelling and calculation methods*, 2006
[19] Zurek S., Marketos P., Meydan T., and Moses A.J. 'Use of a novel adaptive digital feedback for magnetic measurements under controlled magnetizing conditions'. *IEEE Trans. Magn.* 2005, vol. 41(11), pp. 4242–4249
[20] Harrison C., and Anderson P. 'Characterization of grain-oriented electrical steels under high DC biased conditions'. *IEEE Trans. Magn.* 2016, vol. 52 (5), Art. No. 2001604.
[21] Werner E. Austria Patent 191015, 1949

[22] Pfutzner H., and Krismanic C. 'The needle method for induction tests: sources of error'. *IEEE Trans. Magn.* 2004, vol. 40(3), pp. 1610–1616

[23] Yamaguchi T., Senda K., Ishida M., Sato K., Honda A., and Yamamoto T. 'Theoretical analysis of localised magnetic flux measurement by needle probe'. *J. Phys. IV France* 1998, vol. 4(PR2), pp. 717–720

[24] Loisos G., and Moses A.J. 'Critical evaluation and limitations of local flux density measurements in electrical steel'. *IEEE Trans. Magn.* 2001, vol. 37(4), pp. 2755–2757

[25] De Wulf M., Dupre L., Makaveev D., and Melkebeek J. 'Needle probe techniques for local magnetic flux measurements'. *J. Appl. Phys.* 2003, vol. 93 (10), pp. 8271–8273

[26] Loisos G., and Moses A.J. 'Demonstration of a new method for magnetic flux measurement on the interior of a magnetic material'. *Sens. Actuators A* 2003, vol. 106, pp. 104–107

[27] Senda K., Kurosawa M., Ishida M., Komatsubara M., and Yamaguchi T. 'Local magnetic properties in grain oriented electrical steel measured by the modified needle probe method'. *J. Magn. Magn. Mater.* 2000, vol. 215, pp. 136–139

[28] Enokizono M., Tanabe I., and Kubota T. 'Localized distribution of two dimensional magnetic properties and magnetic domain observation'. *J. Magn. Magn. Mater.* 1999, vol. 196, pp. 338–340

[29] Basak A., Rowe D., and Anayi F. 'Thin film sensors for magnetic flux and loss measurements'. *J. Magn. Magn. Mater.* 1996, vol. 160, pp. 197–198

[30] Ilo A., Pfutzner H., Bangert H., and Eisenmenger-Sittner C. 'Sputtered search coils for flux distribution analyses in laminated magnetic cores'. *J. Phys. IV France* 1998, vol. PR2, pp. 733–736

[31] Shilyashki G., Pfutzner H., and Huber C. 'Interlaminar magnetic flux assessment of a transformer core measured by an extra-thin printed foil detector'. *IEEE Trans. Magn.* 2017, vol. 53(9), Art. No. 8400706.

[32] Pfutzner H., and Mulasalihovic E. 'Thin-Film technique for interior magnetic analysis of laminated machine cores'. *Przegląd Elektrotechniczny* 2009, vol. 85(1), p. 43

[33] Mazurek R., Marketos P., Moses A.J., and Belgrand T. 'Development of sensors for measurement of localised flux density in transformer cores'. *Sensor Lett.* 2013, vol. 11(1), pp. 56–58

[34] Ball D., and Lorch H. 'An improved thermometric method of measuring local power dissipation'. *J. Sci. Instrum.* 1965, vol. 42, pp. 90–93

[35] Yamamoto T., and Nozawa T. 'Effects of tensile stress on total loss of single crystals of 3% silicon-iron'. *J. Appl. Phys.* 1970, vol. 41(7), pp. 2981–2984

[36] Moses A.J., and Phillips P. 'Some Effects of stress in GOSS-oriented silicon-iron'. *IEEE Trans. Magn.* 1978, vol. 14(5), pp. 353–355

[37] Hall M., Thomas O., Smith H., and Anderson P. 'Equivalence of measurements on soft magnetic materials in the UK and measurements for operational conditions'. *IEEE Trans. Instrum. Meas.* 2011, vol. 60(7), pp. 2275–2279

[38] Brown D., Holt C., and Thompson J. 'Influence of compressive and tensile stresses at various temperatures on some magnetic properties of transformer laminations'. *Proc. IEE* 1965, vol. 112(1), pp. 183–188

[39] Kai Y., Tsuchida Y., Todaka T., and Enokizono M. 'Influence of biaxial stress on vector magnetic properties'. *IEEE Trans. Magn.* 2014, vol. 50(4), Art. No. 6100204

[40] Inoue H., and Okabe S. 'Magnetic properties of grain oriented electrical steel in model transformer under direct current -biased magnetization'. *J. Appl. Phys.* 2014, vol. 115(17A), pp.332–335

[41] Miyagi D., Yoshida T., Nakano M., and Takahashi N. 'Development of measuring equipment of dc -biased magnetic properties using open type single sheet tester'. *IEEE Trans. Magn.* 2006, vol. 42(10), pp. 2846–2848

[42] Takada S., Yasuda T., and Sasaki T. 'Magnetising properties of an oriented sife sheet core in biased magnetizing states at power frequencies'. *J. Magn. Magn. Mater.* 1996, vol. 157, pp. 455–456

[43] Barbisio E., Bottauscio O., Chiampi M., and Ragusa C. 'Analysis of AC magnetic properties in SiFe laminations under DC biased magnetisation'. *Physica B* 2004, vol. 343, pp. 127–131

[44] Marketos P., Moses A.J., and Hall J. 'Effect of DC voltage on AC magnetisation of transformer core steel'. *J. Electr. Eng.* 2010, vol. 41(11), pp. 4242–4249

[45] Harrison C., and Anderson P. 'Characterisation of grain orientated electrical steels under high DC biased conditions'. *IEEE Trans. Magn.* 2016, vol. 52 (5), Art. No. 2001604.

[46] Zurek S. *Characterisation of Soft Magnetic Materials Under Rotational Magnetisation*. CRC Press, Boca Raton, FL, 2017

[47] Tumanski S. 'Investigations of 2D parameters of electrical steels'. *Proceedings of 1&2-Dimensional Magnetic Measurement and Testing*, 2000, pp. 185–193

[48] Boon C., and Thompson J. 'Alternating and rotational power loss at 50 c/s in 3% silicon-iron sheets'. *Proc. IEE* 1965, vol. 112(11), pp. 2147–2151

[49] Kedous-Lebouc A., Zouzou S., and Brissonneau P. 'On the magnetization processes in electrical steel in unidirectional and rotational field'. *Proceedings of 1st International Workshop on Magnetic Properties of Electrical Sheet Steel under Two Dimensional Excitation*, 1992

[50] Nencib N., Kedous-Lebouc A., and Cornut B. 'Performance evaluation of a large rotational single sheet tester'. *J. Magn. Magn. Mater.* 1996, vol. 160, pp. 174–176

[51] Moses A.J., and Thomas B. 'Measurement of rotating flux in silicon iron laminations'. *IEEE Trans. Magn.* 1973, vol. 9(4), pp. 651–654

[52] Tumanski S. 'Which magnetizing circuit is suitable for twodimensional measurements'. *PTB Bericht*, vol. PTB-E-81, 2002

[53] Ding X., Ren S., Xiong Y., Chen F., and Xu J. '2-D magnetic properties measurement system for electrical steel sheets considering laminated direction mechanical stress'. *IEEE Trans. Magn.* 2017, vol. 53(10), Art. No. 6100511

[54] Enokizono M., Suzuki T., Sievert J., and Zu J. 'Rotational power loss of silicon iron sheet'. *IEEE Trans. Magn.* 1990, vol. 26(5), pp. 2562–2564

[55] Fiorillo F., and Rietto A. 'The measurement of rotational losses at I.E.N.: use of the thermometric method'. *Proceedings of 1st International Workshop on Magnetic Properties of Electrical Sheet Steel under Two-Dimensional Excitation*, 1992

[56] Gorican V., Hamler A., Kribernik B., and Jesenik M. '2-D measurements of magnetic properties using a round RSST'. *Proceedings of 6th International Workshop on 1&2 Dimensional Magnetic Measurement and Testing*, 2000

[57] Zurek S., and Meydan T. 'Rotational power losses and vector loci under controlled high flux density and magnetic field in electrical steel sheets'. *IEEE Trans. Magn.* 2006, vol. 42(10), pp. 2815–2817

[58] Hasenzagl A., Weiser B., and Pfutzner H. 'Novel 3-phase excited single sheet tester for rotational magnetization'. *J. Magn. Magn. Mater.* 1996, vol. 160, pp. 180–182

[59] Pfutzner H. 'Rotational magnetization and rotational losses of grain oriented silicon steel sheets'. *IEEE Trans. Magn.* 1994, vol. 30(5), pp. 2802–2807

[60] Zurek S., Marketos P., Anderson P., and Moses A. 'Influence of digital resolution of measuring equipment on the accuracy of power loss measured in Epstein frame'. *Przegald Elektrotechniczny* 2007, vol. 83, pp. 50–53

[61] Sievert J., Ahlers H., Birkfeld M., *et al.* 'European intercomparison of measurements of rotational power loss in electrical steel sheet'. *J. Magn. Magn. Mater.* 1996, vol. 160, pp. 115–118

[62] Maeda Y., Shimoji H., Todaka T., and Enokizono M. 'Rotational power loss of magnetic steel sheets in a circular rotational magnetic field in CCW/CW directions'. *J. Magn. Magn. Mater.* 2008, vol. 320, pp. 567–570

[63] Zurek S., and Meydan T. 'Errors in the power loss measured in in clockwise and anticlockwise rotational magnetisation. Part 1: Mathematical study'. *IEE Proc.: Sci. Meas. Technol.* 2006, vol. 153(4), pp. 147–151

[64] Ragusa C., Zurek S., Appino C., and Moses A. 'An intercomparison of rotational loss measurements in non-oriented Fe-Si alloys'. *J. Magn. Magn. Mater.* 2008, vol. 320(20), pp. 623–626

[65] Sievert J. 'Reproducibility and standardization aspects of AC measurement methods under one- and two-dimensional magnetization conditions'. *Proceedings of the 6th International Workshop on 1&2-Dimensional Magnetic Measurement and Testing*, 2000

[66] IEC. *IEC TR 62581, Electrical steel – methods of measurement of the magnetostriction characteristics by means of single sheet and Epstein test specimens*, 2010

[67] Birss R., Keeler G., Pearson P., and Potton R. 'A capacitive instrument for the measurement of a large range of magnetostriction at low temperatures and high magnetic fields'. *J. Phys. E: Sci. Instrum.* 1978, vol. 11(9), pp. 928–934

[68] George W., Holt C., and Thompson J. 'Magnetostriction in grain-oriented silicon-iron'. *Proc. IEE – Part A: Power Eng.* 1962, vol. 109(43), pp. 101–108

[69] Brownsey C., and Maples G. 'Theoretical considerations of magnetostriction in silicon iron'. *Proc. IEE* 1966, vol. 113(11), pp. 1855–1858

[70] Simmons G., and Thompson J. 'Properties of grain oriented silicon iron – Part 1: Use of ceramic displacement transducers for the measurement of magnetostriction'. *Proc. IEE* 1968, vol. 115(12), pp. 1835–1839

[71] Banks P., and Rawlinson E. 'Dynamic magnetostriction and mechanical strain in oriented 3% silicon iron sheet subject to combined longitudinal and transverse stresses'. *Proc. IEE* 1967, vol. 114(10), pp. 1537–1546

[72] Greenough R., and Underhill C. 'Strain gauges for the measurement of magnetostriction in the range 4K to 300K'. *J. Phys. E: Sci. Instrum.* 1976, vol. 9, pp. 451–454
[73] Moses A.J., and Davies D. 'Influence of compressive stress on the magnetic properties of commercial (110)[001] oriented silicon iron'. *IEEE Trans. Magn.* 1980, vol. 16(2), pp. 454–460
[74] Anderson P., Moses A.J., and Stanbury H. 'Evaluation of magnetostriction measurement techniques'. *Studies in Applied Electromagnetics and Mechanics.* Vol 13: *Non-Linear Electromagnetic Systems.* IOS Press, 1998, pp. 341–344
[75] Yabumoto M., Arai S., Kawamata R., and Mizokami M. 'Recent development in grain-oriented electrical steel with low magnetostriction'. *J. Mater. Eng. Perform.* 1997, vol. 6(6), pp. 713–721
[76] Stanbury H. 'Magnetostriction effects at angles to the rolling direction in grain-oriented 3% silicon steel'. *J. Magn. Magn. Mater.* 1982, vol. 26(1), pp. 47–49
[77] Anderson P., Moses A.J., and Stanbury H. 'Assessment of the stress sensitivity of magnetostriction in grain-oriented silicon steel'. *IEEE Trans. Magn.* 2007, vol. 43(8), pp. 3467–3476
[78] Nakata T., Takahashi N., Nakano M., Muramatsu K., and Miyake M., 'Magnetostriction measurements with a laser doppler velocimeter'. *IEEE Trans. Magn.* 1994, vol. 30(6), pp. 4563–4565
[79] Nakase T., Nakano M., Fujiwara K., and Takahashi N. 'Measuring system for magnetostriction of silicon steel sheet under AC excitation using optical methods'. *IEEE Trans. Magn.* 1998, vol. 34(4), pp. 2072–2074
[80] Hirano M., Ishihara Y., Harada K., and Todaka T. 'A study on measurement of magnetostriction of silicon steel sheet by laser displacement meter'. *J. Magn. Magn. Mater.* 2003, vol. 254, pp. 43–46
[81] Ban G., and Janosi F. 'Measuring system and evaluation method of DC and AC magnetostriction behaviour to investigate 3.2% SiFe GO electrical steels'. *J. Magn. Magn. Mater.* 1996, vol. 160, pp. 167–170
[82] Benda O., and Klima J. 'Magnetostriction measurement using speckle interferometry'. *J. Electr. Eng.* 1997, vol. 48(8), pp. 133–136
[83] Squire P., and Gibbs M. 'Fibre-optic dilatometer for measuring magnetostriction in ribbon samples'. *J. Phys. E: Sci. Instrum.* 1987, vol. 20(5), pp. 499–503
[84] Squire P. 'Magnetomechanical measurements of magnetically soft amorphous materials'. *Meas. Sci. Technol.* 1994, vol. 5, pp. 67–80
[85] Klimczyk P. 'Novel Techniques for Characterisation and Control of Magnetostriction in G.O.S.S'. *PhD Theses*, Cardiff University, 2012
[86] Somkun S. 'Magnetostriction and Magnetic Anisotropy in Non-Oriented Electrical Steels and Stator Core Laminations'. *PhD Theses*, Cardiff University, 2010
[87] Enokizono M., Kanao S., and Shirikawa G. 'Measurement of arbitrary dynamic magnetostriction under alternating and rotating field'. *IEEE Trans. Magn.* 1995, vol. 31, pp. 3409–3411

[88] Hasenzagl A., Weiser B., and Pfützner H. 'Magnetostriction of 3% SiFe for 2-D magnetization patterns'. *J. Magn. Magn. Mater*. 1996, vol. 160, pp. 55–56
[89] Somkun S., Moses A.J., Anderson P., and Klimczyk P. 'Magnetostriction anisotropy and rotational magnetostriction of a nonoriented electrical steel'. *IEEE Trans. Magn*. 2010, vol. 46(2), pp. 302–305
[90] Smith H. 'Considerations when measuring loss on-line vs Ep/SST in a laboratory'. *Confidential Report*, 2017
[91] Havlicek V. 'On-line AC magnetic measurements at higher velocities of strip movement'. *J. Electr. Eng*. 1997, vol. 50(4), pp. 11–13
[92] Havlicek V., and Mikulec M. 'On-line testing device using the compensation method'. *Phys. Scr*. 1989, vol. 39(17), pp. 513–515
[93] Porter C., Janosi F., Kochman T., Albini L., and Fenaille B. 'EUR 21932: Online measurement of magnetic properties of electrical steel strip'. *European Commission*, 2005
[94] Beckley P. 'Continuous power loss measurement with and against the rolling direction of electrical steel using non-enrapping magnetisers'. *IEE Proc*. 1983, vol. 130(6), pp. 313–321
[95] Khanlou A., Moses A.J., Meydan T., and Beckley P. 'A non- enwrapping single yoke online magnetic testing system for use in the steel industry'. *IEEE Trans. Magn*. 1992, vol. 28(5), pp. 2479–2481
[96] Khanlou A., Moses A.J., Meydan T., and Beckley P. 'A computerised on-line power loss testing system for the steel industry, based on the RCP compensation technique'. *IEEE Trans. Magn*. 1995, vol. 31(6), pp. 3385–3387
[97] Iranmnesh H., Beckley P., and Moses A. 'Investigation of a double sided non-enwrapping "with" and "against" power loss testing system based on the 'flux injection' principle'. *J. Magn. Magn. Mater*. 1992, vol. 112, pp. 74–76
[98] ASTM A773/A773M. *Standard test method for direct current magnetic properties of low coercivity magnetic materials using hysteresigraphs*, 2014
[99] IEC. *IEC 60404-4, Magnetic materials – Part 4: Methods of measurement of d.c. magnetic properties of iron and steel*, 2008
[100] Anderson P. 'A universal dc characterisation system for hard and soft magnetic materials'. *J. Magn. Magn. Mater*. 2008, vol. 320(20), pp. 589–583
[101] Jiles D., Atherton D., Lassen H., Noble D., DeVette J., and Astle T. 'Microcomputer based system for control of applied uniaxial stress and magnetic field'. *Rev. Sci. Instrum*. 1984, vol. 55(11), pp. 1843–1848
[102] Strnat R., Mall M., and Masteller S. 'Precision and accuracy study on measurement of soft magnetic properties using DC hystersigraphs'. *IEEE Trans. Magn*. 2007, vol. 43(5), pp. 1884–1887
[103] Squire P., Sheard S., Carter C., and Gibbs M. 'Digital M-H plotter for low coercivity metallic glasses'. *J. Phys. E: Sci. Instrum*. 1988, vol. 21, pp. 1167–1172
[104] Anderson P. 'Constant dB/dt DC characterisation through digital control of magnetic field'. *Przeglad Elektrotechniczny* 2011, vol. 87(9), pp. 77–78
[105] Garcia J., and Rivas M. 'A quasi-static magnetic hysteresis loop measurement system with drift correction'. *IEEE Trans. Magn*. 2006, vol. 42(1), pp. 15–17

[106] IEC. *PD IEC/TR 62797, International comparison of measurements of the magnetic moment using vibrating sample magnetometers (VSM) and superconducting quantum interference device (SQUID) magnetometers*, 2013

[107] IEC. *IEC 60404-7, Magnetic materials. Part 7: Method of measurement of the coercivity of magnetic materials in an open magnetic circuit*, 1982

[108] Drake A., and Hartland A. 'A vibrating coil magnetometer for the determination of the magnetization coercive force of soft magnetic materials'. *J. Phys. E: Sci. Instrum.* 1973, vol. 6, pp. 901–902

[109] Ahlers H., Ludke J., and Sievert J. 'Coercivity meter with earth's magnetic field compensation'. *J. Magn. Magn. Mater.* 1996, vol. 160, pp. 187–188

[110] Takada S., Yasuda T., and Sasaki T. 'Influence of demagnetization frequency on magnetic properties of electromagnetic steel sheets'. *Appl. Electromag. Mater.* 1995, pp. 191–199

[111] Bertotti G. 'General properties of power losses in soft ferromagnetic materials'. *IEEE Trans. Magn.* 1988, vol. 24(1), pp. 621–630

[112] Beckley P. 'New initiatives in the assessment of surface insulation resistance of coatings on electrical steels'. *J. Mater. Eng.* 1990, vol. 12(1), pp. 47–50

[113] Hamzehbahmani H. 'Development of Novel Techniques for the Assessment of Inter-Laminar Resistance in Transformer and Reactor Cores'. *PhD Theses*, Cardiff University, 2017

[114] IEC. *IEC 60404-11, Magnetic materials – Part 11: Method of test for the determination of surface insulation resistance of magnetic sheet and strip*, 2012

[115] Marion Pera M., Kedou-Lebouc A., Waeckerle T., and Cornut B. 'Characterization of SiFe sheet insulation'. *IEEE Trans. Magn.* 1995, vol. 31(5), pp. 2408–2415

[116] Beckley P. *Electrical Steels: A Handbook for Producers and Users*. European Electrical Steels, Newport, UK, 2000

[117] Moses A., Patel H., and Williams P. 'AC barkhausen noise in electrical steels: influence of sensing technique on interpretation of measurement'. *J. Electr. Eng.* 2006, vol. 87(8), pp. 3–8

[118] Moses A., Patel H., and Williams P. 'AC barkhausen noise in electrical steels: influence of sensing technique on measurement'. *Proceedings of International Conference on Magnetic Measurements*, Zavazna Poruba (Slovakia), 2006, pp. MO–01

[119] Patel H., Moses A., and Williams P. 'The dependence of barkhausen noise measurement on data acquisition parameters'. *Proceedings of the 9th International Workshop on 1 & 2 Dimensional Magnetic Measurements*, 2006, pp. 68–69

Chapter 13
Background to modern electrical steels

The demand for 'electrical steel' first appeared in the 1880s with the development of electric motors, dynamos (generators) and transformers. However, the understanding of electro-magnetism had started many years earlier with the invention of the battery (Allessandro Volta, 1800), the generation of a magnetic field from electric current (Hans Christian Oersted, 1820) and the invention of the electromagnet (William Sturgeon, 1825).

The first rotating device driven by electromagnetism was built by the Englishman Peter Barlow in 1822 (Barlow's Wheel). Michael Faraday discovered electro-magnetic induction in August 1831. Just a few months later, he successfully conducted the experiment that demonstrated the relationship between magnetism and motion. He passed a magnet in and out of a cylindrical helix wound with wire and showed that electricity was produced. As early as 1837 it was reported by William Sturgeon that the use of a core made of divided iron wires as opposed to a solid core, reduced eddy current losses in induction coils.

The years 1885 until 1889 saw the invention of the three-phase electric power system and the three-phase cage induction motor was built first by Michael Dolivo-Dobrowolsky in 1889 [1,2].

The first a.c. power system that used the 'modern' transformer in the United States was in Great Barrington, Massachusetts in 1886. Earlier forms of the transformer were used in Austro-Hungary 1878–1880s and from 1882 onward in England. Lucien Gaulard used an a.c. system for the revolutionary Lanzo to Turin Electrical Exposition in 1884 (Northern Italy). In 1891 Mikhail Dobrovolsky demonstrated three-phase transformers in the Electro-Technical Exposition at Frankfurt, Germany.

Early transformer, generator and motor developers soon realised that cores constructed from solid iron resulted in prohibitive eddy current losses, and their subsequent designs mitigated this effect with cores consisting of bundles of insulated iron wires and, later, laminated sheets of steel. Therefore, the demand for the first 'electrical steel' became apparent.

13.1 History and development of electrical steels

Even at the start of the manufacture of 'electrical steel' (i.e. steel for specific use in the cores of electrical machines) it was clear that the properties required were different from those needed for steels in use for most other applications. Therefore, a development process started whereby the electrical and magnetic properties of these steels were targeted, sometimes in conflict with the requirements for other more normal uses of steel and sometimes complicating the manufacturing process significantly. That being said, as the relative demand for electrical steel was, and still is, small compared to world-wide steel demand; often improvements to electrical steels and the manufacturing process have followed those of general steel production.

13.1.1 Laminations

The company history of Joseph Sankey & Sons Ltd in the United Kingdom records that [3]

> About 1886 it was becoming evident that armatures for dynamos would in future be made from charcoal sheet iron instead of solid wound cores.

The company had been contacted by a London iron merchant, Robert Jenkins, in December 1886, who had done some research into the possibility of making dynamo armatures from charcoal sheet iron instead of solid wound cores and consulted Siemens the electrical engineers. He thought that Sankey's existing machines could be adapted to manufacture electrical laminations. Figure 13.1 shows an original advertisement for laminations produced by the Robert Jenkins Company.

In 1887 Jenkins persuaded Siemens to place an order for electrical stampings in the name of the Harold and Jenkins Company. It is believed that the first

Figure 13.1 Robert Jenkins advertisement for stampings for electrical equipment from Grace's Guide Industrial History of the United Kingdom [4]

Figure 13.2 Photograph of a 1886 commercial transformer showing the laminated core [5] (Reproduced by permission of the Minerals, Metals and Materials Society)

stampings were made from the scrap centres of rims of the new bicycles which were coming into vogue at that time. That was the beginning of the *Electrical Laminations* business in the United Kingdom.

The development of transformers also occurred throughout this time period. The first transformers manufactured in the 1880s had cores made from high grade wrought iron but soon thin laminated sheets, as shown in Figure 13.2, became used to reduce the eddy current problem.

13.1.2 Increased resistivity

In the early 1880s Robert Hadfield had started work on SiFe alloys containing from 1.5% to 5% silicon, in Sheffield, UK. It took several years of research with the help of Barrett and Brown before the remarkable improvement in electrical and magnetic properties of these *silicon steels* were publicly presented in papers published between 1900 and 1902 [6]. Figure 13.3 shows the improvement in both permeability and loss of silicon–iron of 1900 over 1885 wrought iron and 1900 Swedish charcoal iron. The permeability of the SiFe can be seen to be double that of the other materials and its loss is substantially reduced.

With a large interest shown by electrical engineers, Hadfield then looked for partners that would be able to produce the thin sheets of silicon steel needed by the emerging markets. In 1906, Sankeys and the Hadfield Steel Foundry Company of Sheffield entered an arrangement whereby Sankeys were granted an exclusive licence to obtain or manufacture special electrical steel referred to as *stalloy* as a registered brand. In 1909, Sankeys sub-licensed John Lysaght to roll sheets under Hadfield patents at its Orb Works in Newport, South Wales. The Company also granted the Brymbo Steel Company in Wrexham, North Wales, a licence to make steel bars and blooms under the Hadfield patent on condition that these would then be sold to Lysaghts for rolling into sheet. Figure 13.4 shows a 1914 advertisement for Stalloy.

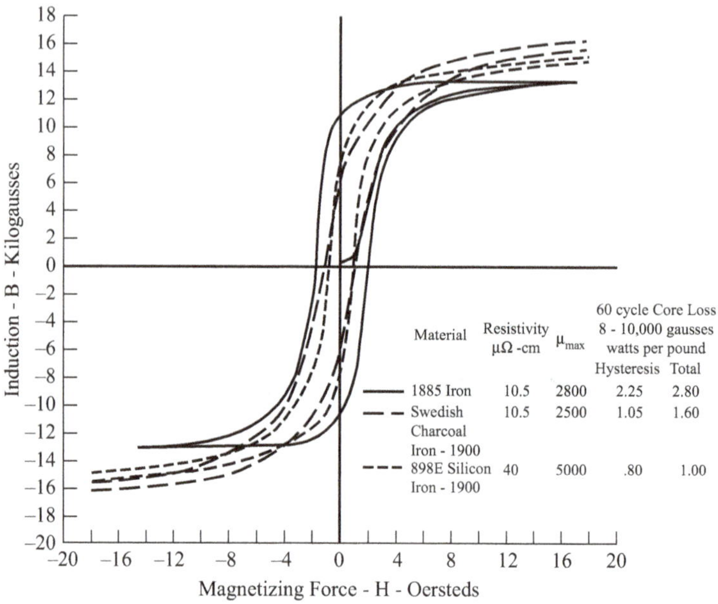

Figure 13.3 Showing improvement in B–H curves of silicon–iron compared to wrought iron and Swedish charcoal iron [5] (Reproduced by permission of the Minerals, Metals and Materials Society)

Figure 13.4 1914 advertisement by the Sankey Company for Stalloy published in Grace's Guide Industrial History of the United Kingdom [4]

Thus began the technology of specialised (silicon containing) electrical steel making. These sheets were produced by hot rolling and folding packs of sheets together, a complex and physically demanding process as outlined in Figure 13.5 [7].

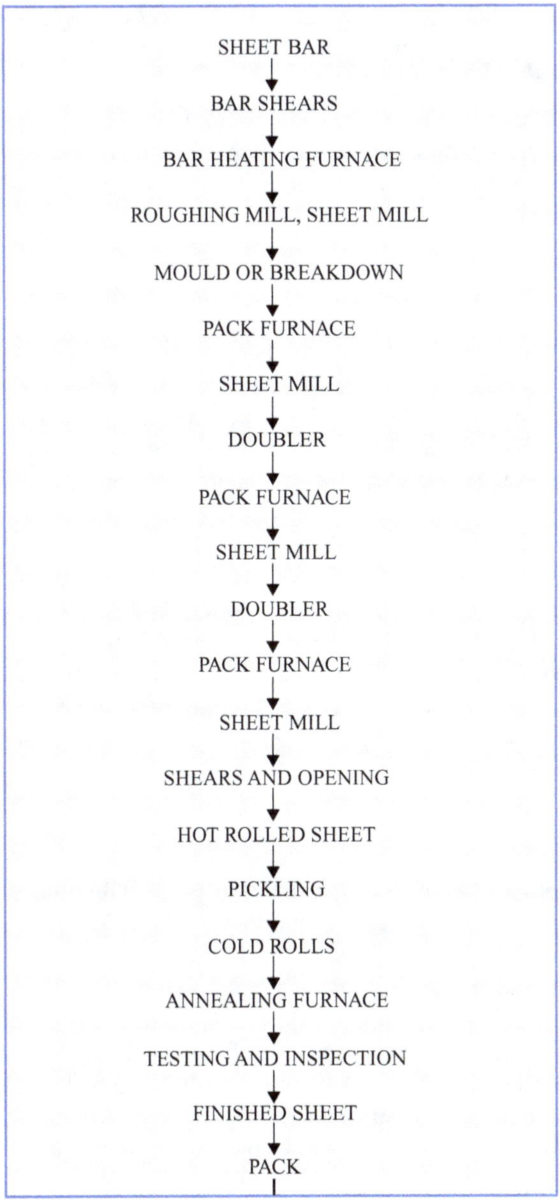

Figure 13.5 Hot rolling process for silicon steel (amended from [7])

Figure 13.6 Photograph showing the manual handling of sheets of silicon–iron at the pack furnace and the despatch area at Orb works in the middle of the twentieth century (from the History of Orb Works, with permission from Cogent Power Ltd)

As can be seen, the hot rolling of silicon steel was a long and complicated process although it lasted well into the late 1960s in the United Kingdom and even into the 2010s in other parts of the world. The following photographs in Figure 13.6 show the pack furnace, where the sheets of silicon–iron are reheated after a number of rolling stages. Note the physical, manual nature of the process requiring two men to move the sheets. The right hand photograph shows the despatch area where the finished sheets were packed and boxed prior to shipping.

The addition of silicon reduced hysteresis loss, increased permeability and increased resistivity, thus reducing the eddy current loss. However, the presence of silicon had the disadvantage that the steel became brittle and hard so that the quantity of silicon had to be limited to not more than 4.5%. The Stalloy produced at a thickness of around 0.35 mm by this hot rolling process had specific total loss values of around 7 W kg^{-1} at 1.5 T, 50 Hz [8]. Figure 13.4 shows an advertisement for the Sankey products.

13.1.3 Purification

In the period between the early 1900s and the 1930s, Glumlich in Germany [9] and Yensen in the United States [10] worked on the effect of impurities on the magnetic properties of iron and SiFe alloys to produce improvements in the cleanliness of commercial electrical steels of the period. This, together with increased understanding following Weiss's work on the domain theory of magnetism produced significant improvements in both loss and permeability. Figure 13.7 shows a comparison of commercial silicon iron sheet from 1903 and 1930 (along with 1934 grain oriented (GO) silicon steel) [5]. It can be seen that losses had more than halved and permeability more than doubled in the silicon iron over the 27-year period.

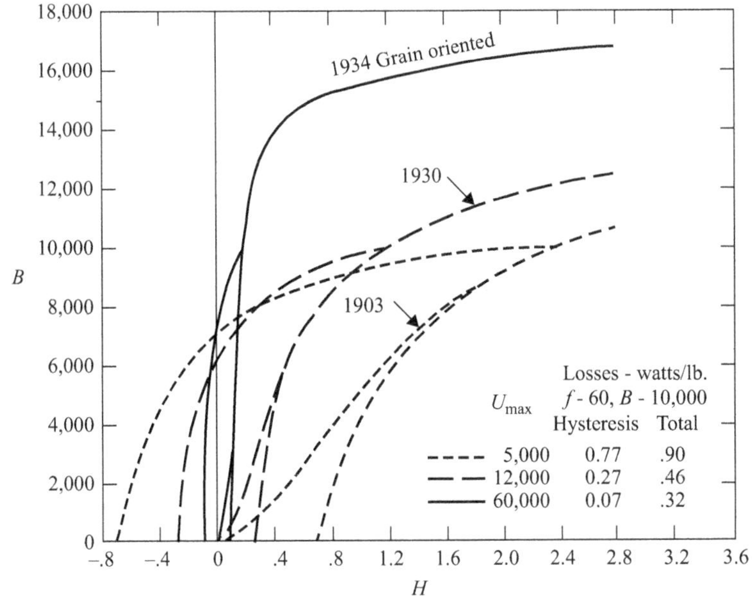

Figure 13.7 Showing improvement in B–H curves of commercial iron silicon sheet from 1903 and 1930 (along with 1934 GO silicon steel) [5] (Reproduced by permission of the Minerals, Metals and Materials Society)

13.1.4 Grain size

In a US patent from 1936, Robert Reno of the Wheeling Steel Corporation which were manufacturing electrical steels at its Beech Bottom works in West Virginia at the time, states [11]

> The improvement of the electrical properties of the steel by addition of silicon has been well established in the art and it is also well known that the watt loss of the silicon steel is essentially dependent upon the kind of treatment which the material undergoes. It is also recognised that the size of the grain structure of the material is of the greatest influence on the electrical properties and is dependent upon the treatment to which the material is subjected.

This is demonstrated in the example of a 1% Si steel which goes on to show a process that produces a hot rolled coil, which is then continuously pickled, cold rolled on a three-stand tandem mill, batch annealed and then skin passed in a range of 1%–10% reduction before a final batch anneal shown in Figure 13.8. The results show loss improvements along with increased final grain size within a range of

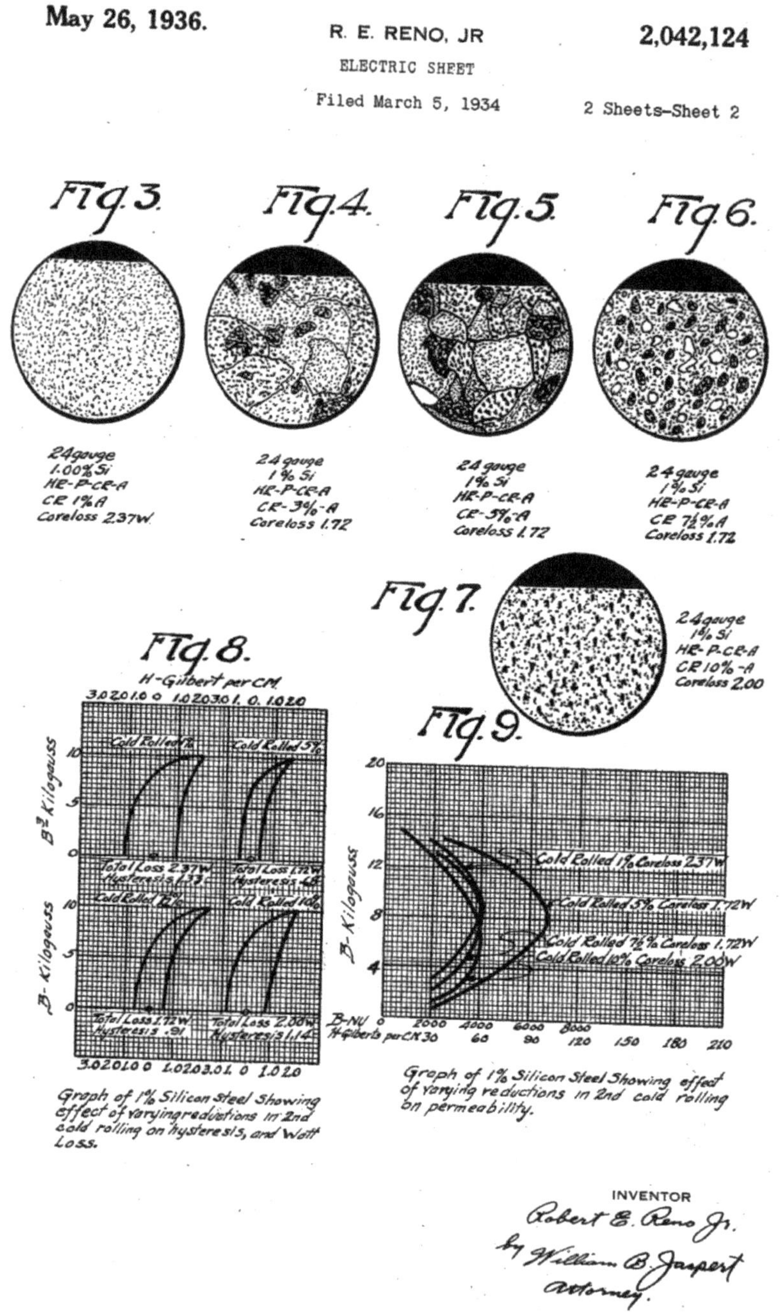

Figure 13.8 *Figures from US patent 2042124 showing the variation in grain size after a range of cold rolling thickness reductions and subsequent annealing and the corresponding improved hysteresis loss and permeability*

critical cold reductions (3%–5%). If figures 4 and 5 as shown in the patent page (Figure 13.8) are compared with figures 3 and 6, the improved hysteresis loss and permeability associated with the larger grain size in figures 8 and 9 can be seen. This is amongst the first cold rolled non-oriented (NO) electrical steel produced in the world, alongside Armco's Special Electric and Electric grades as shown in the 1932 catalogue of Armco Electrical Sheet Steel grades [12].

General steel processing improvements were key in enabling these developments in electrical steels, both in terms of impurity control during steelmaking and the advent of the industrial cold rolling process in the United States in the 1920s and 1930s.

13.1.5 Crystal orientation

It had been recognised in the early 1920s that the silicon steel crystals were themselves anisotropic, but it was not until 1934 that a commercial process was patented by Norman Goss [10] which gave rise to a GO electrical steel. Although at the time Goss had not actually recognised that the material he produced was in fact 'oriented', this is clearly shown in claim 13 of his patent. Richard Bozorth examined and proved conclusively that the Goss material did have a strong $\{110\}<001>$ texture [8]. Bozorth was working at Bell Laboratories in New York at the time and later wrote 'Ferromagnetism', which has become a hugely respected standard reference work.

The first commercial GO cold-rolled steel was produced by the Armco Steel corporation in the United States in 1939 and had a thickness of 0.32 mm and a loss of 1.5 W kg^{-1} at 1.5 T, 50 Hz [13]. The huge improvement over existing electrical steels is clearly shown in Figure 13.7.

GO electrical steel was then industrialised through the 1940s by the Armco Steel Company (and its predecessors) [14] and subsequently licensed to electrical steel producers world-wide through the 1950s.

This Armco GO material continued to be the best available electrical steel until the invention of an improved variant of the Goss orientation. In 1965 the Nippon Steel Corporation announced the production of high permeability GO silicon steel and commercial production began in 1968 [15]. At flux densities of 1.7 T its permeability was three times higher than the best cold-rolled GO steel at the time. The low losses of these steels were largely due to a reduction of around 40% in the hysteresis loss brought about by improved grain orientation.

13.1.6 Coatings

Initially 'coatings' on electrical steels were oxides produced in the hot rolling production process. Although these oxide layers were useful in increasing surface insulation between the stacked layers of laminations in the cores of electrical machines the quality was inherently variable and had issues with the hardness and therefore the abrasiveness to punching tools in lamination manufacture. Early lamination producers were able to offer a variety of insulation as indicated by the text at the bottom of the advertisement shown in Figure 13.4, but this would have been applied after the sheet was cut or the stampings produced.

With the evolution of the GO steel process and the requirement to use a high temperature batch annealing separator a magnesium silicate coating was formed and utilised. These coatings again imparted some degree of surface insulation and also a small reduction in losses due to the applied tension effects from the coating, although this was not necessarily known at the time.

It was not until the advent of cold rolled NO steel and the improvements in control of the continuous annealing atmosphere that reduced the unwanted oxide layers on the surface of electrical steels that it became possible to apply final coatings for insulation and cosmetic purposes as part of the steel cold rolling process. As early as 1932 the Armco Steel Corporation stated [12]

> All grades of Armco Electrical Sheets can be supplied from the mill with an insulative coating designed to increase resistivity of core structures and provide superior punching qualities.

Baldwins Limited in the United Kingdom, in 1939, offered a range of insulations on full sized sheets or finished laminations that included *kaoline*, a baked, china clay-based coating, *varnish*, guaranteed to withstand transformer oil at 180 °C, *cellulose*, oil resisting and suitable for stampings not exposed to a great temperature rise and, finally, *paper* insulation as an alternative to the above [16].

Armco also developed a phosphate-based coating that was applied on top of the 'glass film' produced on the surface of GO electrical steel after high temperature anneal. Further enhancements of this final coating were developed from the late 1930s to the mid-1940s, improving the surface insulation and tension applied by the coating [17].

13.2 Metallurgical requirements and control

Through the early development of electrical steels, it became apparent which chemical and physical requirements were desired to give rise to the best magnetic properties. Some of these were easier, cheaper or more practical to achieve and therefore a process of development focused on the following material features which has continued from those early days until today.

13.2.1 Thickness

Given that the eddy current component of power loss is proportional to the square of the lamination thickness it is clear that thinner electrical steels give rise to lower overall losses. However, there are practical and cost limitations into how thin electrical steels may be rolled. Production costs increase as the thickness is reduced, for example a halving of the thickness results in a doubling of the length and, therefore, unless continuous annealing speeds can be doubled the productivity is reduced significantly. Also, the thinner the electrical steel is rolled the more likely there are to be rolling difficulties and increased scrap produced during production, thus reducing yield and increasing costs. At some point the magnetic benefit of thinner electrical steel is lost, as the stacking factor reduces and the assembled core contains less ferromagnetic material. For the above and many other practical reasons, most electrical steels produced lie in the thickness range of 0.65 mm to 0.18 mm. Usually, the thicker materials are of lower quality, higher loss grades that are cheaper to produce. The thinner, more expensive grades are only used if the cost of the material can be justified by the equipment design specifications.

13.2.2 Chemical composition

There are two main requirements of the chemistry of electrical steels in its final form:

- increased electrical resistivity which is beneficial in reducing eddy current losses and
- high purity, free of unwanted chemicals that would give rise to inclusions and precipitates in the steel microstructure that would interact with domain wall motion and thereby increase hysteresis loss and reduce the permeability.

The main alloy addition to increase electrical resistivity is, of course, silicon and as the costs of production increase as more silicon is added there is again a cost versus benefit balance to be calculated. At the lower, cheaper end of the range of grades of electrical steel, as little as 0.3% Si could be added; this would increase up to a maximum of 3.4% Si in the best grades.

The actual level of silicon is restricted by the brittleness it causes; 3.4% Si is the practical limit. Above this the cold rolling required to produce the required thickness is so difficult as to render the production impractical. However, further increases in resistivity can be and are achieved by adding aluminium in addition to silicon in the best NO grades. In these grades aluminium up to approximately 1.5% can be added to the 3.3% Si steels. In the low Si grades other alloying elements such as manganese and phosphorus can also marginally improve the resistivity [18].

To produce a 'clean' inclusion or precipitate-free steel, low levels of impurity elements are required. This can be achieved in a variety of ways given today's steelmaking technologies, although, as above, the cost versus benefit always needs to be evaluated. Unwanted impurities such as carbon, sulphur and nitrogen in the fully processed NO range of electrical steels can be reduced to low levels (<0.005%) during the secondary steelmaking process. For the semi-processed NO steel grades, where the carbon is useful up until the final *customer anneal*,[1] the carbon level is reduced to <0.005% by the use of a decarburising atmosphere. In GO steel grades where carbon, nitrogen and sulphur are required as part of the process to produce the oriented 'Goss grains' the levels of these elements are reduced by various annealing stages in the later part of the process route of the electrical steel production.

For all the various grades mentioned above, the requirement is to have these very low levels of impurities in the final NO or GO product.

[1]This is an anneal which is carried out after laminations have been stamped from the skin passed cold rolled electrical steel supplied to the customer by the electrical steel producer.

13.2.3 Grain size

Generally, in most normal applications, a large grain size is required to reduce power losses. Such grain sizes are much larger than is considered normal in the field of steel metallurgy. Typical grain sizes in NO grades are in the range 50 microns to 200 microns. Such large grains are achieved in practice via two methods: first by a final high temperature anneal,[2] via a simple grain growth mechanism. Alternatively the use of a critical cold reduction in the range 3% to 10% prior to a final anneal is employed. This final anneal could either be in-house at the electrical steel producer stage for fully finished grades or at the laminator/stamper stage in the case of semi-finished grades.

For GO steels the production route utilises a 'secondary recrystallization' mechanism which results in massive grains of 5 mm to 20 mm diameter in the final product. Indeed, this grain size and subsequent domain size is a little too large to give the lowest loss and an additional process known as domain refinement can be added at the end of the production route. This reduces the losses by sub-dividing the large grains using, for example, laser scribed lines at a spacing of 5 mm to 10 mm perpendicular to the rolling direction.[3]

13.2.4 Crystal orientation

Although electrical steels are divided into NO and oriented classes, it is almost impossible to produce a totally random crystal texture in the so-called NO steels. The directionality of the hot and cold rolling processes imparts a degree of anisotropy in the crystal structure whether it is required or not. Indeed, it could be argued that a totally random texture is not wanted even in rotating machines such as motors and generators. What is required are good and uniform magnetic properties when magnetised at any angle in the plane of the steel sheet. The best magnetic properties are achieved by reducing the level of any <111> crystal directions within the plane of the sheet and increasing the level of <100> directions. Although these are both difficult to do in practice, it is certainly possible to improve the magnetic properties of NO materials by careful control of the various annealing and rolling stages throughout the production process.

In GO material with an extreme final crystal structure, all the grains are aligned in, or close to, one specific orientation. This is the so-called *Goss* orientation named after the inventor of the process. The Goss orientation is in fact {110}<001>, that is a cube on edge with a <001> direction in the rolling direction. This gives rise to very good magnetic properties when magnetised along the rolling direction, but poorer properties away from that direction. Therefore, this material is ideal for use in electrical equipment where the flux can be aligned in the preferential direction, such as transformers.

[2]This is usually in the range from 900 °C to 1050 °C.
[3]See Section 14.3.1.

13.2.5 Coatings

Coatings on electrical steels were initially required to insulate the sheets in the stacked cores of machines to prevent inter-laminar eddy currents although, over the years, it became clear that other aspects of coatings were important. As mentioned in Section 2.10.2 in Volume 2 of this book, tension produced by the coatings on GO material helps to reduce the domain wall spacing and thereby reduce total power loss and provide some degree of protection against harmful compressive stresses that are produced in the manufacturing processes of electrical machines. Particularly, in NO steels the automated process and rapid punching of laminations from the steel strip benefits enormously from coatings that 'lubricate' the punching operation. This results in considerable cost savings for lamination producers via less tool wear/ replacement and better edges and longer production runs.

In general terms, coatings can be divided into three main types: organic, inorganic and mixed organic/inorganic. Virtually all coatings on GO products are inorganic, phosphate or phosphate/chromate based and relatively thin and hard. They provide high and guaranteed surface insulation levels that are a key requirement for transformer manufacturers. On NO steels all three types of coatings and variations in applied coating thickness are available depending on the final customer requirement.

References

[1] https://www.eti.kit.edu/english/1376.php (Accessed 02/02/18)

[2] http://www.engineering-timelines.com/how/electricity/electricity_02.asp (Accessed 10/12/17)

[3] http://www.historywebsite.co.uk/Museum/metalware/sankeys/sankey01.htm (Accessed 01/06/17)

[4] https://www.gracesguide.co.uk (Accessed 03/03/17)

[5] Bechtold J., and Wiener G. 'The history of soft magnetic materials'. *Proceedings of the Sorby Centennial Symposium on the History of Metallurgy*. Cleveland, Ohio, 1963, pp. 501–19

[6] Barrett W.F., Brown W., and Hadfield R.A. 'Electrical conductivity and magnetic permeability of various alloys of Fe'. *Sci. Trans. R. Dublin Soc.* 1900, vol. 7, pp. 67–26

[7] Stanley H.H. 'Modern manufacturing methods for the production of steel sheets'. *J. Inst. Prod. Eng.* 1949, vol. 28, pp. 373–93

[8] Bozorth R. 'The orientation of crystals in silicon iron'. *Trans. ASM* 1935, vol. 23, pp. 1107–111.

[9] Glumlich E., and Goerens P. 'Magnetic properties of Iron-carbon and iron-silicon alloys'. *Trans. Faraday Soc.* 1912, vol. 8, pp. 98–14

[10] Yensen T.D. *Process of treating magnetizable material.* US Patent 1358810, 1920

[11] Reno R. *Electric sheet.* US Patent 2042124, 1936

[12] Armco Electrical Steel Catalogue, 1932, Middletown, Ohio, USA

[13] Goss N. *Electrical sheet and method and apparatus for its manufacture and test*. US Patent 1965559, 1934
[14] Hodnette J., and Horstman C. 'Hipersil a new magnetic steel and its use in transformers'. *Westinghouse Eng.* August 1941, pp. 52–6
[15] Taguchi S., Sakakura A., and Takashima H. *Process for producing single-oriented silicon steel sheets having a high magnetic induction*. US Patent 3287183, 1966
[16] Baldwins Limited, Product Catalogue, 1939, Wilden Works, Stourport-on-Severn, Worcestershire, England
[17] Gifford C. *Production of silicon steel sheet stock having the property of high surface resistivity*. US Patent 2501846, 1950
[18] Lyudkovsky G., Rastogi P., and Bala M. 'Non-oriented electrical steels'. *J. Metals* 1986, vol. 38(1), pp. 18–26

Chapter 14

Production of electrical steels

Current production methods of electrical steels follow the process routes used in typical steel sheet product process, starting with the production of the liquid steel from either the basic oxygen steelmaking process (BOS) using liquid iron produced in blast furnaces as the main constituent, or electric arc steel making where the main constituent is steel scrap produced by recycling. There is no particular advantage for electrical steels in using either route.

14.1 Chemical composition

The final composition of the steel is determined during the steelmaking steps and the secondary steelmaking. Take as an example a 3% Si non-oriented (NO) grade manufactured via the BOS route. The liquid iron produced by the blast furnace reduction of iron ore by carbon is a relatively impure alloy of iron containing up to 4% carbon, manganese in the range 0.5%–2%, phosphorous up to about 1% and sulphur 0.05%. This chemical composition would be unsuitable for any steel product let alone one that required good soft magnetic properties. During the BOS processing the liquid iron is poured into a large vessel and oxygen is blown into the liquid at supersonic speeds. This together with additions to the liquid reduces the impurity levels considerably. A typical finished composition for a 3% Si NO grade at this stage would be 0.3%Si, 0.1%C, 0.3%Mn and 0.04%S. This steel chemistry is still far from that required, the silicon levels now need to be increased along with aluminium while the carbon and sulphur levels need to be significantly reduced. These changes are made in the secondary steelmaking processes.

There are many different possible secondary steelmaking processes, and these vary from steelmaking site to site dependent on the specific requirements of the types of steels made at each site. The size and scale of the equipment used is difficult to imagine for those who have not seen a steelmaking facility, with liquid steel produced in batches of, typically, 150–300 tonnes. At the end of the process the steel will have the desired chemical composition and will be delivered to the continuous casting (concasting) unit at the correct temperature. Note the temperature of the liquid metal at each step is equally as important as the composition. The steel must remain liquid for sufficient time not only to allow the complete melt to be cast into slab but also be of a temperature that can cool sufficiently to form a solid wall to avoid liquid metal breakouts during continuous casting.

The liquid steel is poured into the concasting unit and great care is needed to shroud the liquid stream with an inert gas to prevent any unwanted oxygen pick up. The speed of the casting process is determined by the cross sectional area of the slab, typically 1,150 mm wide by 250 mm thick, and a balance between the cooling of the liquid steel in the ladle and the wall thickness and strength of the solidifying slab as it travels through the copper, water cooled, moulds of the concasting machine. Concasting speeds vary primarily with slab width and thickness in the range 2–3 metres per minute, however higher silicon grades are usually cast at speeds 10%–20% slower than mild steel grades due to the solidification rate and hot ductility differences. The slabs are then cut to lengths as dictated by the reheat facilities available in the next step of the production route and the final coil size required.

14.2 Hot rolled coil production

The slabs produced from the concasting unit now have to be reheated up to temperatures in the range 1,150 °C–1,400 °C, dependent on the grade and type of electrical steel being produced. This is carried out in a slab reheating furnace. Usually today such furnaces are walking beam furnaces where the slabs are slowly moved by a lifting (walking) process through a direct gas heated furnace. The time in the furnace is typically from one to three hours with the aim to get the slab up to a uniform temperature for a set time (soak time) prior to discharge for the hot rolling process.

The soak time and temperature is dependent on the steel composition and the electrical steel grade being produced. Obviously, the temperature has to be high enough to allow the hot rolling mill to reduce the thickness to the desired final gauge required. Typically, electrical steels would be reduced from 250 mm to 2.3 mm. In most, if not all NO grades, as low a reheat temperature as practical is desired as this helps reduce the number of fine precipitates of nitrides and sulphides in the final hot rolled coil. Conversely, for most traditional grain oriented (GO) grades, a high reheat temperature is required as, in this case, nitrides and sulphides need to be taken into solid solution in the reheat furnace so they can be finely precipitated during the hot rolling process. These precipitates, typically in the size range 50–100 nm diameter, are vital to control grain growth in the cold mill processes that follow the production of the hot rolled coil.

The high reheat temperature required for these traditional GO grades is expensive both in energy terms and also in terms of damage/wear to reheat furnaces. To overcome these issues, in recent years, a new evolution of GO material has become more and more widespread. These newer GO processes are referred to as 'low slab reheat' and are reheated to traditional mild steel reheat temperatures in the range 1,150 °C–1,250 °C dependent on their actual chemistry and subsequent cold mill processing. Note that alternative mechanisms must be used to provide the required precipitates needed to control grain growth in the high temperature annealing stage during cold mill processing.

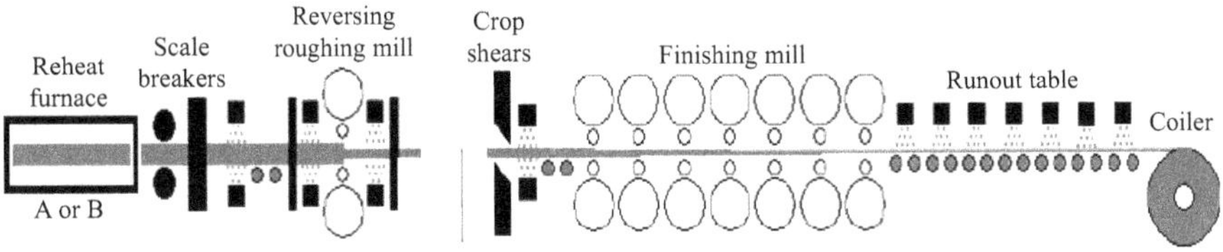

Figure 14.1 Example of processing steps at a hot rolling mill

Following reheating, slabs are discharged from the furnace and then enter the hot mill. Although there are a number of different configurations of hot rolling mill possible, most are divided into similar sections. First, there will be a descaling section to remove some of the surface oxides formed during the reheating process, this is usually carried out by a small degree of rolling and high-pressure water sprays. Following descaling the slab is 'roughed' into a transfer bar usually of the order of 20–25 mm thick. During the rolling process, the deformation that occurs is known as plain strain and, essentially, the width of the slab remains constant, so as the thickness is reduced the length of the slab increases. Every time the thickness is halved the length is doubled. The roughed transfer bar is then fed into the finishing mill, which consists of six or seven hot mills in a line, where the thickness is reduced in steps down to approximately 2.0–2.5 mm final gauge.

On exiting the final hot mill finishing stand the strip runs down a runout table where water cooling is applied before coiling. The hot rolled coil is then allowed to cool and then transported to an electrical steel cold mill site for further processing. An example of a hot rolling mill schematic is shown in Figure 14.1.

As the slab is being rolled, the length is increasing rapidly, and the speed of the slab in the final stands is well over 50 mph (22.5 ms^{-1}) and the original slab lengths of, for example, 8 m will have produced a coil close to 1 km long. In order to produce a hot rolled coil that will give the required magnetic properties after the cold mill processing, key temperatures through the hot mill need to be carefully controlled. These are the slab reheat temperatures and soak time, exit roughing temperature, entry finishing temperature, exit finishing temperature and coiling temperature. Again, different steel chemistries and grades require different actual aim temperatures, but these temperatures can be controlled by the ingoing temperature, degree of cooling applied and the speed of the hot rolling process itself.

For electrical steels, the output of a hot mill is a coil of steel of the selected grade chemistry usually of the order of 20 tonnes, and, typically, 2.3 mm thick and 1,150 mm wide.

14.3 Cold mill processing

14.3.1 Grain oriented electrical steel

Electrical steel cold mill sites are often specific for electrical steel processing and usually have equipment that is different from most other steel cold mills. The hot mill coil edge quality of GO material is usually poor and rough. This gives rise to the need for a continuous edge side trim operation early in the cold mill processing. This side trim operation is either carried out on a bespoke heavy gauge side trim line or as part of the first annealing and pickle line. Side trimming is done using rotary shear blades which cut the strip as the coil is continuously fed through the line. Typically, 15–25 mm is side trimmed off each edge. Often at this line, it is also possible to cut out damaged lengths of hot rolled coil and reweld the undamaged parts back together.

Following the side-trimming operation the coil then passes to a continuous anneal and pickling line. Here the strip is uncoiled and fed through a furnace to heat the strip to a soak temperature in the range 900 °C–1,150 °C, dependent on the type and grade of electrical steel. The time at temperature is short, in the range 30 s–150 s. The strip is then cooled at a controlled rate, again cooling rates being determined by the actual grades being produced. The strip then passes into a descaling section where the surface oxides produced by the hot rolling and latest annealing operation are removed via both mechanical (shot blasting with steel shot) and chemical (acid pickling) processes. The acid used varies with steel producers although hydrochloric acid (HCl) is now the most common, with typical concentrations of 15%–20% HCl at a temperature of 90 °C.

The strip is then coated with a thin rust inhibitor or is lightly oiled before being recoiled. Note that for this anneal–pickle process and any subsequent continuous anneal process it is important to maintain a constant process speed during the anneal. To make this possible, a reservoir of strip is built up both before and after the process sections so that, when the coil end is reached on uncoiling or recoiling, the front or back of the line can stop while the strip stored allows the process section to continue until the next coil is welded to the end of the last. The entry section is then restarted and run slightly faster than the process section speed to rebuild the 'stored' strip. These strip storage systems are usually in the form of multiple loops of strip at either end of the line. These are shown in Figure 14.2 that depicts a GO steel process route at Cogent Power's Orb Works in Newport, UK.

After removal of the surface oxides the coils are ready for the cold rolling process. Here, the strip at a thickness of 2.3 mm is rolled to either an intermediate thickness or its final thickness, again dependent on the grade being produced. In the example shown in Figure 14.2, a single stage cold rolled GO material, the strip would be rolled to a final gauge of 0.23 mm to 0.35 mm in five passes on the reversing mill. That is, the strip is rolled back and fore and five steps are taken to get to the final thickness.

This type of cold rolling (on a four high reversing mill) is commonly used for rolling silicon bearing electrical steels although it is a much slower process than is used for normal mild steels. For the latter, tandem mills, a line of mill stands rather like the hot rolling mill, are used. There are two main reasons for this. First, silicon bearing steels work-harden quickly and the yield strength becomes very high requiring cold rolling mills with considerable load capabilities (more than would normally be designed into a tandem mill stand). Second, silicon steels are also much more likely to snap during cold rolling and delays caused by such

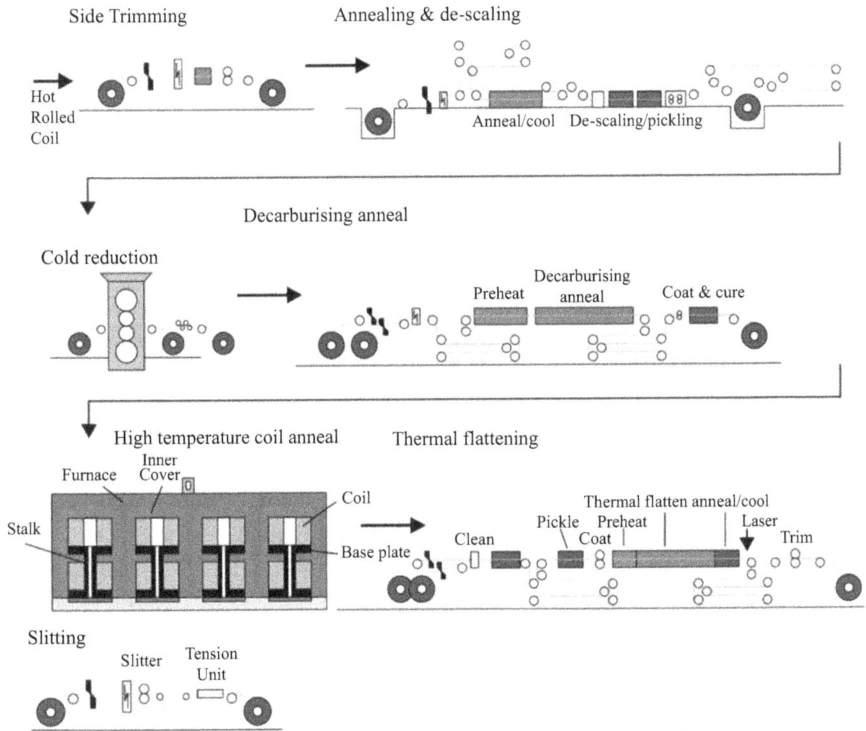

Figure 14.2 Example of a GO steel cold mill process route (Courtesy of Cogent Power Ltd)

strip breaks can take many hours to unravel on a tandem mill but only minutes on a single stand mill.

After cold rolling the next process is a continuous anneal, on a decarburising line. Such a line is essentially a long refractory tunnel furnace, approximately 125 m long, through which the strip is passed. The atmosphere inside the tunnel furnace is very carefully controlled, both to remove carbon from the steel strip and at the same time form a controlled mixture of silica (SiO_2) and fayalite (Fe_2SiO_4) on the surface of the steel. The atmosphere used is mixed hydrogen/nitrogen and water vapour, with the partial pressure of water vapour to hydrogen very tightly controlled. During the process, the strip is heated to 850 °C for approximately 90 s at temperature which causes primary recrystallization of the cold worked steel producing a fairly uniform, single phase, ferritic grain size of 10–20 μm.

Note that for a 3% Si steel with very low carbon levels there is no ferrite to austenite transformation in the steel. After the decarburisation process the carbon level in the steel has been reduced to <0.004%.

At the end of the cooling section on the decarburising line there is a coating unit where a slurry of MgO in water is applied to the strip surfaces and then dried.

The strip is then recoiled before passing on to the high temperature batch anneal process (HTCA).

The next process is a batch anneal process during which the coil is annealed in coil form; not uncoiled and run through a furnace. Here, coils are stacked and heated slowly in a dry hydrogen/nitrogen atmosphere to 1,200 °C; this is again very unusual in normal steel processing. It is during this process that the exceptionally large 'Goss' grains[1] are produced to form the GO steel product. Although there a few different GO production routes now in use throughout the world all of them need this high temperature batch anneal process (HTCA) to produce the 'Goss grains'. During this process, the primary grain structure produced in the decarburising furnace is held in place by a large number of fine precipitates, either sulphides, nitrides or mixtures of both. These precipitates pin the grain boundaries and prevent movement and thereby hold the grain size to below that of the equilibrium grain size that would normally grow as the temperature rises above between 950 °C and 1,000 °C. At these temperatures, the fine precipitates start to break down and the constituents are either taken into solid solution or removed from the steel by oxidation and other surface reactions. As these grain boundary movement inhibitors are removed 'Goss' grains {110}<001> start to grow at the expense of most other grains in the structure due to their grain boundaries being more mobile than others. Once this grain growth has started, a small number of the Goss grains, now with a size advantage, will continue to grow and consume many thousands of nearby grains, resulting in a massive and very unusual grain size for any steel. The process is sometimes referred to as 'secondary recrystallisation'.

Figure 14.3 shows a photomicrograph of a MnS-based conventional grain-oriented (CGO) steel after the decarburisation process with Cu–Mn–S precipitates of the order of 30–70 μm pinning a grain boundary. Figure 14.4 shows a micrograph cross section of a laboratory interrupted high temperature anneal. Samples were heated in a furnace to simulate the heating rates and atmosphere of the HTCA process and then removed and rapidly cooled at defined points in the annealing cycle where the precipitates have grown sufficiently for some 'Goss' grains to grow extensively [1]. When this secondary recrystallization process is complete there are virtually no fine primary grains left and all the grains are close to a 'Goss' orientation, {110}<001>.

At a temperature in the range from 1,000 °C to 1,050 °C, dependent on the inhibitor precipitates used, the secondary recrystallisation process is complete, but the annealing cycle is continued up to 1,200 °C to de-sulphurise and de-nitride the steel completely (i.e. $<0.001\%$ of each). If this is not achieved, then on cooling the sulphides and nitrides could re-precipitate and increase the hysteresis component of loss by interfering with domain wall movement during the magnetisation cycle.

In addition to forming the 'Goss' grain structure and removing S and N from the steel the HTCA process also produces a forsterite (Mg_2SiO_4) coating on the surface of the steel. This forsterite layer is formed by the reaction of the fayalite and silica layer produced in the decarburising anneal and the MgO coating applied at

[1]See Section 2.4 for a description of Goss grains.

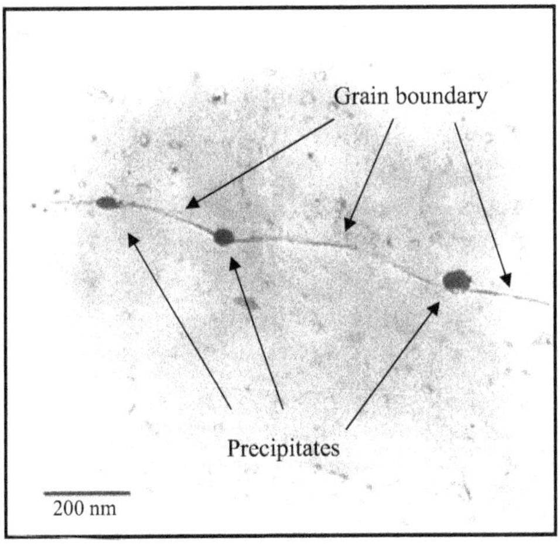

Figure 14.3 *Example of Cu–Mn–S precipitates of the order 30–70 μm pinning a grain boundary in MnS-based CGO after the decarburising anneal (Adapted from [1])*

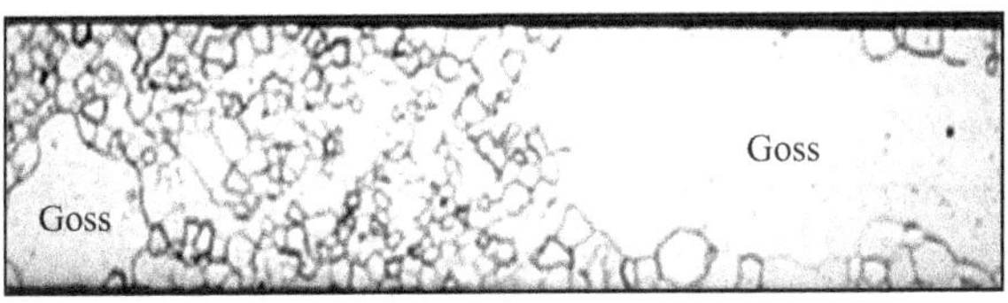

Figure 14.4 *Cross section of a MnS-based CGO grain structure from an interrupted HTCA (sample size is 0.27 mm × ≈ 1.00 mm) (Adapted from [1])*

the end of that process. The MgO also acts as an annealing separator during the HTCA process, preventing the steel surfaces from sticking together. For historical reasons, the crystalline Forsterite layer is referred to as 'glass film'. This 'glass film' layer contributes to the properties of the GO steel in two ways. First, it produces a small but useful surface insulation level. Second, it provides a degree of surface tension that holds the surface of the steel in tension, thereby improving the domain structure and therefore reducing the power loss.

After the HTCA process is completed, with the coils slowly cooled in a dry hydrogen atmosphere, the coils are unpacked and prepared for the next process of thermal flattening. After annealing a coil at high temperature the coil will have a degree of damage to the bottom edge (where the coil will have deformed under its own weight) and there will be little residual strength in the coil form. In short, the material, as it is uncoiled, has extremely poor shape and has a high level of residual stress.

The next stage of processing on the thermal flattening line corrects both these issues by a continuous anneal at 850 °C for around 30 s, with a small elongation (<0.2%) of the strip at temperature followed by a stress relief anneal. During this process, an additional coating is also applied to the steel surface with a liquid coating mix applied by a roll coating system prior to the thermal flattening furnace. These coatings further improve the surface insulation and tension characteristics mentioned previously and are usually phosphate-based and inorganic in nature. At this stage the strip can also be side trimmed again to remove poor edges using rotary shear blades in a similar process to that used on the hot rolled coil described previously in this section. It is also possible to 'domain refine' the steel at the exit of this line, that is, to use a laser system to scribe parallel lines transversely across the strip with a spacing of 5–10 mm. These produce a fine line of stress near the surface of the steel without damaging the effectiveness of the final coating. This domain refinement process can also be done on separate production lines following the thermal flattening unit.

At the end of the thermal flattening line the final magnetic properties of the GO material can be measured and assessed for the first time.

14.3.2 Non-oriented electrical steel

NO electrical steel grades can have a wide range of starting hot rolled coil chemistries dependent on the final magnetic grade required and the actual steel producer. NO magnetic grades are numerous and generally are classified by the total power loss at 1.5 T, 50 Hz, in the range 2 W kg^{-1} to 10 W kg^{-1}. There are also standard thicknesses of 0.35 mm, 0.50 mm, 0.65 mm and 1.00 mm. (See pages 8–11 in [9] for an example of grades available.) Individual electrical steel producers use different chemical compositions to achieve a range of magnetic properties. This usually results in typical silicon levels in the range 0.3%–3.3%. The most common chemistries are 1.3%, 2.4%, 3.0% and 3.3% Si.

As most electrical steel grades have significantly more silicon than normal steels the hot mill coil edges are usually rougher. This gives rise to the need for a continuous edge side trim operation early in the cold mill processing. Usually, any electrical steel with more than 1.3% silicon requires this side trimming operation. This enables subsequent cold rolling to proceed without major difficulties. As is the case with GO material, hot band side trimming necessary on any NO material is usually carried out on a bespoke heavy gauge side trim line, or as part of the first annealing and pickle line.

After coil preparation, the hot band steel coil is processed on an anneal and pickle line (as per the GO steel process described in Section 14.3.1). Some cases such as the poorer, higher loss, final grade material with the lowest Si levels do not require hot band annealing and the annealing furnace will be turned off or set at low temperature during the processing of these coils. For the better quality, lower loss grades the steel will be heated to temperature in the range 900 °C–1,150 °C dependent on the final grade requirements and steel chemistry. After processing through the anneal pickle line the steel has been descaled and prepared for cold rolling.

The pickled coils are usually cold rolled to final thickness (0.35 mm, 0.50 mm, 0.65 mm or 1.00 mm). However, in some cases the steel is first rolled to an intermediate thickness usually 5%–10% thicker than the required final gauge if the material is to be sold as a semi-finished (not finally annealed) grade. Again, this cold rolling is usually carried out on a single stand reversing mill, as is the case for GO material. However, the lower silicon grades are much like 'normal' mild steels and can be rolled on much faster tandem mills if the electrical steel manufacturer has those available.

After cold rolling to final gauge the coils are processed on a continuous annealing line in a controlled non-oxidising atmosphere up to temperatures in the range 900 °C–1,150 °C, again depending on the magnetic requirements and the steel chemistry. Often at the end of such final annealing lines there are, in-line, coating units that can apply coatings as required.

Coating on NO steels can vary depending on the requirements of the steel user. Coatings can be fully organic resins to fully inorganic phosphate or even chromate based with variations containing both. Coating thickness can also vary at the request of the customer, from ≈0.7 μm per side to ≈5 μm per side.

A measurement and inspection system is placed at the end of the production line where the magnetic properties are recorded and physical inspection of the coil takes place. Samples will also be taken from the coil ends and standard test samples cut for testing and grading as per the international standards for measurement and specification (see Chapter 12 of Volume 2 of this book).

In the case of a coil that has been rolled to an intermediate gauge, the coil will still be processed through a continuous anneal line (often the same line) but the annealing temperatures are usually lower (800 °C–850 °C). The coil is then returned to the cold rolling area where it will be cold rolled to final gauge (i.e. given a small final cold rolling of 5%–10%). In this case the coil is sold in the semi-finished state (sometimes referred to as non-finally annealed or NFA (not finally annealed) material).

For semi-finished material, the final magnetic properties have not yet been developed. These are only achieved following a final customer anneal which is usually conducted after stamping of the laminations and the building of the cores of motors, transformers or ballasts have been completed. The anneal is then a batch type anneal where bundles of laminations or cores are annealed at about 800 °C for approximately an hour at temperature. In order for the electrical steel producer to test and certify these steels, samples will have been cut from the coil after final cold rolling and these will have been annealed in a similar way and according to agreed standards. The magnetic properties of samples are measured, and the results reported in a test certificate. The small final cold rolling, together with the final batch anneal, results in a larger grain size than that found in fully processed grades of the same chemistry. This process is known as critical extension and the grain size produced is approximately two to three times larger than that in a fully processed steel. Typical grains sizes would be 150–200 μm compared to 50–100 μm.

Higher silicon grades of NO are produced by JFE Steel Corporation in Japan. These are produced by starting with a conventional 3% Si NO and diffusing Si into the surface using chemical vapour deposition of silicon tetra chloride. The final product contains up to 6.5% silicon. The process is quite complex compared to the usual steel manufacturing processes [2], but it does produce a range of NO materials that have specific uses at higher than normal magnetisation frequencies [3]. These are discussed further in Section 3.1 of Volume 2 of this book.

14.4 Final property assessment

As discussed in Section 14.3, at the end of the thermal flattening line for GO material or the final annealing line for NO material, it is usual to have an on-line continuous magnetic measurement system. This measures and records the power loss of the steel along the whole length of the coil. Together with other automated and manual inspection systems the 'quality' of the coil produced will be assessed and graded at this point. As there are no internationally agreed standards for on-line magnetic measurement, samples are taken from the beginning and end of each coil or sub-coil taken off the thermal flattening or final anneal line. The samples are then cut to size and tested according to international standards and the final magnetic properties are certified. In addition, tests for ductility, surface insulation, coating adhesion and the assessment of edge quality, surface appearance and flatness/shape are carried out at this stage and recorded on the test certificate for the coil.

14.5 Future development

The key drivers for electrical steel development in the coming years will be the demand for improvements in the performance of the electrical machines (transformers, motors, generators, etc.) and the costs of production of both the electrical steel and the electrical machines.

14.5.1 Grain oriented electrical steel

GO electrical steels, primarily used in the cores of electrical transformers account for less than 0.2% of the world's steel production, with GO material volumes of the order of 2.2–2.4 million tonnes per year. However, this very special steel is vital to the electrical supply industry and is a key requirement to meet the growth in electricity use which is forecast globally in the future. Although GO is a technically specialised steel, to all intents and purposes it is traded as a commodity product and the price it governs is largely controlled by supply and demand in the market place.

With the expected rise of generating capacity, a large number of new transformers spanning generation, power and distribution will be required. In addition, there is also a small, but not insignificant, requirement to replace older transformers throughout the industrialised world. The estimated growth in transformer demand is of the order of 3%–4% year on year for the period 2016–2026 according to Goulden Reports in the World Market for Transformers report 2016–2026 [4].

Recent processing improvements from a variety of manufacturers have changed the traditional GO steel production routes and there is now a new range of material with magnetic properties with a distribution that runs from a good conventional material to a good high permeability material. Figure 14.5 shows the range of magnetic properties currently found in GO material produced from traditional routes – Figure 14.5 (left hand figure) – and from the new process routes – Figure 14.5 (right hand figure) [5]. This has resulted in better conventional grades being manufactured than were previously possible, falling between the traditional conventional and high permeability grades. Indeed, demand for some of the current

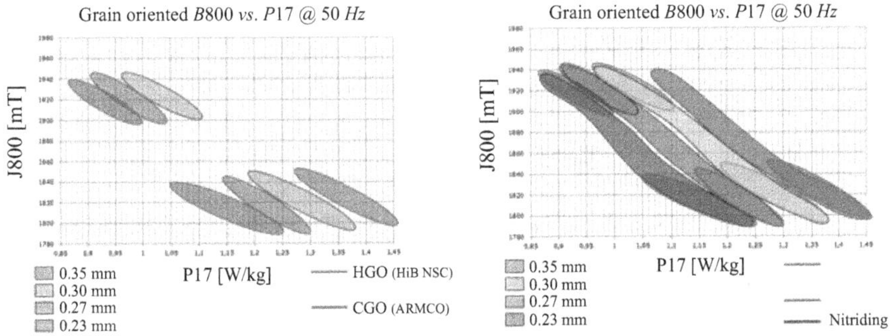

Figure 14.5 Traditional and new process GO loss versus peak induction curves (Courtesy of Rina Consulting, C.S.M. SpA)

conventional grades has all but disappeared, to the extent that a re-evaluation of the international standards for measuring the magnetic properties of GO steel GO magnetic properties is already under consideration [6].

These processing improvements largely involve the move away from 'traditional' GO steel technology using grain growth inhibitors controlled by the hot mill production process to the use of a nitriding stage in the cold mill processing. This nitriding is often referred to 'acquired inhibition' and is usually carried out during or at the end of the decarburising process. This change has reduced the requirements during slab reheating, prior to hot rolled coil production and therefore allows for more latitude in the processes required up to that point. For example, in conjunction with nitriding in the cold mill production it is possible to use thin slab casting or to significantly reduce the slab reheat temperatures in more normal concasting (200–250 mm thick). Thin slab casting produces a slab only 30–60 mm thick which is fed immediately from the casting mould into a shortened hot rolling mill. This, therefore, avoids the need for a large slab reheating furnace, which is replaced by a tunnel furnace through which the thin slab runs and then on into the equivalent of a hot mill finishing stage, this time avoiding the need for hot mill roughing equipment [5]. Such changes have a significant manufacturing cost benefit in addition to the improvements in total power loss associated with the increased availability of high permeability grades over conventional grades.

The impact of such reductions in power loss has important consequences of carbon dioxide mitigation during the lifetime of any transformers built using these better grades of core steel as the no-load loss is reduced and the efficiency of the transformer is improved.

New efficiency standards will drive the use of the most cost effective core material and GO steel manufacturers will continue to drive the reduction of the power losses of these materials. To summarise, there is a change in the market towards improved conventional GO steel having magnetic properties close to traditional high permeability grades becoming readily available.

In addition, more high permeability material is now being produced using the newer processes and demand for these grades will increase. This will enable improvements in transformer efficiencies to continue without the requirements for any radical changes to the materials used or the manufacturing methods beyond those discussed, thereby maintaining stability in the market place.

Efforts are also continuing to improve the quality and consistency of the final coatings, with the aim of increasing the amount of tension applied by such coatings, as this reduces power losses by reducing domain size [7]. Also, many of the existing coatings on GO material use hexavalent chromium (Cr_{VI}) compounds in their coating mixes. Such chemicals are known to be harmful to the environment, but maintaining the current tension levels, the stability and even the curing properties when the hexavalent chromium is removed has proven to be far from straight forward. Although some manufacturers now produce non-Cr_{VI} coatings on GO material, some are not successful and work continues in this area.

There are attempts to produce domain refined grades that are able to maintain the loss improvement achieved by domain refining after annealing. This is of use where small wound core type transformers are produced which include an anneal after the core is wound. The vast majority of domain refined material is produced by laser scribing as indicated in Section 14.3.1. When material of this type is annealed (heated to remove manufacturing stress, for example) the loss improvement from domain refining disappears and loss increases back to the level it was prior to the domain refinement. This is because the laser lines produce a thin band of stress at the surface of the steel and this stress is removed by annealing. In order to produce a domain refinement that can withstand an anneal, usually thin shallow grooves have to be made on the surface of the steel [8], although the surface insulation can often be reduced by such processes. Work continues in this area to improve the final product and the ease of manufacture.

14.5.2 Non-oriented electrical steels

In many ways the future development of NO electrical steels is more complex than that of GO steels because of the many existing grades and range of properties already available. In simple terms, machine manufacturers and designers would like the lowest loss, highest permeability material with mechanical properties that allow for ease, and therefore cheapness, of manufacture – all at the lowest cost.

As discussed earlier, lower losses can be achieved by increased resistivity via additional alloying additions (silicon, aluminium and manganese for example) but virtually all the elements added reduce the permeability at the operating induction ranges for the vast majority of machines. In addition, increased alloy additions increase the manufacturing costs.

Lower losses can also be achieved by reducing the thickness of the steel, but this again increases production costs for both the electrical steel supplier, the producer of the laminations and the machine core assembly. Improved grades, known as high permeability or HP grades can be produced by either improving the crystallographic texture for a given alloy content or by increasing the final grain size for a given alloy content. Three such examples can be seen in [9]. Although such grades are few in number compared to the standard NO grades. There can also be some issues with changes in mechanical properties associated with a high permeability variant of a given standard loss grade, with lower mechanical strength and hardness if lower silicon content steels and/or steels with a larger grain size are used. This can cause production problems for some machine manufacturers.

As was the case for GO material, the use of hexavalent chromium in the manufacturing process has become an environmental issue. For NO material, under the Waste Electrical and Electronic Equipment (WEEE) directive for the sale of products in the European Union any coatings applied to NO steels now have to be free from hexavalent chromium or its use during the manufacturing process [10].

There are also developments for more specific applications – for example, in motors used in electric car drives or ultra fast vacuum cleaners, the cores are subject to higher magnetising frequencies and faster rotational speeds. The need for thin but also higher mechanical strength steels has emerged [11,12]. Metallurgical mechanisms that increase the tensile strength of soft magnetic steels often damage the magnetic properties and, again, compromises between the competing requirements must be balanced. Also, in these areas it is important to note that when considering using electrical steels at frequencies higher than the normal power frequencies of 50 Hz to 60 Hz, some of the usual relationships between metallurgical factors and magnetic properties can change. For example, as the magnetisation frequency is increased the optimum grain size for lowest loss is reduced. Again, it's most important to consider the actual operating conditions applied to core steels in use and not simply rely on standard data that usually only refers to properties at magnetising frequencies of 50 Hz to 60 Hz.

Other examples in specific applications are steels for the cores of direct driven wind turbines are discussed in Chapter 13 in Volume 2 of this book. In such an application, due to the lower magnetisation frequency in the core, a thicker than standard, relatively low alloyed, steel would perform more efficiently. Yet again, this emphasises the importance of the design engineers in fully understanding the steels that are used in their products.

References

[1] Jenkins K., and Lindenmo M. 'Precipitates in electrical steel'. *J. Magn. Magn. Mater.* 2008, vol. 320, pp. 2423–2429

[2] http://www.jfe-steel.co.jp/en/products/electrical/catalog/f1e-002.pdf (Accessed 03/03/18)

[3] http://www.jfe-steel.co.jp/en/research/report/021/pdf/021-04.pdf (Accessed 15/02/18)

[4] http://www.gouldenreports.org/ccs/ccs_library/201701211051360.Goulden_Transformers_2016_brochure.pdf?url_overview=201701211051360.Goulden_Transformers_2016_brochure.pdf (Accessed 14/03/18)

[5] Fortunati S., Cicale S., Schneider J., Franke A., and Kawalla R. 'Developments in the field of electrical steels over the last years'. *Proceedings of the 7th Conference on Magnetism and Metallurgy, WMM16*, 2016, pp. 28–54

[6] Yamaguchi H. *Proposal for introduction of domain-refined grades and addition of high permeability grades on grain-oriented material*. The Japanese National Committee for IEC/TC68, 2013/10/01
[7] Goel V., Anderson P., Hall, J., Robinson F., and Bohm S. 'CrAlN coating to enhance the power loss and magnetostriction in grain oriented electrical steel', *AIP Adv*. vol. 6, Art. No. 055924, 2016
[8] http://www.jfe-steel.co.jp/en/research/report/021/pdf/021-02.pdf (Accessed 11/12/17)
[9] https://cogent-power.com/cms-data/downloads/Cogent%20NO%20brochure%202016.pdf (Accessed 10/12/17)
[10] WEEE – Waste from Electrical and Electronic Equipment (Directive 2002/96/EC)
[11] http://www.jfe-steel.co.jp/en/research/report/021/pdf/021-03.pdf (Accessed 10/03/18)
[12] https://www.tatasteeleurope.com/static_files/Downloads/Energy%20and%20Power/Electrical%20steel/electrical%20steels%20for%20low%20carbon%20vehicles.pdf (Accessed 10/03/18)

Chapter 15
Amorphous and nano-crystalline soft magnetic materials

15.1 Amorphous materials

Amorphous materials were discovered at the California Institute of Technology (Caltech) in the 1960s where it was found that rapid cooling from the liquid state to a solid could suppress the nucleation and growth of crystals to make many types of glassy materials. Thousands of compositions were produced with countless applications in mind. The new class of amorphous metallic glasses soon became referred to as the *dream material* or the *material of the century*.

Appropriate combinations of constitutive elements can lead to magnetic products which may be mechanically hard, corrosion resistant, radiation resistant and sometimes, also, with high mechanical strength and ductility.[1] The magnetic material is produced commercially in the form of ribbon much thinner than any electrical steel, so eddy currents are naturally low under a.c. excitation. This, combined with inherent large resistivity, leads to low losses. No grain boundaries are present to impede domain wall motion, so hysteresis losses are low and very high permeability can be achieved.

The potential applications of such soft amorphous materials were initially treated with a lot of scepticism, but magnetic ribbon was first produced commercially in 1975. This was the basic binary alloy, $Fe_{80}B_{20}$.[2] It was soon recognised that amorphous ribbon was indeed an exciting new class of engineering material which not only could replace conventional soft magnetic materials in many applications but could also satisfy magnetic requirements in entirely new applications [1].

15.1.1 Production of amorphous magnetic materials

The first soft magnetic amorphous materials were produced in 1967 [2]. Most early alloys were very brittle but some had encouraging magnetic properties. One such

[1]Early commercial, soft magnetic, iron-based amorphous ribbon was brittle to the extent that flakes of material could break off the ribbon during core assembly. Today's products do not normally suffer from this problem.

[2]This composition is expressed in atomic percentages. This is normal for amorphous and nanocrystalline compositions in contrast to compositions of crystalline alloys, such as electrical steels, which are normally expressed in *weight* percentages. This practice is adopted in this book.

composition was $Fe_{80}P_{13}C_7$ whose saturation magnetisation is around 0.7 T, its coercivity is approximately 240 Am^{-1} and its Curie temperature is 320 °C. It is interesting to note that Dr Pol Duwez and his colleagues discovered the amorphous material by accident [3], just as was the discovery of the soft magnetic properties of silicon steel by Robert Hadfield and his co-workers over 60 years earlier.

There are many ways of producing amorphous magnetic materials [1]. The biggest challenge is to convert a liquid melt to a solid ribbon extremely quickly so no grains can be formed and therefore the ribbon retains an amorphous state. Typically, solidification from the melt must be cooled at a rate in the range 10^4 °C s^{-1} to 10^6 °C s^{-1}.

Almost all soft magnetic amorphous material is produced by rapid solidification technology (RST) to cast from the melted alloy to form a continuous ribbon or wire[3] [4]. Figure 15.1 shows a schematic representation of the manufacturing route. The required composition is melted and a stream of liquid metal is sprayed under high gas pressure onto the surface of a rapidly rotating copper drum where it immediately cools by conduction and rapidly solidifies. It spins off at high speed to be coiled as a thin ribbon ready for use or further processing. The water cooled drum extracts heat very quickly so that the produced ribbon solidifies fast enough to prevent the formation of any grains. In principle, RST is very simple, but in practice many factors have to be successfully brought together to produce a ribbon that has consistent magnetic, physical and dimensional properties.

In the jet casting process illustrated in Figure 15.1, the molten material is squirted from a distance at a velocity of the order of 30 ms^{-1} onto the outer surface of the wheel. The jet of molten material is inherently unstable which restricts the consistency and width of the resulting ribbon. Economic large scale industrial production was made possible by replacing this by the planar flow process where

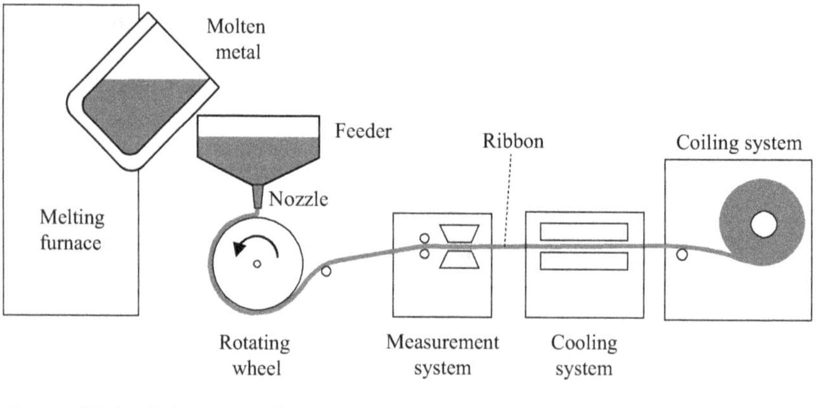

Figure 15.1 Schematic of a basic route for production of amorphous ribbon

[3]Soft magnetic amorphous wires are used mainly for magnetic sensor applications.

the molten metal flows through a special casting nozzle very close to the wheel. This produces a stable puddle of molten metal which solidifies to produce a uniform strip [5]. Introduction of the planar flow process led to an almost tenfold drop in the price of Fe-based ribbon over the first ten years or so of commercial production [6].

In principle, similar steps are used in the commercial production of modern Fe-based amorphous ribbon as shown Figure 9.2 in Volume 2 of this book.

The planar flow technology enables a wide range of compositions to be economically produced as uniform strip up to over 200 mm wide over a thickness range from around 20 μm up to 40 μm [6]. Thinner strip is difficult to produce uniformly so if wound into toroidal magnetic cores, the packing factor can be unacceptably low. It is difficult to produce homogenous thick amorphous strip because heat cannot be extracted quickly enough from its centre to avoid partial crystallisation at the surfaces. This would degrade the magnetic properties.

Soft magnetic amorphous material can be produced in powder form to press as small compacted powder cores claimed to have very good d.c. magnetic properties [7].

15.1.2 Composition

It is extremely unlikely that any composition will be found which simultaneously has a high degree of all the attributes of amorphous magnetic materials, so commercial alloys have been developed to focus on specific combinations of properties for particular applications. Various compositions have evolved for use as distribution transformer cores, magnetic sensors and high frequency devices. These have been fine-tuned, as far as possible, to minimise inherent harmful characteristics such as brittleness, low magnetic saturation, high stress sensitivity, low Curie point and low maximum operational temperature.

All compositions need to include a glass forming metalloid element such as phosphorus, boron or carbon. Ternary amorphous alloys contain the glass forming metalloid, a transition element, such as iron, cobalt or nickel to make the material ferromagnetic, and a non-metallic third element, such as silicon, to provide thermal stability [8]. Today, the commercial soft magnetic amorphous materials are based on the general composition $TM_{100-x}M_x$ where TM is the transition metal with possible additions such as chromium, molybdenum or niobium and M is the glass forming element [9].

Iron-, cobalt- and nickel-based alloys are produced commercially. The Fe-based material is cheapest and has the highest saturation magnetisation. It is made in far greater quantities than the others and used mainly in magnetic cores for power equipment, mostly distribution transformers. The Co-based alloys have low magnetostriction so they have very low stress sensitivity; also they have very good high frequency characteristics. For example, one very thin laboratory-produced, near zero-magnetostriction $Co_{67}Fe_4B_{14.5}Si_{14.5}$ alloy has higher permeability and lower losses than commercial Mn–Zn ferrites at magnetisation frequencies from d.c. to 10 MHz [10]. The NiFe amorphous alloys have intermediate properties.

Of course, all of the materials are magnetically diluted because of the need for the non-magnetic metalloid, so their saturation magnetisation is lower than that of crystalline counterparts. The raw materials are generally abundant. They have inherently low energy footprints because energy is needed only in the casting and coiling process. No hot or cold rolling or subsequent high temperature annealing, used in the production of most crystalline alloys, is needed. Because of this, the energy consumed in the production of commercial Fe-based ribbon is only around 20% of that needed to produce GO silicon–iron.

By its nature, amorphous material is metastable, so in the early days of commercial production it was feared by some potential users that its magnetic properties would deteriorate before the end of the planned working life of some products. However, suitable heat treatment, including magnetic annealing, ensures that stable properties can be expected over the lifetime of commercial materials operating within the temperature guidelines provided by manufacturers [9,11].

15.1.3 Magnetic structure

Because Fe-based amorphous metals are very thin, and have relatively high resistivity, their magnetic characteristics are generally very different to those of electrical steels. However, the underlying reason for the differences is their lack of magnetocrystalline anisotropy. This has a dominant role in determining easy directions of magnetisation and hence the domain structure of electrical steels. The anisotropy also has an indirect effect on the stress sensitivity of the magnetic properties as discussed in Section 15.1.5.

The high magnetocrystalline anisotropy of electrical steels makes domains remain along <100> directions at all times (unless under an extremely high stress or high applied magnetic field). The static domain structure in amorphous material is mainly determined by magnetoelastic anisotropy and some forms of in-built anisotropy.

The result of stress, shape anisotropy and other induced anisotropy, such as that produced by magnetic annealing, normally gives a ribbon a uniaxial anisotropy K of the form

$$K = K_u \mathrm{Sin}^2\theta \left(\mathrm{Jm}^{-3}\right) \tag{15.1}$$

where K_u is a constant and θ is the angle between the magnetisation direction and an easy axis, normally along the length of a ribbon. The anisotropy does not appear to be very well understood, it is certainly not quantified. Very short range, perhaps atomic pair, ordering, may be a significant factor [9].

The near zero magnetocrystalline anisotropy of amorphous materials makes the properties of many compositions very sensitive to heat treatment, particularly magnetic annealing.[4] Figure 15.2 shows the ranges of *B–H* loops obtainable by applying an annealing magnetic field at angles to the longitudinal direction of a ribbon. A longitudinal field normally produces a square *B–H* loop giving very high permeability, whereas transverse field annealing gives a linear, sheared loop, with

[4]Magnetic annealing is the term used for holding a magnetic material in a sufficiently large d.c. or a.c. magnetic field within a certain temperature range to induce a uniaxial anisotropy due to atomic ordering. This anisotropy remains at room temperature in certain magnetic materials.

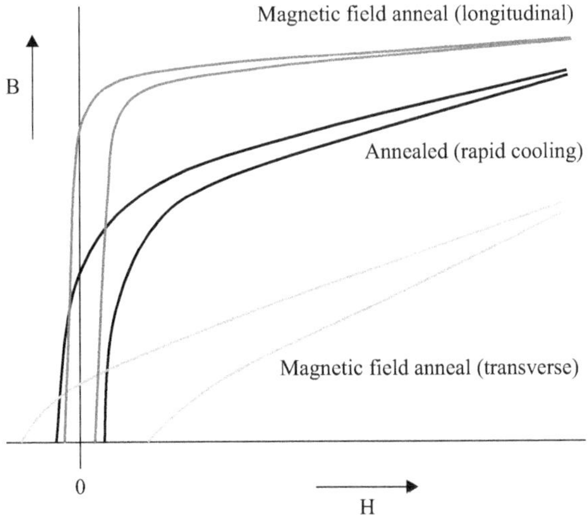

Figure 15.2 B–H loops obtained in an amorphous magnetic material due to different annealing conditions

lower permeability, together but also low eddy current loss. The magnetic annealing mechanism is believed to be similar to that of Permalloy[5] which has very low magnetocrystalline anisotropy [9].

Partly because of the uniaxial anisotropy, domain structures such as those shown in Figure 15.3 are formed after annealing an amorphous material for different lengths of time. There appears to be some form of random uniaxial anisotropy which manifests itself as a wandering of the easy axis [12]. It is possible that a difference between surface and internal magnetisation, not of the form which occurs in electrical steels, also has a strong influence on the magnetic properties [13].

Figure 15.4 shows surface and internal domain structures in as-cast, Fe-based ribbon [14]. The surface structure was observed using the magnetic bacteria technique described in Section 4.7 and the internal structure was deduced using a laser magneto-optic technique in which the internal domain walls were made to oscillate so that their positions could be detected. The surface structure is a maze pattern of domains around 5 μm wide whereas the inter-structure comprises main domains, with 180° walls, 1.0–2.5 mm wide, parallel to the longitudinal direction of the ribbon. These structures[6] can be likened to a random version of the pattern found on the surface of GO steel subjected to compressive stress along its RD.

Magnetisation, up to a high flux density, of these structures along the longitudinal direction is assumed to occur by wall motion of the main internal domains.

[5]See Section 16.2.

[6]At first sight the structures in Figures 15.3(a) and 15.4(a) might look different although they are observed on similar materials. However, the images have different scales and the structures are actually quite similar.

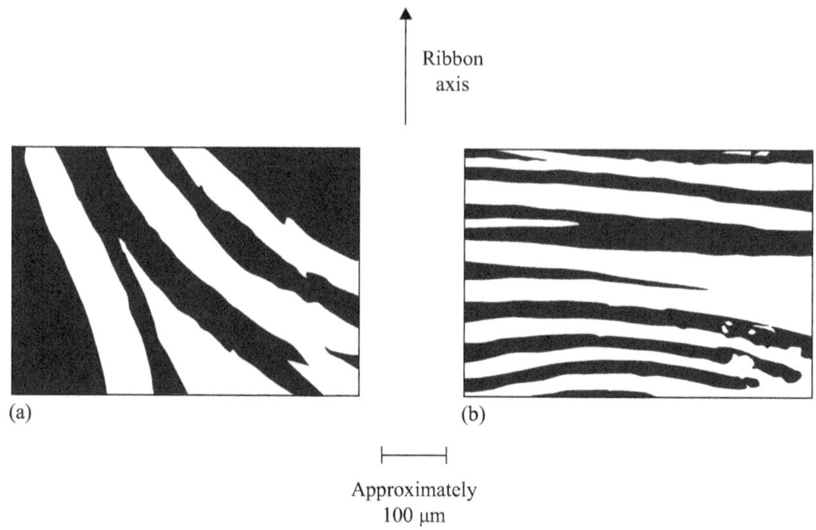

Figure 15.3 SEM images of domains on the surface of a Fe-based amorphous ribbon after annealing at 400 °C for (a) 15 min and (b) 60 min (Figure 4 from [9] reproduced under free licence CC-BY-4.0)

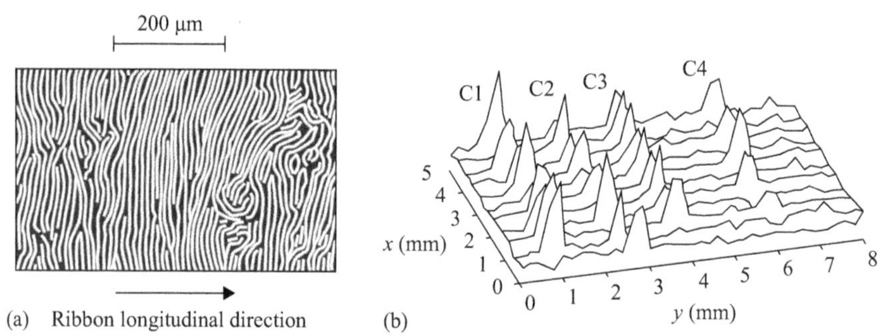

Figure 15.4 Static domain patterns occurring in Fe-based amorphous ribbon (a) surface pattern and (b) position of internal slab domains over the same region (Figure 1 in [14] reproduced under free licence CC-BY-4.0 and modified)

Hence dynamic losses can be assumed to occur due to mechanisms very similar to those occurring in GO electrical steel.

Very similar internal structures occur in as-cast Co-based and Ni-based materials [14].

The loss mechanisms in amorphous material are not yet fully understood. The domain structures are very irregular, as can be seen in the examples in Figure 15.3. The loss is often thought of as having three components, static hysteresis loss, classical eddy current loss and excess loss, just as in crystalline materials. The power loss has a large classical eddy current component [9] so, if other components are negligible, the minimum loss is

$$P_{cl} = \left(\pi B_p f t\right)^2 / 6\rho \left(\mathrm{W m^{-3}}\right) \tag{15.2}$$

However, when such an analysis is carried out in practice using measured data, an anomaly factor of 50, or even higher, can be found [15]. This raises the question whether the analysis has any real physical foundation since P_{ex} (the excess eddy current loss) is so large. New ways of calculating or defining these loss components seem necessary, perhaps not just in amorphous materials.

15.1.4 Coatings and surface treatment

Normally, the surface roughness and a surface oxide layer provides sufficient surface resistance to avoid inter-turn eddy currents in magnetic cores wound from amorphous ribbon. However, eddy currents may flow in random paths between adjacent turns causing extra losses and lower permeability if the packing factor[7] of a core is high, and the ribbon surface is smooth. In practice, the simplest way of overcoming this is to apply an insulating coating to the ribbon. Such a coating has to be thin, since it adds a non-magnetic volume to a core. The coating on GO SiFe is around 5 μm thick so around 3% of the volume of a core is non-magnetic. If the same thickness coating was applied to a 30 μm thick amorphous ribbon, it would immediately make around 30% of the core volume non-magnetic which, normally, would be unacceptable.

Silica-based coatings in the range 50–200 nm have been shown to reduce the excess component of loss in toroidal cores with only a small effect on the packing factor [16]. It also tends to make surface resistivity more uniform, leading to low core to core variation in magnetic properties.

A technique which has proved to be successful in reducing eddy currents at magnetising frequencies in the MHz region, where surface resistivity becomes a greater issue, is to anneal in an oxygen atmosphere to produce a well-controlled, thin oxide layer. The treatment increases the high frequency permeability of amorphous cores by over 50% [17].

Narrowing of the width of 180° domains in large grain GO SiFe by surface treatment improves magnetic properties for reasons explained in Section 2.5 of Volume 2 of this book. The relative width to length ratios of main domains in *Fe–B–Si* alloys can be relatively large so the material can benefit from a suitable domain refinement treatment. Mechanical scribing at 90° to the ribbon length has been shown to reduce losses by 15%–20% at power frequencies [18].

[7]This term is usually applied to the difference between the measured and calculated thickness of a stack of laminations, or the build-up height of a toroidal core. It is also referred to as the stacking factor or the lamination factor. Its value is calculated from the density of a material and its nominal thickness. The coating thickness is not precisely taken into account in the calculation.

However, the exciting apparent power (VA kg^{-1}) can increase significantly unless optimum treatment conditions are satisfied.

Pulsed-laser scribing has proved effective on $Fe_{80}B_{12}Si_8$ ribbon, provided the scribing speed and pulse rate is carefully controlled [19]. Another domain refinement laser treatment using micro hardness indentations in $Fe_{73.5}Cu_1Nb_3Si_{13.5}B_9$ ribbon produced an increase in static hysteresis loss, due to surface domain wall pinning, but the overall loss was reduced by about 5% over a frequency range from 40 Hz to 50 kHz due to reduction of the dynamic component of loss [20].

Until now, the benefit of the small improvements achievable by surface treatment seem out-weighed by extra production cost and increased packing factor so the magnetic benefits of coatings and surface treatment have not been exploited commercially.

15.1.5 Stress sensitivity

As in all magnetic materials, the stress sensitivity of amorphous ribbon depends on the value of the magnetoelastic energy in comparison to all other forms of free energy which might be present. The magnetoelastic energy itself is proportional to the product of magnetostriction and the stress. The magnetostriction, hence the stress sensitivity, varies widely in amorphous materials. Some compositions have been developed to have a high sensitivity to stress for magnetoelastic sensor applications [21].

The stress sensitivity of magnetoelastic sensor material is usually assessed in the form of small toroidal cores which are relatively easy to stress. It is not so easy to apply in-plane compressive stress to a single ribbon.

The stress sensitivity of losses and magnetostriction of $Fe_{78}B_{13}Si_9$ (Metglas[8] 2605 S2) is shown in Figure 15.5 [22]. A compressive stress of 0.5 kg m^{-2}

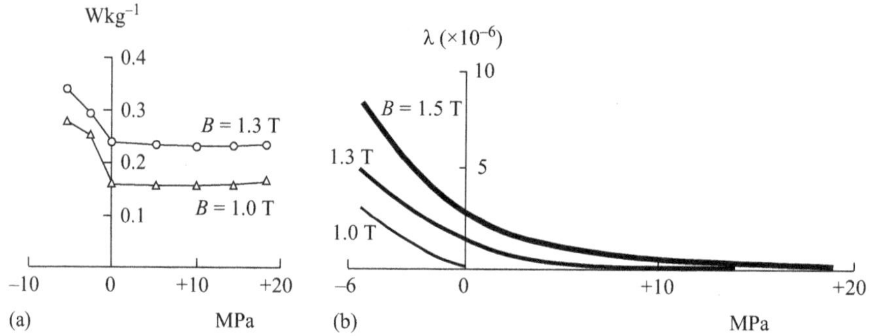

Figure 15.5 Effect of stress on (a) power loss and (b) saturation magnetostriction in field annealed Fe-based amorphous ribbon at 50 Hz magnetisation frequency (Figures 4 and 5 in [22] reproduced under free licence CC-BY-4.0)

[8]Metglas is the trademark adopted by Allied Corporation Inc. when it developed and produced the first commercial products. The company has evolved into Metglas Inc., a Hitachi Group member, based in Conway, South Carolina, USA.

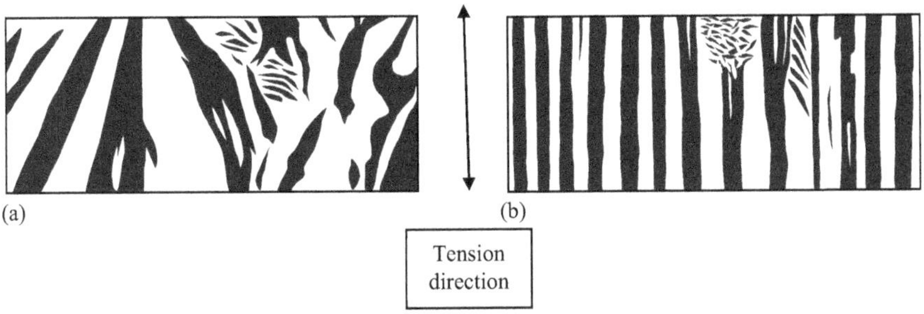

Figure 15.6 Effect of tensile stress on static domain structure of as-quenched $Co_{80}Fe_{20}$ alloy ($\lambda_s < 0$) (a) 35 MPa and (b) 100 MPa (Figure 1 in [25] reproduced under free licence CC-BY-4.0)

(4.9 MPa) applied along the ribbon axis increases the loss by 50%–100% according to the induction level. The magnetostriction also increases significantly. Tensile stress causes losses to fall a little, as expected in positive magnetostriction alloys.

An earlier study of the effect of longitudinal compression on the magnetostriction of a single ribbon of Ni-based $Fe_{40}Ni_{40}P_{14}B_6$ (Metglas 2826) found that the magnetostriction increases by around 30% under a compressive stress of 5 MPa. The stress-free value of the magnetostriction was believed to be reduced by the shape anisotropy of the very thin ribbon [23]

As stated earlier, the stress sensitivity characteristics of an amorphous material depends very much on its composition. Application of tensile stress along the ribbon axis of Ni-based $Fe_{40}Ni_{40}P_{14}B_6$ (Metglas 2826) in the as-cast state can reduce the losses [24] but, after annealing, tensile stress does not reduce the total loss as can be seen in Figure 15.5(a). This is because the stress refines the domain structure of the annealed ribbon to give wide domains whose axes are parallel to the ribbon axis. It has also been shown that in Fe-based amorphous material the main domain walls are *frozen* in place and are not changed spontaneously by increasing tension. Implications of this phenomenon are not widely reported, although it has been shown to occur in GO silicon–iron [24].

Even in near-zero magnetostriction material, stress can have a significant effect on the domain structure as shown in Figure 15.6. This shows that a high stress applied along the ribbon axis of a Co-based amorphous material (with a very low positive magnetostriction) can orientate domains perpendicular to the axis [25].

Bending stress occurs in amorphous material, particularly when wound into a toroidal core. The difference in stress on the convex and concave surfaces of a bent ribbon cause different domain structures near either surface dependent on whether the magnetostriction is positive or negative. Such domain structures observed in Fe-based and Co-based materials have been correlated with changes in coercivity [25–27]. Careful annealing removes much of this stress.

It has been shown that magnetic properties of wide (>10 mm) toroids of *Fe–B–Si* amorphous material are independent of diameter whereas the properties of narrow toroids deteriorate with increasing diameter [28]. The reason is attributed to a dimensional-dependent stress-induced anisotropy due to the winding stress present during heat treatment.

15.1.6 Magnetostriction

The difference between magnetostriction in amorphous materials and equivalent crystalline materials has not been explained. Theories are based on local magnetoelastic effects which cannot be used to predict compositional dependence but are claimed to be useful for understanding temperature and field dependencies [29]. The physics of such mechanisms is outside the scope of this book.

Saturation magnetostriction values of some compositions are given in Tables 15.4 and 15.5. The *engineering magnetostriction* is the most relevant parameter as far as power applications are concerned. An example of its stress sensitivity is given in Figure 15.5(b) [22].

15.1.7 Consolidated Fe-based amorphous material (POWERCORE)

Because amorphous ribbon is so thin, it is difficult and expensive to use in laminated cores. Many methods of dealing with this problem have been tried in order to exploit the benefits of Fe-based material in stacked transformer cores. One approach was the introduction in the late 1980s by Allied Signal Inc. of so-called POWERCORE® strip. The product is no longer available but it is instructive to review some of its history relevant to its application in transformer cores.

Table 15.1 compares the properties of one grade of POWERCORE with Fe-based ribbon and a good quality GO silicon-steel of the time [30]. The properties of POWERCORE are similar to those of the ribbon although its losses and exciting

Table 15.1. Comparison of typical physical and magnetic properties of magnetic core materials for application in power transformers [30]

Property	**Metglas 2605 S2 ribbon**	**POWERCORE strip**	**GO silicon–iron**
Thickness (μm)	20–30	120	270
Specific gravity	7.20	7.20	7.65
Resistivity (μΩm)	1.3	–	4.8
Saturation magnetisation (T)	1.58	1.58	2.00
Curie temperature (°C)	415	415	746
Young's modulus (GPa)	180	150	≥ 120
Lamination factor (%)	≤ 90	90	95–97.5
Thermal conductivity ($Wm^{-1}C^{-1}$)	8	9+	25
Remanent induction (T)	1.3	1.2	1.3
Coercive force (Am^{-1})	1.6	3.2	6.4
Exciting power at 1.4 T, 60 Hz ($VA\ kg^{-1}$)	0.25–0.45	0.5–0.8	1.2 (typical)
Loss at 1.4 T, 60 Hz	0.15–0.25	0.25–0.28	0.9 (typical)

power are higher. The magnetic properties of the best grades of GO silicon–iron today are better than those in the table but they are a useful comparison.

POWERCORE is referred to as *consolidated strip*. It was formed simply by bonding together six layers of Fe-based ribbon. The bonding procedure was not published in detail but it was a spot bonding process using diffused silicon oxide to give a random structure which made the material act neither as a rigid sheet nor as a collection of individual ribbons [30]. Full bonding of the ribbon surfaces would lead to an anomaly factor of around 300 whereas it is 5 to 10 when no bond is present [31]. Not knowing the physical nature of the bond made loss separation analysis difficult. Lack of such knowledge might have restricted development of the material [31,32].

The material used to be pre-annealed to optimise its magnetic properties. It had to be handled more carefully than steel laminations because delamination could occur and, being so brittle, pieces of material could break off from the edges of a ribbon. Obviously small pieces of material present in a transformer tank could cause catastrophic circumstances. This, as well as the need for precise dimensioning of cores, has proved a major obstacle to the usage of POWERCORE [33].

The stress sensitivity of POWERCORE was higher than that of the ribbon itself [31]. The loss could increase by 25% at a stress of just of over 0.2 MPa [32]. The high stress sensitivity has been shown to result in the building factor of a three-phase core being around 2.0, which is far higher than that in an equivalent core assembled from GO silicon–iron [34]. Similar findings were obtained on three-phase cores assembled from POWERCORE with different types of joints [35]. Interestingly, building factors of 1.65 and 2.00 were found for cores with 90° butt-lap corner and T-joints and with 45° mitred corner and 45°/90° T-joints respectively, contrary to the trend found in cores assembled from GO silicon–iron. This is presumably related to completely different flux density distributions in the joints assembled from the two materials.

Because of its inherent hardness, POWERCORE strip could not be cut accurately or cost-effectively using conventional processes. Tool life could be increased by heating the material during the cutting process to reduce its hardness.

An advantage of POWERCORE was that its magnetic properties in the plane of the sheet were far more isotropic than those of GO silicon–iron which lead to very low rotational losses, as shown in Figure 15.7 [30].

The isotropic nature of POWERCORE strip makes the flux distribution within the T-joints of stacked three-phase cores very different from that in joints assembled from GO silicon–iron. Figure 15.8 shows the localised loci[9] of the flux density in a T-joint[10] of a core assembled from POWERCORE strip. If this is compared with that of a typical distribution in the T joints of a GO steel core (such as shown in Figure 5.44 in Volume 2 of this book), the degree of rotation at most points

[9] An explanation of the meaning of loci presented in this way is given in Section 5.9 of Volume 2 of this book.

[10] It might be noticed that this core has an unusual 5 mm yoke stagger, but this does not affect the distribution to any noticeable extent.

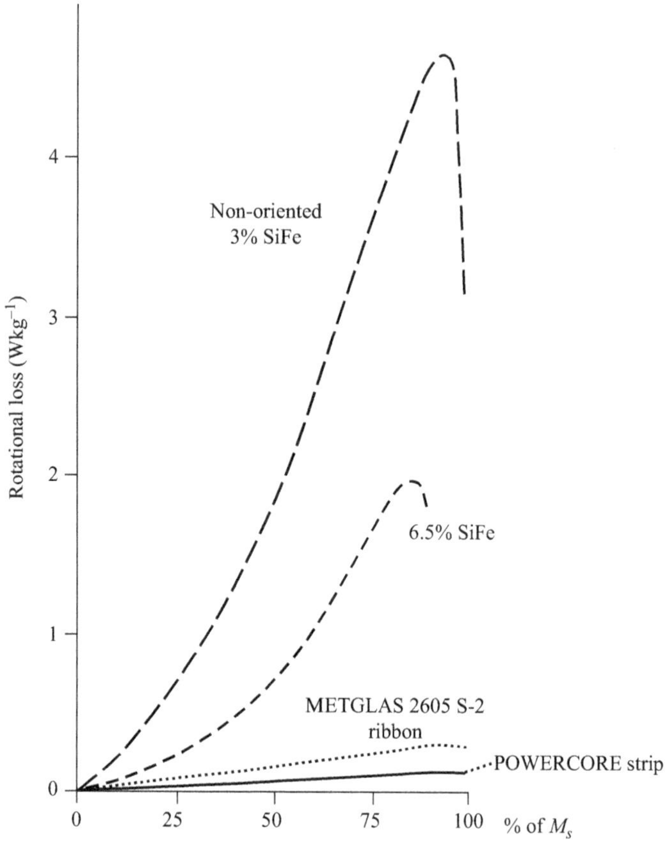

Figure 15.7 Comparison between the rotational loss of electrical steel and Fe-based amorphous material at 60 Hz magnetisation. Magnetisation scale is plotted as a percentage of the saturation magnetisation of each material (Figure 3 in [30] reproduced under free licence CC-BY-4.0 and modified)

within the core is greater in the POWERCORE because of its low anisotropy, which itself is believed to be the cause of a lower ratio of rotational loss to unidirectional a.c. loss in general [37].

POWERCORE ceased being manufactured in the early 1990s, possibly due either to the technical deficiencies outlined in this section or to slow take up in the transformer market.

15.1.8 Bulk amorphous material

At the time POWERCORE was abandoned, some researchers were focusing on producing *thick* Fe-based amorphous ribbon which could be used in stacked transformer cores. Commercial amorphous ribbon is around 20–30 μm thick and

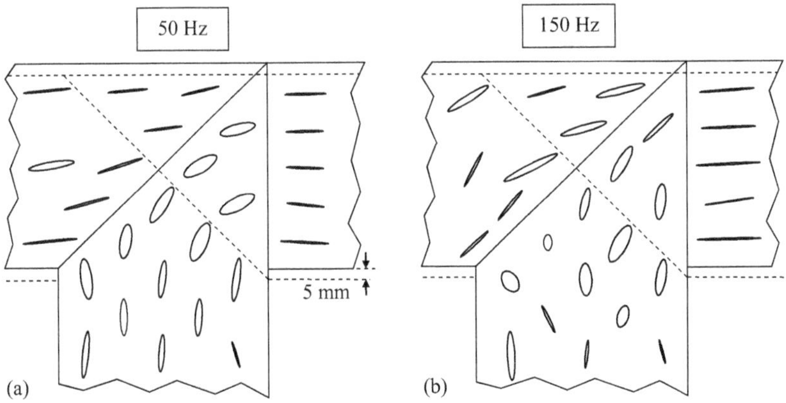

Figure 15.8 Variation of localised flux distribution in a T-joint of a three-phase core assembled from POWERCORE strip (50 Hz, 1.35 T) (a) 50 Hz component and (b) 150 Hz component (Figures 4 and 5 in [36] reproduced under free licence CC-BY-4 and modified)

the thinnest high grade electrical steels are around 100 μm thick. Laboratory produced $Fe_{80}Si_{6.5}B_{12}C_{1}$ amorphous material about 60 μm thick has been found to have surprisingly low losses when built into laboratory scale, single phase, stacked cores. The no-load loss and exciting power was about 30% less than in any amorphous cores [38]. However, no progress appears to have been made in the development of thick ribbon or consolidated amorphous strip.

Until the early 1990s it was believed that solidification rates of 10^5 °C s^{-1} to 10^6 °C s^{-1} were needed to produce amorphous materials with useable magnetic properties. However, by 1995, mainly as the result of pioneering research by Professor Akihisa Inoue's group at Tohoku University in Sendai (Japan), the first *bulk* amorphous magnetic materials were being reported in the United States and Japan. These were produced using a mould casting technique [39].

The innovation which makes this possible is the inclusion of elements which extend the super-cooled liquid region to obtain a glassy structure at a far lower cooling rate than possible with conventional compositions. At least three elements with atomic size mismatch of at least 12% and large negative heat of mixing[11] are needed [40].

An important property is the very high resistivity which means bulk amorphous materials are not restricted to d.c. and low frequency applications but some are usable up to the MHz region [41]. Materials have been produced in rods up to 2.5 mm diameter and up to at least 30 mm long while retaining relatively high saturation magnetisation and low coercivity [42].

Different shapes and sizes seem feasible so the search continues for compositions with magnetic and mechanical properties with potential engineering

[11]The heat of mixing is the enthalpy liberated or absorbed from a substance upon mixing elements or compounds.

Table 15.2. Magnetic properties of a range of compositions of bulk amorphous materials [40]

Composition	M_s (T)	H_c (Am^{-1})	μ	$\lambda_s \times 10^{-6}$
$Fe_{56}Co_7Ni_7Zr_{10}B_{20}$	0.96	2.0	19,100	10
$Fe_{56}Co_7Ni_7Zr_8Nb_2B_{20}$	0.75	1.1	25,000	13
$Fe_{61}Co_7Ni_7Zr_8Nb_2B_{15}$	0.85	3.2	12,000	14
$Fe_{56}Co_7Ni_7Zr_8Ta_2B_{20}$	0.74	2.6	12,000	14

applications. Table 15.2 shows some properties of bulk amorphous alloys whose coercivities and permeabilities are respectively lower and higher than equivalent *Fe–Si–B* ribbon [40].

The disadvantages of bulk amorphous alloys are slow manufacturing, high cost and low saturation magnetisation. However, some materials do have relatively high values of M_s, e.g. $Fe_{69}Co_{18}B_{13}$ has a saturation magnetisation of 1.9 T, but it appears to be unstable [43].

The potential of bulk amorphous soft magnetic materials is uncertain. Newer compositions can be accurately made as larger components in shapes which could find applications in new types of high performance electrical machine, but this is still at the laboratory stage [44,45].

15.2 Nano-crystalline magnetic materials

Only a few families of soft magnetic materials have low anisotropy combined with low magnetostriction necessary to give very soft magnetic properties. These are the Permalloys (Chapter 17), Sendust (Chapter 17), MnZn ferrites (Chapter 18), Co-based amorphous alloys (Chapter 15) and, today, the Fe-based nano-crystalline alloys.

Nano-crystalline materials, commonly referred to as *nano-materials*, are becoming preferred to NiFe crystalline alloys and amorphous ribbon for a growing range of applications. They can be magnetically hard or soft. They are referred to as nano-materials because of their 10–30 nm scale grain size and the alignment of their magnetic moments via atomic exchange interactions over a range of a few nano-metres.

Nano-materials with soft magnetic properties were discovered over 30 years ago [46]. The soft magnetic nano-materials can have low losses at high magnetising frequencies, high resistivity, high saturation magnetisation and low magnetostriction compared to many other materials. Also, their *B–H* loop shapes can be tailored by magnetic annealing. This exciting range of materials fills the gap between amorphous and crystalline soft magnetic materials and is already established in high frequency electronic devices. Nano-materials are being considered for use in a range of power engineering applications.

15.2.1 Production of nano-magnetic material

Nano-materials can be produced in several ways. The most suitable process for developing soft magnetic properties is by controlled crystallisation from an amorphous state.

A simplified version of the process is outlined in Figure 15.9. First, a suitable composition is rapidly solidified to produce an amorphous phase by the RST process. The ribbon thickness, ≈ 10 μm, and width, typically up to 50 mm, are controlled by the wheel speed and nozzle design. Next, the ribbon is annealed at around 550 °C in an inert atmosphere to produce precipitates, in this case copper particles, which act as nuclei for the formation of α-FeSi nano-grains. The final result is an

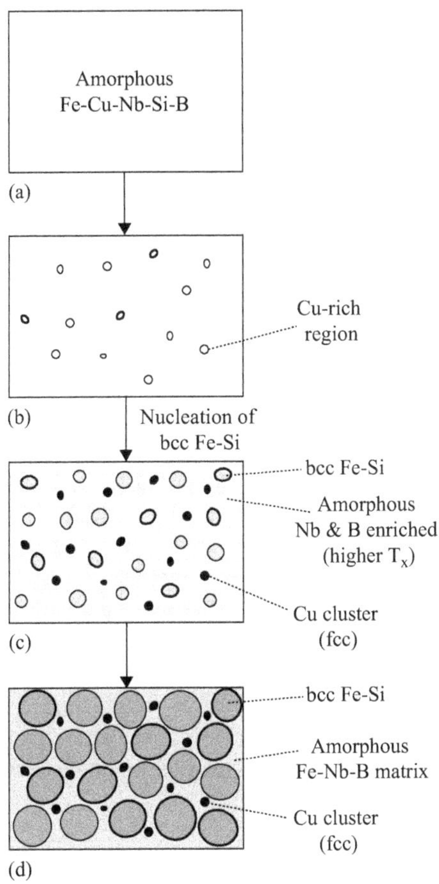

Figure 15.9 Schematic diagram of the development of a nano-crystalline alloy from an amorphous material: (a) as quenched amorphous state, (b) change during early stage of annealing, (c) initial stage of crystallisation and (d) final nano-crystalline state

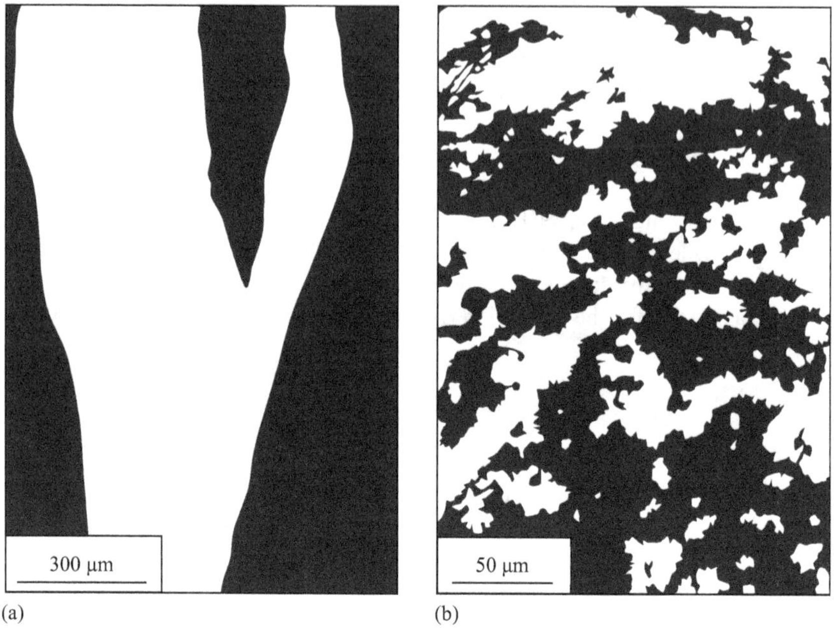

Figure 15.10 Typical domain pattern on the surface of nano-crystalline alloy $Fe_{73.5}Cu_1Nb_3Si_{13.5}B_9$ at (a) room temperature and (b) 320 °C (part of Figures 1 and 8 in [48] reproduced under free licence CC-BY-4.0)

amorphous phase which occupies perhaps 80% of the volume, together with the SiFe nano-grains and a small number of copper clusters. The amorphous state in the resulting nano-material is different to that of the starting amorphous composition.

The two-phase structure gives low coercivity, and a.c. losses lower than equivalent amorphous alloys. Further heat treatment in a magnetic field can be carried out to vary the *B–H* loop shape. The heat treatment needs to be carefully controlled to avoid embrittlement.

Composition can be varied widely. For example, the boron content can be increased at the expense of silicon to give better glass forming qualities but the magnetic properties will deteriorate. More iron can be included to increase the saturation magnetisation, but this can cause other magnetic properties to deteriorate [47] unless the composition is fine-tuned.

Figure 15.10 shows typical domain structures in an iron-based nano-crystalline material [48]. The room temperature structure is very similar to that occurring in similar compositions of amorphous material, again having non-linear domain walls due to the very low anisotropy. Above the Curie point of the fully amorphous phase, the fine structure shown in Figure 15.10(b) is present. This verifies the difference in Curie temperatures of the fully amorphous phase and the equivalent amorphous phase in the nano-material matrix. In general, T_c of the starting, as-cast, fully amorphous phase is in a range from 250 °C to 300 °C whereas it increases to between 600 °C and 750 °C in the nano-crystalline state [48].

As with amorphous alloys, nano-materials can be produced in compacted powder form, but its potential is still not clear.

15.2.2 Magnetic properties

Magnetic domains in nano-materials span several grains since domain wall motion is not impeded very much by the very small grains. Magnetic properties are broadly similar to those of Co-based amorphous ribbon but with far higher saturation magnetisation. The coercivity drops rapidly with decreasing grain size below around 100 nm diameter when the width of the domain walls is comparable with the grain size. This is opposite to what is found in fully crystalline, larger grain, materials. This is illustrated in Figure 15.11 [49]. This is analogous to that reported in grain oriented (GO) electrical steels where precipitates with diameters less than 30 nm have little effect on the coercivity [50]. However, the mechanism in materials with nano-metre size grains is based on the so-called random anisotropy model which is explained elsewhere [51,52].

The magnetic properties of crystalline materials are very dependent on the anisotropy. The best magnetic properties generally are obtained when the anisotropy is close to zero. In nano-materials the local crystalline anisotropy is averaged out because the grain size is lower than the magnetic exchange length.[12]

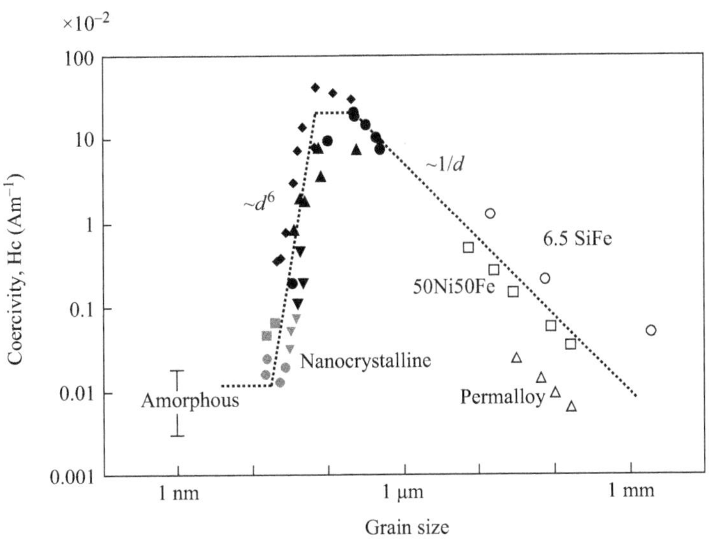

Figure 15.11 Variation of coercivity with grain size in crystalline, amorphous and nano-crystalline soft magnetic materials (Figure 1.1 in [49] reproduced under free licence CC-BY-4.0)

[12]The exchange length is effectively the region over which no deviation occurs between adjacent dipoles. This is mentioned just to emphasise how small the grains are in nano-materials.

The magnetostriction of a nano-crystalline alloy is effectively the volume weighted sum of that of the amorphous and crystalline phases. Therefore, to obtain near to zero magnetostriction in Fe-based alloys, the negative magnetostriction of the crystalline SiFe phase needs to be offset by the positive magnetostriction of the amorphous Fe-based matrix.

The thermal stability is of interest because of the high volume of amorphous phase in nano-crystalline materials. However, its aging resistance is higher than that of similar fully amorphous alloys, so continuous operation at 150 °C is acceptable in most commercial alloys [49].

The magnetic properties of nano-materials tend to deteriorate with increasing temperature, but not so much as to stop the materials from being usable at normal device operating temperatures. For example, the relatively high Curie point of $Fe_{73.5}Cu_1Nb_3$-$Si_{15.5}B_7$ means that, at 500 °C, M_s only drops to 50% of its room temperature value. Its saturation magnetostriction is around $+0.1 \times 10^{-6}$ at room temperature; it reaches a peak of 0.3×10^{-6} at 200 °C and drops to zero at 400 °C [49].

The random anisotropy tends to fall with temperature. This would be expected to cause an improvement in permeability, but the initial increase in magnetostriction with temperature offsets this so the permeability drops a little. This can be compensated for by appropriate choice of annealing conditions and alloy additions to give better high temperature performance.

A significant benefit of the Fe-based nano-materials is their relatively high saturation magnetisation. Figure 15.12 shows how low magnetostriction nano-crystalline materials have much higher saturation magnetisation than groups of low magnetostriction, high permeability amorphous and nickel–iron alloys. Alloys with M_s of over 1.8 T have been developed more recently with correspondingly low losses of around 0.3 W kg^{-1} at 50 Hz, 1.5 T [53]. If the iron concentrations can be increased further, it is feasible that this could eventually rise close to that of high silicon electrical steels and at the same time have far lower losses and higher permeability. However, whether costs and other important properties can be comparable remains to be seen.

15.2.3 Coating and surface treatment

The magnetic properties of nano-materials can be improved by suitable surface treatment in a similar way as for fully amorphous materials. For example, the application of a suitable varnish or epoxy resin to form interlaminar insulation in tape wound cores of pre-annealed $Fe_{73.8}Ni_5Cu_{0.6}Nb_{2.6}Si_9B_9$ reduces losses at magnetising frequencies above 20 kHz, although it causes an increase in loss at low frequency. A more effective treatment is by vapour deposition of silicon oxide which results in a modest drop in loss at 1 kHz [54].

15.2.4 Stress sensitivity

The stress sensitivity depends on the magnitude of the magnetostriction which varies widely in nano-materials. Compositions such as FINEMET $Fe_{73.5}Si_{13.5}Nb_3Cu_1B_9$ have very low magnetocrystalline anisotropy and relatively high magnetostriction,

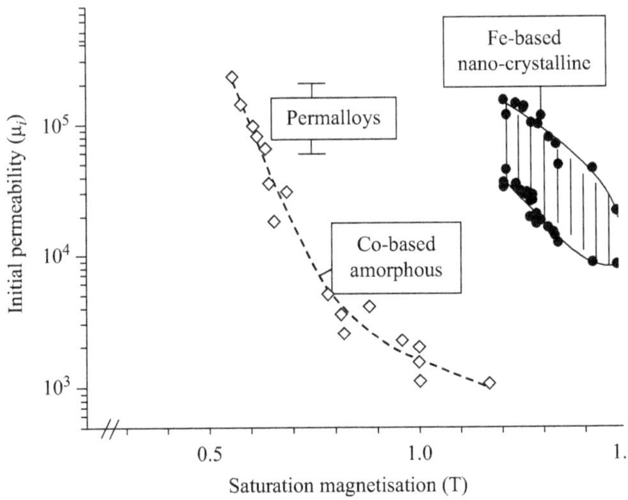

Figure 15.12 Comparison of the initial permeability versus saturation flux density ranges of high permeability materials: (a) Permalloys, (b) Co-based amorphous material and (c) Fe-based nano-crystalline tape (reproduced from Figure 4.1 in [49] under free licence CC-BY-4.0 and modified)

so some magnetic properties, such as the *B–H* loop shape, are very stress sensitive. Such materials are used in magnetic stress sensors [55]. Others, have near zero magnetostriction so they have low stress sensitivity and find applications in power conversion.

15.3 General properties of amorphous and nano-materials

15.3.1 Families of amorphous materials

Typical magnetic properties of alloys within the three generic groups of amorphous materials are compared in this section. Table 15.3 summarises some general properties of the generic groups of amorphous materials.

The general properties of some compositions of the Fe-based group make them very attractive for some power applications at, or close to, 50/60 Hz magnetising frequencies. The other families are more useful for operation at higher frequency, so comparing them at a fixed frequency or induction level is interesting, but not very useful, for assessing possible application performance.

15.3.2 Commercial materials

Well-established commercial manufacturers of amorphous ribbon and nano-crystalline materials are Metglas Inc., Hitachi Metals and Vacuumschmelze (VAC). Company details and product ranges are given on their web sites [56–58]. Commercial products

Table 15.3. Comparative properties of the three families of amorphous magnetic materials [9]

	M_s (T)	Typical core loss (W kg^{-1})	Saturation magnetostriction	Cost	Applications
Fe-based	1.30–1.80	0.08–0.15 at 1.0 T, 50 Hz	32×10^{-6}	Lowest	Transformers
Ni-based	≈0.90		–	Medium	Shielding, sensors
Co-based	0.55–0.77	0.005 at 0.40 T, 50 Hz	≈0	Highest	HF devices, sensors

from each company are referred to throughout this book. Metglas Inc. is now a part of Hitachi Metals. Generally, they manufacture different compositions aimed at different market applications. Other companies have successfully carried out pilot line production of Fe-based amorphous ribbon (e.g. [59]) so new and alternative products might come to market soon.

It should be mentioned that losses and permeability of nominally the same compositions produced by different manufacturers have been reported to vary considerably [60]. This is said to cause significant variation in performance of cores produced from the materials.[13] This seems to have been a relatively limited study and, of course, commercial products are continually improving.

Today's commercial products have evolved from earlier products, often with the same trade names. So comparisons of performance, for example, between the same grades of electrical steel and Fe-based amorphous material made today and a decade or two ago, must be viewed with caution.

Table 15.4 shows the range of the properties of the commercial amorphous and nano-material currently produced by Metglas Inc. Comprehensive references to some of these are given in reviews such as [9].

Examples of properties of commercial materials taken from VAC and Hitachi web sites are shown in Table 15.5. Figures for each group summarise the range of properties of several alloys which have slightly different compositions and are heat treated in various ways to give the wide range of structure sensitive properties shown.

15.4 High silicon micro-crystalline ribbon

Micro-crystalline 6.5% SiFe ribbon could be considered a precursor of both today's magnetic nano-materials and high silicon electrical steels. Research into

[13]It should be emphasised that where comparisons of products from different producers are made, they are only intended to show generic differences due to physical and compositional characteristics.

Table 15.4. Properties and applications of current range of amorphous and nano-crystalline materials [56]

Trade name and main element(s)	**2605 SA1 (Fe-based)**	**2605S3A (Fe-based)**	**2605 HB1 (Fe-based)**	**2705M (Co-based)**	**2714A (Co-based)**	**2826MB (Fe-Ni-based)**	**Nano-crystalline tape FT-3W (Fe-Si-B-Nb-Cu)**
B_s (Tesla)	1.56	1.41	1.63	0.77	0.57	0.88	1.18
λ_s ($\times 10^6$)	27	20	27	< 0.5	< 0.5	12	< 1
T_c (°C)	395	358	364	365	225	353	
Density (kg m^{-3})	7.18	7.29	7.33	7.8	7.59	7.4	7.3
Resistivity (μΩm)	1.3	1.38	1.2	1.36	1.42	1.38	–
Maximum d.c. permeability (annealed)	600,000	35,000		290,000	1,000,000	800,000	6,000–74,000 at 10 kHz
Loss (Wkg^{-1}) (annealed) at 1.4 T, 50 Hz	0.20	0.005 at 0.3 T, 60Hz	0.2	6.0–7.0 at 0.3 T, 10 kHz	0.005 at 0.3 T, 60 Hz	0.02 at 0.3 T, 60 Hz	–
Thickness (μm)	25	18	25	22	15	29	17
Lamination factor (%)	> 84	> 75	> 84	> 75	> 75	> 75	70
Tensile strength (GPa)	1.2	1–2	2.1	1–2	1–2	1–2	–
Young's modulus (GPa)	110	100–110	120	100–110	100–110	100–110	–
Maximum operating temperature (°C)	150	150	–	90	90	125	–
Typical applications	Distribution and power transformers; motors; current transformers	Current transformers; fault protection devices; high frequency cores	Distribution and power transformers; motors; current transformers.	Shielding; sensors; high frequency cores; magnetic switches	Sensors; shielding; high frequency transformers; magnetic amplifiers	Sensors; high frequency cores	High frequency devices; d.c.–d.c. converters; switched mode power supplies; high permeability cores for telecoms and protection devices

Table 15.5. Range of properties found in alloys within the families of amorphous ribbon and nano-materials and manufactured by VAC and Hitachi Metals

Groups by trade name	VITROVAC Amorphous alloys (Co-based)	VITROPERM Nano-material (Fe-based)	FINEMET Nano-material (Fe-based)
B_s (T)	0.55	1.2	1.23
λ_s ($\times 10^6$)	< 0.2	< 0.5	< 1
T_c (°C)	210	> 600	570
Density (kg m^{-3})	7,700	7,100	7,300
Resistivity (μΩm)	1.3	1.15	1.2
Loss (W kg^{-1}) (annealed)		< 100 at 0.3 T, 100 kHz	250–600 Wm^{-3} at 0.2 T, 100 kHz
Thickness (μm)	17–25	20	18
Tensile strength (GPa)	1,000–1,700	1,000–2,700	
Young's modulus (GPa)	100–159	100–150	
Maximum operating temperature (°C)	80–120 depending on conditions	120–180 depending on conditions	
Typical applications of Fe-based alloys[a]		Switched mode converters and power supplies, battery chargers, solar technology, motor inverters, high frequency transformers, common mode chokes, field sensors, shielding, magnetic amplifiers, current transformers, small motors, high frequency inductors and fault protectors	

[a]These do not all apply to both nanocrystalline and amorphous materials.

rapidly quenched high silicon–iron and high aluminium–iron alloys was pioneered by Professor Kenji Narita and co-workers in Kyushu University, Japan in the 1970s.

It can be produced by RST, similar to that used for producing an amorphous structure, starting with rapid quenching of high purity Fe and Si followed by high temperature annealing to produce a relatively ductile ribbon, 0.06–1.5 mm thick. The grain size can be as low as 1 μm. This is 100 times higher than that in nano-material ribbon but around 100 times less than that of typical fully annealed non-oriented electrical steel.

A minimum loss occurring at 6.7% silicon is believed to be due to a coercivity contribution which disappears at this composition when the magnetostriction is zero, but, at the same time, mechanical hardening and embrittlement in the high silicon alloys is due to a partial phase ordering effect [61,62].

Ribbon produced in the laboratory can have satisfactory mechanical properties combined with good low frequency magnetic properties [63]. The hysteresis loss increases rapidly with reducing thickness below around 50 μm but a corresponding reduction in classical eddy current loss offsets this considerably [64].

Probably the *best* magnetic properties found in high silicon micro-crystalline ribbon have been obtained in 4.8% Si material. Losses in 80 μm thick ribbon as low as 0.56 W kg^{-1} at 1.7 T, 50 Hz have been obtained under tensile stress [65].

More recently, larger scale production of 6.5% Si strip, 20–40 μm thick up to 30 mm wide has been claimed to be successful [66,67]. Development quantities of the strip have far lower loss at high magnetising frequency than that of the same compositions produced by other techniques.

High silicon content micro-crystalline ribbon has not yet been developed to its full potential. It can have close to zero magnetostriction, high saturation magnetisation and low losses which can compete with other ribbon materials at high frequency. Its permeability can be high, for example B_{800}[14] can reach 1.86 T [65]. However, laboratory production starts with expensive very high purity iron and silicon. For low to medium frequency power applications, it might have been superseded by the high silicon strip produced commercially in relatively large quantities by chemical vapour diffusion techniques (see Section 3.1 in Vol. 2 of this book).

15.5 Applications of amorphous and nano-crystalline ribbons

Just ten years after the first production of Fe-based amorphous ribbon, over 1,000 small wound amorphous cores had been produced for US utilities [68] by Allied Signal. By the turn of the century an estimated 10% of all new small distribution transformer cores in the United States were using Fe-based amorphous metal [42].

Today, Fe-based amorphous ribbon is well established in 50/60 Hz distribution transformer cores in some parts of the world. Over 4 million such cores had been installed by 2015 by which time the capacity of Metglas Inc had increased to 1,00,000 tonnes per annum [69]. However, its low magnetic saturation hinders its application in large transformers and in some high frequency power cores such as output filter inductors in traction system converters where the low saturation can lead to instability. It is also used to increase the profitability and output of some wind turbine applications. It is used widely in small inductive components in the kHz regime since its thin gauge and high electrical resistivity reduces the effects of harmonic components of flux on losses. The high stress sensitivity of amorphous ribbon can impair the performance and acoustic noise output of wound cores.

The need for high frequency performance in electrical power applications might not be obvious since the vast majority of soft magnetic materials are used at 50 Hz, or 60 Hz. However, there are two reasons for concern over performance in the presence of harmonic components of magnetisation. First, the presence of low

[14]Flux density at 800 Am^{-1}.

frequency harmonics (generally up to the 9th) are increasing in power distribution systems as discussed in Section 13.11 in Volume 2. Second, high frequency harmonic components (in the kHz range) are generated in power electronic control, conversion systems and PWM[15] voltage feeds to electrical machines. These consequences of the substantial improvement in reliability and power ratings of electronic components made during the last few years, leading to their rapidly growing exploitation, are higher harmonic components of flux density in applications where magnetisation at close to sinusoidal 50 Hz operation used to be considered the norm.

Tables 15.4 and 15.5 include some general application areas of amorphous and nano-crystalline materials.

It is difficult to use amorphous material in motors. However, by introducing innovative, although complicated, assembly methods, the material has been used in stators of high speed d.c. motors running at 70,000 rpm producing losses considerably less than when using conventional electrical steel [70]. Other prototype motors using amorphous material have been reported (e.g. [71–75]) but in the foreseeable future they will only find uses in very niche, high performance applications where first cost is not a priority.

The wide operating frequency range of Fe-based amorphous materials, and potentially nano-material, is their main strength in new and potential applications in the renewable energy generation and control sector, as well as electrical energy-saving equipment in general [76,77]. Nano-materials potentially can enable new, energy-saving applications in the electricity generation and distribution sectors where they might make a significant impact.

Toroidal cores of $Fe_{81.8}Cu_{1.0}Mo_{0.2}Si_4B_{14}$ nano-material, over a kilogram in weight, have been shown to have a loss of 2 W kg^{-1} at 0.2 T, 10 kHz, which is one of the lowest values for any metallic soft magnetic material. This, combined with high saturation magnetisation and low magnetostriction, is claimed to indicate its potential for distribution transformers and inductors for power electronic systems [78].

Another benefit of nano-materials is that, like amorphous ribbons, their energy footprint during the production process is low and potentially their costs should drop substantially as production levels increase.

A barrier to using wideband semiconductors, such as GaN or SiC, to their full switching capability up to 1 MHz, in high frequency power conversion is the poor high frequency properties of present commercial magnetic material. However, losses in commercial nano-materials, thinned to 5 μm and magnetised at 500 kHz have been found to be less than 50% of those of the best equivalent ferrites [79]. This is not a practical solution but it is an indicator that there is scope to extend the high frequency range of nano-materials even further.

Potential applications of some of the other thin gauge materials are shown in Figure 15.13.

[15]Losses can increase considerably under pulse width modulation (*PWM*) voltage excitation. This is discussed in detail in Chapter 7 of Volume 2 of this book.

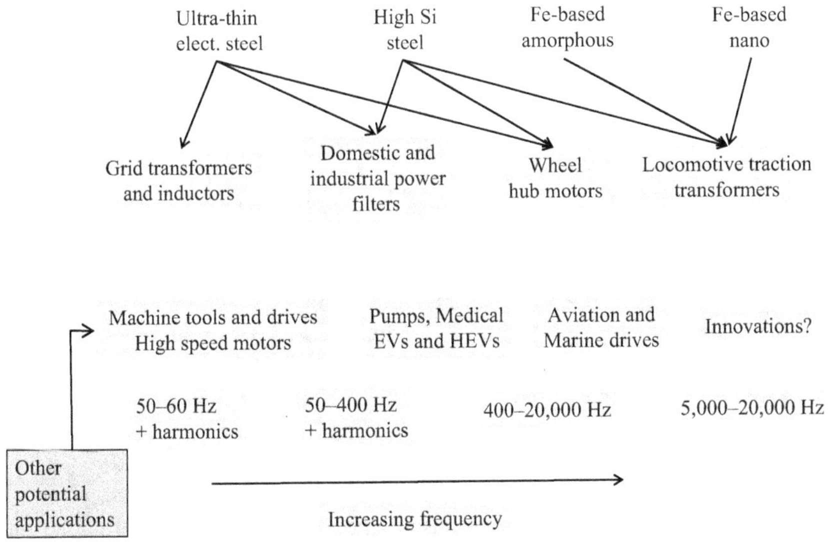

Figure 15.13 Potential applications of thin strip and ribbon over their normal operating frequency ranges

References

[1] Liebermann H.H. 'Manufacture of amorphous alloy ribbons'. *IEEE Trans. Magn.* 1979, vol. 15(6), pp. 1393–1397

[2] Duwez P., and Lin. S.C.H. 'Amorphous ferromagnetic phase in iron-carbon-phosphorus alloys'. *J. Appl. Phys.* 1967, vol. 38(10), p. 4096

[3] Hasegawa R. 'Pol Duwez – an appreciation of his work'. *Amorphous Metals and Semiconductors*, vol. 3 (Eds. Haasen P. and Jaffe R.I.). *Acta-Scripta Metallurgica Proceedings Series*, Pergamon Press, 1986

[4] Moses A.J. 'Steel restructured'. *IEE Rev.* 1994, pp. 11–14

[5] Narasimhan M.C. Continuous casting method for metallic amorphous strips. U.S. Patent 4,221,257, 1980

[6] Fish G.E., and Smith C.H. 'A decade of progress in iron-based amorphous ribbon for electrical power distribution'. Proceeding Symposium on Soft and Hard Magnetic Materials and Applications, Lake Buena Vista, FL, USA, 1986, pp. 7–20

[7] Endo I., Otsuka I., Okuno R., Shintani A., Yoshino M., and Yagi M. 'Fe-based amorphous soft-magnetic powder produced by spinning water atomization process (SWAP)'. *IEEE Trans. Magn.* 1999, vol. 35(5), pp. 3385–3387

[8] Chen H-S., and Warren D.E. Novel amorphous metals articles. U.S. Patent 3,856,513, 1974

[9] Fish G.E. 'Soft magnetic materials'. *Proc. IEEE* 1990, vol. 78(6), pp. 947–969

[10] Beatrice C., Banu N., Ferrara E., and Fiorillo F. 'High-frequency properties of thin amorphous ribbons'. *J. Magn. Magn. Mater.* 2008, vol. 320, pp. e810–e813

[11] Nathasingh D.M., and Liebermann H.H. 'Transformer applications of amorphous alloys in power distribution systems'. *IEEE Trans. Power Deliv.* 1987, vol. 2(3), pp. 843–850.

[12] Buchholtz F., Koo K.P., Dandridge A., and Sigel G.H. 'Easy axis distribution in transversely annealed Metglas 2605 S2'. *J. Magn. Magn. Mater.* 1986, vol. 54–57, pp. 1607–1608

[13] Goser U., and Preston R. 'Mössbauer spectroscopy applied to amorphous metals'. *Topics in Applied Physics and Glassy Metals*. Springer-Verlag, Berlin, 1983, pp. 93–126

[14] Schönhuber P., Pfützner H., Harasko G., Klinger T., and Futschik K. 'Main domains and surface domains in amorphous soft magnetic materials'. *J. Magn. Magn. Mater.* 1992, vol. 112, pp. 349–352

[15] Overshott K.J., and Blundell M.G. 'The variation of power loss and domain wall spacing with applied stress in wide amorphous ribbon material'. *IEEE Trans. Magn.* 1980, vol. 16(5), pp. 1221–1223

[16] Nathasingh D.M., Datta A., and Smith C.H. 'Effects of coatings on the soft magnetic properties of an iron-based amorphous alloy'. IEEE International Magnetics Conference. Hamburg, Germany, 10–13 April 1984

[17] Fukunaga H., Ooae K., Nakano M., Kakehashi H., Ogasawara H., and Ohta Y. 'Improvement in high frequency magnetic properties of amorphous cores by thin surface oxide layers on ribbons'. *J. Phys. IV France* 1998, vol. 8, pp. 83–85

[18] Tan K.S. 'Amorphous ribbon domain refinement by scribing'. *IEEE Trans. Magn.* 1986, vol. 22(3), pp. 188–191

[19] Johnson M.J., Chen R., Jiles D.C., and Ramanan V.R. 'Reducing core losses in amorphous $Fe_{80}B_{12}Si_8$ ribbons by laser-induced domain refinement'. *IEEE Trans. Magn.* 1999, vol. 35(5), pp. 3865–3867

[20] Weidenfeller B., and Riehemann W. 'Surface treatment of Finemet for reduction of power loss'. *J. Magn. Magn. Mater.* 1996, vol. 160, pp. 287–288

[21] Walimont H., and Boll R. 'Applications of amorphous magnetic materials'. *J. Magn. Magn. Mater.* 1982, vol. 26, pp. 97–105

[22] Sasaki T., Shimomura E., and Yamada K. 'Variation of power losses with stresses in amorphous sheets for power applications'. *IEEE Trans. Magn.* 1987, vol. 23(5), pp. 3587–3589

[23] Moses A.J., and Winter J.R. 'Some effects of compressive stress on an amorphous ribbon'. *J. Magn. Magn. Mater.* 1984, vol.41, pp. 152–154

[24] Klinger T., Pfützner H., and Schönhuber P. 'Stress effects on the domain configuration of amorphous and crystalline soft magnetic materials'. *Int. J. Appl. Electromagn. Mater.* 1994, vol. 4, pp. 265–269

[25] Siemko A., Lachowicz H.K., Moser N., Forkl A., and Kronmüller H. 'Stress dependence of magnetostriction and domain structure in metallic glasses'. *J. Magn. Magn. Mater.* 1990, vol. 83, pp. 171–173

[26] Zusman A.I., Ershov N.N., and Vasil'eva N.P. 'Effect of bending deformation on the magnetic properties of amorphous iron-based alloys'. *Izv. Akad. Nauk SSSR Ser. Fiz.* 1989, vol. 53(4), pp. 659–661

[27] Saito A., Konditi D.B.O., Yamamoto K., Tamura T., and Ebisutani K. 'The magnetic properties of bent amorphous ribbons'. *Int. J. Appl. Electromagn. Mater.* 1992, vol. 3(1), pp. 43–46

[28] Luborsky F.E., and Becker J.J. 'Strain-induced anisotropy in amorphous alloys and the effect of toroid diameter on magnetic properties'. *IEEE Trans. Magn.* 1979, vol. 15(6), pp. 1939–1945

[29] Lachowicz H.K., and Szymczak H. 'Magnetostriction of amorphous magnetic materials'. *J. Magn. Magn. Mater.* 1984, vol. 41, pp. 327–334

[30] Norin E.M. 'POWERCORE© amorphous metal consolidated strip'. *Proceedings of Symposium on Soft and Hard Magnetic Materials and Application,* 1987, pp. 21–28

[31] Pfützner H., Schönhuber P., Erbil B., Harasko G., and Klinger T. 'Problems of loss separation for crystalline and consolidated amorphous soft magnetic materials'. *IEEE Trans. Magn.* 1991, vol. 27(3), pp. 3426–3432

[32] Nathasingh D.M., and Liebermann H.H. 'Transformer applications of amorphous alloys in power distribution systems'. *IEEE Trans. Power Deliv.* 1987, vol. 2(3), pp. 843–850

[33] Schönber P., Pfützner H., and Erbil B. 'Comparison of POWERCORE and lased scribed HI-B steel model cores under practical working condition'. *Anal. Fisica* 1990, vol. 86, pp. 96–98

[34] Basak A., Moses A.J., and Yasin M.R. 'Effect of stacking method and clamping on distribution transformer built with amorphous magnetic material'. *IEEE Trans. Magn.* 1991, vol. 27(6), pp. 5196–5198

[35] Passaglia R., and Zannella S. 'Stacked transformer cores using amorphous POWERCORE strip'. *IEEE Trans. Magn.* 1991, vol. 27(6), pp. 5265–5267

[36] Daut I., and Moses A.J. 'Some effects of core building on localised losses and flux distribution in a three-phase transformer core assembled from POWERCORE strip'. *IEEE Trans. Magn.* 1990, vol. 26(5), pp. 2002–2004

[37] Arabi M.A., and Moses A.J. 'An improved method for measuring rotational power loss in electrical steels'. *J. Magn. Magn. Mater.* 1984, vol. 41, pp. 227–229

[38] Sato T., Yamada T., and Ozawa T. 'Core loss of thick amorphous alloy ribbons'. Nippon Steel, Technical Report *No. 39*, January 1989, pp. 57–64

[39] Inoue A., Makino A., and Mizushima T. 'Ferromagnetic bulk glassy alloys'. *J. Magn. Magn. Mater.* 2000, vol. 215/216, pp. 246–252.

[40] Inoue A. 'Bulk amorphous alloys'. In *Amorphous and Nanocrystalline Materials* (Eds. Inoue A., Hashimoto K.). Springer-Verlag, Berlin Heidelberg, 2001, Chapter 1, pp. 1–51

[41] Hasegawa R. 'Design and fabrication of new soft magnetic materials'. *J. Non-Crystal. Solids* 2003, vol. 329, pp. 1–7.

[42] Makino A., Kubota T., Chang C., Makabe M., and Inoue A. 'FeSiBP bulk metallic glasses with high magnetization and excellent magnetic softness'. *J. Magn. Magn. Mater.* 2008, vol. 320, pp. 2499–2503

[43] Moses A.J. 'Opportunities for exploitation of magnetic materials in an energy conscious world'. *Interdiscipl. Sci. Rev.* 2002, vol. 27(2), pp. 100–113
[44] Inoue A., and Louzguine D.V. 'Bulk nanocrystalline and nanocomposite alloys produced from amorphous phase'. *Nanostructured Metals and Alloys: Processing, Microstructure, Mechanical Properties and Applications* (Ed. Whang S.H.). Elsevier R.V, 2016, pp. 152–177
[45] Piccin R., Tiberto P., Chiriac H., and Baricco M. 'Magnetic properties and power losses in Fe-Co-based bulk metallic glasses'. *J. Magn. Magn. Mater.* 2008, vol. 320, pp. e806–e809
[46] Yoshizami Y., Oguma S., and Yamaguchi K. 'New Fe-based soft magnetic alloys composed of ultrafine grain structures'. *J. Appl. Phys.* 1988, vol. 64, pp. 6044–6046
[47] Yoshizawa Y. 'Magnetic properties and applications of nanostructured soft magnetic materials'. *Scr. Mater.* 2001, vol. 44, pp. 1321–1325
[48] Schäfer R., Hubert A., and Herzer G. 'Domain observation on nanocrystalline material'. *J. Appl. Phys.* 1991, vol. 69, pp. 5325–5327
[49] Herzer G. 'Nanocrystalline soft magnetic alloys'. *Handbook of Magnetic Materials*, vol. 10 (Ed. Buschow K.H.J.). Elsevier Science B.V., 1997, Chapter 3
[50] Jenkins K., and Lindenmo M. 'Precipitates in electrical steels'. *J. Magn. Magn. Mater.* 2008, vol. 320, pp. 2423–2429
[51] Herzer G. 'Grain size dependence of coercivity and permeability in nanocrystalline ferromagnets'. *IEEE Trans. Magn.* 1990, vol. 26(5), pp. 1397–1402
[52] Franco V. 'Soft magnetic nanocrystalline alloys: materials and models'. *Mater. Matters* 2016, vol. 11(4), pp. 104–107
[53] Ohta M., and Yoshizawa Y. 'New high-B_s Fe-based nanocrystalline soft magnetic alloys'. *Jpn. J. Appl. Phys.* 2007, vol. 46(2), pp. L477-L479
[54] Szynowski J., Kolano R., Kolano-Burian A., and Polak M. 'Reduction of power losses in the tape-wound FeNiCuNbSiB nanocrystalline cores using interlaminar insulation'. *IEEE Trans. Magn.* 2014, vol. 50(4), Art. No. 6300704
[55] Szewczyk R., Bieńkowski A., Salach J., Fazakas E., and Varga L.K. 'The influence of microstructure on compressive stress characteristics of the FINEMET-type nanocrystalline sensors'. *J. Optoelectron. Adv. Mater.* 2003, vol. 5(3), pp. 705–708
[56] www.metglas.com/products/magnetic_materials/. (Accessed May 2018)
[57] www.vacuumschmetze.com/en/products.html. (Accessed May 2018)
[58] www.hitachi-metals.co.jp/e/corp/imdex.html. (Accessed May 2018)
[59] Wu C.-Y., Lin K.-J., Cheng Y.-T., *et al.* 'Development of amorphous ribbon manufacturing technology'. China Steel, Technical Report *No. 27*, 2014, pp. 28–42
[60] Sixdenier F., Morand J., Salvado O.A., and Bergogne D. 'Statistical study of nanocrystalline alloy cut cores from two different manufacturers'. *IEEE Trans. Magn.* 2014, vol. 50(4), Art. No. 2003604

[61] Fiorillo F., Ferrara E., Ferrando L., *et al.* 'Magnetic losses and mechanical properties of Fe-4 to 7.8 wt% Si rapidly quenched alloys'. *IEEE Trans. Magn.* 1997, vol. 33(5), pp. 3802–3804

[62] Ruiz D., Ros-Yanez T., Vandenberghe R.E., De Grave E., De Wulf M., and Houbaert Y. 'Magnetic properties of high silicon steel with variable ordering obtained through thermomechanical processing'. *J. Appl. Phys.* 2003, vol. 93, pp. 7112–7114

[63] Enokizono M., Teshima N., and Narita K. 'Magnetic properties of 6.5 percent silicon-iron ribbon formed by a melt spinning technique'. *IEEE Trans. Magn.* 1982, vol. 18(5), pp. 1007–1013

[64] Fiorillo F., Ferraro E., Ferrando L., and Appino C. 'Magnetic losses and mechanical properties of Fe-4 to 7.8 wt% Si rapidly quenched alloys'. *IEEE Trans. Magn.* 1997, vol. 33(5), pp. 3802–3804

[65] Arai K.I., Ohmori K., Satoh T., and Yamashiro Y. 'Very low core loss high silicon-iron ribbons'. *IEEE Trans. Magn.* 1987, vol. 23(5), pp. 3221–3226

[66] Liang Y.F., Wang S., Li H., Jiang Y.M., Ye F., and Lin J.P. 'Fabrication of Fe-6.5%Si ribbons by melt spinning method on large scale'. *Adv. Mater. Sci. Eng.* 2015, Art. ID. 296197

[67] Tsuya N., Arai K.I., Ohmori K., Shimanaka H., and Kan T. 'Ribbon-form silicon-iron alloy containing around 6% silicon'. *IEEE Trans. Magn.* 1980, vol. 16(5), pp. 728–731

[68] 'High efficiency amorphous core distribution transformer shipped to 90 utilities'. Press release by Allied Signal (Allied)/General Electric, Hickory, NC, USA, April 1986

[69] Metglas Inc. Amorphous metals distribution transformers. www.metglas.com 2016

[70] Kolano R., Krykowski K., Kolano-Burian A., Polak M., Szynowski J., and Zackiewicz P. 'Amorphous soft magnetic materials for the stator of a novel high-speed PMBLDC motor'. *IEEE Trans. Magn.* 2013, vol. 49(4), pp. 367–1371

[71] Kolano R., Kolano-Burian A., Krykowski K., *et al.* 'Amorphous soft magnetic core for the stator of the high-speed PMBLDC motor with half-open slots'. *IEEE Trans. Magn.* 2016, vol. 52(6), Art. No. 2003005

[72] Dems M., and Komeza K. 'Performance characteristics of a high-speed energy-saving induction motor with amorphous stator core'. *IEEE Trans. Industr. Electron.* 2014, vol. 61(6), pp. 3046–3055

[73] Silveyra P., Xu P., Keylin V., DeGeorge V., Leary A., and McHenry M.E. 'Amorphous and nano composite materials for energy-efficient electric motors' *J. Electron. Mater.* 2015, vol. 45(1), pp. 219–225

[74] Wang Z., Enomoto Y., Ito M., *et al.* 'Development of an axial gap motor with amorphous metal cores'. *IEEE Trans. Ind. Appl.* 2011, vol. 47(3), pp. 1293–1299

[75] Denis N., Inous M., Fujisaki K., Itabashi H., and Yano T. 'Iron loss reduction in permanent magnet motor by using stator core made of nanocrystalline magnetic material'. *IEEE Trans Magn.* 2017, vol. 53(11), Art. No. 8110006

[76] Schwenk H., Beichler J., Loges W., and Scharwitz C. 'Actual and future developments of nanocrystalline magnetic materials for common mode chokes and transformers'. *PCIM Europe*, Nurenberg, Germany, May 2015, pp. 209–216

[77] Kolano R., Kolano-Burian A., Polak M., and Szynowski J. 'Application of rapidly quenched soft magnetic materials in energy-saving electrical equipment'. *IEEE Trans. Magn.* 2014, vol. 50(4), Art. No. 2004804

[78] Ohta M., and Hasegawa R. 'Soft magnetic properties of magnetic cores assembled with a high B_s Fe-based nanocrystalline alloy'. *IEEE Trans. Magn.* 2017, vol. 53(2), Art. No. 2000205

[79] Kulkarni S., Li D., Wang N., *et al.* 'Low loss magnetic thin tape films for off-line power conversion'. *IEEE Trans. Magn.* 2014, vol. 50(4), Art. No. 2004404

Chapter 16

Nickel–iron, cobalt–iron and aluminium–iron alloys

16.1 Introduction

It was pointed out in Chapter 1 that nickel–iron (Ni–Fe) alloys and cobalt–iron (Co–Fe) alloys are well established, high value soft magnetic materials which have been developed for niche applications requiring magnetic properties which cannot be satisfied by silicon steels. They have evolved in response to market demand, in the case of the Ni–Fe alloys, mainly for very high permeability, purposed-shaped *B–H* loops and good high frequency performance, and in the case of the Co–Fe alloys, mainly for high saturation magnetisation, and the capability to operate over a wide temperature range whilst maintaining very good mechanical properties. The aluminium–iron (Al–Fe) alloys are not so common.

Magnetic properties of compositions within each alloy group vary widely. Properties, such as permeability, of alloys in each family group can vary by several orders of magnitude. The compositions around 50%Co/50%Fe, 80%Ni/20%Fe, 50%Ni/50%Fe and 30%–40%Ni/balance iron are commercially important. Production of Co–Fe and Ni–Fe magnetic alloys varies a lot from year to year but shows a gradually upward trend. It is in the range from around 150,000 to 25,000 tonnes per annum [1].

Small amounts of a tertiary element, most commonly vanadium (V), chromium (Cr), copper (Cu) and molybdenum (Mo) are sometimes added to the binary alloys in each group to enhance particular magnetic or mechanical properties.

As with iron and silicon–iron, Ni–Fe and Co–Fe compositions can be produced by powder technology for solid parts [2]. These are only of specialised interest and are not discussed in this book. To the non-expert, the range of trade and group names of Ni–Fe and Co–Fe alloys can be confusing. Common names such as Permendur, Hiperco, Mumetal, Permalloy, Supermalloy and Perminvar refer to close families of alloys. Magnetic properties vary substantially according to the heat treatment as well as small compositional differences between alloys within each group.

Typical values of the magnetic properties of some of the more common alloys are given later in this chapter, but it should be stressed that published values vary widely, partly due to the high sensitivity to heat treatment parameters. Physical and magnetic data, even for specific alloys, should only be treated as a guide.

Metallurgical defects and impurities are particularly harmful to magnetic properties of most Ni–Fe and Co–Fe alloys so the starting alloying elements must be extremely pure, and any contaminates picked up during the manufacturing process must be as low as possible.[1] Non-chemical defects such as vacancies, dislocations and internal stress also have to be minimised by careful heat treatment to ensure that they do not degrade the properties of the commercial alloys [3,4]. Chemical defects, such as non-magnetic compound precipitates of S, O, P, C, Al and Mg must be converted into volatile oxides which can be removed by melting and vacuum refining and/or by solid state refining at a very high temperature in a pure hydrogen atmosphere [5]. The low impurity levels must be maintained during the normal production route of hot rolling, intermediate annealing and cold rolling to final thickness. These close controls are necessary using specialised production equipment which can be adjusted to meet the magnetic requirements of individual alloys [6]. This chapter opens with a brief discussion of some properties of elemental nickel, cobalt and iron which are relevant to the magnetic properties of their magnetic alloys. The chapter continues with an overview, mainly of the Ni–Fe and Co–Fe families. Readers wanting to delve more deeply into the complex structures and wide ranging properties of these alloys need to refer to more specialised text books and reviews, e.g. [7–10].

16.2 Iron, cobalt and nickel

Some characteristics of the individual elements are included here since they are useful in understanding aspects of the magnetic properties of their alloys.

16.2.1 Iron

Pure iron can be produced by laboratory processes such as electrolysis, reduction of pure iron oxide or decomposition of carbonyl.[2] Commercial grades of *pure* iron all contain some impurities, the largest of which are always carbon, sulphur and phosphorus. Published data on the magnetic properties of pure iron vary widely because of its strong sensitivity to even very low impurity levels.

Iron has a body-centred cubic crystal lattice with the six <100> axes being preferred directions of magnetisation. This can be deduced from the positive values of K_1 and K_2 given in Table 16.1. Iron takes on a face-centred crystal form above about 700 °C, but this has no practical importance with regard to the magnetic properties of its alloys. The resistivity of pure iron is very low so laminations are subject to high eddy current loss under a.c. magnetisation and it has no application

[1]Obviously a similar comment could be made on Si–Fe alloys. The degree of degradation or inconsistency of magnetic properties due to impurities always needs to be weighed up against the practical difficulties and costs of producing cleaner alloys. In the case of the very high added-value Ni–Fe and Co–Fe alloys, it is justified to optimise the performance by minimising the presence of these impurities.

[2]Carbonyl iron is a grey powder produced by chemical decomposition of purified iron pentacarbonyl which is a pale yellowish liquid. Iron carbonyl $Fe(CO)_5$ itself is formed by reaction of iron with carbon monoxide.

Table 16.1 Comparison of some physical properties of iron, nickel and cobalt

	Iron	Nickel	Cobalt
Density (kg m^{-3})	7.9	8.9	8.9
Melting point (°C)	1,539	1,455	1,495
Resistivity at 20 °C (μΩm)	0.097	0.069	0.090
Temperature coefficient of resistance (°C^{-1})	0.006	0.006	0.00,007
Young's modulus (GPa)	210	200	209
Magnetocrystalline anisotropy constant K_1 (Jm^{-3})	44,000	−3,400	400,000
Magnetocrystalline anisotropy constant K_2 (Jm^{-3})	21,000	5,300	200,000

Table 16.2 Comparison of some magnetic properties of iron, nickel and cobalt

	Iron	Nickel	Cobalt
Saturation polarisation at 20 °C	2.15	0.61	1.76
Curie temperature (°C)	770	358	1,127
Magnetostriction constant λ_{100} (με)	20	−46	–
Magnetostriction constant λ_{111} (με)	−20	−25	–
Maximum relative permeability	5,000–10,000 (commercial material)	600 [11]	250 [12]
Coercivity (Am^{-1}) (commercial material)	50–250	60	800

in transformer cores. It does have limited commercial use in the form of solid components for d.c. applications.

The structure sensitive[3] magnetic properties of pure iron are very dependent on heat treatments which affect the state of impurities. Typical variations can be found in [13].

Table 16.2 summarises some important magnetic properties of elemental iron, cobalt and nickel. Again, structure sensitive properties vary significantly according to the purity of the test material. For example, the relative permeability of pure commercial iron quoted in the table is at least 35 times lower than that measured on high purity polycrystalline laboratory produced material. The difference in coercivity values is even higher [14].

16.2.2 Nickel

Only a few per cent of the world production of nickel is used in magnetic applications. Pure nickel is produced from nickel carbonyl ($Ni(CO)_4$) via the well-known Monde process. The element has a face-centred cubic atomic lattice at all

[3]Intrinsic magnetic properties are temperature and composition dependent. Extrinsic (structure sensitive) properties such as permeability also depend on heat treatment, stress, etc.

temperatures and it has preferred directions of magnetisation along its eight <111> crystal axes. The hardest directions are the six <100> axes. However, its magnetocrystalline energy is not so dominant as that in pure iron. The striking features of nickel compared to iron are its low Curie point and low saturation polarisation.

16.2.3 Cobalt

Cobalt is a brittle, hard material usually extracted from iron or copper rich ores by initially smelting in an electric furnace followed by chemical processing, heat treatment and electrolysis. It has a hexagonal lattice structure up to around 400 °C to 500 °C when it transforms to a cubic form, although the two phases co-exist over a wide temperature range. It has very high values of anisotropy constants compared to those of iron and nickel. The preferred direction of magnetisation in cobalt is along the *c*-axis (see the crystal form in Figure 2.13(d)) and, because of the high magnetocrystalline anisotropy, magnetisation is far more difficult along any other direction.

The strong anisotropy in cobalt is illustrated by the fact that a field of around 800 kAm^{-1} is required to saturate a crystal along its *b*-axis or *c*-axis compared with around 50 kAm^{-1} in its preferred direction. The corresponding fields required to saturate along the hard and easy axes in iron are around 50 kAm^{-1} and 8 kAm^{-1} and in nickel around 20 kAm^{-1} and 2.5 kAm^{-1}, respectively.

The saturation magnetostriction of pure cobalt of around 60×10^{-6} is the highest of any element. No values of magnetostriction constants of cobalt are included since published values vary widely, perhaps due to the difficulty in interpretation of results from hexagonal crystals.

16.3 Nickel–iron alloys

The Ni–Fe family of soft magnetic materials is probably the most complex and diverse of all soft magnetic material. Their structure sensitive magnetic properties vary widely, not just with composition, but also with processing, heat treatment and ambient temperature. Figure 16.1 shows the variation of resistivity of the binary Ni–Fe family over the complete alloy range from 100% iron to 100% nickel. The corresponding variation of saturation polarisation (at 20 °C) is shown in Figure 16.2. It should be pointed out that the dip in the J_s characteristic is not so prominent at lower temperatures or when the material is more fully saturated. However, the lowest saturation is obtained in materials with around 30% nickel content and at the same time the resistivity is highest. Unfortunately for a.c. applications, high resistivity does not coincide with high J_s.

Figures 16.3 and 16.4 show the variation of crystal anisotropy constant K_1 and the magnetostriction constants λ_{100} and λ_{111} with nickel content. It can be seen in Figure 16.3 that the state of atomic order affects the magnetocrystalline anisotropy

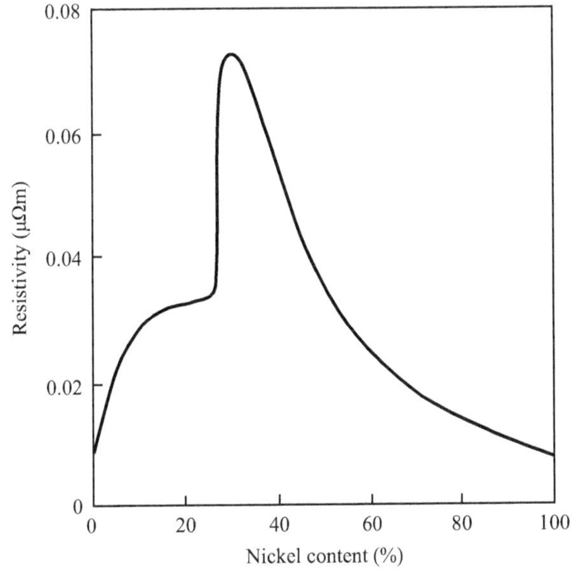

Figure 16.1 Variation of resistivity at 20 °C with nickel content in binary Ni–Fe alloys

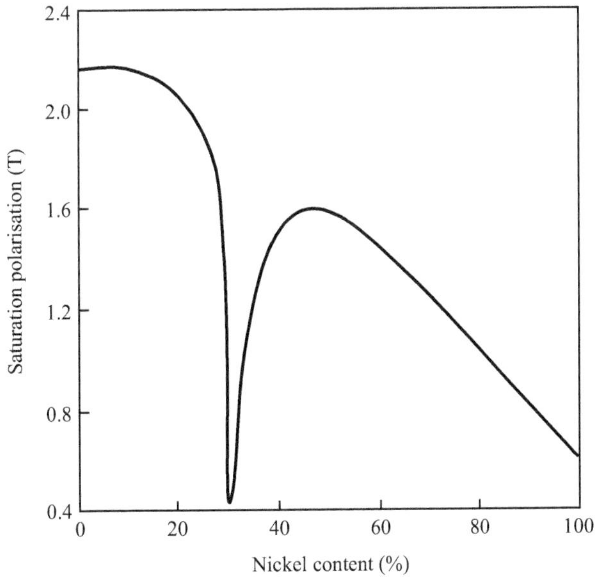

Figure 16.2 Variation of saturation polarisation at 20 °C of binary Ni–Fe alloys with nickel content

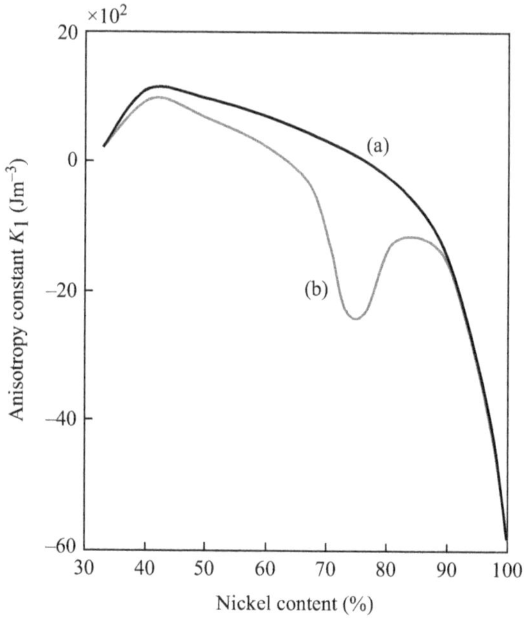

Figure 16.3 Variation of first crystal anisotropy constant, K_1, at 20 °C with nickel content in binary Ni–Fe alloys (a) as-quenched state (disordered) and (b) after slow cooling (ordered)

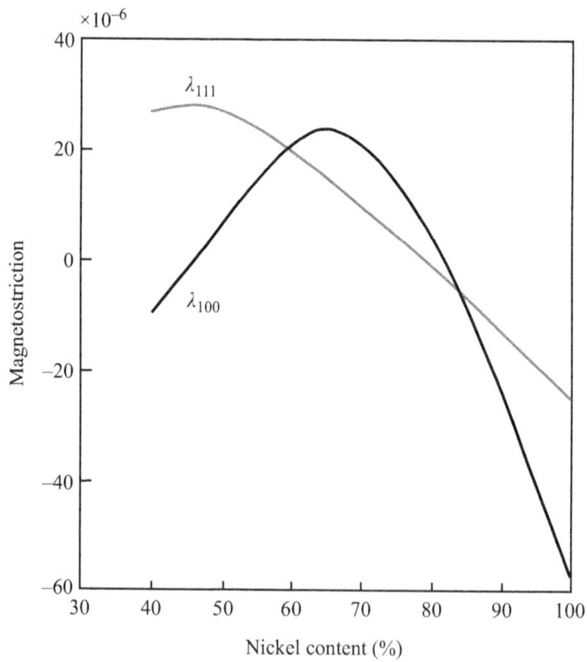

Figure 16.4 Variation of magnetostriction constants of Ni–Fe alloys with nickel content

which, in turn, is likely to affect structure sensitive magnetic properties. The curves in Figure 16.4 are for material quenched after a high temperature anneal to avoid atomic ordering.[4]

The wide potential range of properties can be exploited by using suitable processing and additional alloying elements to produce alloys with more than around 30% nickel content with very applicable magnetic properties. The reason for the difficulty of exploiting lower alloy content, which of course infers lower raw materials costs, is the complex form of the phase diagram[5] of the Ni–Fe alloys and the very low Curie point of those with low nickel contents.[6]

To explain the impact of the phase diagram of the Ni–Fe alloys we need to first consider important allotropes of iron, i.e. the desirable α phase, which is magnetic, and the non-magnetic γ phase. The proportion of these, and other, phases in a given composition is dependent on the temperature. The γ phase exists during high temperature heat treatment. At room temperature the α phase is the stable form but sufficient time is needed during cooling for the γ to α phase conversion to take place for the material to be as magnetic as possible. This process is facilitated by appropriate heat treatment, and perhaps alloy addition, to change the phase diagram. However, the outcome of this is that transformations in alloys below around 30% Ni are difficult to control so alloys with higher nickel content have to be optimised for industrial application.[7]

Figure 16.5 shows the variation of initial permeability with nickel content. The significant effect of heat treatment, in this case the rate of cooling after the final high temperature anneal, over a portion of the alloy range strongly stands out. However, again the absolute values vary significantly according to sample preparation. The variation of maximum permeability follows a similar trend with nickel content and, again, peaks occur in compositions around 40%–50% nickel and at around 80% nickel. However, the magnetic properties of previously produced commercial binary Ni–Fe alloys were limited by the impossibility of optimising the basic parameters shown in Figures 16.1–16.4 at particular nickel contents.

The key to the strong effect of heat treatment in the alloys with around 80% nickel is that quenching inhibits the natural tendency for ordering of the nickel and iron atoms during cooling. The alloy must be disordered in order to obtain a low value of K_1, together with low magnetostriction, hence the need for the quench

[4]Ordered in this context means that the Ni and Fe atoms are arranged in a regular array repeated throughout the material. Disordered means the atoms are randomly distributed. Structural magnetic properties are different in the two states. Ordering can occur also in Co–Fe and Si–Fe alloys.

[5]Material scientists will be very familiar with phase diagrams of alloys which are very important in relation to the preparation and properties of all metallic materials. The phase diagrams of the Ni–Fe system are discussed in detail in publications such as [14,15].

[6]The Curie point of binary compositions around 30%Ni70%Fe is in the range 50 °C to 200 °C. This gradually increases with increasing nickel content up to around 400 °C to 500 °C for 50% nickel and up to around 600 °C for the 70% nickel range.

[7]Iron alloys with up to around 5% nickel are actually feasible since only the α phase is present during conventional heat treatment conditions and no phase conversion occurs.

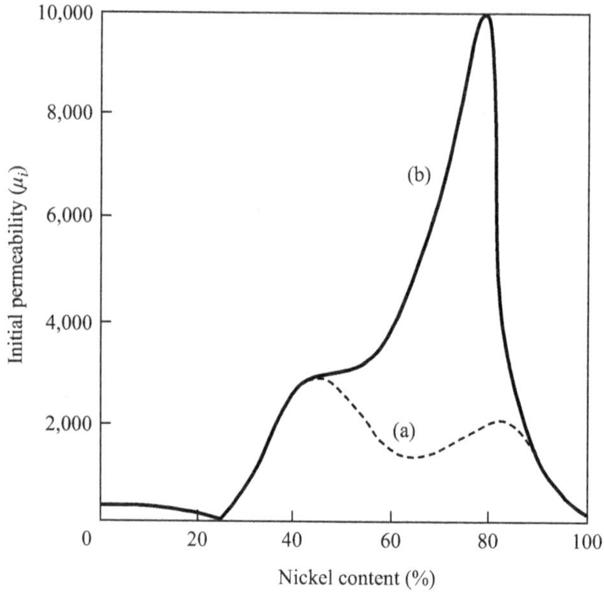

Figure 16.5 Typical variation of initial permeability with nickel content in binary Ni–Fe alloys at 20 °C after (a) slow cooling and (b) rapid cooling (Figure 1 in [16] reproduced under free licence CC-BY-4.0 and modified)

from a high temperature. The quench itself can set up harmful stresses. However, introducing small quantities of extra alloying elements can avoid the need for the quench.

Magnetic phase changes and atomic ordering in general in the Ni–Fe alloys are quite well understood and can be used to explain the trends in magnetostriction, anisotropy and saturation induction characteristics of the alloys. The large variation in magnetic properties of Ni–Fe alloys in the range 30% Ni to 80% Ni can be broadly explained in terms of changes of their crystal anisotropy and magnetostriction with composition [16].

It is beyond the scope of this book to discuss this further, but full details of the metallurgical processes involved can be found in dedicated works such as [14,16].

It is interesting to look briefly at the evolution of commercial Ni–Fe alloys. John Hopkinson, while Professor of Electrical Engineering at King's College in London, was probably the first to carry out comprehensive studies of the Ni–Fe alloys, while at the same time carrying out his more famed work on electrical machines. He discovered the unusual characteristics of the 25% and 78% Ni–Fe alloys [17]. Later, Robert Hadfield and co-workers, renowned for their discovery of silicon–iron, also carried out much research on Ni–Fe alloys and confirmed that

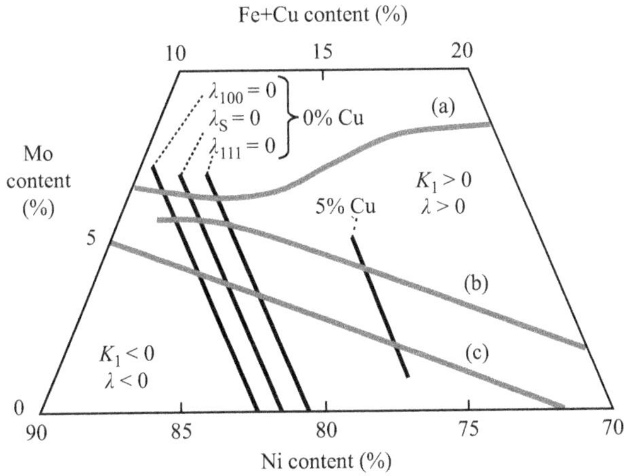

Figure 16.6 Compositions giving simultaneous $K_1 = 0$ and $\lambda = 0$ in high nickel content Permalloys after (a) slow cooling {10^{-3} °C min^{-1}}, (b) medium cooling rate {1 °C min^{-1}} and (c) quenching {10^3 °C min^{-1}} (a) is in an ordered state and (c) is in a disordered state (Figure 25 in [14] reproduced under free licence CC-BY-4.0 and modified)

some alloys, with up to 31% nickel, had high permeability at low induction. This gave the impetus for the search for other Ni–Fe alloys with good magnetic properties [18]. This early work included attempts to find alloys with high saturation polarisation, to quantify the influence of impurities and to determine the low field values of permeability and hysteresis of wide ranges of alloys [19].

The barrier in the applicability of the binary Ni–Fe alloys to obtaining optimum resistivity and initial permeability for the same nickel content is overcome by adding a third element, such as chromium, to increase the resistivity and shift its peak position [14]. Following this discovery, the benefits of adding various alloying elements to control the balance between optimised mechanical or magnetic properties of the Ni–Fe alloys became widely exploited.

Figure 16.6 [14] shows an example of the complex combination of variables which leads to the anisotropy constant K_1 becoming zero in some Permalloys alloys. Curves (a), (b) and (c) are examples of how different combinations are obtained using different cooling rates to produce different levels of ordering. When exploiting these factors to obtain the desired combinations of magnetostriction and anisotropy, it is important neither to produce a negative effect on resistivity or saturation magnetisation, nor to introduce other harmful factors, such as internal stress.

The first commercial Permalloy was a binary 78% Ni alloy, but today the Permalloy family comprises a wide range of alloys with other elements included to enhance their properties.

Table 16.3 Percentage elemental contents in some commercial Permalloys [14]

Main magnetic characteristic	Ni	Fe	Mo	Cu	Cr
High μ	77	13.5	4	5	–
High μ_m	80	14.5	5	–	–
Low μ	76	16.0	–	5	2.5

Another well-known commercial family is the *Mumetals* [20] developed at the Telegraph Construction and Maintenance Company[8] in the United Kingdom. Mumetal, like Permalloy, has continually been developed. Today it is still based close to the 78% nickel composition typically as 77%Ni16%Fe5% Cu and 2% Cr or Al. Table 16.3 gives some common compositions of commercial Permalloys with high nickel content. Their magnetic properties vary considerably, even within the narrow 78% to 80% Ni range.

Properties of the Ni–Fe alloys can be tuned quite precisely over a wide range to suit particular applications. To achieve the necessary high purity in commercial alloys, they are produced from very pure constituent elements. The commercial materials are produced as strip up to typically 250 mm wide and from 0.025 mm to 1.5 mm thick. Sometimes a thin insulating coating, a few microns thick, is applied to reduce the need for extra insulation between strips or turns within toroidal cores and also to prevent inter-laminar sticking during high temperature annealing.

Impurity levels must be very low to achieve the very high permeability. This is achieved by melting the raw materials in a vacuum induction melting furnace. Next, hot and cold rolling to final thickness is carried out followed, in turn, by high temperature heat treatment in a hydrogen atmosphere. The heat treatment reduces the impurity level and develops the required magnetic properties during cooling at a closely controlled rate. The sulphur and oxygen levels are each less than around 0.001 weight % in high grade NiFe alloys.[9]

The main characteristics of the commercial Ni–Fe alloys can be summarised as:

Compositions around 73%–83%Ni

These have the highest permeability of all Ni–Fe alloys because of the disappearance of magnetostriction and K_1 but at the expense of M_s being only around 0.8 T. Hysteresis loss is low, but low resistivity results in high eddy current losses. The simultaneous very low anisotropy and magnetostriction is achieved by adding small amounts of elements such as Mo, Cu or Cr. Their ultimate properties are very dependent on heat treatment, particularly the rate of cooling after high temperature annealing.

Compositions around 36% and 45%–50%Ni

These have the highest value of M_s in the Ni–Fe family, but at the cost of lower permeability than the 80%Ni alloys. Of course, lower nickel content makes these

[8]This is now Telcon Metals.

[9]The sulphur and oxygen levels in high grade electrical steel are less than 0.003%.

Table 16.4 Typical properties of alloys close to the 35%, 50% and 80% Ni compositions

Material	36%–40% Ni	45–55% Ni	75–81% Ni
B_s (Tesla)	0.2–0.8	1.6	0.8–1.0
Resistivity (μΩm)	0.85	0.45	0.15–0.65
Density (kg m^{-3})	8,150	8,250	8,800
Curie temperature (°C)	60–150	450	350
Saturation magnetostriction (με)	+22	+25	0 to +1
D.C. coercivity (Am^{-1})	16–40	6–12	0.2 to 2.0
Maximum relative permeability	30,000	40,000–100,000	24,000–500,000
Initial permeability	5,000	5,000–10,000	6,000–250,000

alloys cheaper. They can be processed to give reasonably isotropic in-plane magnetic properties and *B–H* loops can be made to be quite rectangular which is important in some electronic component applications. Some alloys, particularly in the 50% nickel regime, benefit from magnetic annealing [14].

Compositions around 30%Ni
These have higher resistivity than the other Ni–Fe compositions, but at the expense of lower permeability, lower M_s and a Curie point as low as 50 °C to 150 °C. Low losses, due to their high resistivity, makes them candidate materials for high frequency cores.

Table 16.4 summarises typical properties of commercial alloys in the three main composition ranges. Again, it should be stressed that values can vary widely according to heat treatment and the influence of additions of other elements such as Mo or Cu. Indeed, the properties of some compositions depend critically on obtaining precise control of heat treatment dwell times and temperatures, cooling rates and the exact composition of the annealing atmosphere.

It will be noted that the range of values of structure sensitive properties given in Table 16.4 is very wide. Some of the reasons for this have already been touched upon. The texture formed as a result of cold rolling can have a large effect on the shape of the *B–H* loop. The information in the table simply gives an idea of the range of possibilities for different applications.

Depending on the composition, magnetic properties can be sensitive to stress and temperature.

16.4 Perminvar

The Perminvars are commercial tertiary alloys incorporating the three main ferromagnetic elements with widely varying compositions centred around 45%Ni30%Fe25%Co and 70%Ni23%Fe7%Co. Perminar was developed to obtain an alloy with high permeability at low field for undersea cable shielding. Its initial permeability is only of the order of 500 and maximum permeability is only 2,000–4,000 but it has a linear *B–H* curve over a wide field range. M_s is 1.5 T at room temperature and the Curie temperature is around 600 °C. A disadvantage is its low resistivity of around 1800 μΩm, but today small additions of Cr or Mo are included in commercial alloys to considerably improve on these properties. Comprehensive details of the properties of the Perminvar family are given in [21].

16.5 Cobalt–iron alloys

The soft magnetic Co–Fe alloys have completely different compositions and structures to the cobalt steels used as permanent magnets. A high carbon content is required to produce the high coercivity which gives the permanent magnetic properties whereas low carbon is essential to produce soft magnetic properties.

Figure 16.7 shows the variation of room temperature saturation magnetisation of the binary Co–Fe alloys. The extremely high values of M_s, in the 30% to 50% Co range – far higher than any other known material – has been the incentive to develop alloys around this composition for magnetic cores needing high power densities. The variation of resistivity with cobalt content follows a complex pattern [22] as shown in Figure 16.8 and its value near to 50%Co is four times less than that of grain oriented silicon–iron.

Magnetostriction constants and anisotropy constants of binary Co–Fe alloys are shown in Figures 16.9 and 16.10, respectively. Just like the Ni–Fe alloys, Co–Fe alloys are very susceptible to atomic ordering particularly near the 50%Co composition. The degree of order does not appear to affect magnetostriction but the anisotropy constants can swing from $+1 \times 10^4$ Jm^{-3} to -1×10^4 Jm^{-3} from the fully ordered to the fully disordered states [23]. The curves in Figure 16.10 reflect the effect of the disordered and partially ordered states.

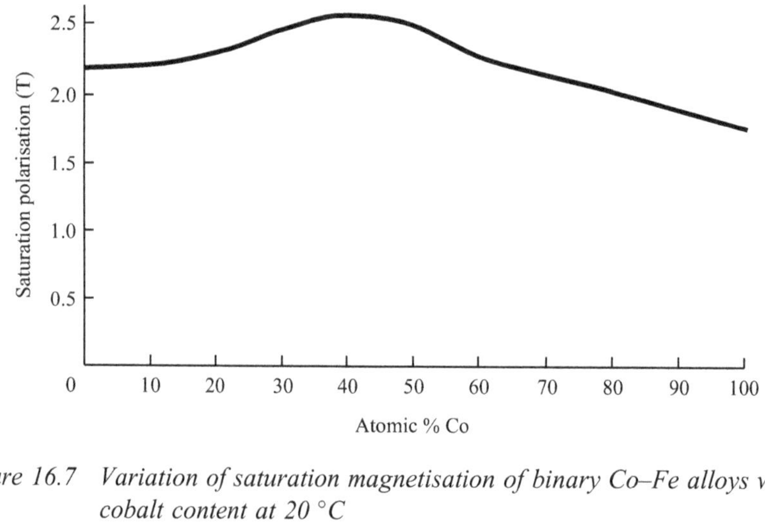

Figure 16.7 Variation of saturation magnetisation of binary Co–Fe alloys with cobalt content at 20 °C

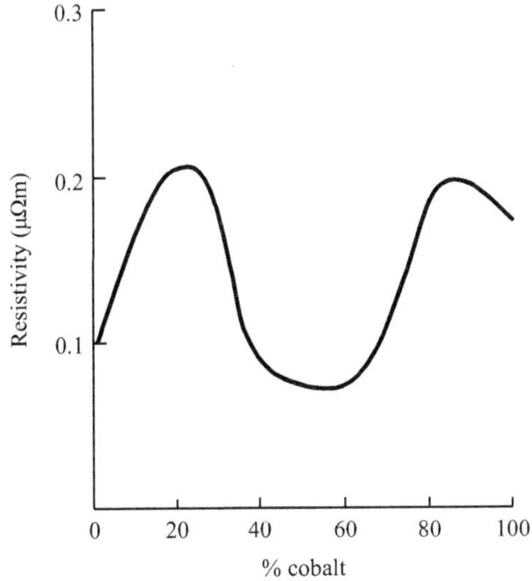

Figure 16.8 Variation of resistivity of binary Co–Fe alloys with cobalt content at 20 °C (Figure 6.3 in [22] reproduced under free licence CC-BY-4.0 and modified)

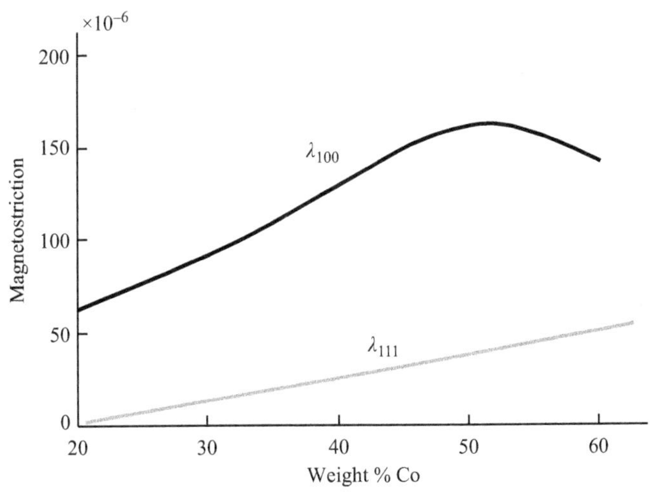

Figure 16.9 Variation of magnetostriction constants of binary Co–Fe alloys with cobalt content at ambient temperature

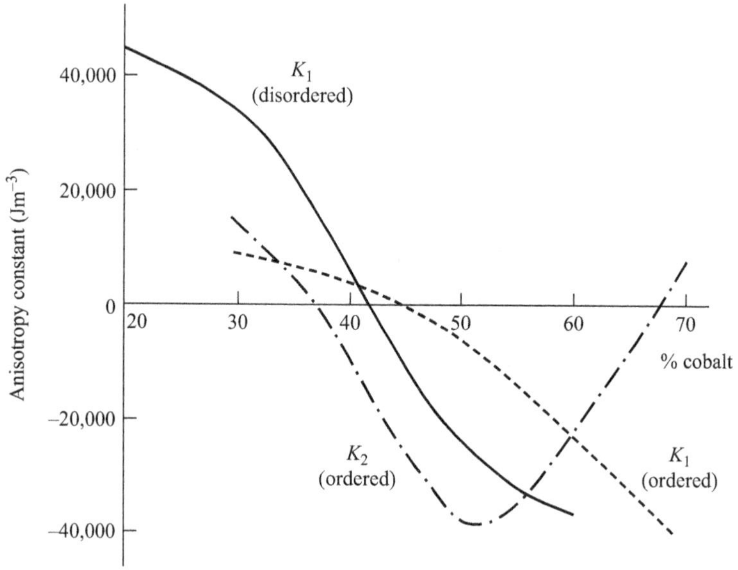

Figure 16.10 Variation of magnetocrystalline anisotropy constants K_1 and K_2 of binary Co–Fe alloys with cobalt content at 20 °C showing the effect of ordering (Figure 8 in [14] (disordered) reproduced under free licence CC-BY-4.0 and data on p. 570 in [24] (ordered))

A practical effect of the sign and magnitude changes of the anisotropy constants K_1 and K_2 is that the easy directions of magnetisation change from <100> to <111> if the cobalt content is increased to over about 46% [25].[10]

The magnetostriction constants of the alloys near 50%Co are among the highest of all materials and unfortunately show opposite trends to those of the magnetocrystalline constants which pass through zero at around this composition. Despite this, the binary 50%Co50%Fe alloy has a relatively high permeability at high flux density [26].

The commercial potential of the 50%Co alloy was recognised and patented by G. W. Elmen at the Bell Laboratories and named *Permendur* [27]. Although commercial development was impeded by the difficulty in producing satisfactory mechanical properties in the binary alloys, this marked the birth of modern ternary Co–Fe alloys.

Figure 16.11 shows how the initial and maximum permeabilities of Co–Fe binary alloys vary widely with cobalt content and also to a lesser, but still significant degree, according to prior heat treatment [28]. The peaks in high induction permeability at around 50% and 90% cobalt are difficult to exploit practically in the

[10]The composition at which this occurs depends on the state of order and temperature. The polycrystalline alloys normally have some anisotropy induced by rolling so the change of easy direction will change the magnetic properties. Little seems to be known about the practical effect of the phenomenon.

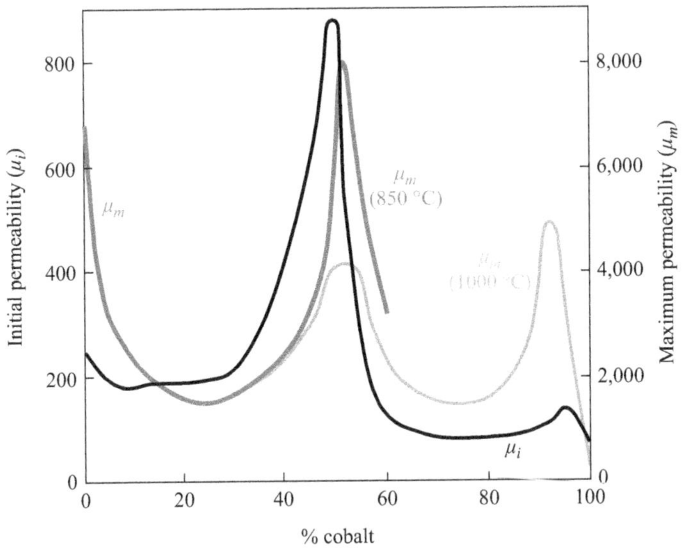

Figure 16.11 Variation of initial and maximum permeability of Co–Fe binary alloys after annealing at 850 °C and 1,000 °C (reproduced from Figure 6.8 in [28] under free licence CC-BY-4.0)

binary alloy because of the fundamental difficulty in producing it in strip form. In the atomically ordered state, within the binary 30% Co to 70% Co range, the alloys are very brittle and cannot be cold rolled. A solution to the ductility problem was found to be to add vanadium to the binary Co–Fe alloy [29]. It was found that this significantly improves the ductility of alloys with up to 90% Co in the disordered state, the only constraint being that the vanadium content should be below 3% to maintain good magnetic properties.

Another important benefit of adding vanadium is that the resistivity increases roughly in proportion to the vanadium content. The resistivity of the 50%Co50%Fe alloy increases substantially from under 0.1 μΩm to 0.26 μΩm by the addition of 2% vanadium whereas the saturation polarisation at room temperature only drops by around 0.1 T.

Unfortunately the 50%Co alloy (even with added vanadium) has phase transformations which have restricted its commercial development. A phase diagram of the binary Co–Fe system is shown in Figure 16.12 [30]. At ambient temperature, the ordered body-centred cubic FeCo (α') is the stable state as shown. The α' phase goes through an order–disorder transformation, at about 730 °C, into the α phase and above around 950 °C the non-magnetic face-centred cubic γ phase forms. After heating to a higher temperature, the γ, or mixed $\alpha + \gamma$ phase, reverts to the ordered α' phase. However, this structural change generates defects detrimental to magnetic permeability which cannot be completely eliminated by a stress relief anneal so there is a temperature close to 900 °C which should not be exceeded during heat treatment [23]. This means that purifying at a very high temperature, which can be

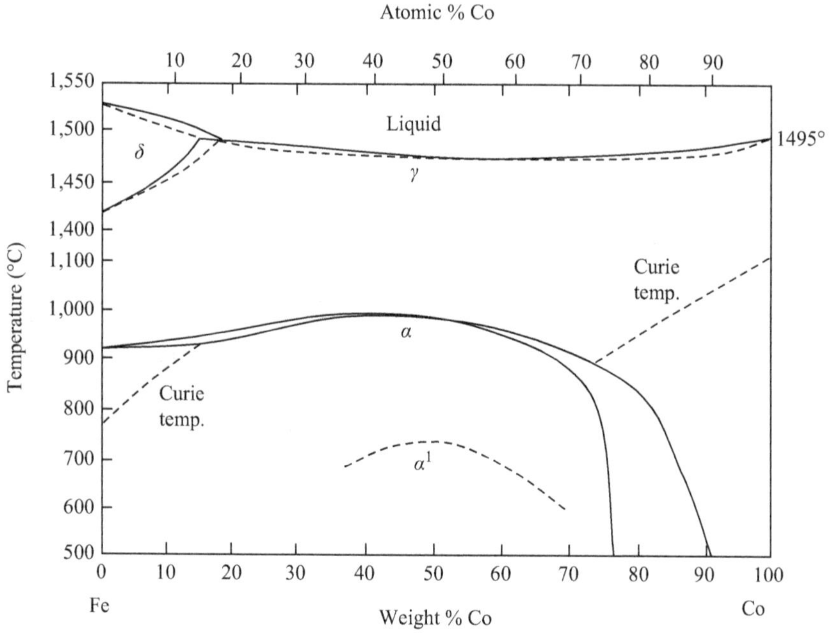

Figure 16.12 Phase diagram of the binary Co–Fe system (Figure on p. 1186 in [30] reproduced under free licence CC-BY-4.0)

performed on Ni–Fe and Si–Fe alloys, is not possible. This problematical $\alpha \leftrightarrow \gamma$ phase change, is a reason why such extremely pure raw materials are needed to obtain good magnetic properties. Also, on cooling, ordering occurs rapidly and this drastically degrades the mechanical and magnetic properties [31].

The brittleness introduced in this way is partly attributed to chemical ordering, although it is recognised that the mechanism is far more complex, perhaps involving hydrogen embrittlement at grain boundaries and other phenomena during heat treatment [32]. The simplest way of creating sufficient ductility in the 2%V49%Co49%Fe alloy is to quench from a high temperature. Its effectiveness depends on good control of the quenching process, which must be carried out at more than 6000 °C s^{-1} [23]. Of course, such rapid quenching is liable to introduce lattice strains and other inhomogeneities which degrade magnetic properties.

Originally the use of 2%V49%Co49%Fe alloy was restricted by its poor ductility of just 1% to 2% elongation after final anneal. However, with improved purity, higher strength can be achieved, with a relatively minor impact on magnetic properties, by heat treatment in the 680 °C to 700 °C range. The effect on magnetic and mechanical properties after heat treatment at around 700 °C is shown in Figures 16.13 [33] and 16.14 [34]. It can be clearly seen that the best mechanical properties are achieved after the lower temperature anneals and best magnetic properties after the higher temperature anneals.

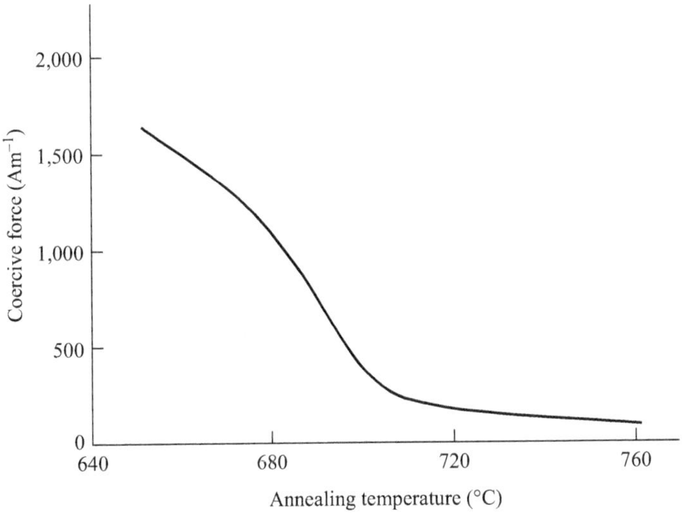

Figure 16.13 Variation of coercive force in Fe-Co-2V after heat treatment within the temperature range 640 °C to 760 °C (Figure 1 in [33] reproduced under free licence CC-BY-4.0 and modified)

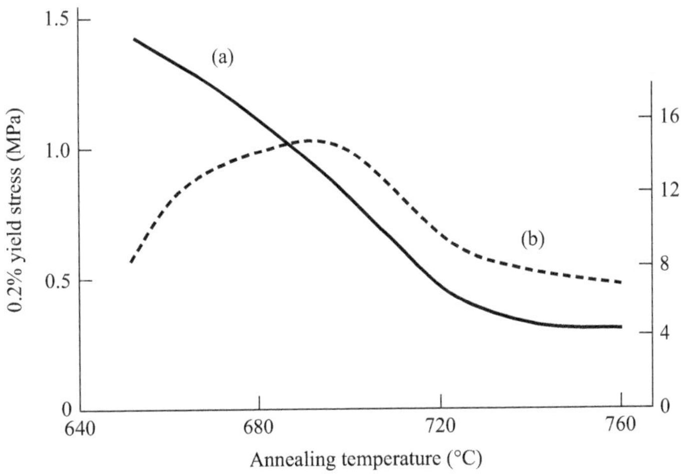

Figure 16.14 Variation of (a) ultimate tensile stress and (b) elongation of Fe-Co-2V after heat treatment within the temperature range 640 °C to 760 °C (Figures 1 and 2 in [35] reproduced under free licence CC-BY-4.0 and modified)

Table 16.5 Typical intrinsic properties of binary Co–Fe alloys compared with Si–Fe and Ni–Fe alloys

Material (approx. %, balance is Fe)	23% Co	35% Co	50% Co	96% Co	3% Si–Fe	49% Ni–Fe	78% Ni–Fe
Saturation polarisation (T)	2.40	2.45	2.42		2.03	1.60	1.08
E_k (Jm^{-3})	380,000	130,000	−160,000		360,000	34,000	<−11,000
λ_{100} ($\times 10^{-6}$)	90	134	148		27	7	12
λ_{111} ($\times 10^{-6}$)	8	20	37		–5	27	2
Resistivity, (μΩm)	0.16	0.11	0.04	0.23	0.45	0.35	0.15
Density (kg m^{-3})	7,980	8,040	8,180		7,660	8,220	8,590
Curie temperature (°C)	940	980	960	1,040	740	520	600

Sensitivity of the properties to annealing temperature falls significantly if a small amount of nickel is included in the 2V-Co-Fe alloy [34,36]. Addition of around 5% Ni improves the mechanical properties, giving elongation of up to around 20%, double that of 49%Fe49%Co2%V, without serious loss of magnetic performance. Substitution of around 0.35% tantalum (Ta) for vanadium increases the ductility of the binary alloy and it increases M_s by around 0.1 T, although a.c. losses increase [36]. Alternatively, the addition of a similar amount of niobium (Nb) inhibits grain growth during processing and final heat treatment which results in improved mechanical and magnetic properties.

Table 16.5 compares properties of the main groups of *binary* Co–Fe alloys with 3% silicon steel and Ni–Fe alloys. As with the Ni–Fe alloys presented in Section 16.3, some of the values quoted are averages of figures from a range of sources so they should only be treated as a guide. Likewise, the comparison with the Si–Fe and Ni–Fe is only included to give an impression of the wide range of properties generally found in the traditional crystalline alloys.

The commercial alloys are mainly based on compositions around 50% and 35% Co together with small additions, mainly of vanadium, nickel or chromium. They are produced in a similar manner to the Ni–Fe alloys mentioned in Section 16.3. Their properties are, of course, very different from the equivalent binary alloys. Detailed information can be obtained from suppliers' Internet sites but the general characteristics of the commercial Co–Fe alloys are summarised here:

Compositions around 50%Co

These invariably contain up to around 2% vanadium and combine high saturation magnetisation with high permeability. Subject to suitable heat treatment, they can have excellent mechanical properties. They must be annealed at high temperature in hydrogen and cooled at an appropriate rate to develop grain growth for optimum magnetic properties. The critical effect of these heat treatments on the *B–H* characteristics of a typical alloy is shown in Figure 16.15. This shows the extremes of

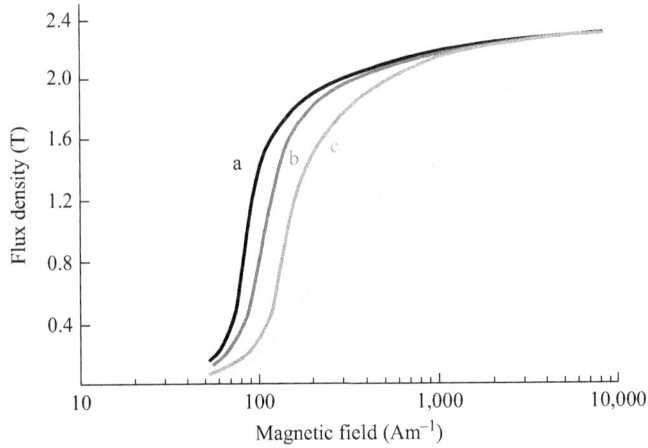

Figure 16.15 Typical variation of d.c. B–H curves of commercial Permendur with heat treatment temperature: (a) 850 °C, (b) 800 °C, (c) 720 °C and (d) 960 °C (Figure 5 in [36] reproduced under free licence CC-BY-4.0)

final anneals between 720 °C (which produces reasonable a *B–H* curve but good mechanical properties) and 950 °C (which produces poor mechanical and magnetic properties). Established trade names include Vanadium-Permendur, Permendur-2V, VACODUR, Permendur-49 and 2V-Permendur (now renamed Hiperco 50).

Compositions around 35%Co
These have high saturation and are generally used as solid parts. Alloys might contain other elements such as chromium to increase the resistivity but losses are high compared to the 50% alloys.

Compositions around 24%Co
Because they contain less cobalt, these are the cheapest of the Co–Fe family. They are very ductile but their permeabilities are relatively low.

Compositions around 90% to 95%Co
These very high Co alloys are not very common but they have a very high Curie point around 1040 °C and can be used at temperatures up to at least 950 °C.

Table 16.6 compares magnetic properties of laboratory produced Co–Fe alloys compared with commercial electrical steels.

Table 16.7 compares typical properties of the binary alloy and modern commercial alloys containing additional elements. The data for the binary alloys are extracted from [11] and the Permendur data are the average of values given in data sheets of products from three manufacturers. The main point to note is that the permeability and coercivity of the two groups do not differ much, whereas the large difference in resistivity would lead to much higher losses in the binary alloy. In general, the permeability and a.c. losses of many binary Co–Fe alloys are

Table 16.6 Comparison of magnetic properties of laboratory produced Co–Fe alloys with those of typical electrical steels (0.3 mm thick laminations) [37]

Alloy	d.c. properties			a.c. properties		
	μ_{max}	H_c (Am^{-1})	B_r/B_m	B_{10} (T)	μ_{max}	Loss at 1.5 T, 50 Hz ($W\ kg^{-1}$)
96%Co4%Fe	12,200	28	0.5	1.41	7,500	3.2
50%Co50%Fe	12,000	64	0.7	2.15	9,000	2.6
25%Co75%Fe	5,000	120	0.5	1.60	5,000	5.0
NO SiFe (M19 grade)	10,000	20	0.6	1.45	8,000	2.8
GO SiFe (M5 Grade)	65,000	8	0.8	1.80	25,000	1.0

Table 16.7 Typical properties of binary 50% Co–Fe and an equivalent current commercial alloy

	Binary Co–Fe [11]	Modern Permendur
M_s (T)	2.45	2.25–2.35
Resistivity (μΩm)	0.04	0.35–0.45
Curie Temperature (°C)	980	940–950
Coercivity (Am^{-1})	100–200	35–600
Initial Permeability μ_i	900	800

Table 16.8 Comparison of loss of 49%/49%/2% alloy (annealed to obtain optimum magnetic properties) with a typical non-oriented Si–Fe (both 0.35 mm thick)

	1.5 T		2.0 T	
	50 Hz	400 Hz	50 Hz	400 Hz
49%/49%/2V%	1.5	30	2.5	60
Typical NO SiFe (M250-35A)	2.35	41	–	–

comparable with mid-range non-oriented silicon steel but obviously far more expensive. However, binary Co–Fe alloys are no longer produced commercially.

Table 16.8 shows a.c. losses of 49%Co49%Fe2%V Permendur compared with a mid-range non-oriented silicon steel measured at room temperature. As implied earlier in this section, such limited comparisons give a reasonable indication of the relative merits, but they do not show a complete picture. As already seen, the Co–Fe–V alloys can be heat treated to give a wide range of properties according to what mechanical properties are needed. Also, the silicon–iron alloy would most frequently be used at a magnetising frequency of 50/60 Hz and peak flux density of 1.5 T, whereas the Co–Fe–V alloy is more often used at 400 Hz and at flux densities of 1.8 T and higher.

Detail of the characteristics under rotational magnetisation is given in Chapter 8.

Tables 16.9 and 16.10 summarise properties[11] of commercial alloys produced by Vacuumschmelze [38]. These are examples of the wide range of alloys commercially available today. Other manufacturers produce similar ranges of alloys with broadly similar properties.

The temperature dependence of properties of Co–Fe alloys is discussed in Chapter 18.

16.5.1 Stress dependence of magnetic properties of Co–Fe alloys

The magnetic properties of Co–Fe alloys can be very stress sensitive. The room temperature sensitivity of the permeability of 27% and 50% Co-based alloys to externally applied compressive stress along the magnetising direction is shown in Figure 16.16 [39]. The permeability drops substantially on application of a relatively low stress of 5 MPa but the rate of degradation drops at higher stress. The loss of 27% Co–Fe and 50% Co–Fe under tensile stress has been found to fall by 30% but, under some circumstances, it rises to the un-stressed value at a high stress of 200 MPa. At the same time it shows a strong dependence on the magnetisation angle with respect to the RD due to the material texture [40].

The mechanisms of the stress sensitivity are very similar to those outlined in Chapter 11 but, as far as the authors are aware, no attempts to model the effects have been reported.

The stress sensitivity of the Co–Fe–V compositions is of most commercial interest because of its widespread use in high speed, high output rotating machines for aerospace applications. The stress sensitivity characteristic of the loss is very dependent on heat treatment as can be seen in Figure 16.17 [41]. The magnetic properties are best after the higher temperature anneals for the reasons discussed in Section 16.5.[12] The shaded bands represent the spread of characteristics measured in batches of the same material annealed under nominally the same conditions. Similar characteristics have been reported under compression after high temperature annealing [39] but the characteristics after low temperature annealing have not been verified.

In practice, building stress in electrical machine cores is often randomly distributed in the plane of the laminations so the characteristics found under unidirectional stress might not be the best indicator of the stress sensitivity in practice.

[11]Where a range of values is given, the lower H_c/higher μ_m values are those obtained after heat treatment to obtain optimum magnetic properties.

[12]The relatively low flux density is used as a convenient reference for directly comparing all the materials, because those annealed at low temperature could not be magnetised at high flux densities. However, similar trends are found at high flux density in the materials annealed at the higher temperatures.

Table 16.9 Nominal magnetic properties of alloys produced in strip form by one manufacturer today[a] [33]

Trade name	Composition	B_s [b] (T)	λ_s(με)	Typical d.c. coercivity (Am−1)	d.c. maximum relative permeability	Loss (W kg^{-1})			
						1.5 T		2.0 T	
						50 Hz	400 Hz	50 Hz	400 Hz
VACOFLUX 48	Fe49Co49V2	2.35	+70	35	18,000	1.5	30	2.2	58
VACOFLUX 50	Fe49Co49V2	2.35	+70	50	15,000	1.6	31	2.5	60
VACODUR 49	Fe49Co49V2+Nb	2.35	+70	50–110	8,000–15,000	1.6–3.0	31–42	2.5–5.0	60–78
VACODUR 50	Fe49Co49V2+Nb	2.33	+70	110–160	5,000–8,000	3.0–4.5	42–53	5–7.7	78–93
VACODUR S Plus	Fe49Co49V2+Ta+Zr	2.30	+70	120–600	1,100–6,000	3.0–11.5	43–109	5–22	74–208
VACOFLUX 17	Co17Cr2+Mo Fe residual	2.27	+25	100	3,500	3.8	54	7	88

[a] Where two values are shown, one is typical value after anneal to optimise magnetic properties and the other after anneal to optimise mechanical properties.
[b] Note that B_s here is the saturation magnetisation, saturation polarisation values are a little lower, e.g. for VACOFLUX 48 $J_s = 2.30$ T, $B_s = 2.35$ T.

Table 16.10 Resistivity and typical mechanical properties of some commercial Co–Fe alloys[a] [33]

Trade name	**Resistivity (μΩm)**	**Yield stress (MPa)**	**Tensile strength (MPa)**	**Elongation at fracture (%)**
VACOFLUX 48	0.42	190	220	2
VACOFLUX 50	0.42	250	550	6
VACODUR 49	0.42	210–390	400–720	4–8
VACODUR 50	0.42	390–450	720–820	8–10
VACODUR S Plus	0.42	400–800	800–1,200	6–8
VACOFLUX 17	0.41	250	450	32

[a]Where a range of values is given, the higher values are obtained after heat treatment to produce optimum mechanical properties.

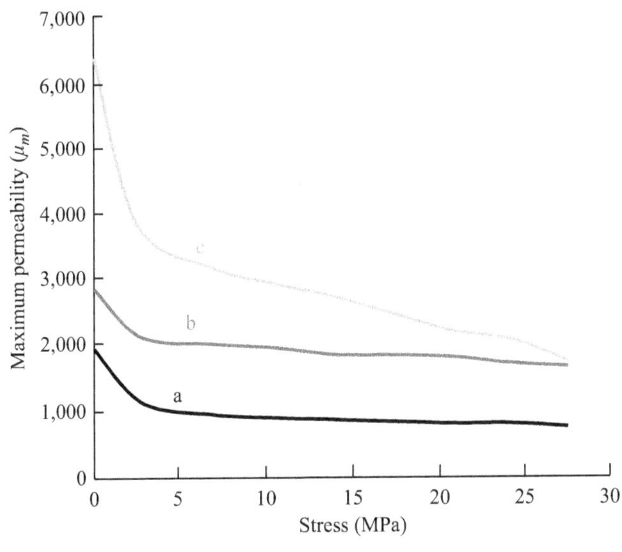

Figure 16.16 Effect of compressive stress on the maximum permeability of three commercial Fe-Co alloys: (a) Hiperco 27, (b) Hiperco 50 and (c) Hiperco 50 HS[13] (Figure 1 in [39] reproduced under free licence CC-BY-4.0 and modified)

Measured and modelled values of permeability show that when only a small uniaxial compression is applied along the RD of 49%Fe49%Co2%V the permeability falls [42]. This is as would be expected based on the loss characteristics after high temperature annealing shown in Figure 16.17. However, when equal values of biaxial tension or compression were applied to sheets of the material the decrease

[13]Hyperco 50 HS is annealed to obtain high mechanical strength. The Hyperco 50 has the same composition but is annealed to optimise the magnetic properties.

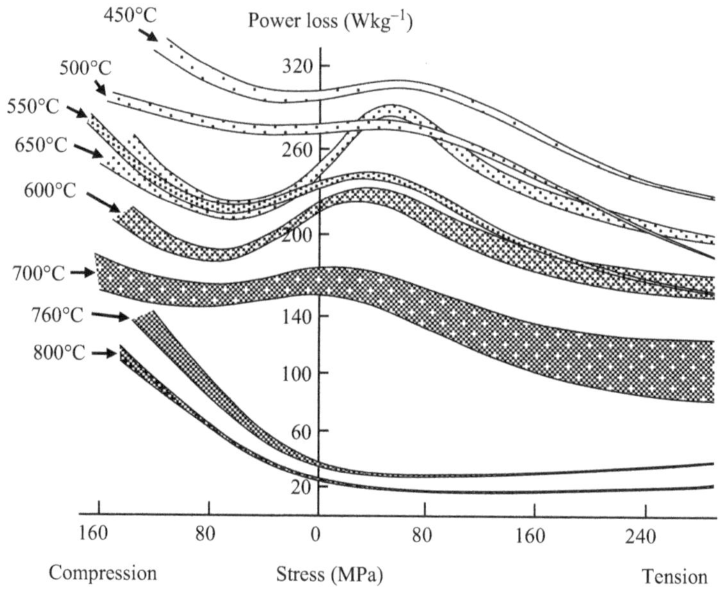

Figure 16.17 Stress sensitivity of power loss of commercial Co–Fe–V alloy magnetised at 1.4 T, 400 Hz after annealing at 450 °C, 500 °C, 550 °C, 600 °C, 650 °C, 450 °C, 760 °C and 800 °C (Figure 2 in [41] reproduced under free licence CC-BY-4.0 and modified)

in permeability is far less than under uniaxial compressive stress [42]. The authors attributed the effect of biaxial stress to the combination of low magnetocrystalline anisotropy together with high magnetostriction. It is interesting to note that similar biaxial stress characteristics have been found [43] in grain oriented SiFe, which, of course, has high magnetocrystalline anisotropy with relatively low magnetostriction.

16.6 Aluminium–iron alloys

Binary Al–Fe alloys, referred to as the Alperms, also known as Alfenol alloys have interesting properties in the 10%–16% Al (balance iron) range. The composition with 13%Al is most common and is called *Alfer*. It is produced in strip and ribbon form and is used in magnetoelastic sensors because of its very high saturation magnetostriction.

Depending on the composition, Alperms can have a coercivity of less than 5 Am^{-1} and a maximum permeability of over 50,000. The resistivity can be more than 100 μΩm, four times that of 3% SiFe and double that of Permalloy, so they are used in small scale, high frequency, applications. However, the Curie points of the Alperms are of the order of 400 °C. Their main disadvantages are susceptibility to surface oxidation, due to high Al content, and the fact that aluminium itself is far more expensive than silicon.

The most well-known aluminium–iron (or more strictly Al–Si–Fe) alloy is Sendust which was developed at Tohoku University, Japan in the late 1930s as an alternative to Permalloy for use in electronic applications. Because of its brittleness, it is normally produced as a metal powder composite of around Fe85%Si9% Al6%. Its relative permeability, of the order of 150,000, does not compete with the best Permalloys but its relatively low coercivity of 5 Am^{-1}, moderate M_s of around 1.0 T, near zero magnetostriction and magnetocrystalline anisotropy makes it attractive in cases where a cheaper, lower performance product is acceptable.

Adding small quantities of aluminium to silicon–iron promotes grain growth, increases resistivity and reduces potential problems with embrittlement, so this ternary alloy is used instead of electrical steel for some applications.

16.7 Applications

16.7.1 Ni–Fe alloys

The market for magnetic strip with high initial permeability grew rapidly in the 1920s with the increasing demand for shielding materials for the emerging telecommunications industry, particularly undersea telegraph cables. Since the first cables were laid in the 1860s, it was known that signal distortion in long conductors reduced the signal speed. Complete shielding could improve this considerably but the permeability of the existing iron-based strip materials was not high enough. This led to the drive to attempt to increase the permeability of Ni–Fe alloys. This led in turn to the development by Elmen of Permalloy, based on Ni78%Fe22% and patented in 1916 [19].

The submarine cable application of Permalloy is only of historic interest, but it created the foundation for the wider development of high permeability magnetic materials for now long-established applications in shielding, sensors and high frequency electronic devices.

Today, one of the biggest application areas for crystalline Ni–Fe alloys is in magnetic cores of residual current detectors (RCDs) whose key requirements are high permeability, and either a flat *B–H* loop or a rounded loop depending on their particular method of operation. However the increasing material performance demands for a.c. and d.c. sensitive RCDs might be better satisfied by nanomaterials (see Section 16.2), which will probably take a large proportion of these emerging markets.

44%NiFe has even been successfully used as rotating machine cores for the experimental *Solar Impulse* long range aircraft where very low losses and high saturation magnetisation were needed.[14] Composite Ni–Fe cores can be wound from two or more types of material to combine their magnetic characteristics. For example, by inter-winding strips with different magnetic properties, improved

[14]Shown at presentation of [10].

magnetic characteristics can be obtained due to the effect of combining high permeability and high saturation magnetisation materials.

16.7.2 Co–Fe alloys

Due to their high cost, ternary Co–Fe alloys are used in specialised applications where low volume or weight is essential or where high temperature operation is required. They are mainly used in aerospace, defence and medical applications including a.c. generators, electromagnet pole pieces, relays and actuators.

The key parameters in stator and rotor laminations of high power density electrical machines are high saturation magnetisation and high mechanical strength maintained over a very wide ambient temperature range. Many Co–Fe–V alloys have been developed over years specifically with these objectives in mind and it is still their most important application. In some rotor cores, high temperature mechanical creep deformation under high rotor centrifugal stress is an important characteristic to which some Co–Fe alloys are less susceptible [44].

A practical combination of good a.c. magnetic properties in stator laminations and high strength rotor laminations punched from the same high saturation magnetic alloy is difficult to achieve. The simplest solution is to anneal stator and rotor laminations at different temperatures to optimise magnetic and mechanical properties separately. Apart from the practical inconvenience and cost of the extra heat treatment, the dimensional stability of laminations becomes an issue. Heat treatment of alloys with around 50% Co, causes a permanent anisotropic decrease in length of around 0.06% which is dependent on the annealing temperature [45]. This has to be taken into account in precision lamination design. An alternative solution, where less demanding performance characteristics are needed, is to anneal at a very precise temperature, somewhere around 700 °C, in order to obtain a compromise final product with adequate magnetic and mechanical properties for rotor and stator laminations. It can be seen from the characteristics shown in Figures 16.13 and 16.14 that, by annealing at a temperature on the steep part of the curves, a compromise between optimum magnetic and mechanical properties is possible. However, it is possible for batches of nominally the same composition to have slightly different annealing characteristics which creates a challenge to optimise annealing conditions without significant trial anneals resulting in possible wastage of expensive material.

The world-wide drive to the more electric aircraft (MEE) stimulated the search for high strength soft magnetic materials for motors, generators and magnetic bearings. New engine configurations call for stator and rotor operation up to above 400 °C with a high temperature yield strength of 700 MPa to 800 MPa for rotor laminations.

The desire to use integrated generators mounted inside the MEE aircraft engine can push present Co–Fe–V alloys to their limits [46]. An even greater capability to operate safely and predictively at 540 °C with a yield stress of up to 1000 MPa is being suggested to give electrical designers scope for innovating the electrical system even further. Such demanding requirements are being called for in other prospective automotive applications, magnetic bearings and high speed drives [47,48].

Incremental developments in Co–Fe alloys are not likely to satisfy the anticipated, even more demanding needs of next generation high speed machines, but neither does any other existing soft magnetic material. Co–Fe alloys containing other additions are promising [31] so at least incremental improvements should continue and, perhaps, the required yield stress performance is within the capability of Fe–Co–2V alloys [49]. It is claimed that a complete manufacturing change to a continuous annealing production process could lead to such high strength material with low magnetic losses [48].

Innovative approaches such as reinforcing the Co–Fe-based alloys with ceramic or silicon carbide might lead to better combinations of mechanical and magnetic properties. Tensile strengths of over 1000 MPa have been achieved in some such materials at 600 °C [46]. Further development might eventually result in a new family of advanced materials, perhaps containing magnetic friendly reinforcing materials, such as carbon nanotubes, capable of withstanding these severe demands. They will come at a very high financial cost, but they may provide solutions which are impossible to arrive at with conventional soft magnetic materials. No completely new alloy or processing route seems to be on the horizon at present. However, fundamental breakthroughs in magnetic materials have been made spontaneously in the past, so user demands might result in another breakthrough in the future.

References

[1] Gehrmann B., and Grosskopf R.-D. 'The market for nickel and cobalt base alloys'. *16th Conference on Soft Magnetic Materials*, Dusseldorf, Germany, Verlag Stahleisen, 2003, pp. 31–40

[2] Libsch J.F., Both E., Beckman G.W., Warren D., and Franklin R.J. 'Iron-cobalt and iron-cobalt-nickel alloys'. *Trans. AIME* 1950, vol. 188, pp. 287–296

[3] Adler E., and Pfeiffer H., 'The influence of grain size and impurities on the magnetic properties of the soft magnetic alloy 47.5% NiFe'. *IEEE Trans Magn.* 1974, vol. 10(2), pp. 172–167

[4] Pfeiffer F., and Kunz W. 'The influence of grain structure and non-magnetic particles on the magnetic properties of high-permeability Ni–Fe alloys'. *J. Magn. Magn. Mater.* 1977, vol. 4(1–4). pp. 214–219

[5] Couderchon G. 'Improvement in 80 Ni Permalloys through solid state refining'. *IEEE Trans. Magn.* 1978, vol. 14(5), pp. 524–526

[6] Handley G.W., Major R.V., and Waters B.H.C. 'Aspects of production of nickel-iron and cobalt–iron magnetic materials'. *Proceedings on Soft Magnetic Materials 2*, European Physical Society, Cardiff Conference, April 1975, pp. 84–90

[7] Pfeifer F., and Radeloff C. 'Soft magnetic Ni-Fe and Co-Fe alloys – some physical and metallurgical aspects'. *J. Magn. Magn. Mater.* 1980, vol. 19, pp. 190–207

[8] Boll R. (Editor). *Soft Magnetic Materials: Fundamentals, Alloys, Properties, Products, Applications: The Vacuumschmelze Handbook*. Heyden & Son Ltd, 1979

[9] Sundar R.S., and Deevi S.C. 'Influence of alloying elements on the mechanical properties of FeCo-V alloys'. *Intermetallics* 2004, vol. 12(7–9), pp. 921–927

[10] Volbers N., 'Soft magnetic crystalline NiFe and CoFe alloys status and developments'. *Proceedings of the International Conference on Magnetism and Metallurgy*, Part I, Dresden, Germany, 12th–14th June 2018, pp. 43–58

[11] Bozorth R.M. *Ferromagnetism*. IEEE Press, Piscataway, NJ, 1993, p. 870

[12] Couderchon G., and Tiers J.F. 'Some aspects of magnetic properties of Ni-Fe and Co-Fe alloys'. *J. Magn. Magn. Mater*. 1982, vol. 196–214, pp. 196–214

[13] Bozorth R.M. *Ferromagnetism*. IEEE Press, Piscataway, NJ, 1993

[14] Bozorth R.M. *Ferromagnetism*. IEEE Press, Piscataway, NJ, 1993, p. 54

[15] Heck C. *Magnetic Materials and Their Applications*. Butterworth & Co, London, 1974

[16] Bozorth R.M. 'The Permalloy problem'. *Rev. Mod. Phys.* 1953, vol. 25(1), pp. 42–48

[17] Hopkinson J. 'Magnetic properties of alloys of nickel and iron'. *Proc. R. Soc. Lond.* 1890, vol. 48, pp. 1–13

[18] Barrett W.F., Brown W., and Hadfield R.A., 'Researches on the electrical conductivity and magnetic properties of upward of one hundred alloys of iron'. *J. Inst. Electr. Eng*. 1902, vol. 31, pp. 674–729

[19] Yensen T.D. 'Magnetic properties of ternary Fe-Si-C alloys'. *Trans. Am. Inst. Electr. Eng*. 1924, vol. 43, pp. 145– 75

[20] Smith W.S., and Garnett H.J. *Nickel alloys of high magnetic permeability*. British Patent 223972 (Appl. 25/8/23)

[21] Bozorth R.M. *Ferromagnetism*. IEEE Press, Piscataway, NJ, 1993, pp. 160–180

[22] Bozorth R.M. *Ferromagnetism*. IEEE Press, Piscataway, NJ, 1993, p. 193

[23] Josso E. 'Iron-cobalt-vanadium alloys: a critical study of phase diagrams in relation to magnetic properties'. *IEEE Trans. Magn.* 1974, vol. 10(2), pp. 161–165

[24] Bozorth R.M. *Ferromagnetism*. IEEE Press, Piscataway, NJ, 1993, p. 570

[25] Chin G.Y., and Wernick J.H. *Ferro Magnetic Materials*, vol. 2 (Ed.Wohlfarth E.P.), North-Holland Pub Co., 1980, pp. 55–188

[26] Ellis W.C. 'Study of physical properties of electrolytic Co and its alloys with Fe'. *RPI Bull., Eng. Sci. Ser.* 1927, vol. 16, pp. 1–5

[27] Elmen G.W. *Magnetic material and applications*. U. S. Patent No. 1739752 (Appl. 28/12/27)

[28] Bozorth R.M. *Ferromagnetism*. IEEE Press, Piscataway, NJ, 1993, p. 197

[29] White J.H., and Wahl C.V. *Workable magnetic compositions containing principally iron and cobalt*. U.S. Patent 1862559 (Appl. 14/8/31)

[30] Nishizawa T., and Ichida K. *Binary Alloy Phase Diagrams*, vol. 2 (Ed. Massulski T.B.) ASM, Metals Park, OH,1990, p. 1186

[31] Sundar R.S., and Deevi S.C. 'Soft magnetic FeCo alloys: alloy development, processing, and properties'. *Int. Mate. Rev.* 2005, vol. 50(3), pp. 157–192

[32] Chen C.W. 'Soft magnetic cobalt-iron alloys'. *Cobalt* 1964, vol. 22, pp. 3–21

[33] Thornburg D.R. 'High-strength high-ductility cobalt-iron alloys'. *J. Appl. Phys.* 1969, vol. 40(3), pp. 1579–1580

[34] Major R.V., Martin M.C., and Branson M.W. 'The development of cobalt-iron soft magnetic alloys with improved mechanical properties for use in electrical generators'. *IEE Conference Proceedings on Advances in Magnetics*, September 1976, pp. 53–56

[35] Major R.V., and Samadian V. 'Physical metallurgy and properties of a new high saturation Co-Fe alloy'. *J. Mater. Eng.* 1989, vol. 11(1), pp. 27–30

[36] Thomas B. 'The influence of nickel on the magnetic and mechanical properties of Co-Fe-V alloy'. *IEEE Trans. Magn.* 1980, vol. 16(2), pp. 444–454

[37] Couderchon G., and Josso E., 'Magnetic Co-Fe alloys for systems operating at very high temperature'. *Proceedings of 3rd International Conference on Soft Magnetic (SMM3)*, Bratislava, Czechoslovakia, September 1977

[38] Soft magnetic CoFe alloys: VACOFLUX and VACODUR. www.vacuum schemeltze.com (Accessed 17th February 2017)

[39] Turgut Z., Horwath J.C., Huang M.-Q., and Fingers R.T. 'Effect of Ta addition on mechanical and magnetic properties of FeCoSiBTa alloys'. *J. Appl. Phys.* 1999, vol. 88, pp. 6692–6694

[40] Turgut Z., Huang M.Q., Horwath J.C., Hinde R., Kubicki J., and Fingers R.T. 'Effect of tensile stress and texture on magnetic properties of FeCo laminates'. *IEEE Trans. Magn.* 2004, vol. 40(4), pp. 2742–2744

[41] Moses A.J., and Thomas B. 'Stress sensitivity of the a.c. magnetic properties of Permendur after various heat treatments'. *IEEE Trans. Magn.* 1976, vol. 12(2), pp. 103–106

[42] Hubert O. 'Influence of biaxial stresses on the magnetic behaviour of an iron-cobalt sheet- experimental and modelling'. *Przeglad Elektrotechniczny* 2007, vol. 83(4), pp. 70–77

[43] Moses A.J., and Phillips P.S. 'Some effects of stress in Goss-oriented silicon-iron'. *IEEE Trans. Magn.* 1978, vol. 14(5), pp. 353–355

[44] Fingers R.T., and Rubertus C. 'Application of high temperature magnetic materials'. *IEEE Trans. Magn.* 2000, vol. 36(5), pp. 3373–3375

[45] McCunn T.H. 'Dimensional stability of Vanadium Permendur'. AIP Conference Proceedings, Number 5, 1971, pp. 1534–1538

[46] Godfrey S. 'Development of fibre-reinforced soft magnetic metal-matrix composite materials'. *Developments in Magnetic Materials, UK Magnetics Society Seminar*, 25th April 2005

[47] Dorninger F. 'High strength electrical steel'. *5th Proceedings of WMM'12*, June 2012, pp. 56–59

[48] Waeckerlé T., Batonnet R., Wéry. T., and Petit F. 'Fully processed Fe-Co soft magnetic laminations for high-speed electrical machines'. *IEEE Trans. Magn.* 2014, vol. 50(4), Art. No. 2002606

[49] Fingers R.T., Coate J.E., and Dowling N.E. 'Mechanical properties of iron-cobalt alloys for power applications'. *Proceedings of the Intersociety Energy Conversion Engineering Conference*, 1997, vol. 1, Paper 97364, pp. 563–568

Chapter 17
Consolidated iron powder and ferrite cores

17.1 Background

Magnetic components have been produced by compacting powders for many years. They are sometimes, known as *dust cores*. The most well-known is probably Sendust which was produced from powder partly because its composition was too brittle for production as strip using conventional routes. More recently, iron and SiFe bulk components have been developed and commercialised for applications requiring mass produced, cheap, mechanically strong and dimensionally accurate net shape cores for magnetic components.

These iron and silicon iron compressed components are commonly referred to as *soft magnetic composite* (SMC) cores or consolidated powder cores. General advances in powder metallurgy have resulted in their applications in many sectors, particularly as functional materials, replacing laminated cores for small motors and actuators in the automotive and aerospace industries.

The potential industrial importance of magnetically soft ferrites was recognised in the early 1900s[1] when they were referred to as magnetic insulators. Commercial exploitation began to accelerate almost 50 years later, by which time their fundamental physics and technology had become understood based on the Néel theory of magnetism. It is interesting to note that, at about the same time, another accidental discovery in magnetic materials was made.[2] In 1950, researchers at the Phillips Research Labs in the Netherlands were preparing semiconductor materials and, by mistake, produced barium ferrite ($BaFe_{12}O_{19}$), which they found to be strongly magnetic. The material was handed over to magnetics researchers and within 2 years *Ferroxdure* permanent magnet was being commercially produced [1].

Strong characteristics of the soft ferrites are extremely high resistivity, wide operating frequency range, good temperature stability and low cost. Disadvantages are low saturation polarisation, poor thermal conductivity, brittleness and low tensile strength. The most well-known, and unique, property of soft ferrites is

[1]In fact a US patent was filed in 1886 by Fritts proposing the production of magnetic cores from an iron and rubber composite. This might have been the precursor of flexible permanent magnets. If applied using modern powders and processing it could be possible to produce very low magnetostriction components this way but we have not investigated the history or this a.c. possibility!

[2]Of course it was found by accident that 3% SiFe can have good soft magnetic properties and also amorphous material was discovered by accident as mentioned in Chapter 15.

retention of useable magnetic properties into the MHz magnetisation regime. The average resistivity of soft magnetic ferrites is around 10^4 Ωm, or 10^{12} times higher than that of metallic alloys which suggests we need not consider eddy currents. However, some ferrites, for example magnetite (Fe_3O_4), have far lower resistivities so high frequency eddy currents are possible.

Commercial materials are based on the MnZn and NiZn families which complement each other over a wide magnetizing frequency range. The resistivity of the NiZn ferrites is an order of magnitude higher than the MnZn ferrites so they are used more in the very high frequency range which is of little interest for power applications so they are briefly covered in this book. The MnZn ferrites are used widely up to 30 kHz magnetisation frequency and are an alternative to metallic magnetic alloys at lower frequencies. Comprehensive details of the full soft ferrite family can be found elsewhere (e.g. [2,3]).

17.2 Consolidated iron and SiFe powder cores

17.2.1 Production

SMC components can be made by a standard powder metallurgy processing route. Figure 17.1 shows factors which have an influence on their magnetic properties. Powder particles of up to around 400 μm are initially mixed with a lubricant and perhaps other organic additives. The lubricant, or binder, is added to obtain high density during the compaction process and it is later removed by annealing.

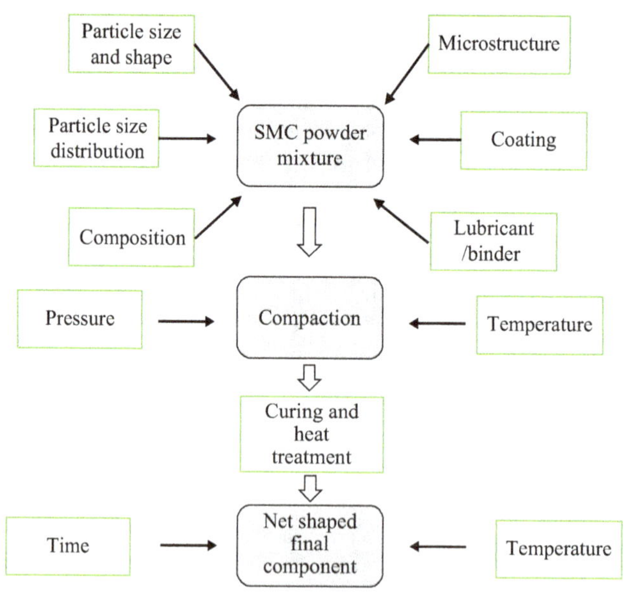

Figure 17.1 Outline of processing route for producing an SMC including factors affecting the magnetic properties

The size and shape of the powder particles are critical to the production process and final magnetic properties. Particles might be coated with an insulating material to reduce eddy currents which occur during a.c. magnetisation. Sometimes this is called a bonding process, although no actual bonding agent is used.

The powder mix is compressed at say 500 MPa to 800 MPa which compacts and compresses the particles. The compact is cured to remove the organic materials and then perhaps, heat treated typically at 500 °C, to improve the magnetic properties. The final product is a near net shape component with close dimensional tolerance. There are many ways of compaction which have differing effects on the magnetic properties [4].

The magnetic and mechanical properties of this ready-to-use component depend on processing parameters as well as starting with a highly compressible powder mix. The final density basically determines the saturation polarisation and the insulation layer's thickness and integrity control the permeability, coercivity and a.c. losses.

17.2.2 SMC compositions for power applications

SMC components can be made from any suitable elemental or compound powders which can be compacted by a conventional powder metallurgy (PM) route. A wide range of SMC products is commercially available. Selection can be based mainly on the magnetising frequency required in the application. The most common compositions are simply iron and silicon–iron although other iron-based alloys are also produced commercially. The iron composite can be an alternative to silicon–iron alloys since it is able to operate at a higher flux density over the upper range of the usable frequency range of silicon–iron. It is also an alternative in the lower frequency range to MnZn ferrites, offering higher flux density.

Normally, SMCs can be operated at temperatures of up to 150 °C to 200 °C.

17.2.3 Magnetic properties

The magnetisation process and magnetic properties of an SMC component, with a given composition, depends very much on the interface between adjacent compressed coated particles.

Figure 17.2 shows this region before and after treatment. Before compaction the particles effectively have a double coating of the insulation layer and the binder. After processing, the binder, which could be an epoxy resin, has mainly been removed and the compaction has reduced the number of voids as in Figure 17.2(b).

Eddy currents can flow between particles under a.c. magnetisation if short circuits occur between them where the insulating coating does not fully cover the surfaces. This increases the loss and reduces the permeability. A consistently thick coating satisfies the insulation need, but increases the non-magnetic volume of the component, thereby reducing its effective permeability and saturation polarisation. In practice, the coating is made to be as thin as is practically possible, but this might lead to localised, randomly distributed electrical short circuits between particles.

It is not necessary to use insulated powder if an SMC is to be used under d.c. magnetisation. The magnetic volume is increased so the material has higher

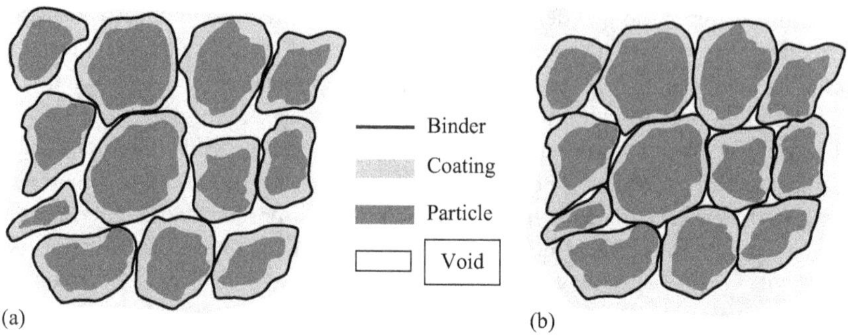

Figure 17.2 *Schematic representation of the structure of SMC particles: (a) before compaction and (b) after compaction (not to scale)*

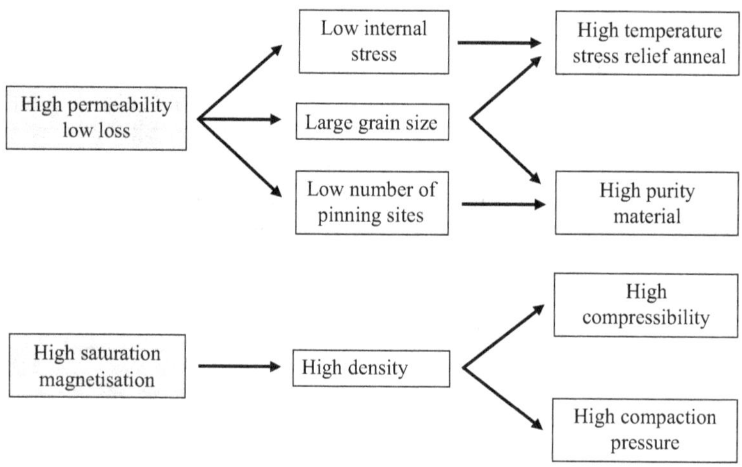

Figure 17.3 *General requirements for optimum a.c. magnetic properties in SMC cores*

permeability then when coated powder is used and its saturation magnetisation can be close to that of the bulk components of the particle material. Of course, in some systems magnetic components operating under nominally d.c. magnetisation can be subjected to localised a.c. fields.

Requirements for optimum a.c. permeability and loss differ from the requirements for high saturation magnetisation. Some general requirements for a given material composition are shown in Figure 17.3.

Components produced by conventional PM processes have isotropic mechanical properties. Correspondingly, SMCs have isotropic magnetic properties so they are able to carry flux equally easily in all directions. This gives scope for device topologies not feasible when using strip wound or laminated cores.

Table 17.1 Typical characteristics of some soft magnetic composites [5,6]

	Somaloy[a] 500	**Somaloy 700**	**2.5% SiFe**	**3% SiFe**	**6.5% Si**	**2.5% Si0.45%P**	**49%Fe49% Co2%P**
Density (kg/dm^3)	7.4	7.45	7.4	7.5	7.4	7.5	7.8
J_s (T)	1.95	2.02	1.96	1.95	1.76	1.95	2.2
μ_m	650	350	9,000	8,000	16,500	8,000	3,200
B_r (T)	0.25	0.06	1.0	1.0	1.2	1.2	1.4
H_c (Am^{-1})	250	70	40	46	2.6	50	240
Resistivity (μΩm)	>80	90	50	55	90	45	50
Source/ comments	[5] Iron powder, small grain	[5] Iron powder, large grain	[6]	[6]	[6]	[6]	[6]

[a]Somaloy is a recognised trademark of Höganäs.

Table 17.1 shows properties of some commercial SMCs. These are *typical* values of each composition. These can vary substantially according to processing. For example, the resistivity of the Somaloy 700 range of products varies from 70 μΩm to 1000 μΩm [5].

The a.c. loss of iron powder cores is two to three times that of the best equivalent non-oriented (NO) steels at 50 Hz magnetisation. The permeability and coercivity are structure sensitive so they depend on purity, internal stress and porosity. Impurities such as C, O and N above their solubility levels are also harmful. The losses are similar to those of low grade NO steel at 400 Hz; above that the losses are lower in the powder core. Generally speaking, Fe-based SMCs compete with low grade electrical steels best at frequencies around 1 kHz, whereas for SiFe-based SMCs this rises to around 5 kHz.

Table 17.2 compares typical losses of NO steels with an iron-coated SMC [7]. In this case the eddy current loss is simply obtained by subtracting the d.c. hysteresis loss from the total measured loss. The loss components follow the expected trends in the NO steels, the hysteresis loss falling with increasing component thickness, but the eddy current loss is not proportional to f^2 as is expected from the classical loss separation discussed in Section 7.8. This is partly because the excess loss has not been included in the analysis and partly because the classical approach to eddy current analysis is not appropriate for the reason given in Section 18.2.4.

The hysteresis component of loss dominates at relatively low magnetising frequencies in an SMC. It is far higher than in electrical steel because of the effects of the compaction stresses which cannot be removed by the heat treatment. At high frequency the loss of the SMC is less than that of poorer grade, thick, NO electrical steel. Of course, the loss performance has to be weighed up against material cost and the cost savings capitalising on the use of the net shaped SMC component

Table 17.2 Comparison between losses of typical mid-range NO electrical steels of different strip thickness with those of a Fe-based SMC (P_e – eddy current loss, P_h – hysteresis loss and P_{total} – total loss) [7]

Designation	Power loss at 1.0 T (W kg^{-1})								
	50 Hz			400 Hz			1,000 Hz		
	P_e	P_h	P_{total}	P_e	P_h	P_{total}	P_e	P_h	P_{total}
NO 0.2 mm	0.5	2.5	3.0	4.0	11	15	25	25	50
NO 0.3 mm	0.6	2.0	2.6	10	10	20	80	25	105
NO 0.5 mm	1.4	1.8	3.2	23	10	33	160	25	185
Fe-based SMC	0.6	6.0	6.6	13	40	53	47	100	147

which can often be produced in a single step compared to the many steps needed in production of the same component using laminations.

The structure sensitive magnetic properties of SMC cores are stress sensitive, although not as much as the equivalent solid material. It has been found that the losses of SMC rings and rods increase by up to around 70% when subjected to a compressive stress of 150 MPa [8]. The sensitivity falls at high frequency as the proportion of stress-independent eddy current loss increases.

17.2.4 Loss components in SMCs

The losses in composite cores can be divided into static hysteresis, classical eddy current and excess loss as for laminated steels, but different structures must be taken into account. To identify general trends the simple approach in [7] can be used, but for more accurate and meaningful analysis the eddy current paths must be taken into account. Under a.c. magnetisation, two separate eddy current paths are possible. The eddy currents can circulate within each particle, confined by the insulating coating, but they can also circulate in agglomerates of contacting particles and even in the whole sample depending on the degree of insulation [9].

Surprisingly good agreement has been found between measured losses in SMC cores and calculated values obtained by solving Maxwell's field equations using conventional finite element analysis. A modified material conductivity has been employed which presumably takes account of the eddy currents within the particles [10]. The main conclusion to this is that the physical size of an SMC component has a large influence on the eddy currents. This results in the per unit losses being larger in larger components. This concurs with earlier findings on SiFe SMC cores [11].

The above approach has some physical basis if the insulation is generally poor throughout the core, but, if the insulation resistance is high, global eddy currents cannot flow.

Another approach is to assume that three loss components are present so the total loss P_{total} is given by

$$P_{total} = K_h B_p^{\alpha} f + K_{ex}(B_p f)^{1.5} + K_{cl}(B_p f)^2 \ (\mathrm{Jm^{-3}}) \tag{17.1}$$

Table 17.3 Loss separation constants for a typical SMC compared to those of a typical NO silicon–iron [11]

Material	K_h	α	K_{ex}	K_{cl}
0.35 mm thick, 3% SiFe	0.017	1.87	0.000,64	0.000,07
Somaloy 500 SMC core	0.144	1.96	0.000,32	0

where K_h, K_{ex} and K_{cl} and α are constants for a given flux density. These can be calculated from a series of measurements for a given material and the equation can be used to predict losses under different magnetising conditions. The constants evaluated in this way are far different from values found for SiFe as can be seen in Table 17.3 [12]. The hysteresis coefficient of the SMC core is almost ten times greater than that of the electrical steel, in line with the expectation that the hysteresis loss will be far higher. Of course, the expression is of limited use since a new set of constants are needed for each material. Its main use is in evaluating machine core losses from computed flux density distributions but this is inaccurate in itself for reasons given in Chapter 11 in Volume 2 of this book.

If it is assumed that the eddy currents are restricted to the individual particles then the classical eddy current loss can be written as [13]:

$$P_{cl} = \frac{K_e B_p^2 f^2 d_p}{\rho} \left(\mathrm{Jm}^{-3}\right) \tag{17.2}$$

where K_e is a geometry dependent constant, B_p is the peak flux density in the particle, f is the magnetising frequency, ρ is the electrical resistivity and d_p is the smallest dimension in the plane of the eddy current circuit. It is difficult to determine the values to put into this expression.

The resistivity in (17.2) is that of the particle material itself. This is different from the resistivity of the SMC component produced by compacting the particles. The bulk resistivity depends on random contacts between particles due to poor insulation coverage and cannot be directly used in macroscopic eddy current models unless the random nature of the contact between particles is taken fully into account [14]. This can be treated as a model taking into account the classical eddy current losses and the statistical theory of losses to predict total losses quite accurately under non-sinusoidal flux waveforms from a few hertz up to 10 kHz [15].

Loss depends on the dimensions of an SMC core and, generally, the measured eddy current component increases with increase of cross sectional area [13]. The loss in an SMC toroidal core can depend on the core height. This dependency would be expected to disappear in a perfectly insulated core where the eddy currents are confined within the particles themselves. From measurements on SMC cores, it has been found that the loss can be expressed simply as

$$P_t = K_e f^m \left(\mathrm{Jm}^{-3}\right) \tag{17.3}$$

where K_e is a constant and m has a value between 1.7 and 1.9 [11]. This equation for *eddy current* loss effectively includes excess and classical losses which are proportional to $f^{1.5}$ and f^2 respectively, so a value between the two is to be expected, the closer the exponent m is numerically to 2, the greater is the classical component.

The following expression for eddy current loss is claimed to be applicable to many materials including SMCs, electrical steel and amorphous materials [16]:

$$P_t = 8.4wlf^2B_p^2/\rho n\pi \left(\mathrm{Jm^{-3}}\right) \tag{17.4}$$

where w and l are the width and length of bar domains in the particles, n is the number of domains per unit volume which varies with frequency in accordance with $n \propto f^u$ where the exponent u lies within 0.35 and 0.65. The expression assumes only bar domains are present and that their walls move freely (i.e. no hysteresis occurs). This is a gross simplification of the actual domain structure and wall motion within the particles, but it does show some correlation with measured high frequency losses in highly resistive SMCs cores [11] and implies that the domain wall dynamics play a major role.

Another approach takes the demagnetising factor into account in the derivation of a formula for fast computation of P_h and P_e in SMCs assuming no global eddy currents are present and all the particles are spherical. Effective permeabilities of the particles and the complete composite can be obtained, from which the eddy current loss of an SMC composite can be computed reasonably accurately provided that an effective volume term based on the permeability values is included [17]. A similar approach is to use an equivalent hysteresis model in conjunction with linked macro and micro models for the total losses [18].

Several approaches to loss prediction have been proposed, most of which are claimed to be effective. However, none give a complete solution and claims on accuracy are sometimes based on the inclusion of constants to compensate for assumptions made in the derivations to bring solutions more in line with measurements.

Loss calculations are mainly aimed at evaluation of high frequency eddy current components making the assumption that the eddy current loss dominates and the hysteresis loss is low and constant at a given flux density. None can adequately predict losses at low magnetising frequency when hysteresis dominates. This is not surprising since this has only been partially achieved in the case of homogeneous materials as discussed in Chapter 4 in Volume 2.

17.2.5 Applications of iron-based SMCs

SMCs are being used as cores for many devices including actuators and choke reactors of switched mode power supplies (SMPSs). Many types of prototype motors using SMCs have been reported. These include claw pole, induction, transverse flux and permanent magnet designs. Several are believed to be in large scale commercial production, exploiting the 3D topology in relatively cheap cores with good dimensional tolerance and high temperature stability [19].

Magnetic properties can be tailored for these applications by varying the powder properties and compaction processes. Table 17.4 summarises the performance of

Table 17.4 Range of properties of insulated iron composites produced by Höganäs Corporation [7]

Curie temperature (°C)		>300
Density (kg m^{-3})		7,430–7,600
Induction at 10 Am^{-1} (T)		1.47–1.64
Permeability at 60 Hz		360–550
Coercive field Am^{-1}		246–300
Resistivity (μΩm)		>50 to >1,300
Tensile strength (MPa)		40–90
Loss at 1.0 T (W kg^{-1})	60 Hz	7.3–9.0
	100 Hz	11.2–15.0
	200 Hz	24–31
	400 Hz	52–67
	1,000 Hz	146–245

insulated powder composites produced by one manufacturer. The push appears to be towards easily compatible, oxide-coated SiFe composites produced by reactive atomisation for hybrid vehicle applications where losses as low as 150 kWm^{-3} at 0.1 T, 10 kHz have been reported [7].

A problem for designers seems to be that 3D topologies cannot yet be exploited to the full because core loss models are based on 1D magnetic theory and 1D material constants used in formulating loss expressions are based on 1D measurements which might not be appropriate [20]. This throws doubt on to the accuracy of some advanced design predictions. However, as designers learn to fully exploit the 3D isotropic properties of these materials, and the manufacturing freedom powder metallurgy allows, the performance gains from SMCs will improve further.

A practical drawback is still the lowering of permeability and saturation polarisation caused by the powder insulation and the need in some applications for improvement of the natural strength by additional means such as resins. SMCs are finding far more applications as their magnetic performance continues to improve [21–23]. The relevant standards are produced by the International Electrotechnical Commission Technical Committee TC 51 magnetic components, ferrite and magnetic powder materials [24].

17.2.6 Opportunities for future developments

The main challenge still is simultaneously to obtain high saturation polarisation and low losses over a wide frequency range without impacting on cost or mechanical performance. One avenue is improvements to the powder coating material. For example, the use of plastic-coated particles reduces core loss considerably in the 10–100 kHz frequency range, but it becomes significantly higher than in silicon-steels at up to 1 kHz [25]. Iron oxide coatings are reported to give the possibility of high B_s and relatively low core loss [26].

Nano powder composites have been investigated for many years and it seems they are applicable at very high magnetising frequencies where their magnetic performance is comparable to Sendust or NiZn ferrites. For example, cores produced using iron oxide-coated iron particles 10 to 20 μm in diameter have almost 100% density and resistivity of around 10^{11} times that of the iron particles with constant permeability over a wide frequency in the MHz range [27]. MnZn ferrite coated on laboratory produced SMC particles has been found to increase the permeability and lower the loss at low frequency by making the whole composition a little more magnetic [28].

Compacted coated Fe–Co–V nanoparticles have been incorporated into SMCs and predicted to give extremely soft magnetic properties [29], although this does not appear to have been achieved yet. Whether any of these different coating possibilities can outperform those commercially available today, particularly at low frequency, remains to be seen.

A long term possibility, not yet fully investigated, is the use of graded powders to improve the 3D flux distribution in complex components by control of the localised directional permeability. In theory, an SMC component can be made as an integrated core combining soft and hard magnetic powder regions. There would be many metallurgical and magnetic challenges to overcome, but it could be an ultimate goal for SMC components.

The magnetisation in each particle in an SMC depends on its own demagnetising factor which can be exploited to produce low loss solid components using far larger particles than normal which inherently have lower hysteresis. SMCs produced from such acicular engineered particles, up to 5 mm long, have been demonstrated to have a.c. magnetic properties as good as, or better than, low grade NO steel at low frequency and comparable with better grades at high frequency [30]. A similar approach is to produce high anisotropy by directional ordering of oblate spherical particles to enhance mechanical and magnetic properties [31].

SMCs produced from atomised iron-based amorphous powder have been found to have superior properties to Sendust cores, even under a.c. magnetisation. They are claimed to be commercially viable [32].

3D screen printing has been demonstrated as a way of producing compact copper windings. Multiple materials can be used in one step to produce complete motor components including an SMC stator core saving production time and giving scope for improved machine performance by close control of flux and thermal paths [33].

17.3 Soft ferrites

17.3.1 Basic structure

Ferrites exhibit the phenomenon of *ferrimagnetism* which was briefly introduced in Chapter 3. They are homogenous, ceramic materials comprised of various oxides, the most common of which is iron oxide. A single ferrite has the composition MO. Fe_2O_3 where M is a divalent metal element. Mixed ferrites contain more than one metal element, the appropriate combinations of which have superior magnetic properties to single ferrites.

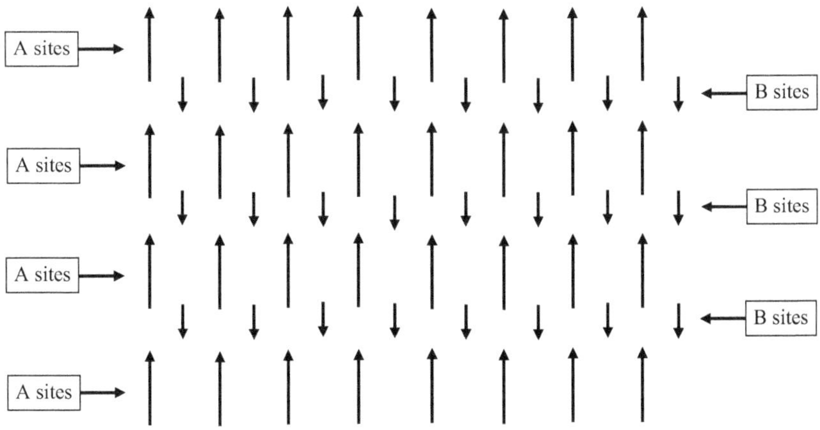

Figure 17.4 Basic arrangement of dipoles on A and B sites

Just like most ferromagnetic materials, the magnetic properties of ferrites are controlled by their magnetostriction and crystal anisotropy constants which determine their magnetic domain structures.

Magnetic ions occupy two types of lattice site in a ferrite. These are referred to as *A* and *B* sites where the dipoles spins are in opposite directions as shown in Figure 17.4. Dipoles at sites *A* and *B* are aligned along a preferred direction of magnetisation but in opposite directions so the net saturation polarisation is less than that of dipoles at the *A* and *B* sites separately.

Domains are formed to minimise the free energy and the presence of an external field encourages magnetisation to build up, initially remaining in a preferred direction, according to the relative dipole moments of the dipoles at the *A* and *B* sites. This ordered structure gives rise to a net spontaneous magnetisation. Increasing the temperature causes the dipole spins to be disturbed until the resultant magnetisation falls to zero at the Curie point, just as in the case of a ferromagnetic material.

As mentioned in Section 17.1, characteristics of only MnZn ferrites are outlined in this section, although some comparisons with other types of ferrite are included to show the wide variation with composition.

17.3.2 Production

Figure 17.5 summarises the process for producing MnZn ferrite. Powder oxides are mixed in the appropriate proportions and ball milled to obtain an initial homogenous mixture of powder sizes, typically 0.5–0.8 μm average diameter. During the process lubricant and binder materials may be added. This is followed by a so-called calcination process[3] where the material is heated to around 1,000 °C to partially decompose the powder, evaporate off volatile impurities and homogenise the mixture.

[3]Calcination is a process of heating to high temperature, sometimes in a controlled atmosphere. It originated in the cement production industry.

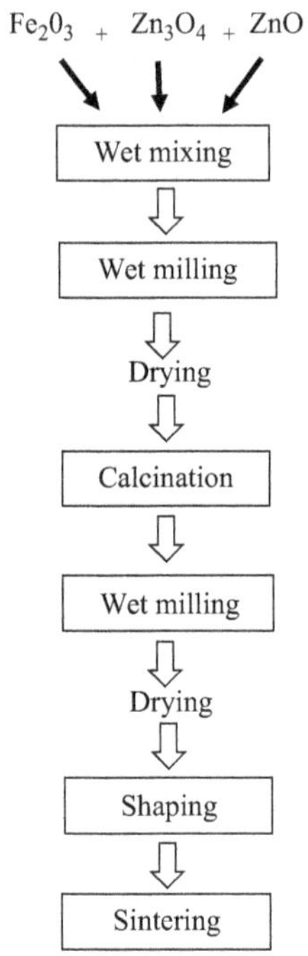

Figure 17.5 Flow chart showing basic route for ferrite production

The mix is then pressed into net shape parts by one of many types of pressing processes [4]. The next, and most critical parts of the process, is sintering to obtain the desired combination of properties by what can be a complex heat treatment cycle. The binder and lubricant are removed during this process. Finishing involves coating or other surface treatment according to the specific needs of an application.

The magnetic quality of the product depends on various microstructural factors, the most important of which are given in Table 17.5. The sintering temperature should be as high as possible to obtain an average grain size greater than 30 μm with a narrow size distribution.

Several methods of producing from nano-powders, made possible by general advances in nano-powder technology, are being considered [35]. It is possible that this new technology, combined with improved sintering processes, might lead to a future family of ferrites with better low frequency performance and improved mechanical properties for power devices.

17.3.3 Magnetic properties

Properties of some single ferrites are given in Table 17.6. The wide range of properties is very obvious. More useful combinations of properties are found in mixed ferrites. Typical properties of MnZn and NiZn ferrites are given in Table 17.7. These vary widely from grade to grade and values quoted in publications vary widely. The structure sensitive properties can be frequency dependent.

Table 17.5 Factors influencing the magnetic and mechanical properties of MnZn ferrite [34]

Chemical composition
Average grain size and distribution
Composition and structure of grain boundaries
Size and distribution of micro cracks
Planar and point defect, dislocations
Pore size and pore size distribution
Lattice structure and phases
Density after sintering

Table 17.6 Properties of some common single oxide ferrites

Ferrite	K_1 **m** ^{3}	λ_s **(με)**	M_s **(T)**	T_c **(°C)**	μ_i
$MnFe_2O$	−2,800	−5	0.5	300	250
$CoFe_2O$	380,000	−110	0.6	585	70
$NiFe_2O$	−6,200	−26	0.34	585	10
$CuFe_2O$	−6,000	−10	0.17	455	70
$MgFe_2O$	−2,500	−6	0.15	440	10

Table 17.7 Typical properties of commercial MnZn and NiZn mixed ferrites

Ferrite	K_1 **(Jm^{-3})**	λ_s **(με)**	**D.C. resistivity (Ωm) [35]**	M_s **(T)**	T_c **(°C) [4]**	**Loss (Wm^{-3}) [35]**	μ_i **[4]**
MnZn ferrite	+50 to −500	≈ 0	5–10	0.5	120–250	60–100 at 0.2 T, 25 KHz	750–15,000
NiZn ferrite	−150		20,000	0.33	90–500	150 at 0.01 T, 3 MHz	10–1,000

As can be seen, the main advantages of ferrites are their very good performance at high frequency, low cost and the possibility of being produced in a wide range of shapes and sizes. Their disadvantages, apart from low saturation polarisation and Curie point, are poor mechanical strength, brittleness and relatively poor temperature stability.

The loss of cores wound from silicon–iron or amorphous ribbon are dependent on geometry [36,37]. Similar trends are found in MnZn toroidal cores, despite them being solid as opposed to the strip wound cores of the other materials [38].

17.3.4 Loss components in ferrite cores

The loss is assumed to be the sum of two components at high magnetising frequency. These are a quasi-static hysteresis component and a dynamic component made up of eddy current loss and a residual loss associated with magnetic relaxation and resonances in ferrites. The hysteresis component P_{hys} is assumed to be affected by microstructure and internal stress, just as in metallic magnetic alloys. The low anisotropy reduces hindrance to domain motion and the low magnetostriction helps minimise the stress sensitivity of the loss.

It is interesting to note that it is recognised in the ferrite designer community that P_h/f is frequency dependent although it is usually assumed constant at low magnetising frequencies [38].

The classical eddy current loss of a ferrite core can be written in a similar way as for metallic alloy thin strips, simply as

$$P_{cl} = K_{cl}f^2B_p^2/\rho \ (\mathrm{Jm^{-3}}) \tag{17.5}$$

where K_{cl} is a constant dependent on the core shape and dimensions.

The residual loss P_r originates from the direct release of energy from precessing electron spins into the crystal lattice (sometimes called spin damping). It can be expressed in terms of frequency dependent real and imaginary parts of the a.c. permeability [39].

The total loss per cycle can be separated into three components as shown in Figure 17.6. There is a strong resemblance to the loss per cycle versus frequency characteristics of metallic alloy strip but different mechanisms seem to be involved.

If the dimensions of a MnZn ferrite are sufficiently small, the classical eddy current component of loss is low. Also, below around 400 kHz, P_r is small so the loss is dominated by the hysteresis component. From this, it appears that P_h is the only component of loss which occurs at power frequencies but, just as with electrical steels, the physical nature of the loss mechanisms are not well understood. For example, when the classical eddy current loss expression in (17.5) is used,

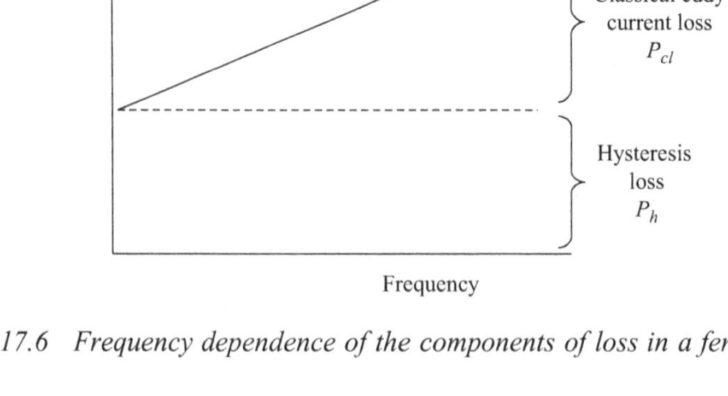

Figure 17.6 Frequency dependence of the components of loss in a ferrite core

limitations in physical analysis and other factors lead to gross differences in modelled losses compared to measured values [40].

The eddy currents are generally constrained within individual grains at low frequency because of the high resistivity, but because the resistivity drops rapidly at frequencies above around 10 kHz global eddy currents flow in a similar manner to that in homogeneous metallic material [41]. This suggests that modelling of high frequency eddy current losses in ferrites is complicated by the spin damping process and the models used today are not appropriate.

Like most other soft magnetic materials, MnZn ferrites are used in applications where the magnetisation is subject to a d.c. bias (or offset). The effect on losses seems to vary a lot from material to material and also with temperature and method in which the bias is applied and measured. Generally, it is found that losses are not affected up to relatively high bias ratios, e.g. a bias of 50 mT has no measurable effect on losses when the a.c. component is varied between 30 mT and 240 mT at 25 kHz to 200 kHz [42]. When the bias field is increased to more than 100 mT in this particular example the loss rapidly increases by an order of magnitude. The mechanism does not appear to be well known but is probably different from that in steels when losses increase immediately on application of a small bias which takes the magnetisation past the knee of the *B–H* curve as described in Section 12.7.

17.3.5 Applications

The unique combination of very high resistivity and low magnetostriction together with low magnetocrystalline anisotropy gives the ferrites their ultra-high frequency performance into the hundreds of GHz microwave region. They can easily be produced in a single step as toroids, rods, planar cores, tubes and other forms for integrating into power electronics and telecommunication systems where they are used in the kHz range.

Increased demand for miniaturisation of power electronic power supplies and other electronic components has escalated the use of soft ferrites, so they are used at increasingly higher frequencies in the kilowatt power range. For example, they are used in output transformers for ultrasonic transducers operating at around 2 MHz to matching transformers operating in the 30 MHz range.

The MnZn ferrites have saturation polarisations up to around 0.5 T so, despite their lower cost and other advantages referred to earlier, they rarely compete with electrical steels in 50/60 Hz applications. However, it is possible that they might find advantages over alternative materials in future applications in devices operating under severe harmonic distortion with a fundamental component of magnetisation around or even below 1 kHz.

References

[1] De Vries M. *80 Years of Research at the Phillips Natuurkundig Laboratorium (1914–1994)*. Amsterdam University Press, 2005, p. 95

[2] Snelling E.C. *Soft Ferrites: Properties and Applications*, 2nd edn. Butterworths, 1988

[3] *Soft Ferrites: A User's Guide*. Magnetic Materials Producers Association, Chicago, IL, 1998
[4] Hultman L. 'Development of SMC materials for motor applications'. *UK Magnetics Society, Seminar: Developments in Magnetic Materials,* Farnborough, UK, 28th April 2006,
[5] http://www.hoganas.com/ (Accessed May 2018)
[6] http://www.sgtec.com/material/silicon-iron (Accessed May 2018)
[7] Narasimhan K. Insulated iron powders, SMC, current state and future possibilities. http://www.magneticsmagazine.com/conference/2016/Presentation/P2PTechnologies_Narasimhan.pdf (Accessed March 2017)
[8] Hori K., Chun-Feng J., Ishihara Y., Todaka T., and Yoshikawa K. 'Magnetic properties of powder magnetic core under the stress'. *Przeglad Elekrotechniczny* 2007, vol. 83(4), pp. 78–82
[9] Lefebvre L.P., and Gelinas C. 'Effect of material insulation and part geometry on the AC magnetic performances of PM soft magnetic composites'. *Advances in Powder Metallurgy and Particulate Materials MPIF*, vol. 7, Princeton, NJ, USA, 2001, pp. 36–50
[10] Nord G., Pennander L.-O., and Jack A. 'Loss calculations for soft magnetic composites'. *16th ICEM Conference, Cracow,* September 2004
[11] Saito T., Takemoto S., and Iriyama T. 'Resistivity and core size dependencies of eddy current loss for Fe-Si compressed cores'. *IEEE Trans. Magn.* 2005, vol. 41(10), pp. 3301–3303
[12] Jansson P., and Jack A. 'Magnetic assessment of SMC materials'. *21st Annual Conference on Properties and Applications of Magnetic Materials,* Chicago, USA, 13–15th May 2002
[13] Gelinas C., and Brydges D. 'Insulated iron powders for automotive applications'. *Society of Automotive Engineers Inc*, vol. 1, 2003, p. 447
[14] Appino C., Bottauscio O., De la Barrière O., Fiorillo F., Manzin A., and Ragusa C. 'Computation of eddy current losses in soft magnetic composites'. *IEEE Trans Magn.* 2012, vol. 48(11), pp. 3470–3473
[15] De la Barrière O., Appino C., Fiorillo F., *et al.* 'Characterization and prediction of magnetic losses in soft magnetic composites under distorted induction waveform'. *IEEE Trans. Magn.* 2013, vol. 49(4), pp. 1318–1326
[16] Sakaki Y., and Imagi S. 'Eddy current losses in polycrystalline and amorphous soft magnetic materials'. *Jpn. Trans. IEE* 1981, vol. A-101(11), pp. 575–581 (in Japanese)
[17] Sato T., Aya S., Igarashi H., Suzuki M., Iwasaki Y., and Kawano K. 'Loss computation of soft magnetic composite inductors based on interpolated scalar magnetic property'. *IEEE Trans. Magn.* 2015, vol. 51(3), Art. No. 8001704.
[18] Fujisaki F., and Yao A. 'Magnetic multiscale model for local eddy current flow in complex material with insulted particles' *IEEE Trans. Magn.* 2017, vol. 53(6), Art. No. 7300204
[19] Jansson P., Persson M., Jack A.G., and Mecrow B.C. 'Powdered soft magnetic materials for medium frequency applications'. Presented at *Soft Magnetic Materials '96*, San Francisco, USA, February 1996

[20] Guo Y., Zhu J., Lu H., Lin Z., and Li Y. 'Core loss calculation for soft magnetic composite electrical machines'. *IEEE Trans Magn.* 2012, vol. 48 (11), pp. 3112–3115

[21] Pennander L.O. 'Progress for soft magnetic composite materials'. *Proceedings of the 7th International Conference on Magnetism and Metallurgy,* Rome, Italy, 13–15 June, 2016, pp. 55–59

[22] Schoppa A., and Delarbre P. 'Soft magnetic composite materials – their magnetic properties and applications' *Proceedings of the 6th International Conference on Magnetism and Metallurgy*, Cardiff, UK, 17–19 June 2014, pp. 112–127

[23] Schoppa A. 'Soft magnetic powder materials – Status of development and applications'. *Proceedings of the International Conference on Magnetism and Metallurgy*, Part I, Dresden, Germany, 12th–14th June 2018, pp. 389–395

[24] IEC. TC 51: Magnetic components, ferrite and magnetic powder materials. www.icc.ch

[25] Goldman A. 'The magnetic applications choice among ceramic, metallic strip, or metal powder cores'. *J. Mater. Eng. Perform.* 1995, vol. 4, pp. 395–400.

[26] Zhao G., Wu C., and Yan M. 'Fe-based soft magnetic composites with high B_s and low core loss by acidic bluing coating'. *IEEE Trans. Magn.* 2015, vol. 51(11), Art. No. 2002904

[27] Sugaya Y., Inoue O., and Kugimiya K., 'Soft magnetic properties of nanostructure-controlled magnetic materials'. *IEEE Trans. Magn.* 1994, vol. 30(6), pp. 4945–4947.

[28] Wu S., Sun A., Xu W., Zhang Q., Zhai F., Logan P., and Volinsky A.A. 'Iron-based soft magnetic composites with Mn-Zn ferrite nanoparticles coating obtained by sol-gel method'. *J. Magn. Magn. Mater.* 2012, vol. 324, pp. 3899–3905

[29] Turgut Z., Fingers R.T., Piehler H.R., and McHenry M.E. 'Microstructural and magnetic observations of compacted FeCoV nanoparticles'. *IEEE Trans. Magn.* 2000, vol. 36(5), pp. 3024–3025.

[30] Bularzik J.H., Krause R.F., and Kokal H.R. 'Properties of a new soft magnetic material for AC and DC motor applications'. *J. Phys. IV France* 1998, vol. 2, pp. 747–753

[31] Namkung M., Bryant R.G., Fox R.L., and Buchman A. 'Magnetic and mechanical properties of molded iron particle cores'. *IEEE Trans. Magn.* 1996, vol. 32(5), pp. 4890–4892

[32] Endo I., Otsuka I., Okuno R., Shintani A., Yoshino M., and Yagi M. 'Fe-based amorphous soft-magnetic powder produced by spinning water atomization process (SWAP)'. *IEEE Trans. Magn.* 1999, vol. 35(5), pp. 3385–3387

[33] Linder M., Bräuer P., and Studnitzky T. 'Utilization predictions for electrical machines with advanced materials and production technologies'. *Proceedings of the 5th International Conference on Magnetism and Metallurgy,* Ghent, Belgium, 20th–22th June, 2012, pp. 291–303

[34] Michalowsky L. ‘Manganese-zinc ferrites: market and potential – further improvement of properties’. Presented at *Soft Magnetic Materials 98*, Barcelona, Spain, 20–22 April 1998

[35] Mirshekari G.R., Daee S.S., Mohseni S. T., et al. ‘Structure and magnetic properties of Mn-Zn ferrite synthesized by glycine-nitrate auto-combustion process’. *Adv. Mater. Res.* 2012, vol. 409, pp. 520–552

[36] Moses A.J., and Ling P.C.Y. ‘Dimensional factors affecting magnetic properties of wound cores’. *Phys. Scr.* 1989, vol. 40, pp. 249–251

[37] Moses A.J., Leicht J., and Fox D. ‘Influence of geometry and wave shape on magnetic properties of magnetic amorphous material’. *J. Magn. Magn. Mater.* 2005, vol. 290–291, pp. 1520–1523

[38] Landgraf F.J.G., Lazaro-Colan V., Leicht J., Janasi S.R., and Lopes M.F. ‘Geometry effect on the magnetic properties of manganese zinc ferrite’. *J. Magn. Magn. Mater.* 2008, vol. 320, pp. e857–e859

[39] Stoppels D. ‘Developments in soft magnetic power ferrites’. *J. Magn. Magn. Mater.* 1996, vol. 160, pp. 323–328

[40] Fiorillo F., Beatrice C., Bottauscio O., and Carmi E. ‘Eddy-current losses in Mn-Zn ferrites’. *IEEE Trans. Magn.* 2014, vol. 50(1), Art. No. 6300109

[41] Zaspalis V.T., Kolenbrander R., Guenther R., and Van der Valk P. ‘Loss mechanisms in modern MnZn ferrites for power applications’. *Proceedings of the 9th International Conference on Ferrites* (Ed. Sohoo R.F.), San Francisco, CA, USA, 2004, pp. 579–584

[42] Baguley C.A., Carsten B., and Madawala U.K. ‘The effect of DC bias conditions on core losses’. *IEEE Trans. Magn.* 2008, vol. 44(2), pp. 246–252

Chapter 18

Temperature and irradiation dependence of magnetic and mechanical properties of soft magnetic materials

All magnetic properties of engineering materials are temperature dependent to varying degrees. This is to be expected since their intrinsic saturation magnetisation, magnetostriction and magnetocrystalline anisotropy are all temperature sensitive. These determine the domain structure within a magnetic material, hence the magnetic properties at any temperature.

Knowledge of the temperature dependence of magnetic properties is important for two reasons. First, if a material can be used safely, and effectively, at a high temperature, the efficiency and rating of many types of electrical machine can be increased. Second, there is no reason why the best choice material for room temperature operation should still be the best option at a different operating temperature.

The range of temperatures over which a magnetic material might be required to operate is normally restricted to established industrial standards and norms. For example, for general industrial usage, magnetic materials are expected to operate over a range from −40 °C to + 100 °C. For military applications, the normal requirement is for satisfactory operation from − 55 °C to + 125 °C and in specialised aerospace applications the range can span from a temperature of near −273 °C up to over 1,000 °C.

The most important high temperature magnetic properties for rotating machines and transformers are saturation magnetisation, power loss, permeability and coercivity. For magnetic materials used in electronic systems the *B–H* loop shape is often the key parameter and for some types of solenoids d.c. magnetic properties are the most important.

In a.c. applications, soft magnetic materials can be thought of as being active thermal materials producing heat as a result of eddy current and hysteresis losses. The core of an electrical machine designed for optimum output power to weight performance is magnetised at the highest possible flux density and magnetising frequency at the expense of higher than normal losses.[1] The result is a high temperature rise which might give rise to an extra cost for cooling. Air, water, oil, hydrogen and liquid metal are used directly or indirectly to cool magnetic cores in

[1]Of course, high power output also implies high copper losses in the electrical windings, but these implications are not considered here.

applications ranging from energy efficient power transformers to rotating machines in nuclear reactor environments.

Mechanical properties can be at least as important as magnetic performance so the suitability of a magnetic material to operate at an elevated (or very low) temperature must also be judged on its retained mechanical strength, thermal stability and creep resistance. For example, the ever growing demand for more compact, high power, rotating machines for use as aircraft engine starter/electricity generators leads to the need for efficient soft magnetic materials with a yield strength of over 800 MPa at a continuous operating temperature of possibly 500 °C [1].[2]

Sample shape and size can influence the temperature characteristics so this must be taken into account when comparing materials. Measurements are made under many different magnetisation conditions, often not fully specified. Frequently, insufficient estimates of errors and experimental uncertainties are given in the literature, so direct comparisons between results from different sources can be difficult to interpret.

Magnetic measurements near the Curie point are inherently difficult to perform. Because the relative permeability is close to unity, accurate control of the flux paths to ensure uniform magnetisation is difficult. Experimental test systems containing dimensionally accurate ceramic components and high temperature insulation need to be carefully designed, set up and operated. There are no standard specifications or guidelines to follow.

For the above reasons, material characteristics included in this chapter are presented to show general trends.

Sections 18.1 and 18.2 summarises the temperature dependence of structure insensitive magnetic properties of important soft magnetic materials. This is followed in Sections 18.3 and 18.4 by a presentation of temperature characteristics of structure sensitive properties of some families of commercial soft magnetic materials. The chapter concludes with sections covering low temperature (Section 18.6) characteristics, approaches to modelling temperature effects (Section 18.5), possible effects of temperature gradients in magnetic cores (Section 18.7) and the general effects of irradiation on magnetic properties (Section 18.8).

18.1 Effects of temperature on structure insensitive magnetic properties

18.1.1 Saturation magnetisation M_s

Thermal agitation disturbs magnetic dipole alignment as shown in Chapter 2 and causes the saturation magnetisation to fall, as shown in Figure 2.2(a), from its maximum value M_0 at absolute zero temperature to disappear completely at the Curie temperature.

Figure 18.1 compares measured and theoretical variation of saturation magnetisation with temperature of iron, cobalt and nickel. Magnetisation and temperature are

[2]The rotors of such machines revolve at tens of thousands of revolutions per minute, so high yield strength of the laminations is essential to avoid the possibility of bursting.

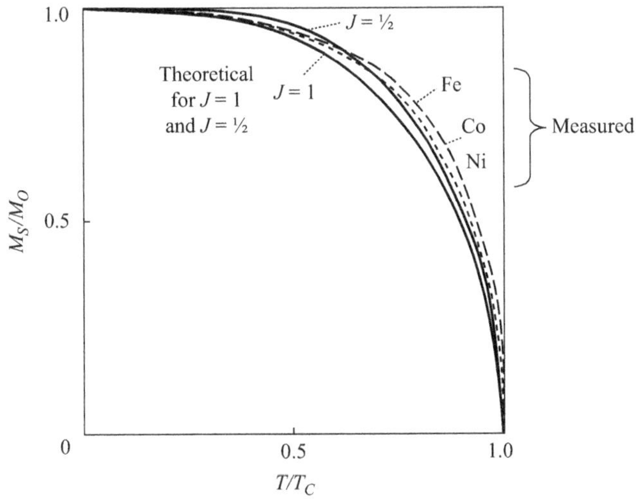

Figure 18.1 Theoretical and measured temperature variation of the saturation magnetisation of iron, cobalt and nickel (Figure 5.2 in [2] reproduced under free licence CC-BY-4.0 and modified)

plotted as ratios of the saturation magnetisation M_0 at 0 K and the Curie temperature T_c, respectively. The measured and theoretical values are very similar. Magnetisation of each material drops from the value at 0 K by around 3% at a temperature equal to half its Curie point, i.e. 522 K, 700 K and 316 K (249 °C, 427 °C and 43 °C) for iron, cobalt and nickel, respectively. This immediately points to the important practical fact that a high Curie point results in a wide effective operating temperature range.

The measured temperature variation agrees closest with a theoretical value calculated using $J = 0.5$ where J is the total atomic angular momentum [2]. In fact, J cannot take the value of exactly one half in iron, cobalt or nickel but an exact analysis is unnecessary here. It is sufficient to say that for practical applications (2.5) and Figure 18.1 give a reasonable approximation to the temperature variation of M_s.

The Curie points and saturation magnetisation of some common materials are given in Table 18.1. By referring to the general form of the temperature characteristics shown in Figure 18.1, an effective operating range of each material can be estimated. Note, however, that values quoted for the nickel–iron and cobalt–iron alloys vary in practice according to the presence of small quantities of additional elements.

18.1.2 Resistivity

The eddy current loss of a soft magnetic material is inversely proportional to its resistivity as shown in Section 7.7. The cause of the resistivity is simply the effect of electrons in the electric eddy currents colliding with atoms in a conducting material so it is not surprising that it is temperature dependent. The change of

Table 18.1 Curie point and temperature coefficient of resistivity of some common soft magnetic material[a]

Material	Saturation magnetisation at 20 °C	Curie temperature (°C)
Iron	2.15	770
Nickel	0.61	358
Cobalt	1.76	1127
GO 3%SiFe	2.04	730
49%Co49%Fe2%V	2.36	932
50%Ni50%Fe	1.50–1.60	500
6.5%SiFe	1.80	700
78%NiFe	0.75	460
25%CoFe	2.40	930
MnZn ferrite	0.35–0.4	180–250 [3]
Fe-based amorphous material	1.56	395
Fe–Cu–Nb–Si–B nano-crystalline material	1.25	570

[a]Some of these vary with exact composition and processing so the values are only to be taken as a guide.

resistivity with temperature is quantified in terms of the temperature coefficient of resistance α which is obtained from the empirical equation

$$\rho_\theta = \rho_0(1 + \alpha_\theta[\theta - \theta_0]) \ (\Omega\text{m}) \tag{18.1}$$

where ρ_θ and ρ_0 are the resistivity at temperature θ and a reference temperature θ_0, respectively. The temperature coefficient of resistivity α_θ is the mean value between temperatures T_θ and T_0.

Equation (18.1) is often referred to as the linear approximation of the variation of resistivity with temperature where α_0 is assumed constant over the temperature range in question.

The magnitude of α_θ for any material is very dependent on the impurity level. It is often quoted as being around 0.005 °C^{-1} for iron which results in a substantial 50% rise in resistivity for a 100 °C temperature rise. It is not possible to calculate the temperature coefficient, but Figure 18.2 shows the temperature variation of resistivity of iron plotted as an average of measurements on iron samples with different impurity levels presented in several publications. It will be noticed that the slope of the characteristic (α_θ) versus temperature curve is not constant.

Table 18.2 shows how the resistivity of iron and nickel vary with temperature and the mean values of α over two temperature ranges [4]. In both cases α varies substantially depending on the temperature range over which it is averaged. It is not easy to find reliable data for variation of α with temperature in electrical steels but it is expected to be similar to that of iron.

In metals and alloys, α is almost always positive meaning that the resistivity increases with temperature. Hence eddy current loss under a.c. magnetisation tends to fall with increasing temperature.

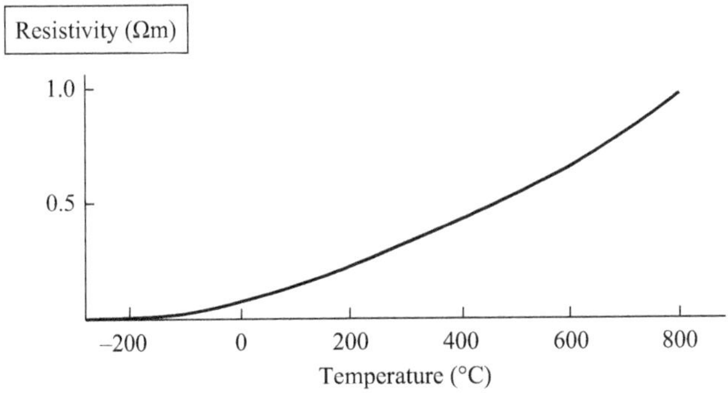

Figure 18.2 Temperature dependence of the resistivity of iron

Table 18.2 Variation of resistivity and the temperature coefficient of iron and nickel

Material	Resistivity (μΩm) [4]			$\alpha(^\circ C^{-1})$			
	−193 °C	20 °C	227 °C	−193 °C to 20 °C range	20 °C to 227 °C range	−193 °C to 227 °C range	Nominal value at 20 °C
Iron	0.7	9.6	23.7	0.042	0.068	0.054	0.005
Nickel	0.6	6.9	17.7	0.038	0.052	0.041	0.006

In ferrites, ρ_θ can be positive or negative, even over different temperature ranges for the same composition [5]. However, because the resistivities of ferrites are extremely high, any change due to temperature is of little consequence.

In amorphous metals, α can be negative and it varies in a complex manner with temperature [6,7]. However, this does not seem to be of any practical importance since the resistivities of iron-based, cobalt-based and nickel-based amorphous materials are far higher than equivalent crystalline alloys, α itself is low so it has little effect on the resistivity at any practical operating temperature, which itself is restricted in value by their low Curie and recrystallization temperatures.

18.1.3 Magnetocrystalline anisotropy constants

Magnetocrystalline energy is often the dominant factor controlling the domain structure and hence the magnetic properties of crystalline materials, particularly electrical steels. The energy arises from the strong coupling of orbital and electron spins of individual atoms with the electric field associated with the crystal lattice. This is difficult to predict in theory so it is sufficient to say here that thermal agitation of the crystal lattice makes the magnetocrystalline anisotropy constants $K_1$and K_2 extremely temperature sensitive.

Table 18.3 shows the temperature variation of the magnetocrystalline constants[3] of iron [8]. K_1 is the dominant term over the whole temperature range and is always positive. K_2 is very low compared with K_1 and is always negative. Equations (3.2)–(3.4) show how the easy direction of magnetisation depends on the magnitude and sign of K_1 and K_2. It can easily be shown from these equations that the <100> directions are easy directions of magnetisation if $K_1 > 0$ and $K_2 > -9K_1$. This condition is satisfied for iron over its entire ferromagnetic temperature range.

Another point to note is that both constants are close to zero at high temperature, hence the material becomes almost isotropic. The magnitudes of the quantities presented in Table 18.3 are only indicative since, even at room temperature, reported values do not agree well. For example, in a summary of results from ten independent investigators, values of K_1 for iron of between 478,000 Jm^{-3} and 525,000 Jm^{-3} and K_2 between −34,000 Jm^{-3} and −59,000 Jm^{-3} have been reported [9].

The temperature characteristics of the magnetocrystalline constants of electrical steels are not well documented but do follow similar trends to those of iron as can be seen in Table 18.3. Generally, in cubic crystalline materials, at ambient temperature, K_2 is far less than K_1 so it is ignored in most analyses [13]. However, it can be seen that, in iron, at high temperatures, they are similar in magnitude so this assumption cannot be applied indiscriminately.

Figure 18.3 shows the temperature variation of the anisotropy constants of cobalt and nickel. In cobalt, both constants gradually drop over the full temperature range (over which cobalt has its hexagonal crystal structure). At low temperature K_2 becomes increasingly negative so that the condition is satisfied which makes

Table 18.3 Temperature variation of magnetocrystalline energy constants of iron and silicon–iron

Temperature (°C)	**K_1 (Jm^{-3}) for iron [8]**	**K_1 (Jm^{-3}) for 3%SiFe [12]**	**Temperature (°C)**	**K_2 (Jm^{-3}) for iron [8][a]**
−253	52,000	42,000	−253	−16,000
−80	49,800	37,000	−100	−15,000
0	47,500	35,300	−10	−13,000
123	39,800	29,000	110	−8,000
200	31,700	23,100	210	−5,000
405	12,000	10,500	410	−3,000
678	8,000	0 (at 610 °C)	450	−3,900

[a]These values are included because they cover the full temperature range but other investigations [10,11] give positive values of K_2 at 20 °C. The orders of magnitude agree and, whether positive or negative, the argument in the text remains the same. Also in [8] positive K_3 terms (referred to in Section 3.6.1) are presented alongside the lower magnitude negative K_2 terms which might go some way towards explaining the different results in [10,11].

[3]The variation of the third anisotropy constant K_3 is shown in [8]. It is not discussed here since it has no practical significance to industrial soft magnetic materials.

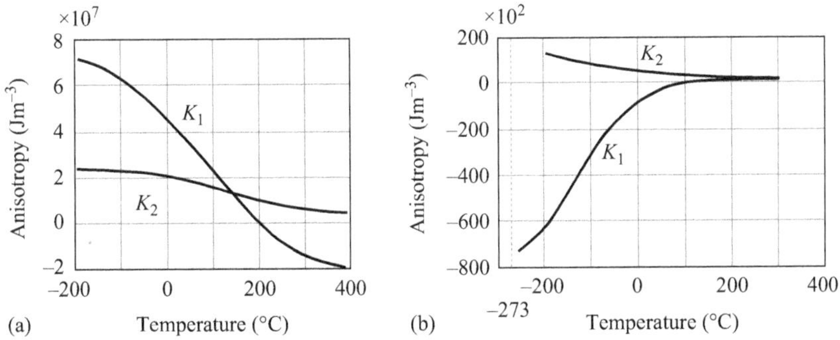

Figure 18.3 Temperature variation of K_1 and K_2 in (a) cobalt and (b) nickel (modified from Figures 12.12 and 12.13 in [15] under free licence CC-BY-4.0 and modified)

<111> the easy direction of magnetisation[4] rather than <100> which is satisfied at higher temperatures. Of course, the values of K_1 and K_2 in nickel are very low at the higher temperatures, so the actual magnetocrystalline anisotropy is small.

18.1.4 Magnetostriction constants

The magnetostriction constants λ_{100} and λ_{111} of many materials are extremely temperature sensitive. They vary in sign and magnitude in ways which are still difficult to explain. Measurement is difficult, which might partly explain a lack of verifiable published values, particularly at high temperatures.

Figure 18.4 shows the variation of λ_{100} and λ_{111} of uncoated, laboratory prepared, SiFe and pure iron samples. The temperature variation of both constants of the iron and 1%SiFe follow similar trends over the full temperature range as might be expected, but the λ_{100} characteristics of the high silicon alloys are quite different. The measurements, made using high temperature strain gauges, are subject to inaccuracies due to the doubt over the magnetisation state of the samples as well as the experimental uncertainties [16] so such characteristics are only a guide. However, what is evident is that the addition of silicon causes λ_{111} to decrease over the entire temperature range. No conclusive explanation for these changes has been reported.

The a.c. magnetostriction characteristics of commercial grain oriented (GO) 3%SiFe up to 710 °C have been used to help produce a model based on the temperature dependence of magnetic domain structures to estimate the temperature variation of λ_{100} [17]. Predictions using the model agree with the characteristics of the 3.85%SiFe single crystal shown in Figure 18.4 up to around 300 °C, but agreement is poor at higher temperatures. This is attributed partly to the effect of the coating on the commercial material and the possibility of unintentional magnetic annealing affecting the characteristics at the higher temperatures.

[4]This condition is introduced in Section 3.6.

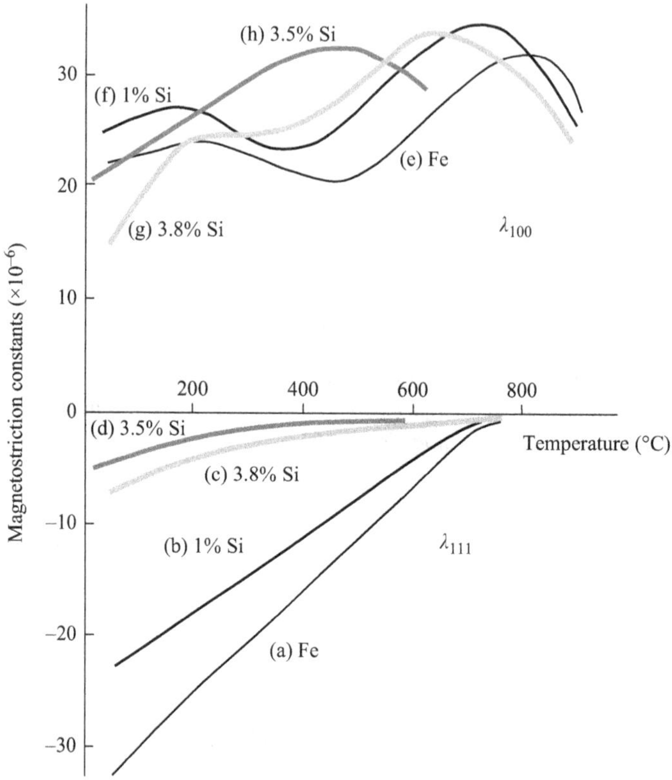

Figure 18.4 Temperature variation of magnetostriction constants in 0% to 3.8% SiFe alloys – (a)–(d) λ_{111}; (e)–(h) λ_{100}: (a) Fe [16], (b) 1%SiFe [16], (c) 3.8%SiFe [16] (d) 3.5%SiFe [14], (e) Fe [16], (f) 1%SiFe [16], (g) 3.8%SiFe [16] and (h) 3.5%SiFe [14] (produced from figures in [14] and [16], reproduced under free licence CC-By-4.0 and combined)

18.2 Effect of temperature on permeability, coercivity and losses

Coercivity, permeability and losses depend on the intrinsic magnetic properties discussed in Section 18.1. Therefore, if the loss of a given material falls over a particular temperature range, it is common to assume that coercivity also falls and permeability increases. However, this is not necessarily the case.

For example, consider the coercive force. In crystalline materials, H_c is a function of λ_s/M_s or K_1/B_s [18]. The effect of λ_s on the magnetic properties of SiFe alloys at ambient temperature in general is low compared to the effect of K_1, so the K_1/B_s relationship dominates. Since K_1 falls quicker than λ_{100} and, hence, λ_s with temperature, at elevated temperature, H_c is expected to decrease as is found to be the case. At the same time, the temperature coefficient of permeability may be either positive or negative depending on whether magnetisation occurs mainly by domain wall motion or by rotation of the magnetisation vector, i.e. the permeability might, or might not, follow the opposite trend to that of H_c depending on the magnitude of the magnetising flux density and frequency.

Of course the relative magnitudes of the anisotropy and magnetostriction constants play a major role in determining domain structures whose variation with an applied magnetic field in turn determines the values of H_c, μ, λ_s and losses. However, if the effect of temperature on these properties is considered, it soon becomes apparent that the process is so complex and difficult to predict accurately.

18.3 The d.c. and a.c. properties of silicon steels at elevated temperatures

Figure 18.5 shows a typical set of d.c. magnetisation curves of GO 3%SiFe measured over the temperature range from 20 °C to 700 °C [18]. The gradual fall of high field induction with increasing temperature is to be expected from the reduction in M_s shown in Section 18.1.1. It might not be so obvious from the figure that, at low magnetisation, the flux density increase with field becomes far more rapid as the temperature is increased, i.e. the high temperature permeability increases. This is clearer in Figure 18.6 in which it can be seen that the field at which the permeability is maximum decreases as the temperature increases and the

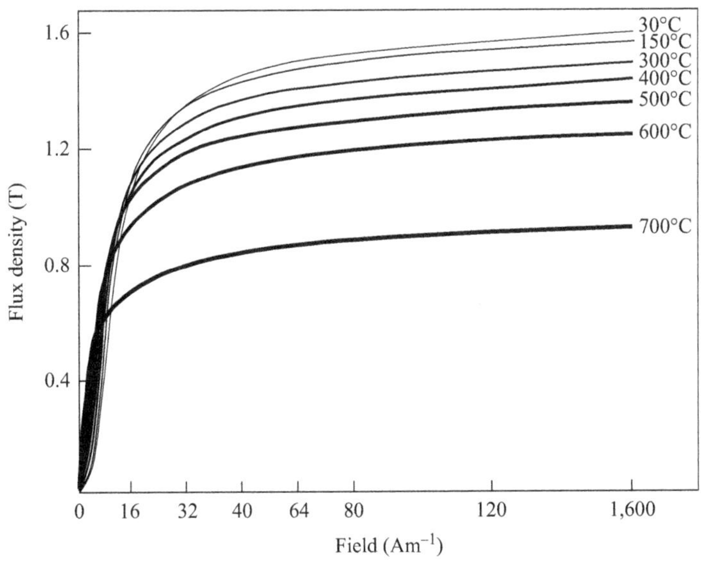

Figure 18.5 Temperature dependence of d.c. magnetisation curves of GO 3%SiFe (Figure 1 in [18] reproduced under free licence CC-BY-4.0 and updated)

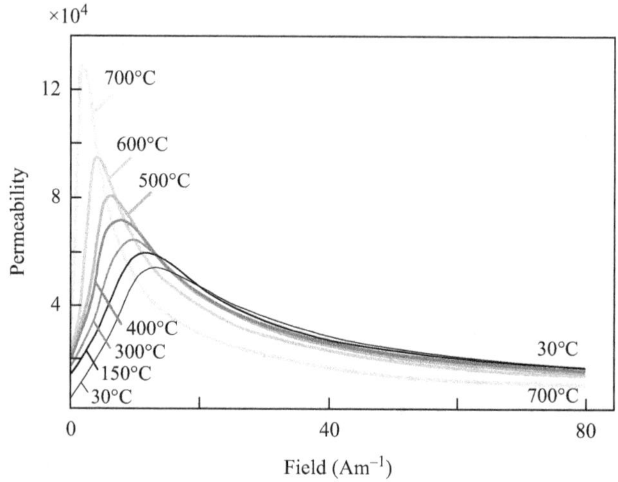

Figure 18.6 Temperature dependence of d.c. permeability on magnetising field in GO 3%SiFe (Figure 2 in [18] reproduced under free licence CC-BY-4.0 and updated)

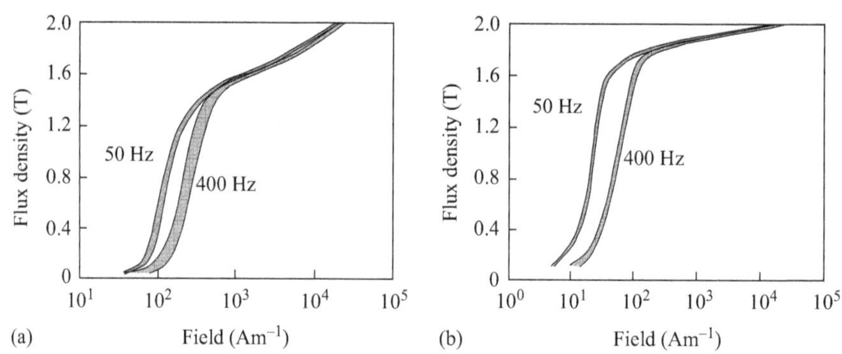

Figure 18.7 Temperature dependence of a.c. B–H curves of electrical steel magnetised at 50 Hz and 400 Hz: (a) 0.50 mm thick NO SiFe and (b) 0.3 mm thick GO SiFe (Figure 1(a) and (b) in [20] reproduced under free licence CC-BY-4.0 and modified)

maximum permeability itself increases with temperature. This is associated with the drop in K_1 helping to annihilate harmful lancet domains. The degree of the annihilation increases with improved orientation, thinner material and the presence of tensile stress [19].

The temperature characteristics of the a.c. magnetisation curves of typical grades of non-oriented (NO) SiFe and GO SiFe at 50 Hz and 400 Hz are compared in Figure 18.7 [20]. The *B–H* characteristics are measured over a temperature range from 20 °C to 200 °C and the spread of the characteristics is simply shown by the

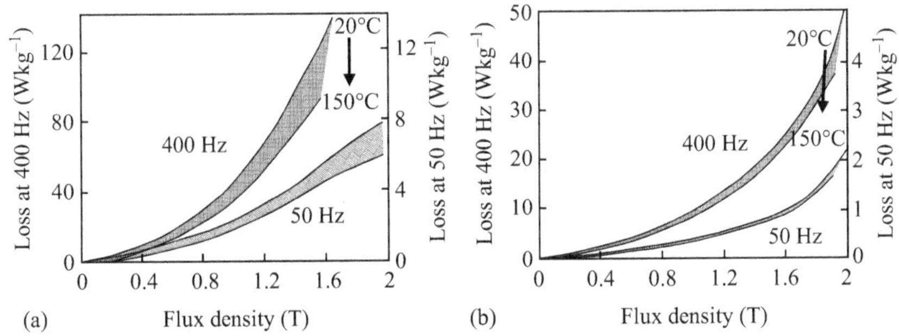

Figure 18.8 Temperature dependence of a.c. loss characteristics of electrical steel magnetised at 50 Hz and 400 Hz: (a) 0.50 mm thick NO SiFe and (b) 0.3 mm thick GO SiFe (Figure 1(c) and (d) in [20] reproduced under free licence CC-BY-4.0 and modified)

envelopes. Over this relatively narrow temperature range, the characteristics are similar, slightly more so at the lower magnetising frequency. Closer inspection shows that the characteristics of both materials improve marginally with increasing temperature, the effect being a little greater in the NO steel. The variation follows the same trend as that of the d.c. characteristic plotted over a wider temperature range in Figures 18.5 and 18.6.

Figure 18.8 shows the corresponding temperature variation of losses in the same NO and GO materials whose *B–H* curves are shown in Figure 18.7 [20]. The loss of both materials falls appreciably with increasing temperature. The percentage drop in the NO steel is much higher than in the GO material. At 400 Hz, 1.5 T, the loss of the NO steel drops by around 25% as the temperature is increased from 20 °C to 200 °C and at 1.5 T, 50 Hz it drops by around 15%. The corresponding figures for the GO material are 12% and 8%, respectively. The probable reason for this difference is the greater benefit of the increase in resistivity with temperature in the thicker NO steel where a higher proportion of its loss is due to eddy currents.

Similar a.c. loss and permeability characteristics have been observed in NO SiFe but over a narrow temperature range of 25 °C to 100 °C [21]. At 2.0 T, 50 Hz the loss falls by around 10% at the upper temperature which is attributed to the increase in resistivity. At low flux density the permeability increased as expected by around 10%, but at high flux density it dropped by 30% which can be attributed to thermal agitation.

The power loss per cycle of a typical grade of conventional GO 3%SiFe over the temperature range from 20 °C to 530 °C at 1.5 T, 50 Hz is shown in Figure 18.9 [12]. These all have the well-known shape similar to that shown in Figure 6.11 so the approximate analysis into eddy current and static hysteresis components of loss can be carried out to obtain characteristics such as those shown in Figure 18.10.

The percentage fall in hysteresis loss per cycle shown in Figure 18.10 is far lower than that of the dynamic losses. However, direct measurement of static

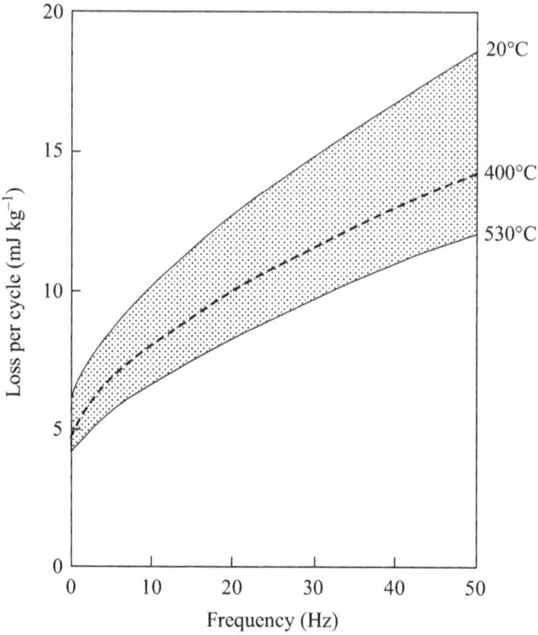

Figure 18.9 Temperature dependence of total loss per cycle in conventional GO 3%SiFe over the range from 20 °C to 600 °C when magnetised at 1.5 T, 50 Hz (Figure 2 in [12] reproduced under free licence CC-BY-4.0 and modified)

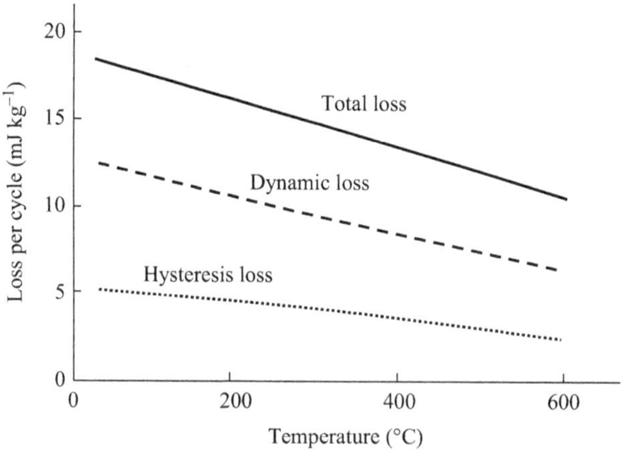

Figure 18.10 Temperature dependence of static and dynamic losses in conventional GO 3%SiFe at 1.5 T, 50 Hz (Figure 3 in [12] reproduced under free licence CC-BY-4.0 and modified)

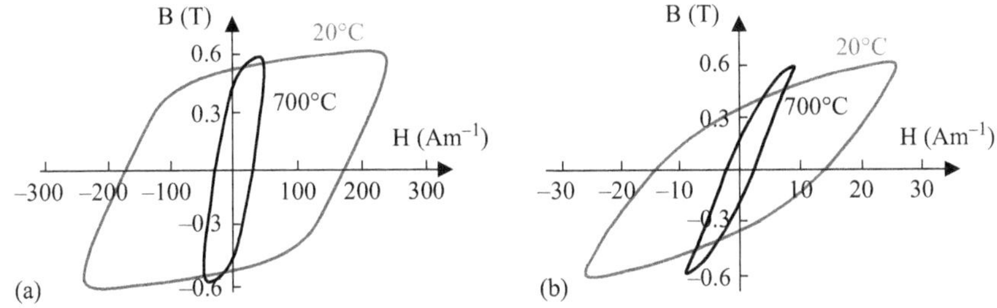

Figure 18.11 B–H loops at room temperature and 700 °C of (a) cold rolled sheet steel and (b) NO 6.5%SiFe (50 Hz, 0.6 T) (Figure 5 in [24] reproduced under free licence CC-BY-4.0)

hysteresis has shown that it is independent of temperature in the range 20 °C to 200 °C in both NO and GO steel [20]. Quoted ratios of hysteresis loss at ambient temperature to that at a high temperature depend to some extent on the measurement method. It is reported that P_h falls by up to 80% in high purity cobalt–iron alloys and 50% in GO 3%SiFe when the temperature is increased from 20 °C to 400 °C [22], whereas another report concludes that in GO silicon–steel it is independent of temperature [23].

Remarkable changes to *B–H* characteristics occur at high temperature. Figure 18.11(a) shows a.c. *B–H* loops of a 1.0 mm thick cold rolled steel sheet and a 0.1 mm thick sheet of commercial NO 6.5%SiFe [24]. In each case a substantial drop in both hysteresis and classical eddy current components of power loss with temperature is associated with the 70% drop of H_c.

Of course, the very high temperature characteristics of electrical steel shown in Figure 18.11 are mainly of academic interest. Its magnetic properties are still usable at 500 °C to 600 °C although its mechanical properties are poor at these temperatures [25]. In many applications it is necessary to restrict the normal operating temperature to less than around 200 °C to avoid long term deterioration of electrical insulating materials.

The surface coating on GO SiFe is applied partly to create a beneficial tensile stress in the strip in order to reduce the degradation of loss, permeability and magnetostriction due to externally applied mechanical force. The effectiveness of this coating drops with increasing temperature even when no permanent damage to the coating occurs.[5]

The a.c. magnetostriction of GO 3%SiFe also drops significantly with increasing temperature as shown in Figure 18.12 [17]. The increase occurring up to around 380 °C is followed by the sharp decrease, which does not appear to correspond to changes in λ_{100}. This, is again probably strongly associated with the effect of temperature on the properties of the coating.

[5]Fuller discussion of the role of the coating on GO steel can be found in Section 2.10.2 in Volume 2 of this book.

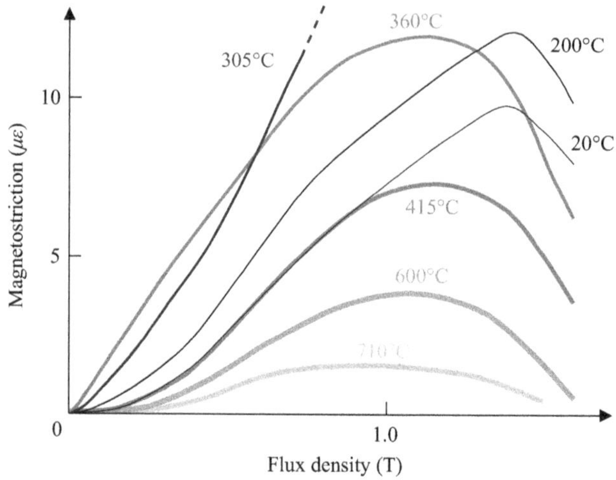

Figure 18.12 Variation of 100 Hz component of magnetostriction with temperature in GO 3%SiFe when magnetised sinusoidally at 1.0 T, 50 Hz (Figure 2 in [17] reproduced under free licence CC-BY-4.0)

18.4 Temperature dependencies of magnetic properties of various material

A comparison of temperature characteristics of four materials from common groups of soft magnetic materials is shown in Table 18.4 [22]. Numerical values are representative of each group, but not necessarily the best found in practice. One expected trend is the small drop in saturation magnetisation with increasing temperature in the three materials with high Curie points. Note that the permeability of all the materials increases with temperature. The hysteresis loss drops in each case, contradicting other claims that it should be independent of temperature. The conclusion from this comparison is that, as expected, the Co–Fe–V alloy is the most attractive material for power cores operating at up to 450 °C because of its all-round suitability [22].

Thermal stability, included in Table 18.4, refers to an irreversible change in magnetic properties which can occur in all magnetic materials subjected to either cycling from room temperature to a moderately high temperature (e.g. around 300 °C) or continuous operation at a higher temperature.

The very high Curie temperatures of the members of the CoFe-based family leads to them having the best high temperature performance of all the soft magnetic materials. Figure 18.13 shows typical *B–H* curves of 24% Co and 49% Co alloys at temperatures up to 950 °C [26]. Both materials maintain high B_s at high temperatures, but in each case there is no consistent trend due to the complex metallurgy of these alloys such as the ordered to disordered transformation in the 49% Co alloy at around 700 °C.

Table 18.4 Change in values of typical properties of common materials with an increase in temperature from ambient to 400 °C [22]

	B_s at 24 °C	T_c (°C)	Permeability at 24 °C	H_c at 24 °C	Ratio of parameters at 400 °C to their values at 24 °C			Thermal stability
					B_s	Permeability	Hysteresis loss	
50/50 NiFe	1.60	480	7,650	0.12	−0.54	+0.50	−0.33	Poor
GO 3%SiFe	1.97	730	3,380	0.30	−0.92	+1.00	−0.50	Average
Supendendur (Co–Fe–V)	2.30	940	4,360	0.18	−1.0	+1.00	−0.85	Average
Permendur Co–Fe–V	2.30	940	3,353	0.18	−0.94	+0.99	−0.80	Very good

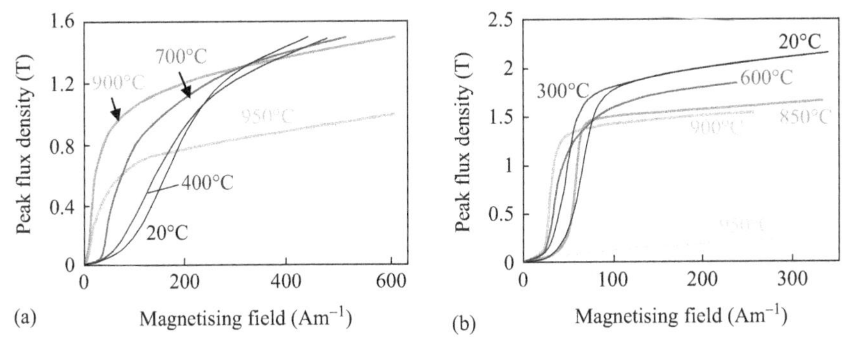

Figure 18.13 Temperature variation of a.c. magnetisation curves (50 Hz) of (a) a 24Co/0.75Cr/balance Fe alloy and (b) a 49Co/49Fe/ 2V alloy (Figure 2 in [26] reproduced under free licence CC-BY-4.0)

Corresponding variation of losses in some CoFe alloys is shown in Figure 18.14 [26]. The samples were magnetised at only 1.0 T, 50 Hz in order to be able to make the comparison over the full temperature range. The phase and order/disorder transformations cause the complicated changes in the 49% Co alloy. The 95% Co alloy has the lowest losses of all the materials at 1000 °C and its saturation magnetisation at this temperature is in excess of 0.9 T. The more common 24%Co0.75%Cr/balance Fe and 49%Co49%Fe2%V alloys exhibit excellent magnetic and mechanical properties up to 900 °C. If an even higher operating temperature is needed, the 49%Co0.35%Ta/balance Fe alloy is suitable for use up to 960 °C and, as stated earlier in this section, the 95%Co5%Fe alloy will perform well over 1000 °C [26].

The main benefit of the Co–Fe alloys is that their properties are not seriously degraded over a wide operating temperature range. Figure 18.15 shows how important physical and magnetic properties of a 27%Co binary alloy and a Co–Fe–V alloy (Hiperco 50A) also remain usable at temperatures several hundred degrees higher than with other alloys [27]. Both alloys retain 90% of their room temperature

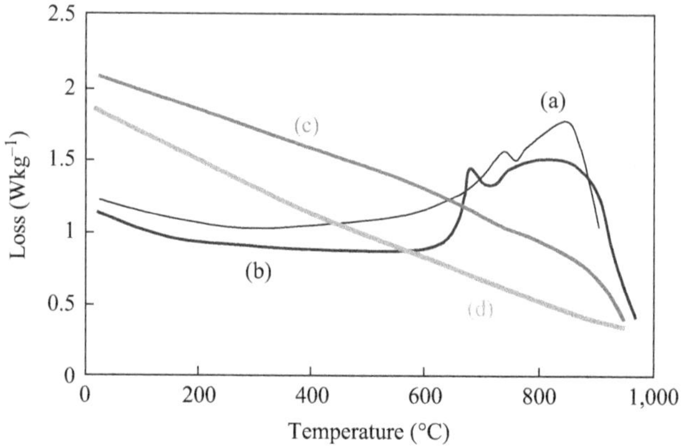

Figure 18.14 Temperature dependence of losses at 1.0 T, 50 Hz of CoFe alloys: (a) 0.35 mm thick 49%Co49%Fe2%V, (b) 0.20 mm thick 49% Co0.35%Ta/balance Fe, (c) 0.25 mm thick 24%Co0.75%Cr/ balance Fe and (d) 0.13 mm thick 95%Co5%Fe (Figure 5 in [26] reproduced under free licence CC-BY-4.0)

induction at 600 °C as well as a useful drop in coercivity. Overall, both alloys are claimed to be adequate in operation up to 450 °C [28].

Depending on the application, several of the materials already discussed in this chapter are good candidates for operation at elevated temperature. Less common materials which can be used to 600 °C are members of the Alfenol[6] alloys [28].

The magnetic properties of soft ferrites are very temperature sensitive. M_s decreases by as much as 50% as the temperature rises from 20 °C to 100 °C [29]. Increasing the temperature also decreases the remanence and coercive force.

The Curie point of amorphous alloys falls normally in a temperature range from 350 °C to 420 °C and the recrystallisation temperature, when they revert to a magnetically poorer crystalline structure, lies between around 410 °C and 550 °C. As a result, depending on their composition, the normal maximum continuous operational temperature is within the range from 90 °C to 150 °C.

The focus of investigations of temperature effects in amorphous magnetic materials has been on high temperature stability, relaxation mechanisms and ageing [30]. However, one study [31] concluded that the loss of POWERCORE fell linearly by around 1% per 10 °C up to around 140 °C, which is very similar to that found in GO steel. However, the exciting power almost doubled at 140 °C compared with its normal room temperature value; a far higher increase than that found for GO steel. Of course, magnetisation and loss mechanisms are very different in the two materials and it is not known to what extent the temperature sensitivity of POWERCORE depends on its consolidated form.

[6]See Section 16.6 for a short description of the Alperms.

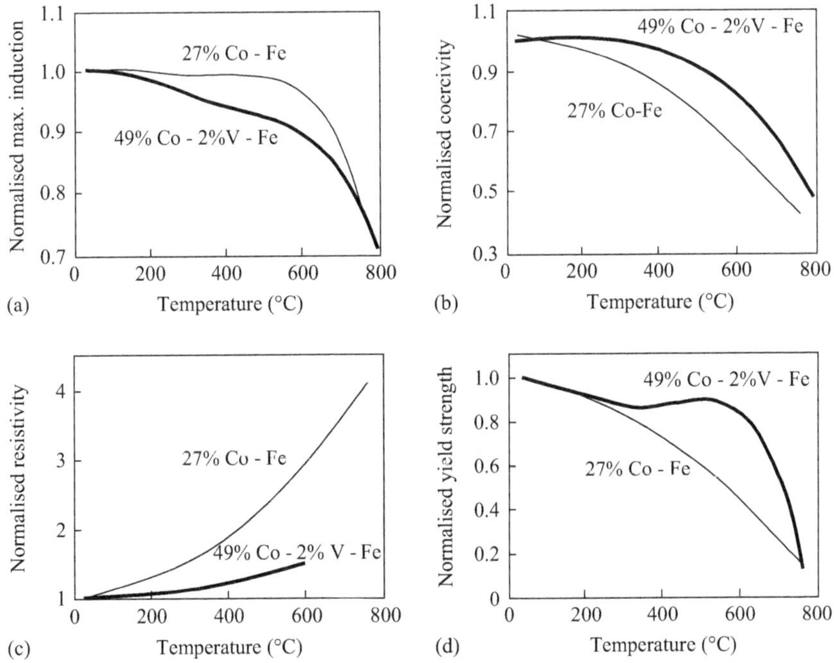

Figure 18.15 Normalised temperature dependence of (a) saturation magnetisation, (b) coercivity, (c) resistivity and (d) yield stress of two commercial CoFe alloys (reproduced from various figures in [26] under free licence CC-BY-4.0)

It has sometimes been suggested that even operating at a temperature only a little above normal, might cause a slow deterioration of properties due to crystalisation of the amorphous structure. However, no noticeable change takes place. For example, it has been proved that some iron-based amorphous material alloys are stable at a continuous operating temperature of 100 °C for at least a 30-year life span [32].

Soft magnetic nanocrystalline alloys have an amorphous phase as described in Section 16.2 so their temperature dependencies would be expected to be generally similar to amorphous alloys. The actual temperature dependence is far more complex than in single phase materials. It is strongly controlled by the processing, the composition of each phase and the evolution of the microstructure [33]. So the expectation is that, as the temperature approaches the Curie point of the amorphous phase, a serious degradation of the properties occurs. However, if grains are sufficiently small and only a small volume transform to crystallites, superparamagnetic behaviour can occur leading to high temperature performance, perhaps up to 650 °C, which is not possible in normal nanocrystalline or amorphous alloys [34]. The saturation magnetisation of commercial FINEMET alloys drops by less than 10% by increasing the temperature from 0 °C to 150 °C leading to a very small drop in permeability over the same operating range [35].

18.5 Modelling high temperature performance

Structure-sensitive a.c. and d.c. magnetic properties depend on the balance between the influence of anisotropy and magnetostriction whose temperature dependencies is complex so it is not surprising that no generic model has been proposed for predicting factors such as *B–H* loop shape or losses.

Some empirical methods of predicting temperature variation of *B–H* curves and losses of soft magnetic materials have been produced. A logical approach is to take an existing model, which has been shown to work well at ambient temperature, and modify it to take account of temperature sensitive parameters to predict *B–H* loops for example for use in computational electromagnetic solutions for electrical machines [36]. The approach has been shown to produce good agreement between measured and predicted high temperature *B–H* curves of a Ni50/Cr9/balance Fe alloy, but poor prediction of minor loop shapes was attributed to inaccuracies associated with the representation of high temperature anhysteretic curves[7] needed in the model [37].

A simple approach to predicting the temperature dependence of losses in electrical steel is based on introducing a factor to take account of the temperature change in resistivity into any one of the common loss separation expressions given in Chapter 7. If the loss is separated into dynamic and static hysteresis components, and it is assumed that the hysteresis loss per cycle is independent of temperature, then a temperature coefficient k can be included to account for the temperature dependence of resistivity to arrive at the following expression for the total loss at temperature θ [38]:

$$P_\theta = k_h f B_p^x + \frac{k'_{ec}}{1 + k(\theta - \theta_0)} f \int_0^T \left(\frac{dB}{dt}\right)^2 dt \; \left(\text{Jm}^{-3}\right) \tag{18.2}$$

The first term is an experimental value of static hysteresis. θ_0 is a reference, or base, temperature, k is obtained by curve fitting,[8] B_p is the peak flux density and k_h and x are static hysteresis coefficients (assumed constant at a given flux density). k'_{ec} is an eddy current coefficient given by

$$k'_{ec} = \frac{t^2}{12D\rho(\theta)} \tag{18.3}$$

where t is the lamination thickness, D is its density and $\rho(\theta)$ is the resistivity at the measurement temperature given previously in (18.1). Figure 18.16 shows the close agreement between the loss occurring in NO SiFe predicted by using the model and measured values over a wide range of magnetising frequencies and flux densities at 200 °C [38]. The band at each frequency represents the spread between predicted

[7]The anhysteretic magnetisation curve is a single valued *M–H* relationship of an ideal material free of internal defects, etc. which cause the normal d.c. hysteresis loop. The concept is used widely by material modellers.

[8]The constant k is actually the temperature coefficient of resistance α defined in (18.1).

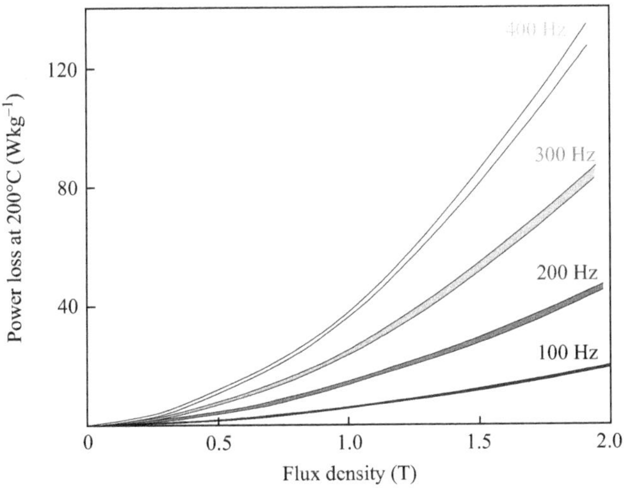

Figure 18.16 Predicted and measured variation of iron loss with flux density over the frequency range from 100 Hz to 400 Hz in NO SiFe at 200 °C (Figure 5 in [38] reproduced under free licence CC-BY-4.0 and modified)

and measured values. The model works equally well for GO SiFe. It is really a curve fitting technique from which parameters can be obtained and used to produce *B–H* or *loss-B* characteristics. These, in turn, are used in time stepping *FE* (finite element) electromagnetic field solvers to predict magnetic core performance at elevated temperature. Good accuracy should not be expected over a wide temperature range because the coefficient k can only be assumed to be constant over a narrow temperature range and other assumptions made in deriving (18.2) may not be valid at high temperatures. The method seems to work adequately, but it is questionable whether it is related to any true physical magnetic phenomena.

However, similar approaches of loss separation and fitting of model coefficients from experimental measurements also work well for NO electrical steel. This shows that hysteresis loss and classical eddy current loss are less temperature sensitive than the excess loss [39]. However, this approach predicts that the total loss drops linearly between 20 °C and 230 °C by around 10% to 20%, the highest decrease occurring in grades with high silicon content.

It is important to minimise the size of ferrite cores to allow them to operate at higher frequencies whilst also minimising losses, so power converter designers are continually demanding better modelling of high temperature magnetic characteristics. Scalar hysteresis models[9] have been successfully used to predict the change in loop shape with temperature, but the temperature coefficient of resistivity is treated as a constant [29].

[9]See Chapter 4 of Volume 2 of this book.

In general, modelling the hysteresis core losses under more complex, practical magnetisation conditions, incorporating the effect of temperature fluctuation, is very challenging. Enhanced Presach models taking temperature dependence into account, are claimed to be capable of predicting behaviour to within 20% over a wide range of temperatures and magnetising frequencies [40].

18.6 Magnetic properties at cryogenic temperatures

Niche applications are growing for magnetic cores of electrical equipment operating at cryogenic temperatures, defined here as lower than −150 °C. Magnetic characterisation at low temperature has not attracted as much attention as high temperature characterisation because of the previous limited industrial interest and the expectation that magnetic properties are only a little different to those at normal ambient temperature.

It can be shown from (2.5), or by a rough estimate from Figure 18.1, that M_s is only a few per cent higher at 0 K than at room temperature in iron, cobalt and nickel alloys because of their relatively high values of T_C. This trend is found in many materials including SiFe alloys, Permalloys, and 2V Permendur.

As shown in Figure 18.2, the resistivity of iron tends to fall approximately linearly with temperature in the cryogenic zone. This tends to increase the eddy current losses. The anisotropy constant K_1 continues to increases at low temperatures, but at a decreasing rate as shown in Table 18.3. However, the values of K_2 and K_3 are much lower than K_1 over the entire temperature range, so they are neglected in most studies of iron-based alloys [41]. Likewise, in nickel, K_1 becomes even more dominant at low temperature as shown in Figure 18.3.

We are not aware of any quantified reports of the variation of magnetostriction constants of any soft magnetic materials at cryogenic temperatures.

Structure sensitive magnetic properties of SiFe alloys and 2V Permendur change only a little at cryogenic temperatures, whereas in other materials they can become substantially poorer. For example, in pure iron, 49 Permalloy and 4Mo Permalloy the relative permeability falls to around 60%, 25% and 15% respectively of their values at 300 K [41]. Similar adverse trends are found in coercivity apart from in the SiFe and 2V Permendur, where, again, only a small deterioration occurs.

The low temperature dependence of a.c. losses in SiFe alloys is dependent on the magnetising frequency. For example, when GO SiFe is magnetised at 50/60 Hz, at low to medium induction, the loss at 4 K is no more than 10% higher than its value at room temperature, but at 400 Hz to 1,000 Hz it increases by 20% [42]. This is attributed to the fact that at room temperature, P_h is the dominant component of loss at high frequency, whereas at low temperature the low resistivity makes P_e dominant, more so at higher frequency.

The temperature coefficient of resistance of SiFe alloys is very dependent on the silicon content as can be seen in Figure 18.17. At 4 K it falls to around 85% of its value at 300 K in 4.5%SiFe whereas it falls to only 30% of the room temperature value in 0.85%SiFe [42].

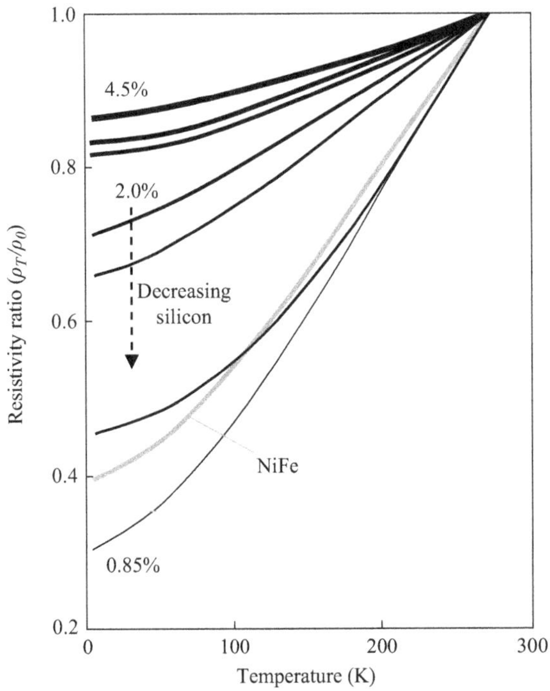

Figure 18.17 Low temperature variation of resistivity of SiFe alloys expressed as ratios of the values at 20 °C (Figure 5 in [42] reproduced under free licence CC-BY-4.0 and modified)

It is anticipated that the low temperature sensitivity of the power loss of 6.5%SiFe to stress will almost vanish due to its near zero magnetostriction constants. The magnetic properties might actually improve at very low temperatures.

The temperature variation of the properties of nickel–iron alloys at cryogenic temperatures is complex due to the fact that the relative magnitudes of the magnetocrystalline and magnetostriction constants vary significantly with composition. The behaviour still seems not to be well understood. Generally the magnetic properties deteriorate with decreasing temperature with both H_c and B_r sometimes increasing to double their room temperature values at 4 K [42]. Hysteresis loss of low grade NO SiFe has been found to increase by around 10% at 77 K compared to the room temperature value at 1.5 T, 50 Hz, but, in better grades, the increase is only marginal [43]; the explanation is not obvious.

18.7 Effect of non-uniform temperature gradients in magnetic core laminations

Temperature gradients are present to various degrees between the inside and surface of a magnetic core and between regions of high and low localised loss.

The core loss can be estimated based on the specific temperature characteristics of the particular core steel. However, the uneven temperature distribution also causes localised stress gradients, partly due to the temperature dependence of magnetostriction, and partly due to differential thermal expansion of the core material.

It has been found that a temperature difference of 10 °C can cause a compressive stress of 10 MPa in regions of a core assembled from GO SiFe laminations. This causes additional losses as well as the magnetostriction to vary by a factor of three from the core centre to its edges [44]. It does not appear to be known whether temperature gradients generally increase core loss or acoustic noise. It seems that today's powerful multi-physics software packages, incorporating thermal and magnetic solvers, might have the necessary capability to quantify the effect.

18.8 Effect of irradiation on soft magnetic materials

Deep space missions requiring innovative power supplies and compact fission energy generation systems have been considered for decades. These need soft magnetic cores capable of operating in a strong irradiation environment. Shielding is generally expensive and space consuming so magnetic materials whose properties are less sensitive to radiation, which ranges from gamma rays to fast and slow neutrons, are desired.

In general, irradiation creates obstacles to magnetic domain motion, hence soft magnetic properties deteriorate. Such obstacles might remain in place indefinitely or the material might revert to its original state over a period of time.

Material evaluation is very costly, time consuming, potentially dangerous and sometimes security sensitive. It is not surprising that results published in the public domain are rather limited.

The characteristics of soft and hard materials irradiated by neutron sources over periods of weeks have been obtained [45]. A simple trend has been found in soft magnetic materials in general. The permeability and coercivity of high permeability materials, including Supermalloy, Mumetal and other NiFe alloys, are seriously degraded, sometimes by orders of magnitude. However, lower permeability alloys, such as SiFe and Co–Fe–V alloys experienced almost no change. No explanation has been proposed.

Radiation damage in high permeability materials is often regarded as being permanent but it can be recovered to a large extent by suitable heat treatment if the material is accessible. The chance of partial recovery shows that irradiation does not cause any irreversible chemical transmutations in these materials. The cause of the magnetic damage can be attributed to the production of material inhomogeneity by vacancies, interstitials or more complex defects.

It has been found that if the temperature is higher than $T_c/2$ greater damage occurs but the result of combinations of different radiation levels and temperatures is difficult to predict [46].

References

[1] Fingers R.T. 'Application of high temperature magnetic materials'. *IEEE Trans. Magn.* 2000, vol. 36(5), pp. 3373–3375

[2] Brailsford F. *Physical Principles of Magnetism*. Van Nostrand, London, UK, 1966. p. 103

[3] Snelling E.C. *Soft Ferrites: Properties and Applications*, 2nd edn. Butterworth, 1988

[4] htpps://en.wikipedia.org/wiki/Electrical_resistivitiesy_of_the_elements (data page). (Accessed January 2017)

[5] Bhavikatti A.M., and Mugli M. 'Comparative studies of some electrical properties of nickel ferrite and cobalt ferrite'. *Indian J. Sci. Res.* 2015, vol. 12(1), pp. 232–234

[6] Dolocan V. 'Temperature dependence of the electrical resistivity in some amorphous ferromagnetic alloys'. *IEEE Trans. Magn.* 1988, vol. 24(2), pp. 1832–1834

[7] Nagel R. 'Temperature dependence of the resistivity in metallic glasses'. *Phys Rev. B* 1977, vol. 16, pp. 1694–1698

[8] Gengnagel H., and Hofmann U. 'Temperature dependence of the magnetocrystalline energy constants K_1, K_2 and K_3 of iron'. *Phys. Stat. Sol.* 1968, vol. 29, pp. 91–97

[9] Bozorth R.M. *Ferromagnetism*. IEEE Press, Piscataway, NJ, 1993. p. 567

[10] Brailsford F. *Physical Principles of Magnetism*. Van Nostrand, London, 1966. p. 143

[11] Bozorth R.M. *Ferromagnetism*. IEEE Press, Piscataway, NJ, 1993. p. 570

[12] Ferro A., Montalenti G., and Soardo P. 'Temperature dependence of power loss anomalies in directional FeSi 3%'. *IEEE Trans. Magn.* 1976, vol. 12(6), pp. 870–872

[13] Bozorth R.M. *Ferromagnetism*. IEEE Press, Piscataway, NJ, 1993. p. 563

[14] Bozorth R.M. 'Directional ferromagnetic properties of metals'. *J. Appl. Phys.* 1937, vol. 8, pp. 575–588

[15] Bozorth R.M. *Ferromagnetism*. IEEE Press, Piscataway, NJ, 1993. p. 568–569

[16] Tatsumoto E., and Okamoto T. 'Temperature dependence of the magnetostriction constants in iron and silicon iron'. *J. Phys. Soc. Jpn.* 1958, vol. 14 (11), pp. 1588–1594

[17] Pike E.C., and Moses A.J. 'Effect of temperature on a.c. magnetostriction in grain-oriented silicon-iron'. *J. Mater. Sci.* 1977, vol. 12, pp. 187–191

[18] Chen C.W. 'Temperature dependence of magnetic properties of silicon-iron'. *J. Appl. Phys.* 1958, vol. 29(9), pp. 1337–1343

[19] Iwata K., Suzuki M., Hashimoto M., *et al.* 'Temperature dependence of lancet domains in grain-oriented Fe-3 % Si steels'. *IEEE Trans. Magn.* 2015, vol. 51(11), Art. No. 2002504

[20] Chen J., Chen Z., Wang D., *et al.* 'Influence of temperature on magnetic property of silicon steel lamination'. Proceedings of the 61st International Conference on Magnetic Materials, New Orleans, USA, November 2016

[21] Nakaoka M., Fukuma A., Nakaya H., Miyagi D., Nakano M., and Takahashi N. 'Examination of temperature characteristics of magnetic properties using a single sheet tester'. *IEEJ Trans. Fundam. Mater.* 2005, vol. 125 (1), pp. 63–68

[22] Pandey R.K., Weichold M.H., and Palmer D.W. 'Materials characterization for high temperature transformers'. *IEEE Trans. Magn.* 1980, vol. 16 (5), pp. 749–751

[23] Bullingham J.M. 'A Study of the Effects of Temperature and Non-Linear B/H Loops in Grain-Oriented Silicon-Iron Transformer Laminations'. *PhD Thesis*, University of London, UK, 1971

[24] Takahashi N., Morishita M., Miyagi D., and Nakano M. 'Comparison of magnetic properties of magnetic materials at high temperature'. *IEEE Trans. Magn.* 2011, vol. 47(10), pp. 4352–4355

[25] Counderchon G., and Josso E. 'Magnetic Co-Fe alloys for systems operating at very high temperature'. 3rd International *Conference on Soft Magnetic Materials (SMM3),* Bratislava, Czechoslovakia, 1977, pp. 93–100

[26] Samadian V., and Major R.V. 'High temperature magnetic characteristics of iron-cobalt alloys'. *IEEE International Magnetics Conference (INTERMAG),* Brighton, UK, 1990

[27] Masteller M.S. 'Soft magnetic materials for hostile environments'. *Gorham Intertech Conference Proceedings, Soft Magnetic Materials 98,* Barcelona, Spain, April 1998

[28] Kueser P.E., Pavlovic D.M., Lane D.H., Clark J.J., and Spewock M. 'Properties of magnetic materials for use in high-temperature space power systems'. Scientific and Technical Information Division, NASA, Report No. SP-3043, 1967

[29] Lu H.Y., Zhu J.G., and Hui S.Y.R. 'Measurement and modelling of thermal effects on magnetic hysteresis of soft ferrites'. *IEEE Trans. Magn.* 2007, vol. 43(11), pp. 3952–3960

[30] Fish G.E., and Smith C.H. 'A decade of progress in iron-based amorphous ribbon for electrical power distribution'. *Proceedings of SympOsium on Soft and Hard Magnetic Materials and Applications*, Lake Buena Vista, FL, USA, 1986, pp. 7–20

[31] Schönber P., Pfützner H., and Erbil B. 'Comparison of POWERCORE and lased scribed HI-B steel model cores under practical working condition'. *Anal. Fisica* 1990, vol. 86, pp. 96–98

[32] Fish G.E. 'Soft magnetic materials'. *Proc. IEEE* 1990, vol. 78(6), pp. 947–971

[33] McHenry M.E., Willard M.A., and Laughlin D.E. 'Amorphous and nanocrystalline materials for applications as soft magnets'. *Prog. Mater. Sci.* 1999, vol. 44(4), pp. 291–433

[34] Willard M.A., Daniil M., and Kniping K.E. 'Nanocrystalline soft magnetic materials at high temperatures: A perspective'. *Scr. Mater.* 2012, vol. 67, pp. 554–559

[35] Nanocrystalline soft magnetic material FINEMET. http://www.hitachi-metals.co.jp/products/elec/tel/pdf/hl-fm9-h.pdf (Accessed September 2017)

[36] Bergqvist A. 'Magnetic vector hysteresis model with dry friction-like pinning'. *Phys. B Condens. Matter*. 1997, vol. 233(4), pp. 342–347

[37] Sixdenier F., Messal O., Hilal A., Martin C., Raulet M.-A., and Scorretti R., 'Temperature–dependent extension of a static hysteresis model'. *IEEE Trans. Magn*. 2016, vol. 52(3), Art. No. 7300104

[38] Chen J., Wang D., Cheng S., Wang Y., Zhu Y., and Liu Q. 'Modeling of temperature effects on magnetic property of nonoriented silicon steel lamination'. *IEEE Trans. Magn*. 2015, vol. 51(11), Art. No. 2002804

[39] Novak G., Kokošar J., Bricelj M., Bizjak M., Steiner Petrovič D., and Nagode A. 'Improved model based on the modified Steinmetz equation for predicting the magnetic losses in non-oriented electrical steels that is valid for elevated temperatures and frequencies'. *IEEE Trans. Magn*. 2017, vol. 53 (10), Art. No. 2001805

[40] Zegadi L., Rousseau J.J., Allard B., Tenant P., and Renault D. 'Model of power soft MnZn ferrites, including temperature effects'. *IEEE Trans. Magn*. 2000, vol. 36(4), pp. 2022–2032

[41] Ackermann F.W., Klawitter W.A., and Drautman J.J. 'Magnetic properties of commercial soft magnetic alloys at cryogenic temperatures'. *Proceedings of Cryogenic Engineering Conference*, University of Colorado, Boulder, USA, 1970, B-3, pp. 46–50

[42] Gniewek J.J., and Ploge E. 'Cryogenic behaviour of selected magnetic materials'. *J. Res. Natl. Bur. Stand. Sec. C: Eng. Instrum*. 1965, vol. 69C(3), pp. 225–236

[43] Miyagi D., Otome D., Nakano M., and Takahashi N. 'Measurement of magnetic properties of nonoriented electrical steel sheet at liquid nitrogen temperature using single sheet tester'. *IEEE Trans. Magn*. 2010, vol. 46(2), pp. 314–317

[44] Niwa H., Yanase S., Hashi S., and Okazaki Y. 'Magnetic properties of grain-oriented magnetic steel sheets under temperature distribution'. *J. Magn. Magn. Mater*. 2003, vol. 254–255, pp. 376–378

[45] Gordon D.I. 'Magnetic cores and permanent magnets in hyper-environments'. *Proceedings of the Institute of Environmental Sciences*, Illinois, USA, 1960, pp. 205–228

[46] Bowman C.L. 'Design issues for using magnetic materials in radiation environments at elevated temperatures'. *Proceedings of the Nuclear and Emerging Technologies for Space*, Albuquerque, NM, USA, 25–28 February 2013, paper 6801

Index

www.ingramcontent.com/pod-product-compliance
Lightning Source LLC
LaVergne TN
LVHW011518200926
848484LV00001B/63
* 9 7 8 1 7 8 5 6 1 9 7 0 0 *